HYDRAULICS OF DAMS AND RESERVOIRS

HYDRAULICS OF DAMS AND RESERVOIRS

by Fuat Şentürk, Ph.D.

Doctor of Toulouse University
Former Director of the Research Center
of the General Directorate
of the State Hydraulic Works
Ankara, Turkey

Water Resources Publications

For Information and Correspondence:

Water Resources Publications
P. O. Box 260026 • Highlands Ranch • Colorado 80126-0026 • USA

HYDRAULICS OF DAMS AND RESERVOIRS

by

Fuat Şentürk, Ph.D.

Photograph by Fuat Şentürk
Cover Designed by Water Resources Publications

ISBN Number 0-918334-80-2

U.S. Library of Congress Catalog Card Number: 92-62859

To my courageous wife

TABLE OF CONTENTS

FOREWORD

This book, translated from Turkish into English, is an extensively revised, expanded and updated edition of the 1988 publication entitled *Dams Hydraulics* authored by Dr. Fuat Şentürk. He sent me a courtesy copy in 1989 for review. Even though it was written in Turkish, numerous and easily understood figures, tables, photos and equations, helped me realize the significance of this important work. We discussed the possibility of completely rewriting the original book as an English edition in order to make this valuable information available to as many professionals as possible. The idea was to use the examples of realization of dams and reservoirs from all around the world instead of mostly those in Turkey, as was done in the 1988 edition. Then, to improve the content of the book, we discussed expanding the existing chapters, including a relatively condense chapter on hydraulics of reservoirs and adding other chapters. The topic of dam breaches and the resulting flood waves was not included in this edition. It is usually treated as a special subject in separate texts. Dr. Şentürk's positive reaction to these suggestions started his five-year long undertaking in creating a completely new and comprehensive text in English, ultimately resulting in this publication entitled *Hydraulics of Dams and Reservoirs*.

This life-long and successful pursuit in research and development, as well as in laboratory and field practice by Dr. Fuat Şentürk, will contribute to various objectives in the application of the up-to-date hydraulic methods to planning, design, construction, operation, maintenance and renovation of large, medium and small dams and reservoirs and their hydraulic structures. This book is intended to serve both practicing professionals and graduate and special continuing education studies scholars. Both specialists and students will benefit from the many solved problems in this publication which are offered as examples, as well as the exercise problems, with solutions that are given in the separate *Problem Solution Manual*.

For developed countries, where large numbers of dams and reservoirs already exist, continuing their operation and maintenance will require many improvements because of increasing need, possibility of accidents or both. As an example, an estimate was made by the U.S. Corps of Engineers for the United States of America concerning upgrading the safety of large dams to handle large floods. Of the 5,000 largest dams and reservoirs in the U.S., out of an existing total of 55,000 small, medium and large dams and reservoirs, it will be necessary to invest more than 25 billion dollars in the future. At the same time, aging dams and their hydraulic structures will require renovating.

For developing countries, this text is essential for planning, designing, constructing, maintaining and operating new dams. For all other countries, both the building of new dams and the maintenance of existing dams and reservoirs, should be of primary concern. This impressive effort by Dr. Fuat Şentürk will be fully justified and rewarded if this publication helps meet these pressing needs.

Vujica Yevjevich
Professor Emeritus of Civil Engineering
Colorado State University
Fort Collins, Colorado, USA

PREFACE

The idea for the book *Hydraulics of Dams* was conceived in 1974. It was written in Turkish for the General Directorate of Hydraulic Works in Ankara, Turkey and distributed among engineers as an inner-departmental publication. The first edition was comprised of only the actual Chapters 2, 3 and 4. Many of the subjects discussed in different chapters were prepared as technical papers for international conventions such as IAHR & ICOLD. The second edition that was released in 1977, also included Chapters 5, 6, 7, 9, 10, 11 and 12. Originally, it was used as a reference book in Turkey. With recent revisions, it is even more useful as a practical reference guide to different hydraulic processes.

The method of computation of the shear stress on the spillway profile and on the ski jump is used in the design of large dams such as Iril-Emda (North Africa), Aigle (France), Keban, Karakaya (Turkey), etc. Since this method has proven to be successful, it has been in use ever since. The approach described in the computation of the control structure, has also been adopted by many countries worldwide and recognized as a time and labor saving procedure which leads to economically efficient results.

The book considers water in its journey from its entry to the reservoir, in the reservoir, in the approach channel, on the control structure, in the discharge channel and in the downstream river up to a control section. Then the hydraulics of reservoirs is added to its content. To help understand the different concepts, problems are included in the text, with additional problems and solutions given in the *Problem Solution Manual*.

Chapters 1-5 describe spillways that are most often used in engineering practice. Chapter 1 discusses the historical background of dam construction worldwide. Chapter 2 introduces the basics of dam spillways such as spillway design; components; layout; type, and classification. Chapter 3 and 4 analyze the hydraulics of free-flow spillways and gated spillways, respectively; and Chapter 5 discusses the hydraulics of special types of spillways, broad-crested weir spillways, side and lateral spillways, shaft spillways, siphon and compound spillways. Downstream channel hydraulics is discussed in Chapter 6. The forces involved in stability analysis such as static and dynamic loads on spillways are brought up in Chapter 7.

Chapter 8 addresses the hydraulics phenomena in reservoirs such as waves, density currents, sedimentation, etc. Chapter 9 deals with the hydraulics of the bodies of dams. The osculation devices used for measurement purposes and interpretation of measurements are also presented in this chapter, together with hydraulic computations. Chapter 10 discusses feasible solutions associated with the detail design of upstream and downstream cofferdams and diversion tunnels. Different problems concerned with outlet works and their solutions are outlined in Chapter 11.

Preface

Chapter 12 addresses different effects on the degradation, aggradation and local sour in downstream channels at dams.

In general, design engineers use hydraulic laboratories to obtain answers to questions which cannot be treated by theoretical approaches. Chapter 13 will help design and model engineers work together in harmony. Many of the problems which were solved on scale models are today submitted to computer analysis. The result is much more rapidly attained and efficient. Attention to this feature is given in Chapter 13 and in the *Problem Solution Manual*.

The *Problem Solution Manual* in conjunction with *Hydraulics of Dams* will be of benefit to specialists in preparing the final design of hydraulic structures and to students by helping them understand the hydraulic phenomenon that enters into the design of dams. These problems enable engineers to better follow the behavior of hydraulic structures.

In the preparation of this manuscript, I wish to acknowledge the assistance and support received from various individuals and support groups. In particular I wish to thank Dr. A. Şentürk and Dr. E. Demiröz for their assistance in reviewing the manuscript, checking figures and other related details. Also I wish to thank Dr. T. Akar from Istanbul Technical University and Mr. A. Yalçin, Head of the Drafting Office of the Research Center, Ankara, for their assistance with preparation of figures and graphics and Ms. Kantemir for her assistance in the clerical work of the manuscript.

I wish to acknowledge the excellent assistance provided by Mr. D. Groner, PE in reviewing the complete manuscript and the related tables, etc. The preparation of this book was only possible because Mr. D. Groner spent more than one year in review the complete manuscript.

I also am indebted to Dr. V. Yevjevich for his valuable suggestions on the content of the book and the encouraging remarks on different subjects.

Finally I would mention the help of the former directors general and the acting Director General of DSI, Mr. Ö. Bilen for their encouragement in using DSI's realizations in the field of hydraulic engineering.

I would like to mention the name of President Demirel with whom I had the opportunity to collaborate beginning from 1954 up to 1960. At that time he was Director General of the General Directorate of State Hydraulic Works of Turkey; he is today the President of the Turkish Republic. He has conceived the Research Institute I directed from 1954 to 1975. He was the driving force of projects like the Keban, the Karakaya, the Ataturk, etc., and following his example I had the opportunity to work on different hydraulic details of such large dams and include in this book the knowledge I have accumulated during my professional life.

Fuat Şentürk
August 1994

Chapter 1

HISTORICAL BACKGROUND OF DAM CONSTRUCTION

1.1 INTRODUCTION

It is well known that water resources are essential to living organisms, including humans beings. Therefore, numerous attempts have been made to harness the useful aspects of water resources, and to fight against those forces of water which are destructive. The earliest approaches for this latter purpose were apparently made in China some 4000 years B.C. History tells that similar technology for constructions of water resources structures were developed simultaneously in Mesopotamia and in Egypt. In fact, and approximately at the same time, the Egyptians constructed a dam which was located in the vicinity of Memphis City in the Nile Valley. This structure, 15 m high by 450 m long, had a rockfill core with an impervious clay blanket on the upstream side (Legget, 1939).

Engineering skills among the Americans during that period seem to be unknown. Between 4000 B.C. and the 1st century B.C., the centers of scientific activity were in Mesopotamia, Egypt and Anatolia. The region became a natural bridge between an advanced East and the less developed West. Remnants of early hydraulic structures are still visible in Turkey along the western coast of Anatolia. Here, the large aqueducts, tunnels, water retaining structures and cisterns of those times still exist. Several civilizations flourished in this area before the 15th century. Transfer of technology from East to West probably occurred during that millennium, when rivers were improved and bridges, aqueducts and dams were built to be used by the conquering armies (Fig. 1.1).

The innovations in water related structures occurred during the development of the Ottoman Empire. The Ottomans had acquired knowledge for improvement of river channels, construction of bridges, dams and flood control structures (example is the bridge in Fig. 1.2). They knew how to determine the maximum scour around bridge piers. Some hydraulic structures built during this period have lasted 600 years and are still in use in Turkey, Bulgaria, former Yugoslavia, Greece, Africa and the Middle East.

1

Figure 1.1. Aqueduct for water supply of Istanbul.

Figure 1.2. An old Ottoman bridge, still in use in Anatolia.

1.1.1 Worldwide Dam Constructions

A. Early Dams in Persia

Many dams were built in Persia, India, Ceylon, China and other ancient societies during the expansion of technology and other knowledge worldwide. Dams constructed in Iran during that period include:

- 40 m high gravity dam on the Saveh River

- The Bandar-Emir gravity dam

- The Gulistan Dam near Masshad

- The 28 m high and 40 m long Karab Dam.

Some of these structures were repaired in the years 1628-1787 A.D. in the era of Shah Abbas.

B. Early Dams in the Middle East

Many hydraulic structures and particularly dams were built in the Middle East during the 28 centuries between 1600 B.C. and 1200 A.D. Construction techniques were developed in the later centuries with hydraulic engineering recording a prodigious jump during the Renaissance in Europe at the end of the 16th century.

The Mesopotamian hydraulic structures were relatively advanced. The Babylonians built numerous retaining structures in the Tigris Valley such as the Mardook Dam, which was an earth-fill dam reinforced by aquatic plants. The people of Saba built the Marib Dam (1700 B.C.), a 32.50 m high and 3,200 m long, rockfill dam, in the vicinity of their capital city Sabaea. Djizveh and Taif Dams in the same region were also built at the same time. Similarly, it seems that the Queen of Sheba built a dam in Ethiopia during the Solomon times.

C. Early Dams in Spain

Dams were built in Spain, first by Romans and then by Moors. The two best surviving examples of the first Roman dams are to be found near Merida, which are still in use today. The Proserpina Dam (Fig. 1.3), 427 m long and 13 m high, forms a reservoir for irrigational purposes. The Cornalbo Dam, 200 m long and 20 m high, has a larger reservoir (Fig. 1.4). Some of these dams are massive constructions of masonry and in fact it has been claimed that this is the Moors' most important contribution to Spanish technology (Smith, N.A.F.).

Moors were well acquainted with construction of hydraulic structures for many centuries and they carried their knowledge to countries they conquered. This transfer of knowledge took more than 1,000 years. Two examples are the Cardete Dam near Tudela, built in 1220 A.D. and the still

3

surviving Almonacid Dam near Balchite, in Aragon. These structures had massive, straight masonry walls built across a deep gorge. The original height was estimated to be 22 m high with a length of 75 m.

Toward the end of the 14th century, the Almansa Dam was built in the province of Albacete and in 1576 was raised keeping the lower original part as it was. It was 15 m high, estimated to be ~3 m thick at the crest, ~10 m thick at the base and curved in the plan. A similar structure is the Tibi-Alicante Dam built between 1579 and 1594 A.D. on the river Monegre. It had a height of 43 m and a thickness at the crest of 20 m. It is interesting to note that similar dams were built in Iran and in the Ottoman Empire during the same period. Two more dams can also be cited in Spain. Figure 1.4 shows cross sections for the Elche Dam (26 m high) and the Relleu Dam (32 m high and 34 m long),

More sophisticated structures were built in Spain in the centuries that followed. These early dams presented similar characteristics; they were composed of two walls with an impervious fill between them. Although the provision for a spillway or outlet was not well defined, they were equipped with outlet wells to convey water to the user. The Elche Dam was the latest in this series. The date of construction was 1640 and again, the spillways were not incorporated into the dam design.

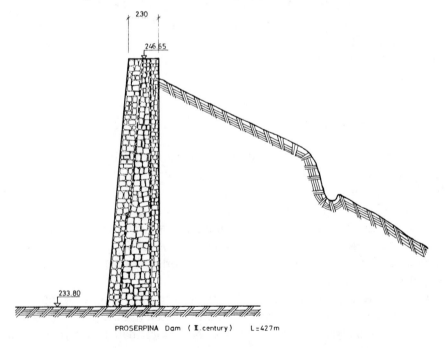

PROSERPINA Dam (II. century)   L = 427 m

Figure 1.3. The Proserpina Dam, 2nd century A.D. L = 427 m.

4

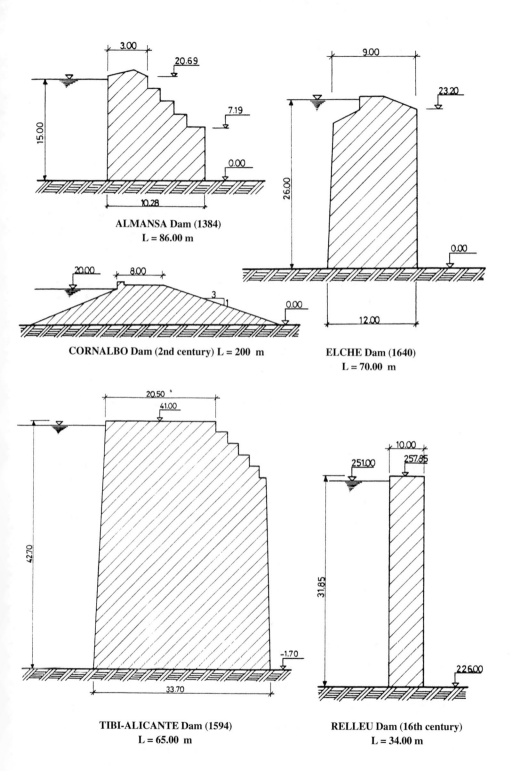

Figure 1.4. Cross sections of five of the early Spanish dams.

D. Early Dams in the Americas

The most developed peoples in the North and South Americas were the Incas, Mayas and Aztecs. The Incas established their empire in the southern regions of today's Peru, Equador, Columbia, Mexico and the Amazonian watershed. Extensive research has not yet been carried out to determine their technical abilities. However, inscriptions recently discovered in caves reveal that they understood well irrigation. They irrigated fertile plains of mountainous regions some 4000 years B.C. The Inca's civilization flourished particularly in 12th century A.D. The Mayas built large religious structures. The Aztecs built the pyramids of Teotihuacan (ceremonial center). These peoples then constructed works to suit their social needs (such as the Mayan irrigation canals and roads in the Yucatan Peninsula; the hydraulic works on the Anahuac Plateau, etc).

Different in-situ investigations led to the discovery of the remnants of hydraulic structures supplying potable water to Cholula from the Iztaccihual volcano. This water supply was shared by the cities of Xoloc, Atlixco and Yanhitalpan. Later on, the Aztecs grew in number and dominated the whole region. They built new hydraulic structures, among which those built by Netzahualcoyotl may be seen today.

In order to prevent the pollution of Lake Tenochtitlan's fresh water by the salt water from Lake Texcoco, a dike was built and named after Netzahualcoyotl. A similar dike was built between the two lakes of Texcoco and Xachimilco by Tizoc.

The first new dams were built in the years following the Spanish conquer of the region. Their characteristics are given in Table 1.1. No spillway was provided at the Yuriraga Dam. In the case of the Saucillo Dam, a special section was provided for flood spills. The idea of spilling excess water became apparent during the 16th century A.D., not only in the Americas, but also in Europe.

TABLE 1.1

Characteristics of Early Dams in Americas

Name	Year of Compl.	River	Type	Height	Crest Length
Yuriraga	1550	Rio Lerma	Gravity	12	-
Saucillo	1730	Arroyo Socillo	Gravity	11	175
Natillas	1760	Rio Grande	Gravity	12	100
Huapaugo	1765	Rio Huapaugo	Gravity	14	840
San Antonio	1765	Arroyo Zarco	Gravity	11	160

The typical cross section of the Saucillo Dam was a 2.80 m thick masonry retaining wall with buttresses used to assure stability.

The Roman idea of two vertical walls for forming the cross section of the dam and the impervious material in between was not used in American dams. These structures were mostly vertical masonry retaining walls.

However, the canal builders of the pre-Inca Peru (Ortloff, C.R., 1988) showed the remarkable hydraulic engineering skill for solving flow problems. A very interesting case is the problem consisting of reducing the erosive power of supercritical flow upstream of an aqueduct in an irrigation canal, which conveys water to an irrigation area in the Intervalley Kingdom of Chimor, Peru. The stream lines in the upstream reach of the canal are visualized in Fig. 1.5.

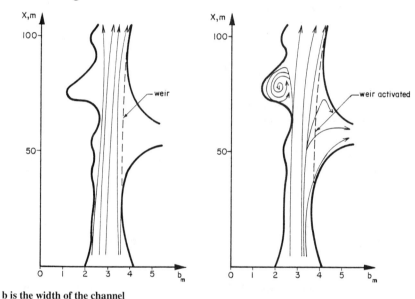

b is the width of the channel

Figure 1.5.   Flow visualization in the irrigation canal of Intervalley, Kingdom of Chimor, Peru.

Assume that the regime of flow is supercritical. During normal irrigation period the flow pattern is shown in the left diagram. The streamflow lines follow approximately the rigid boundaries of the channel. The side weir visible on the right shore is not activated. The aqueduct (not shown in figure), down-stream of the channel, can carry the flow easily.

During the flood time, the water depth increases. The left side groin is activated and a vortex takes place in the small cavity. This vortex directs the flow lines towards the right shore diminishing the wetted area, and increasing the flow depth which activates the overflow weir at the right shore. Eventually a hydraulic jump takes place in this region, as a function of the geometric characteristics of the channel, and the system works as a short structure. This has two effects:

1. The discharge of the right side weir increases, and

2. The discharge of the flow downstream of the jump decreases, enabling the aqueduct to carry the flood water without overtopping, due to the fact that for smaller linear discharge the flow depth in supercritical regime becomes smaller. Ortloff stated that a model experiment confirmed this case.

Effectively, it seems that such an interesting hydraulic achievement can be realized in modern times, but this solution was reached by Incas around 1000 A.D. Engineers showing such a high performance in solving hydraulic problems can also solve the water storage problems for irrigation of their fertile lands.

Unfortunately, the traces of old dams in these regions no longer exist, which would enable one to locate their spillways and discuss their construction. However, the unique hydraulic solution described above can be considered proof of the existence of early dams and spillways conceived by the pre-Inca engineers. A similar example of an advanced irrigational system is in Sichuan, China, near Chengdu. That system is nearly 19 centuries old, and most of the original features are still in use, along with several that were modernized.

1.2 HYDRAULIC STRUCTURES OF ANATOLIAN PENINSULA BEFORE THE OTTOMAN EMPIRE

As a special attention and an instructive example, the case of the ancient hydraulic structures before the Ottoman empire are discussed here in detail, since the author is very familiar with that region and the history of its hydraulic structures. The evidence of three kinds of hydraulic structures can be found in the Anatolian Peninsula, namely irrigation canals, water supply aqueducts and dams. The construction techniques used to build these three kinds of structures were similar. Dams were conceived as two perpendicular masonry walls, with an impervious fill inbetween. The construction technique for aqueducts was exactly the same. The only difference was in the impervious fill which was placed between the two masonry walls. The material used for aqueduct walls was not impervious, it was a simple rockfill. The irrigation canals excavated in the earth vanished during the course of the years, but the remnants of such conveying structures on rocky cliffs still exist. It is known that the water supply system of the ancient city of Side, located at the Mediterranean littoral (Fig. 1.6), was supplied from

Figure 1.6. **The Dumanli Springs (Q_{min} = 23 m³/sec), (Courtesy of DSI).**

the Dumanli spring. This spring was probably the largest karst spring from one outlet in the world (in the minimum around 23 cubic meters per second). Now, the spring is covered by the Oymapinar Reservoir.

Water from the Dumanli Spring was used also to irrigate the fertile plain of the Manavgat River. Similarly, water was conveyed to the other plains along the Mediterranean Sea in the Anatolia.

The small city of Demre was located southwest of Side in the Demre Plain. A tunnel dug through the Demre cliffs to supply the city with water was a spectacular sight. The Demre Plain was irrigated by a system of tunnels and canals from upstream karst springs. The extensive irrigation systems developed and maintained along the coastal regions supported a number of happy and financially prosperous communities.

1.2.1 Dams and Construction Techniques Used by Romans in Anatolia

It is known that Anatolia was the cradle of several past civilizations. Romans, Hellenes, Persians, Byzantines, Huns and finally Turks, particularly the Ottoman Turks, were inhabitants of this peninsula. The Ottoman Empire lasted 600 years and was succeeded by the modern state of Turkey. It is interesting to investigate the engineering skills of these past populations. Unfortunately, systematic research has not been undertaken yet, but it has been possible to identify some of the remains of old engineering activities. Neither the exact construction date nor the period during which they were built have been determined precisely, but they constitute the elements needed to explain the technical understandings of ancient peoples.

A. Purpose of Building Dams by the Romans

Some of the Anatolian dams, assumed to be from the Roman era, are: Çavdarhisar, Böget, Örükaya, Semali and Löştügün Dams. Their construction purpose was identical; they were flood control structures. They were equipped with the bottom orifice to discharge the flood waters. Some of them were also equipped with gates, so that the stored water could be used for irrigation. They were located at the entrance of small meadows and were surrounded by flourishing settlements.

B. Construction Techniques Used by the Romans

Overall, Roman construction techniques were similar for all hydraulic structures. The concept of a spillway had not yet been developed. Dam foundations were shallow. Prevention of damage from flood downstream of the bottom outlet orifice was not provided or, if it existed, it was not adequate. As a result, nothing remains to be seen that might indicate the technique used for construction. Probably a large number of these structures were subject to complete failure so that nothing at all remained. The only survivors were the ones which were constructed in rocky gorges or on a river with an aggrading bed. These earlier dams were formed by two masonry walls with shallow foundations and an impervious fill between them. In the Böget Dam, the main walls were fortified by an upstream impervious fill (Fig. 1.7).

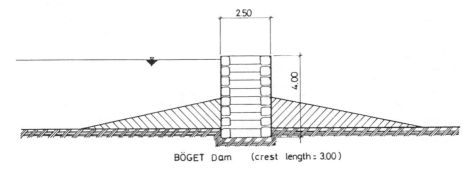

Figure 1.7. The Böget Dam; the crest length = 300 m.

The Çavdarhisar Dam is of a similar construction (Fig. 1.8). The thickness at the crest is 6.00 m. The upstream and downstream retaining walls were cut from the stone masonry. The dimension of stones are 1.00 x 0.60 x 0.60 m. Their weight is around 500 kgs. The Çavdarhisar River has an aggrading bed, so the structure is protected from flood damage. The walls are not tight because of the dry stone masonry. The impervious fill is of a poor quality, causing the dam to leak water.

A similar situation exists in case of the Örükaya Dam. The height of this structure is 16 m, the length at the crest is 40 m and the thickness at the bottom of the valley is 12 m. The mean dimensions of the cut stones are 1.00 x 0.70 x 0.70 m. Some of these stones are up to 2.50 m long and weigh some 3 tons. At the bottom of the dam, a flood discharge sluice can still be seen. This opening is curved at the top and is formed by using a special technique. Observations indicate that the reservoir formed by the Örükaya Dam was also used to store water for irrigation. Gate slots for closing the orifice can still be observed, but the spillway does not exist. It is possible to say that the Örükaya Dam was a relatively recent structure. Consequently, a new technique had been used to fill the joints between the stones. Molten lead was poured into the joints and when solidified, it bonded the stones. Recent investigations conducted by the Turkish National Research Council on this structure did find traces of lead in joints.

Figure 1.8. The Çavdarhisar Dam (Courtesy of Unal Özis).

The construction method used for the Böget Dam was similar to one used in the Örükaya Dam. The maximum height of this dam was only 4 m, its length 300 m. The center line bends downstream. The foundation of this structure is on limestone and an impervious blanket is used to prevent leakage. Spillway does not exist and the downstream discharge canal also does not exist. The dimensions of cut stones used in construction are 0.50 x 0.50 x 2 00 m.

One of the most interesting dams known in Anatolia is the Löştügün Dam (Stark, H., 1957). It is located on the Dövenci River in the northern region of Anatolia. The Dövenci River is a tributary of the Yesilirmak River, one of the largest water courses of the Peninsula. After traversing the limestone area it enters a region covered with loam and clay, and conglomerates as the valley widens. The Löstügün Dam is constructed in this region characterized by a limestone mass, 80 m long and 50 m large at the base which divides the river in two parts. The early dam builders considered this place a proper location for the dam. They did not use the serpentine available in the vicinity. They chose an earth fill structure instead of rockfill. This was an unique example of this kind of construction. The axis of the dam was placed in such a way that the limestone mass was incorporated into the dam fill, dividing the entire structure into the two distinct parts. The southern portion is 70 m long and the remainder is only 30 m long. The total length of the structure is some 130 m long including the limestone block. An obtuse angle exists between the two branches, the apex being on the limestone. The height of the dam is 12 m. Today grasses and trees covering the surface of the earth-fill make it look like a structure of uniform material. The width at the crest is 20 m and the width at the bottom is assumed to be 70 m. The upstream and downstream slopes are equal to each other and are estimated to be equal to 1:2 (1 vertical and 2 horizontal).

In general, the problem of spilling for earth dams was difficult to solve. A bottom orifice in such a case was of some use. The engineers of the Löstügün Dam applied a new solution. This solution was interesting but not sufficient. They used the rocky reach to locate the spillway. This spillway was a circular gallery with a one square meter section area starting at the river bed. Horizontal at the entrance, it extended to a sloping chimney which ended at a level higher than the river bed but lower than the crest downstream of it. The traces of erosion at the entrance showed that this system worked continuously for years with moderate success.

Today the water intake does not exist. It seems that it was located at the left bank of the valley. It is not known when the dam was constructed. The aggradation caused a superelevation of the river bed of approximately 1.00 m. A high flood demolished the left portion of the structure. Today only the right portion exists. A large dam, the Almus Dam was recently built at a location some 10 km upstream from the fertile region of the Löstügün Dam.

1.2.2 Dams Constructed by Urartu

One of the earliest structures on the Anatolian Peninsula is the Faruk Dam (Fig. 1.9a and 1.9b). The Faruk Dam is located at the Eastern region of Turkey near the Great Van Lake. The exact date of construction is not known. It is assumed that it was built by the Urartu Turks (900 - 600 B.C.). The Urartus were very skilled people. They were particularly fond of building hydraulic structures such as dams and irrigation canals (Şamram irrigation canal, 45 km long). The Faruk Dam was composed of masonry walls with impervious material between them. The foundations were shallow. It was equipped with an orifice which can be seen in Fig. 1.9b.

The development of engineering concepts of dam construction and particularly of spillways was rather slow. The idea of spilling excess water and measures to be taken to prevent leakage are rather recent. The following examples show these facts once more.

Figure 1.9a . The Faruk Dam; view from downstream.

Figure 1.9b. The Faruk Dam; bottom orifice.

1.3 DAMS CONSTRUCTED DURING THE OTTOMAN EMPIRE

The Ottoman Empire covered a large part of Europe, the Middle-East and Northern Africa at the end of the 16th century. The Ottoman Turks constructed important engineering works in that part of the world. Numerous bridges, roads, mosques and irrigation systems, including dams and canals, can be found even today in Algeria, Syria, Anatolia, Bulgaria, former Greece, Yugoslavia, etc. The particular techniques they used to build those engineering structures are unfortunately not known. They did not publish their knowledge so that one could be aware of solutions they used for solving different kinds of problems. However the following facts may be cited.

They invented the floating foundations. Some great mosques of Istanbul are floating on mud and move in accordance with the water level in the nearby sea. They knew how to determine the maximum bed degradation in rivers during floods. They placed foundations of many bridges below that level, and in such a way that those structures are in use even today. They invented a particular convection system for aeration of mosques. They built the largest cupola in the world during the 16th century (the Selimiye Mosque, Edirne, Turkey).

1.3.1 Techniques Used and Developed by the Ottoman Turks for Construction of Dams

The method of construction was not very different from the method used by the ancient Anatolian civilizations. However, some improvements were made and they are:

- The Ottomans used dams for water supply purposes.

- Dams were formed by two masonry walls with an impervious fill between them. The difference from the old technique was that a pozzolanic cement was used to bind the stones with the lead used to assure the total impermeability.

- Spillways did not exist on dams. The Ottoman Turks accepted the overtopping and lined the abutments accordingly. The crest and the downstream of dams were lined with marble (Fig. 1.10). For decades many floods were spilled using this system. Yet no major damage occurred. Naturally, such a solution, though spectacular was very

15

Figure 1.10. Crest Composition of an Ottoman Dam (Valde Dam, 1796).

expensive. However, reservoirs were also utilized by the Istanbul citizens as recreation areas and were equipped with various facilities.

The early Ottoman dams are in use even today, and the municipality is in charge of their operation. In recent years spilling facilities have been added to some of these dams. The early Ottoman dams were used to store water for domestic purposes and the special devices were used for delivering water to customers.*

---

* Special intakes were developed by Ottoman Turks and used for centuries. Their definitions for discharge units used, are:

- 1 lüle is the discharge of a tube discharging under hydraulic head of 96mm and having a diameter of 26 mm. It is approximately equal to 40 m^3/h

- 1 lüle = 2 half lüle = 4 kamis (flute) = 8 masura = 16 half masura = 32 needles = 64 crescents.

1.3.2 Characteristics of Ottoman Dams

The characteristics of some of the Ottoman dams built around Istanbul, Turkey are given in Table 1.2.

TABLE 1.2

Characteristics of Ottoman Dams Built for Water Supply of Istanbul

Measures are in meters.

Name of the dam	Name of the river	Dam height	Crest length	Crest thickness	Bottom thickness	Year of constr.
Osman II	Topuzlu	9.91	65.55	5.21	6.86	1684
Ahmet III	Topuzlu	9.41	65.50	6.20	9.44	1720
Mahmut III	Kirazli	11.60	45.50	7.00	9.00	1838
Topuzlu	Bahceköy	13.84	66.30	4.32	7.80	1748
Valde	Bahceköy	11.50	70.30	4.20	5.40	1796
New Dam	Bahceköy	15.45	95.00	6.60	9.49	1839

1.4 RECENT TECHNIQUES USED FOR CONSTRUCTION OF DAMS

New techniques were developed in civil engineering in the application of concrete and the use of earth moving equipment for construction of roads and retaining structures. A similar evolution in engineering of dams also has occurred during this period. A dam was no longer a simple structure, intended to store water in a reservoir. It was to be equipped with a spillway, an intake and a bottom outlet valve. The exact dates of development of these concepts are not known. However, these kinds of appurtenant structures were known to have existed at the end of the 19th century and at the beginning of the 20th century.

The first spillways were free flow structures. A downstream discharge channel carried water to a stilling basin and the restitution of water to the main stream was assured after complete dissipation of energy was achieved. Much research was conducted on these subjects during the decades following the year 1930. The development of hydraulic laboratories and techniques of modeling were also developed during that period.

The progress in technology of concrete, as applied to arch dams, marked the beginning of a more advanced period of dam construction. The dam heights were increased. During World War II the air attacks caused a lot of anxiety. The idea of deviating the air bombs downstream of dams preventing their blast, is at the base of the modern concept of ski jump spillways. During this period, a great deal of research on this type of spillway underlined the works of hydraulic laboratories. Different forms of flip buckets were developed and dam heights were increased accordingly. Based on the newly born technology, dams as high as 200-300 m were built.

Today, dam engineering forms a separate chapter of civil engineering. Thousands of dams have been constructed around the world and hundreds of them are under construction.

Chapter 2

INTRODUCTORY CONCEPTS ON DAM SPILLWAYS

2.1 INTRODUCTION

Spillways are hydraulic structures intended to bypass flood water which cannot be contained in the allotted storage space of reservoirs. They constitute a basic and important element of a dam. In case of a diversion structure the spillway forms the structure itself. Their use is worldwide, but they have only been studied rationally in the second decade of the 20th century.

The design of spillways takes into consideration different factors such as:

- Selection of spillway layout

- Spillway components (approach channel, control structure, discharge channel, terminal structure)

- Hydraulic computations

- Stability studies

- Economic studies.

An acceptable design can only be obtained as a consequence of serious consideration of factors affecting the final choice. This choice must be the safest and the most economical of all possible solutions.

2.2 SELECTION OF SPILLWAY LAYOUT, SPILLWAY TYPE AND CLASSIFICATION OF SPILLWAYS

2.2.1 Selection of Spillway Layout

The selection of spillway layout depends on various factors:

- Type of dam

F. Sentürk

- Amount of excavation and possibility for its use as embankment material

- Stability of foundations and excavations

- Hydraulic conditions

- Overall economy.

A. Spillway Layout for Embankment Type Dams

In embankment dams the spillway can be located in a nearby valley in such a way that the dam body maintains its integrity (Fig 2.1a - Fig 2.1b), or, instead of a surface spillway a shaft or tunnel spillway can be used (Fig. 2.2).

If such a solution is found impractical, the spillway can be placed on a concrete structure incorporated with the earth fill or rockfill embankment (Fig. 2.3). The gravity dam on which the spillway is placed can be a lateral structure as in case of the Keban Dam (Turkey) or a central structure is used as in case of the Escollera Dam (Mexico) to suit geological conditions of the dam site.

B. Spillway Layout for Concrete and Masonry Dams

For gravity concrete and masonry dams, the spillway can be placed on the dam body (Fig. 2.4).

In arch dams, if a free flow spillway is the solution, it can be placed immediately below the crest of the dam and water is discharged freely into the air (Fig. 2.5). If sufficient crest length is not available, the solution can be a gated orifice spillway. In this case the leakage problem which may exist around the gates due to high pressure must be solved (Figs. 2.6a and 2.6b).

2.2.2 Type and Classification of Spillways

The type of spillway is a function of the solution selected for solving the spilling problem:

1. Free flow spillways

2. Gated spillways or controlled crest spillways.

Each of them includes different types of control structures. Table 2.1 summarizes the classifications selected in this book.

River: Ilhan; Type: Earthfill; Height: 50 m; Year of Completion: 1980

Figure 2.1a.   The Asartepe Dam Spillway, Turkey.

River: Kitayama, tributary of Kumano; Type: Concrete arch
Height: 111 m; Year of completion: 1964

Figure 2.1b. The Ikehara Dam Spillway, Japan. (Courtesy of Japanese National Committee on Large Dams).

Figure 2.2. Gate controlled Spillway of the Round Butte Dam placed in the diversion tunnel (USA) (Courtesy of US Committee-Inter. Comm. on Large Dams).

River: Firat; Type: Rockfill and concrete gravity
Height: 207 m; Year of completion: 1975

Figure 2.3. Gravity structure on which the spillway is placed at the Keban rockfill dam, Turkey.

Figure 2.4. Spillway of the Sariyar Dam, Turkey.

Figure 2.5. Freeflow spillway of the Bangala Dam, South Rhodesia (Courtesy of Coyne and Bellier).

Figure 2.6a. **The Kukuan Dam Spillway, Formosa (Courtesy of Coyne and Bellier).**

Figure 2.6b. Leakage from the gate seal.

TABLE 2.1

Classification of spillways

I-Free Flow Spillways

1.1 -   Classification in respect to geometry of control structures
        - Straight drop spillways
        - Special types of control structures.

1.2 -   Classification in respect to the profile selected
        - Creager's profile
        - Second degree parabola profile
        - Third degree parabola profile
        - Compound profile
        - Broad-crested profile.

1.3 -   Side spillways

1.4 -   Shaft spillways

1.5 -   Siphon spillways

1.6 -   Box inlet drop spillways

II-Gated Spillways

2.1 -   Flash boards and stop logs

2.2 -   Rectangular lift gates

2.3 -   Radial gates

Several examples are given to define different types of spillways.

A. Straight-Drop Spillways

They are in general used together with thin arch dams (Fig. 2.7). Flow is discharged freely and its energy is dissipated while dropping through the air into a natural or artificial stilling basin. A certain hazard must be considered in using this kind of spillway. If good aeration of the lower nappe is not

**Figure 2.7. Straight-drop spillway, the Ribou Dam, France
(Courtesy of Coyne and Bellier).**

assured, a pulsating flow can take place on the spillway. In general, groins are used to break the uniformity of the nappe enabling the atmospheric pressure to build up beneath the lower nappe (Couesque Dam, France, Fig. 2.8).

B. Special Types of Control Structures

These types of control structures are extensively used in France and in North Africa. Figure 2.9 illustrates one of these types. This solution is applied particularly for solving problems due to lack of space. The crest length is increased by the use of a special type of structure in the form of a marguerite as shown in Fig. 2.9.

C. Creager Type Profile

This type of profile is extensively used in spillway engineering. A detailed study of Creager profile will be given in Chapter 3. The build-up of subatmospheric pressures can be avoided in this type of profile.

Figure 2.8. The Couesque Dam Spillway, France (Courtesy of Coyne and Bellier).

Figure 2.9. Marguerite type control - the Sarno Dam Spillway, Algeria.

D. Second and Third Degree Parabola Profiles

They are used for controlling subpressures on a gated spillway. They are extensively used in Europe and in the United State (see Chapter 3). Figure 2.10 shows this kind of profile used most often .

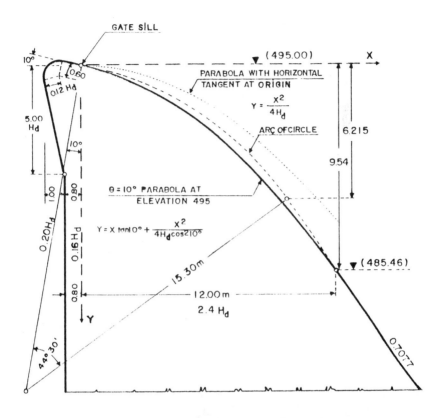

Figure 2.10. Comparison of different profiles.

E. Compound Profile

It is used for facilitating the engineering feasibility of the control structure. Figure 2.11 defines such a profile.

F. Compressed Profile

This type of profile is generally used on weirs.
The nappe drops freely, coming from the crest without adhering to the weir profile (Fig. 2.12).

29

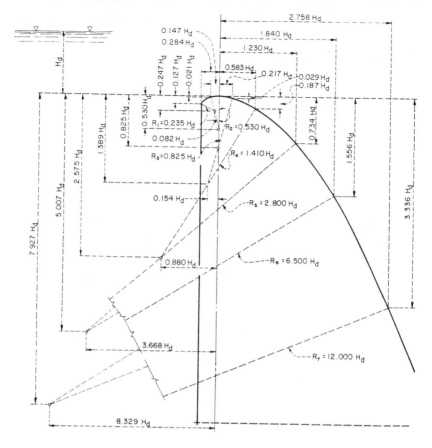

Figure 2.11. Compound profile, USBR, (from the book: "Design of Small Dams").

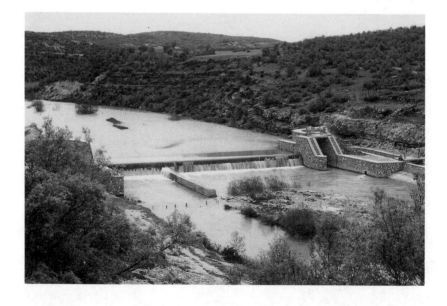

Figure 2.12. Compressed profile, the Kütahya Weir, Turkey.

G. Broad-Crested Profile

This type of profile in general is used for weirs. A complete description of broad-crested weirs is given in Chapter 5.

H. Side Spillways

This solution has been applied in past years for solving layout problems in narrow gorges and in the case of large design floods (Fig. 2.13).

Figure 2.13. Side Spillway - the Porsuk Dam Spillway, Turkey.

I. Shaft Spillways

It is still used in many cases for solving layout problems in narrow gorges and rather small design floods (Fig. 2.14).

J. Siphon Spillways

The siphon spillway is used as an auxiliary spillway in case of small design floods. The self-priming and self-regulating action is sometimes retarded or does not function satisfactorily; for this reason, it is not entirely safe.

Figure 2.14. Shaft Spillway, North Africa, the Sarno Dam Spillway.

K. Box-Inlet Spillways or Tunnel Spillways

Figure 2.15 shows such a spillway. It is used for solving special delivery problems.

Figure 2.15. The Yellowtail concrete tunnel spillway, USA (Courtesy of U.S. National Committee on Large Dams).

L. Gated Spillway

Gates are generally used in large dam spillways which discharge large floods. Figure 2.16 shows such a spillway.

Detailed definitions and characteristics of these structures are given in the following chapters. Their hydraulic computations are given in Chapters 3 and 4. Also, many examples are presented illustrating the application of theoretical formulas and the basis for selection of the proper formula.

Figure 2.16. The Aigle Dam Spillway, France (Courtesy of Coyne and Bellier).

2.3 SPILLWAY COMPONENTS

The components of a spillway are:

*   The approach channel
*   The control structure
*   The downstream canal
*   The terminal structure.

Figure 2.17 shows only the three components of a spillway (without the approach channel).

Figure 2.17. Spillway components, the Atatürk Dam Spillway,
Turkey: 1-Downstream channel, 2-Control structure,
3-Terminal structure (Courtesy of DSI).

2.3.1 Approach Channel

The approach channel is a conduit intended to direct the flow towards the control structure. It is generally used in embankment type dam spillways. When the spillway is placed on the body of a concrete structure it may be eliminated, but if the approach channel is deeply submerged as in the case of Keban Dam, Turkey, helicoidal currents can take place complicating the flow phenomenon. The following features are taken into consideration when designing the approach channel.

A. Upstream Inlet

The design of the upstream inlet may solve two basic problems:

*   The decrease of the entrance head losses

*   The stabilization of the flow in the channel.

The stabilization of the flow in the channel is particularly important. Its effect is not local and it influences flow in the control structure, the discharge canal and terminal structure.

The entrance head losses, if not studied carefully, attain important values where either the spillway cannot evacuate the design flood or the freeboard diminishes dangerously. An interesting example is the upstream inlet of Gökçekaya Dam, Turkey. The original design was tested in the Hydraulics Laboratory of the Research Department, Ankara, Turkey and the head loss at the entrance was found to be equal to 1.40 m.

B. Layout

The layout of the approach channel constitutes the geometrical factor influencing the stabilization of the flow. In general, the channel presents a curvature in plan which causes the flow to pile up towards the convex side of the conduit. Adequate side piers on the control structure may correct this inconvenience, but this solution is costly and must be avoided.

In case of inadequate design or malfunction of the hydraulic structure, the upstream water level increases beyond the limits imposed by the designer and water is discharged in the approach channel by overtopping from the upstream guide wall completely disturbing the stability of the flow.

C. Maximum Velocity Permitted

Criteria do not exist for determining the maximum allowable velocities in the approach channel. However, it is recommended that excess velocity, which causes high head losses and local erosion in the conduit, be avoided. As a maximum value, $4.5 \cong 5.00$ m/sec can be selected. In the example of the Gökçekaya Dam, Turkey, the accepted values are:

$Q_{max} = 5300$ m³/sec $H = 19.50$ m $L = 62.00$ m

$U_{max} = 4.5$ m/sec $U = 2.00 \cong 2.5$ m/sec

2.3.2 Control Structure

The major component of a spillway is the control structure. It is composed of different elements defined as follows.

A. Side Walls

This element constitutes the transition structure. It is formed by plane surfaces as shown in Fig. 2.18 or rounded surfaces as shown in Fig. 2.19.

A selected solution is a function of layout of the spillway, the topographical conditions and the hydraulic conditions.

Figure 2.19. Rounded transition between the approach channel and the control structure of the Hasan Ugurlu Dam (Courtesy of DSI, Turkey).

Figure 2.18. Transition in the downstream channel of the Seyhan Dam Spillway, Turkey.

B. Side Piers

Side piers take the place of side walls when an approach channel does not exist. Figure 2.20 shows the side piers of Keban Dam, Turkey. They are designed to correct the flow entering the spillway. The same idea was also used in Spain for correcting the flow in Aldeadavila Dam Spillway. If the side piers are eliminated, the flow is disturbed over the control structure. This results in higher discharge in some bays and diminished in others. Measurement on scale models has shown head losses as high as 1.5 meters and differences in discharge capacity between central bays and lateral bays attaining some 40%. In some instances, particularly at weirs, the flow does not pass through the center bays in an upstream downstream direction, but instead is reversed so that it enters the upstream pool through the side bays.

Figure 2.20. Side piers of the Keban Dam Spillway, Turkey.

C. Control Devices

The control devices are the central and basic structures of a spillway. They accelerate flow and control it at the crest. It can be placed at the dam body as in the case of the Keban Dam Spillway for example (Fig. 2.3) or in the approach channel as in the case of the Asartepe Dam Spillway, Turkey (Fig. 2.1a). This structure will be studied in detail in Chapters 3 and 4.

D. Piers

Piers are designed to support gates. Hydrostatic forces, wave forces, earthquake forces, etc., are transmitted to foundations by means of piers. The clearance and the geometric form of piers are a function of the gate and flow characteristics. Also the geometric form of piers must be studied together with the geometric form of the control structure (see subsection 4.2.2E). The components of piers are the side gate seal plates, the stop log slots, the ladders, etc. Sometimes the gate operating devices are also placed on the piers.

E. Operating Bridge, Road Bridge

The operating bridge and the road bridge are supported by piers. The gate hoisting device is a separate structure placed on the operating desk. In some instances a road bridge is designed to enable trucks to reach the hoisting device and for general traffic use.

F. Gates

The gates are used to control the water level in the reservoir. During flood time they are partly or completely raised and flood waters are evacuated from the spillway. Different types of gates are used in spillways or weirs. In large dams, radial gates are usually chosen; plate gates, wagon gates, etc., are used for solving different kinds of hydraulic problems for weirs.

2.3.3 Downstream Channel

The excess flood waters are evacuated through the spillway bays first, then through the discharge channel. The width of this channel can be chosen equal to or less than the length of the control structure. If smaller, a transition structure is inserted between the spillway and the channel (Fig. 2.18). Where large floods are expected, a large downstream channel is required. Under such conditions, the spillway piers are extended to the terminal structure. Such a solution has been adopted for the Keban Dam downstream channel. The height of the separating wall must be chosen in such a way that overtopping be avoided during flood time (see subsection 6.2.2). In designing layout and the longitudinal section of the downstream channel, curvatures must be avoided. If a curve is placed in plan, the centrifugal forces cause a pile up of flow opposite to the center of curvature. If the curve is placed in the vertical section, subpressure may occur at the bottom of the channel; both cases are dangerous.

2.3.4 Terminal Structures

There are two type of terminal structures:

1. Structures permitting the energy to dissipate at the downstream water level,

2. Structures permitting the energy to dissipate above the downstream water level.

The first category is known as stilling basin and the second as a ski jump (see Chapter 6).

2.4 BASIC CONCEPTS IN THE CHOICE OF DESIGN FLOOD AND THE OPERATION OF RESERVOIRS

2.4.1 General Consideration

The choice of design flood constitutes the basic factor in design of spillways. In past years many examples of dam failure caused by the choice of an inadequate design flood alerted engineers to reconsider the problem of spillway design. The failure of the Malpasset Dam in France caused more than 500 deaths and the catastrophe of the Vaillont Dam, Italy, about 300. In recent years the failure of the Bend Deresi Dam, Turkey inundated the capital city of Ankara and caused the death of one person and heavy damages. In the USA also, many examples of dam failure can be cited. These failures reflect the importance of a problem which can be resolved by a proper choice of the magnitude of the inflow and the volume of the flood hydrograph.

The design flood is obtained by using the unit hydrographs and the catastrophic rainfalls. Flood routing yields the design floods which can be used for sizing of spillways.

A rather empirical method is also used in Europe. The return periods of flood discharges are computed by means of an acceptable procedure. In general, Gumbel's method is chosen for obtaining the return period curve. The design flood is then chosen from the computed return periods. Table 2.2 summarizes different possibilities used particularly in Turkey.

TABLE 2.2

Return periods which can be selected for spillway design floods

Type of hydraulic structure	Return period in years
Earth dams	Infinity
Rockfill dams	Infinity
Gravity dams	1,000
Arch dams	1,000
Weirs (near populated rural areas)	250
Weirs (rural areas)	100

The apprehension of dam failure is so great that sometimes extreme values for magnitude of the design flood are adopted. Two examples, taken from Turkey are used here to illustrate this fear of the unknown.

1. The maximum recorded flood at Keban Dam was 6,000 m^3/sec for a return period of 45 years. The International Board charged with the review of maximum flood design criteria decided to use 30,000 m^3/sec.

2. The maximum recorded flood at Hirfanli Dam was 2,000 m^3/sec, the magnitude of the design flood was taken equal to 8,000 m^3/sec. (The return period of 2,000 m^3/sec is 60 years).

Naturally these figures were extra safe but required a more costly solution not only for the hydraulic structure but also for all of the downstream water works.

2.4.2 Magnitude of the Outlet Flow

The volume and the magnitude of the design flood hydrograph influence directly the spillway type and dimension. Many solutions exist for routing the flood to obtain the design flood hydrograph. In general, computers and calculators, on a smaller scale, facilitate the use of mathematical solutions instead of graphical ones.

Bresse has written the equation of continuity as follows:

$$Q_i \, \Delta t - Q_o \, \Delta t = \Delta h \, A \qquad (2.1)$$

where

Q_o is the outflow discharge

Q_i is the inflow discharge

Δh is the increase of the depth of reservoir

Δt is the time increment, and

A is the area of the reservoir free surface.

$Q_i = f(t)$ and $A = \emptyset(h)$ can be obtained from in-situ measurements; the time increment is to be chosen by the engineer and the function $Q_o = f_1(h)$ is also to be chosen by the designer. At the beginning of time scale (t_o) and at the end of the interval $(t_o + \Delta t)$, Q_{i1} and Q_{i2} are known. The mean value of the inflow discharge to be considered in Eq. 2.1 is then

$$Q_{im} = \frac{Q_{i1} + Q_{i2}}{2}$$

The surface area A_1 of the reservoir at time t_o and A_2 at time $t_o + \Delta t$ also can be obtained if the depth increment Δh is known, then

$$A_m = \frac{A_1 + A_2}{2}$$

Substituting these values in Eq. 2.1 yields

$$\frac{Q_{i1} + Q_{i2}}{2} \, \Delta t - Q_o \Delta t = \frac{A_1 + A_2}{2} \, \Delta h \qquad (2.2)$$

Assume the increment of the hydraulic head on the spillway is Δh and the head at the beginning of time period h_o. Then the total hydraulic head will be $h_o + \Delta h$ and $Q_o = f(h)$. A trial and error must be applied for solving Eq. 2.2. Δt is chosen and Δh is assumed, then $Q_{im}\Delta t$ is known; $A_m\Delta h$, $Q_o\Delta t$ can be computed. The obtained value of Δh at the end of the trial and error corresponds to the chosen values Δt (Problem 2.02). Figure 2.21 shows the outflow hydrograph.

The top point of the outflow hydrograph is always situated on the inflow hydrograph and

$$Q_d = Q_{o\ max}$$

Q_d is the maximum discharge to be adopted for sizing the spillway capacity.

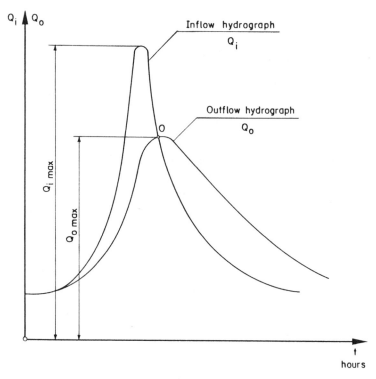

Figure 2.21. The inflow-outflow hydrograph.

2.4.3 Operation of Reservoirs

Reservoir operation constitutes a special branch of hydraulic engineering. Only a general description of it will be given in this section. Good operation is necessary for obtaining efficient use of the available reservoir capacity. Reservoir operation requires the consideration of a competent forecast so that the reservoir is full at the end of the rainy period and empty at its beginning. A median reservoir water level is usually the basis for obtaining the maximum turbine efficiency. These considerations may seem to be contradictory.

The problem can be solved by a comprehensive study of the yield of reservoir which leads to the choice for the volume requirement of the reservoir. Naturally a mathematical model must be developed and efficiently solved to determine the effective yield of the reservoir during many hydrological cycles.

The spillway is equipped with gates to discharge excess water in case a catastrophic flood occurs. These gates, when opened completely could evacuate a catastrophic discharge to the downstream valley causing heavy damages. The danger of hoisting the gates improperly during a normal flood is a problem that must be studied attentively. Such contingencies are studied and discussed in International Forums (ICOLD, Istanbul Congress).

2.5 SELECTION OF SPILLWAY TYPE

The selection of the type of spillway must be done by a combination of the storage capacity of the reservoir, the discharge capacity of the outlet, the type of the dam, the topography, the geological conditions of the spillway site, pertinent factors of hydrology, hydraulics, cost and involved risks.

Risks related to dam failure directly influence the choice of the type of spillway. The outflow must be chosen in such a way that it can be carried by the main water course downstream of the dam without any overtopping which could endanger the population (risk of life) and private property. The solution of the problem is difficult to reach and the data are costly to gather and take a lot of time and effort. The type of dam determined as a result of this study must be checked once more by flood routing and by determining the high water levels downstream of the dam. In case the imposed boundaries are exceeded, the study of type determination must be repeated.

If the determination of the dam type presents more than one solution, the most economical one should be chosen. Once the spillway type is chosen, its cost and the cost of the dam can be evaluated and the combined costs can be compared. A general discussion of factors influencing the choice of the type of spillway follows.

2.5.1 The Selection of Spillway Type as a Function of the Type of Dam

The type of spillway is influenced by the type of dam. The control structure, for example, cannot be placed on the body of an earth fill dam as in the case of a gravity dam. A free flow spillway as used on a thin arch dam cannot be adopted for a gravity dam; it is clear that the type of dam influences the type of spillway.

In case of concrete dams, when the valley is narrow and the hills steep, it may be difficult to locate the spillway near the hill because of lack of space. Then the spillway can be placed on the body of the dam, or a lateral spillway can be chosen. In considering an arch dam, a free fall spillway can be chosen without a downstream discharge canal (Fig. 2.8, the Couesque Dam).

In case of embankment dams and for steep hill sides, the solution can be a shaft spillway or a spillway placed on a lateral valley, if such a valley exists. A more frequently chosen type is a concrete structure with a spillway intercalated between sections of an embankment dam (the Keban Dam).

In case of large floods and narrow valleys, the spillway crest length can be increased by using special types of spillways as shown in Fig. 2.9. A more universal type of long spillway crest is shown in Fig. 2.22.

Figure. 2.22. The spillway of the Mehmet Sumra Dam, Turkey.

An interesting type of spillway is used in Europe for increasing the value of the discharge coefficient. The gates are located to block the entrance of the spillway in such a way that the hydraulic head is greater than the height of the gate which is limited by the dam body. The spillway works as a bottom valve. Naturally the discharge of the spillway is increased but it leads to the important problem of placing tight water seals between adjacent sections (Fig. 2.6a).

2.5.2 Selection of the Type of Spillway as a Function of Topography of the Dam Site

The selection of the type of spillway as a function of dam type generally results in more than a single solution. The exact location of the spillway is determined according to topographical data. Economical considerations take the first place in this determination.

2.5.3 Selection of the Type of Spillway as a Function of the Geology of the Dam Site

The geology of the dam site influences directly the choice of the spillway type regarding:

- Stability of the foundation

- Possibility of the use of excavated material in the dam embankment.

The latter, in particular, influences the combined cost of the dam and the spillway. In many cases the geology of the foundation shows faults. A detailed investigation of these faults may alter the layout of the spillway as in the case of the Keban Dam, Turkey and the Infiernillo Dam, Mexico. If a ski jump terminal structure is chosen, the foundations must then be studied taking into consideration the vibrations caused by deflectors. Geological investigations pertaining to foundations can be grouped as:

- Erosion capability

- Deformations due to vibrations

- Effects of high hydrostatic pressures.

The geologist in charge prepares:

- Geological longitudinal sections and cross sections and geological plan of the spillway site

- Position of the faults in plan and in cross sections

- Variation of ground water levels as a function of time.

  For each particular case a meticulous geological study should be performed.

A. Existence of High Hydrostatic Pressures

High hydrostatic pressures are built up downstream of high dams. Figure 2.23 shows the effect of such pressure downstream of the Keban Dam, Turkey. Heavy grout injections have been made to fight infiltration. In such a case, the foundation rock must be massive, free of faults, and extending deeply into the ground. If the rock does not have these characteristics, a shaft spillway can be chosen. A foundation without faults is difficult to encounter. However, if active faults exist it is recommended that the type of spillway not be changed but the location may be changed. For solving a similar problem at the Keban Dam, the location of spillway was in fact, changed.

Figure 2.23. Leakage through existing rock downstream of the Keban Dam due to high hydrostatic pressure.

B. Existence of Vibration on the Control Structure or on the Terminal Structure

When subatmospheric pressures build up on the control structure, vibrations are immediately generated. Also the terminal structure is often subject to vibrations due to high velocities of flow and the generation of vacuum beneath the lower nappe. In such cases the foundations must be stable and capable of supporting vibrating loads. If not, the type of spillway must be changed and the ski jump avoided. In Keban Dam the ski jump restitution structure was designed at first to deviate the flow by means of concrete deflectors. It turned out that these deflectors regenerated vibrations so heavy for the flip bucket foundation to support. This led to elimination of deflectors.

In the case of stilling basins, vibrations cause compaction of the structure foundation and must also be avoided. The following types of spillway are sources of extra vibrations:

- Shaft spillways
- Siphon spillways
- Drop inlets, and
- Spillways where subatmospheric pressures are regenerated.

C. Spillways with Erodible Foundations

Spillways with erodible foundations are difficult to handle. The foundation rock is weathered or is very deeply located. The foundation material is either loose or is not sound enough. In such cases only straight spillways without any subpressure must be chosen; then an extra compaction due to vibration is avoided. In order to prevent erosion exactly at the toe of the dam structure the flood discharge must be evacuated at a point far downstream (Fig. 2.24).

Basalts and weathered andesites must also be avoided to assure the stability of the foundation in cold regions. Another solution consists of the use of two spillways; one for low pressures and the second for high pressures. The flood discharged by the high pressure spillway must be carried downstream far enough from the dam toe. An example is given in Turkey's Demirköprü Dam. Even such a solution was not enough to prevent deep erosion of the downstream basalts.

2.5.4 Selection of Spillway Type as a Function of Hydrologic Conditions

The magnitude of the design peak of flood hydrograph and its volume have important impact on the design of spillways. A second factor is the reservoir operation. These two factors considered together play an important role in decision making for the determination of the type of spillway. Hydrologic conditions and upstream flow conditions combined are the essential factors for the differentiation between a free flow spillway and a gated spillway. Then the design flood hydrograph is used for sizing the control structure. The (Q_{max}/Q_{100}) ratio is an important factor in this task. For

$$\frac{Q_{max}}{Q_{100}} > 3$$

47

River: Tone; Type: Concrete Arch; Height: 131 m; Year of Completion: 1967

Figure 2.24. The Yagisawa Dam Spillway, Japan (Courtesy of the Japanese National Committee on Large Dams).

an emergency spillway should be considered for economical reasons. The design flood discharge may then be chosen by satisfying the double inequality,

$$Q_{max} > Q_d > Q_{100.}$$

For discharges greater than Q_d the emergency spillway will spill the excess water. The floods of the Firat River at Keban Dam site can be taken as an example.

$$Q_{100} = 9,000 \text{ m}^3/\text{sec}, \quad Q_{max} = 30,000 \text{ m}^3/\text{sec},$$

then

$$\frac{30,000}{9,000} > 3$$

An emergency spillway can be designed to solve the problem. In fact, the first design of this spillway was a fuse-plug together with a gated spillway (Şentürk, F., 1965). The emergency spillway used at the Seyhan

Dam and developed as a fuse-plug allows for hydrologic errors by the designer to be taken into account. The problem is reduced to selecting Q_d, which is, in this case, a function of the reservoir operation. A function of Q_d is also the characteristics of the control structure. The existing relation between Q_d and the coefficient of discharge and the design head are outlined in section 3.3.

In general, the economic life of a reservoir is assumed to be around 100 years. Then the spillway may easily discharge Q, but it is customary to assume

$$Q_d > Q$$

In case that $Q_{max} \cong Q_{100}$ then a free flow spillway can be chosen yielding an economical solution. For $Q_{max} > Q_{100}$ gated spillways are preferred to free flow spillways. Dam failures in all cases must be avoided, so modern designers choose conservative values for the magnitude of design floods.

For free flow spillways the beginning of vibration of the nappe must be determined carefully on design hydrographs. Assume that vibrations begin at time t_1. In this case the duration of vibrations will be $\Delta t = t_2 - t_1$, t_2 being the time at which the vibrations end (Fig. 2.25).

In reality the discharge at which the vibrations end is greater than Q_1, the discharge at the beginning of vibrations. Then Δt is the virtual time duration and the real time duration will be

$$\Delta t = t_2' - t_1$$

If the period of vibrations as shown in Fig. 2.25 is acceptable, then it should be $\Delta t \leq 6$ hours. This inequality is arbitrary. It depends on experience of the designer. A detailed research is not performed on this subject (Partensky,W., 1965). If the inequality is not satisfied then the shape of the ogee must be corrected. The first measure to be taken is to increase Q_1 in this case, then the time increment will decrease.

2.5.5 The Selection of Spillway Type as a Function of Economic Conditions

A. Use of Materials Obtained from Excavations of Foundations of Spillways as Fill Materials

The transport of fill material influences the total cost of embankments of dams. If the excavated material from foundations of the approach channel, the control structure and the downstream channel can be used in the construction of dams, it will contribute to reducing the overall cost. Where the spillway is near the dam the transport cost of material is naturally

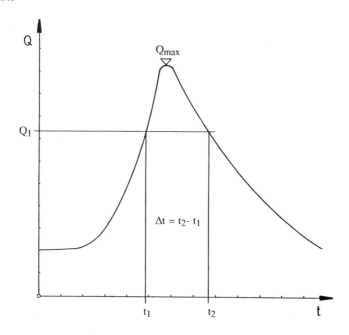

Figure 2.25. The duration of vibrations of a free flow spillway.

reduced. The primary consideration is the quality of the rock or the other suitable excavated material. In case the material is good enough to be used for embankment fill, the cost of the dam is reduced. Where good rock is available in the foundation of spillways, its components will be used as a quarry and so plays an important part in the selection of the type and location of spillway.

B. Economic Study for Free-Flow Spillways

When the technical feasibility of constructing the spillway or the combination of dam and spillway is established, the economic feasibility studies determine the final spillway selection. The best solution will be the most economical among various options without compromising integrity of the structure.

The total cost of spillway is equal to the sum of costs of various elements comprising the main structure. The total cost pertaining to each technically feasible alternatives is computed separately and the variations are plotted on a system of coordinates as shown in Fig. 2.26.

The cost elements forming the total cost are:

1. Cost of hydraulic structure

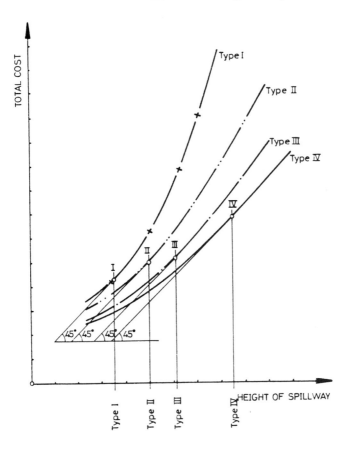

Figure 2.26. Economic feasibility study for spillways.

2. Increase of the total cost due to the increase of the cost of the dam as a function of increased height of spillway

3. Cost of the right-of-way

4. Extra risk cost due to catastrophic floods

5. Variation of the cost of spillway in case bottom valves are also used to evacuate the flood waters.

The cost estimate should be prepared for each technically feasible type and its variation as a function of the spillway height. This is shown graphically in Fig. 2.26. In that figure, the variation of the cost of the four technically feasible types of spillway, are shown. For example, the increase in the rate of change of cost of the type I spillway is smaller to the left of point I than to the right of it. The variation from this point on follows a certain curve defined by points I, II, III and IV. This curve is the locus of optimal points characterizing the variation of cost as a function of various technically feasible types of spillways.

Figure 2.27 shows similar variations taking into consideration not only the spillway but also the dam. The cost variation curves here represent the variation of the combined cost of spillway and dam. Two different types are shown in Fig. 2.27. On the two figures (Figs. 2.26 and 2.27) the total cost is shown along the ordinate axis.

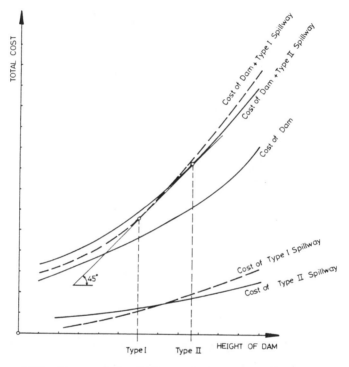

Figure 2.27. Economic feasibility study for the combination of dam and spillway.

The cost of the dam and the spillway combined represents the largest item influencing the total cost. Minor influences are:

- Increase in the coefficient of discharge; this increase yields a decrease in the length of control structure, which reduces the cost of spillway.

- Type of spillway chosen may cause an additional increase in cost of the upstream and downstream features of the dam. The extra cost of right-of-way should be added to the total cost of the structure.

- Discharge capacity of spillway is increased to allow for safety from the unknown factors. This added cost may be added to the total cost of the structure.

- If bottom outlets are used as an extra flood evacuation, the decrease in the design flood may be considered.

The comprehensive study for determining the final, selected type of spillway is the prime request, especially for high dams, because the type of spillway chosen has the major impact on the total cost (Problem 2.05). Even in this case the final decision must be taken by the chief designer.

C. Economic Study of Gated Spillways

If a gated spillway is chosen, the first step to be taken is to finalize the optimization study for gates. This study is usually conducted as follows:

1. The water level in reservoir is assumed constant, so that the hydraulic head will stay constant in the optimization study.

2. The minimum number of gates must be two. The hydraulic computation should be prepared for gates 3, 4, 5, etc; with their respective dimensions computed.

3. The cost estimate relative to gates 3, 4, 5, etc. should also be prepared. Excavation, transportation, civil engineering and gate costs are taken into consideration in preparation of the final total cost estimate.

4. The total cost versus the number of gates is plotted and variations in the corresponding curve studied. This curve will have a minimum point. The number of gates nearest to this point should be chosen for the required number of gates, if no other factor affects this decision.

5. An optimization study should be prepared for each case of determining the number of gates.

2.6 BASIC DATA USED FOR SPILLWAY DESIGN

The basic data used for design of spillways can be grouped as follows:

- General data:

  a - Topographic data

  b - Hydrologic data

  c - Climatic data

    d - Geologic data

    e - Ecological and environmental data.

- Reservoir data:

    a - Reservoir map

    b - Road and public utilities map

    c - Miscellaneous data.

2.6.1 General Data

The nature of these data and the sources from which they may be obtained are given in subsequent sections.

A. Topographic Maps

1. Quadrangle maps, planimetric maps, aerial photographs
2. River plan and profiles
3. National parks and monuments
4. Local specific areas.

These maps may usually be obtained from governmental offices, such as the General Directorate for mapping in Turkey; the U.S. Department of the Interior, Geological Survey and the U.S. Department of the Army, in the United States of America.

In general, topographic data required for the design of spillways are (USBR, 1974):

Local and vicinity maps:

1. Location of the project
2. Location of the existing structures
3. Public utilities such as highways, railroads, airports, etc.
4. Stream gauging, sampling, meteorological, etc., stations.

In general, the scale of the map used in design and survey is chosen as follows:

- Large scale maps: 1/25,000
- Planning maps 1/200 - 1/100.

B. Hydrologic Data

- Stream flow records (daily discharge, monthly volumes, recorded peaks)

These data can be obtained from the U.S. Department of Commerce, U.S. Corps of Engineers in the United States, and from EIE (General Directorate for Electrical Survey) in Turkey:

- Inflow hydrographs;

- Sediment measurements, analysis of dissolved solids;

- Ground water observation records in the vicinity of the dam;

- Water rights.

The reservoir yield, the design flood hydrograph, etc, are computed from these basic data. Project water requirements such as allowances for irrigation and power efficiencies, conveyance losses, reuse of return flow and stream releases for fish must be provided by involved authorities. The dead load storage will be determined by the engineer on the basis of these data.

The use of hydrologic data is outlined in subsequent sections; only the general definitions of these items are given in this chapter.

- The form of design flood hydrograph

This form enables the designer to prepare a rational design for the downstream discharge canal. The three characteristic features of this hydrograph are: its peak, its volume and the time required for attaining the peak.

- The peak of design flood hydrograph

Q_{max} is different and greater than the design flood Q_d,

$$Q_{max} > Q_d$$

Dimensions of a spillway are determined by taking Q_d as the main variable. The control structure may discharge Q_{max} as well by using some restricted values of one of the variables such as the freeboard. This consideration is rejected in the USA practice but accepted in the European countries (and in Turkey as well). For instance, Q_{max} is a discharge which has a return period of 10,000 years. The economic design age of a reservoir is generally considered to be 100 years. The probability of occurrence of such a flood in this interval of 100 years is very small. If Q_{max} happens by bad luck, then it will be discharged by using a certain percentage of the freeboard, and the accompanying damage will be paid by the owner of the hydraulic work. Such consideration can result in greater economy. It will be discussed in detail in Chapter 3 and 4. The spillway of the Kiralkizi Dam is sized according to this principle. The design discharge of the Aslantas Dam in Turkey was 11,000 m^3/sec and the catastrophic discharge, 13,000 m^3/sec. It is possible to find the application of this principle on many spillways in Turkey and in Europe.

- Reservoir operational level

The reservoir operational level is the level chosen for the generation of hydroelectric power. It is assumed constant or almost constant. If the reservoir is used for irrigation also, instead of a single value for the operation level, a minimum and a maximum level are defined. The volume of water for that difference in levels is used for irrigational purposes.

- Characteristics of river bed of the main water course

The characteristics of flow in the main river should be studied in detail. This study is necessary for determining the local scour at the toe of the transition spillway element and the degradation of the bed when sediment is retained in the reservoir (see Chapter 8). The necessary data to be collected are:

1. Planning map of the main river downstream of the dam axis (scale 1/100).

2. Cross sections of the river downstream of the dam axis at intervals of 100 m. These sections may extend to elevations 10 to 20 meters higher than the maximum water level.

3. Soil sampling (at least one sample for each 1,000 m^2).

The exploratory drillings should go down to the sound rock or to a depth recommended by a qualified geologist. Samples obtained from these drillings should be sent to soil mechanics laboratories for the following tests:

- Sieve analysis

- Internal friction angle, Ø

- Cohesion, c

- Shear stress, τ

4. Discharge-elevation curve;

5. Determination of S = f(Q) curve (S, the energy gradient);

6. Sieve analysis of the sediment collected from the surface of the bed;

7. Determination of the coefficient C and the exponent of the function,

$$Q = C \left(\frac{h}{k} \right)^n$$

where k is the relative sand roughness. Generally, it is taken to be equal to D_{65}.

8. Determination of a control section downstream of the dam axis.

Data shown under 1, 2-4-5-6 and 7 above, should be used in hydraulic scale model investigations. Data shown under 2, 3, 4, and 8 should be used in mathematical computation of downstream degradation (See Chapter 12 and Simons, D.B. and Şentürk, F., 1992).

- Ground water observations in the vicinity of dam axis

These observations are of primary importance for the life of a spillway. They are:

1. Variation of ground water table in the vicinity of reservoir and dam axis;

2. Chemical analysis of water from nearby sources;

3. Determination of isotope content of ground water;

4. Determination of isotope content of nearby water sources.

The test results should be used for studying the foundation stability of the dam and spillway. Investigation should also be made for water losses through the eventual underground channels between the reservoir and the eventual springs downstream of the dam.

C. Climatic Data

These data are:

1. Variation of monthly temperature, and rainfall, and the storm intensities;

2. Evaporation rates;

3. Variation of daily temperature;

4. The wind direction and temperature;

5. The ice formation (average starting and ending dates and the variation in thickness).

D. Geologic Data

The geologic data are:

1. The investigation of geologic formations, determination of karstic and thermal caverns, exposed lava, exposed gravel and glacial deposits, etc.;

2. The observation of changes in water table during at least two annual cycles;

3. The determination of evaporates in foundations;

4. The geological report taking all the above data into consideration.

The existence of evaporites in the vicinity of foundations complicates the design of the spillway and the dam. Evaporates were isolated in the foundation of the Dicle Dam in Turkey. A very complicated investigative process was needed to determine:

1. The exact position of evaporites in the vertical plane, passing through the dam axis and vicinity, to determine the evolution of the rock around the dam site.

2. The absorption and the deformation capabilities of evaporites by the in-situ experiments.

3. Determination of the location of the evaporites and where they emerge to the surface by studying aerial photographs.

These investigations took three years and the design was made on the basis of test results. The dam was under construction in 1993.

E. Ecologic Data

The ecologic data are:

1. Impact of the project on wild life,

2. Provision for wildlife protection,

3. Provision for public protection,

4. Fish population to be cultivated in the reservoir,

5. Recreational development.

2.6.2 Reservoir Data

These data comprise the following items:

1. A sound triangulation system for horizontal and vertical level controls for the spillway and dam.

2. Reservoir area to be connected to a specific large scale triangulation system or to the nationwide triangulation system. In Turkey, the latter system is used together with the national coordinate system.

3. A cartographic map to be prepared before the reservoir fills.

4. A map for relocation of roads and railroads.

5. A map for relocation of public utilities.

2.7 RISK ANALYSIS IN SPILLWAY DESIGN

The design analysis criteria for spillways have been changing with time. A review of spillways of old dams gives us pause for reflection. Recent dam failures are an important consideration in the today's spillway design. Thus, a new approach has been chosen in considering the design of spillways: the decision analysis process (Von Thun, J. Laurence, 1987).

It is reasonable to consider this decision analysis process in three phases:

Phase 1:   Hazard analysis

Phase 2:   Risk analysis of alternatives

Phase 3:   Decision on modification.

In the hazard analysis, the results are used to evaluate the nature of the problem. In the risk analysis of alternatives, a quantitative evaluation of the economic risk and life loss risk is prepared. The results obtained are compared qualitatively according to economic risk reduction, life loss risk reduction and the costs and benefits of alternatives. The decision regarding modification is based upon an objective review of qualitative, quantitative, social, environmental, political, organizational and other factors influencing the design of spillways.

2.7.1 Hazards

The hazard analysis must be done for a newly designed spillway in order to assess the downstream impact of various discharge situations. The steps to be taken are summarized in Table 2.3.

TABLE 2.3

Outline of Steps in Hazard Assessment for the Design of Spillways

1.	Perform flood routings to determine the impact on dams as a function of flood level and establish outflow without dam failure.
2.	Identify failure mode and failure thresholds for each particular case.
3.	Consider the dam failure and determine the inundated area.
4.	Compute the consequences as a function of the inundated area.
5.	Evaluate the hazard based on incremental assessment.

2.7.2 Quantitative Risk Analysis

The quantitative risk analysis can be done taking the following steps:

1. Estimating the return period of the potentially damaging loads;

2. Estimating the potential adverse response of the system;

3. Estimating the system response to loads;

4. Estimating the consequences of failure;

5. Determining the risk cost or expected loss.

Loads include the static reservoir load, seismic load and flood load. The risk cost may be partial or total. The partial risk cost is the likelihood of load level for a specific load type multiplied by the likelihood of specific failure mode at given load level for the specific load type, multiplied by property damages produced by specific failure.

The total risk cost is the sum of risk costs for all load levels and all load types pertaining to each particular failure mode.

The comparison of the cost of damage with the total annual risk cost relates the stability of the dam to the likelihood of possible failure. Loads that have damage potential for causing dam failure are to be considered first. To prevent such damage, spillway capacity can be increased and the risk computation recalculated. Similar comparisons must be done by considering separately the static reservoir load, the seismic load, etc. Von Von Thun (1987) recommends the representation shown in Table 2.4 for the variation of different loads and their corresponding risks.

TABLE 2.4

Variation of Different Loads and their pertinent risks

Load Range	Load Range Probability	Potential Failure Mode	Response Probability	Damage Cost / Life Loss Condition	Annual Risk Cost
Step I		**Step II**	**Step III**	**Step IV**	**Step V**
40-60%		Aux. Spill.			
PMF		Failure			
60-80%		Aux. Spill.			
PMF		Failure			
60-80%		Service			
PMF		Spillway			
		Failure			
60-80%		Dam			
PMF		Overtopping			
80-100%		Aux. Spill.			
PMF		Failure			
80-100%		Dam			
PMF		Overtopping			
80-100%		Aux. Spill.			
PMF		Failure			
80-100%		Service			
		Spillway			
Static					
Full Reservoir		Dam Overtopping			
Seismic M 5.5-6.5		" "			
Seismic M 6.5-7.5		" "			
Seismic M 7.0-7.5		" "			

PART II of TABLE 2.4

Loss of life potential-exposure condition	Annual probability of exposure condition occurring

A - People in recreation pursuits within 10 miles (l6 km) of the dam

B - Evacuation related deaths (1 or 2 people)

C - Direct loss of life of up to 50 people for night failure

(8 pm-4 am)

(PMF: Probability of Mode of Failure)

2.8   PROBLEMS

Problem 2.01

Name five examples of each type of dam enumerated in the text, built in your country and give their characteristics.

Problem 2.02

Route the flood. Its characteristics are given in the following data.

Data

1. Characteristics of the inflow hydrograph (Table 1, Problem 2.02)

2. The variation of reservoir surface area in respect to the depth of reservoir is

$$A = 0.00094H^{1.81614} \cong 0.00094H^{1.82} (10^6 \text{ m}^2)$$

$$H = 61.25+h$$

where

H   is the depth of reservoir, and

h   is the depth of water on the spillway crest.

61.25 is the height of the spillway.

TABLE 1, PROBLEM 2.02

Time	Q
day	m³/sec
1	623
2	714
3	812
4	902
5	1002
6	1098
7	2137
8	5550
9	2999
10	1349
11	916
12	826
13	759
14	699
15	640

3. $Q_o = 45.98 \, h^{1.66} \, (m^3/sec)$

$$Q_o = \text{the outflow discharge}$$

4. The gates of the spillway will be fully raised for

$$Q_i = Q_o = 3565 \, m^3/sec.$$

Problem 2.03

Determine on the outflow hydrograph of Problem 2.02 the virtual duration of vibrations assuming that vibrations begin for $Q = 3,800 \, m^3/sec$.

Problem 2.04

The cost estimate pertaining to spillway gates 3, 4, 5, 6, and 7 are given in Table 1, Problem 2.04. Determine the number of gates to be used in the design.

TABLE 1, PROBLEM 2.04

Cost estimate for gates 3-4-5-6 and 7 of a spillway.

Number of the gates	Cost estimate of the structure 10^6 units
3	500
4	350
5	300
6	320
7	400

The units are in the unit money of the country where the dam is to be constructed.

Problem 2.05

The total cost variation of possible three solutions for dam and spillway combinations are given in Fig. 1, Problem 2.05.

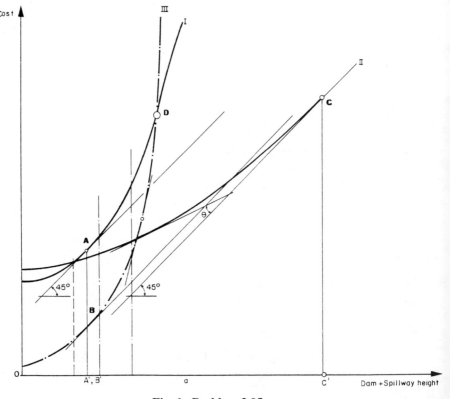

Fig. 1. Problem 2.05.

Compare these solutions and determine for each one the economical dam spillway heights. If the dam spillway height is given, choose the most economical solution.

Problem 2.06

Route the flood, the characteristic of which are:

$$Q_0 = 0.41 \, Lh \, (2gh)^{1/2}$$

where

H is the depth of the reservoir

L = 45 m, is the length of the spilling crest, and

h is the hydraulic head of the spillway.

It is assumed that at time 0, the water surface elevation in the reservoir is either 810 or 870 + h. The volume or the area versus water surface elevation relationships are given in Table 1, Problem 2.06.

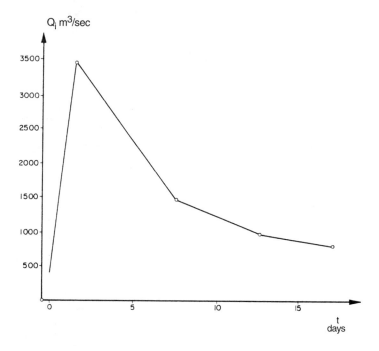

Fig. 1. Problem 2.06. Inflow Flood Hydrograph.

Data

TABLE 1, Problem 2.06

Water Surface Level	Volume 10^6 m^3	Surface Area km^2
840	3705	90
830	2870	78
825	2506	70
820	2175	63
810	1607	51
800	1151	40
790	793	28
780	522	22
775	414	21
770	322	17
760	183	12
750	92	7
740	39	4
730	12	2
720	2	0.8

Determine the peak outflow and the outflow hydrograph. (See Fig. 1, Problem 2.06.)

Chapter 3

HYDRAULICS OF
FREE FLOW SPILLWAYS

3.1   INTRODUCTION

The hydraulic computation for design of spillways is basically a function of its sizing. The designer in charge of hydraulic calculation must understand flow phenomena. The flow of water is not governed by a restricted set of theoretical formulae. Instead, it is subject to many and diversified factors, many of which are based upon the empirical observations rather than purely theoretical deductions.

The first step is a complete understanding of the hydraulic conditions which occur during flood discharges from upstream to downstream.

The hydraulic study of spillways begins with the study of flow taking place in the approach channel, and then continues with the control structure, the downstream channel and the terminal structure. The flow in the main course of the river up to a control section situated far downstream of the dam, can also be considered as a part of the spillway flow.

Different phases of the hydraulics of spillways will be outlined in detail in the following sections.

3.2   HYDRAULICS OF APPROACH CHANNEL

The approach channel is the hydraulic structure conveying the flow to the control structure. The hydraulic computation of the approach channel cannot be done using classical methods. These methods assume that the flow taking place in the channel is uniform, gradually varied or spatially varied; but this flow is quite different from the actual conditions. The only way to solve this problem is to use a hydraulic model study (see Chapter 13). In some cases, the flow can be assumed very near to gradually varied flow; then taking the restrictions imposed by the topography and by the geometry under consideration, the flow in the approach channel can be mathematically simulated.

Model studies show the variation taking place in the flow and the velocities at the bottom of the channel can be measured. These measurements and visual observations suggest the possible amount of erosion that can be expected on the floor of the channel. Then, preventive measures to be taken are developed. Designer and model engineers should work together to determine the scale of the model in such a way that the characteristics of flow can be measured at the selected points and sections.

If the approach channel entrance is symmetrical and the channel is perfectly prismatic from its beginning to its end without any curvature in either horizontal or vertical planes and slopes in the direction of the flow, a mathematical simulation can be useful. When the velocity in the approach channel is quite small, the pile up of the flow towards one side of it can be neglected and the flow assumed gradually varied. The hydraulic computation of such a flow using classical procedure can be achieved.

Usually, the flow in the approach channel is subcritical and gradually varied. Then the determination of the water surface line and energy gradient line begins from a control section where the depth is known or can be computed, and proceeds upstream. The control section downstream of the channel is the control structure. There the depth of flow is known when the discharge Q is given. Then the water surface line should be computed according to the following guidelines:

1. The entrance of the channel is designed in such a way that the counter currents, stagnant water areas and the piling up of flow towards one side of the channel are eliminated.

2. The canal is prismatic for its entire length.

3. Any individual obstacle can cause a counter current and if it exists, it should be removed.

The mean velocity of the flow can be computed by taking the average of the sections. The alignment of the channel should be as close to a straight line as possible.

4. If the discharge of the channel is constant, then the flow is steady.

5. The regime of flow is usually turbulent subcritical. If the regime of flow is turbulent supercritical, a special procedure for determining the water surface line can be developed. (Turbulent supercritical regime is considered only in the case of weirs and intakes in torrential water courses).

6. The flow is mostly uniform or gradually varied.

Under the above conditions it is possible to define mathematically the flow taking place in the approach channel.

The approach channel of a spillway is shown in Fig. 3.1. The length of this channel is shown by L and the control structure is situated at the downstream end of it. Assume furthermore that the discharge is Q; then the hydraulic head of the control device can be computed as:

$$Q = mA\,(2gH)^{1/2} = mBh\,(2gH)^{1/2} \qquad (3.1)$$

where

Q is the discharge in m³/sec or in cfs

A is the wetted area in m² or in square feet

$$H = h + \frac{U^2}{2g}$$

h is the depth of water on the crest of the control structure in meters or in feet, measured at a station clear of the influence region of the control device.

U is the mean velocity in the approach channel in m/sec or in fps

B is the total length of the crest (m or feet) or the width of the channel at section 1-1 and

m is the coefficient of discharge

For small values of U, $U^2/2g$ is negligible, therefore the above relation takes the following form:

$$Q = mBh\,(2gH)^{1/2} = CBh^3$$

$$C = m\,(2g)^{1/2}$$

3.2.1 Hydraulic Computation of the Approach Channel for Uniform Flow

The basic conditions for which the flow is uniform are:

- The channel is prismatical;

- The mean velocity U is constant, which means that any backwater or a draw down effect influences the flow.

In this case any uniform flow formula can be used to determine the characteristics of the flow. If Manning's formula is used:

$$Q = \frac{A}{n}R^{2/3}\,S^{1/2} \qquad (3.2a)$$

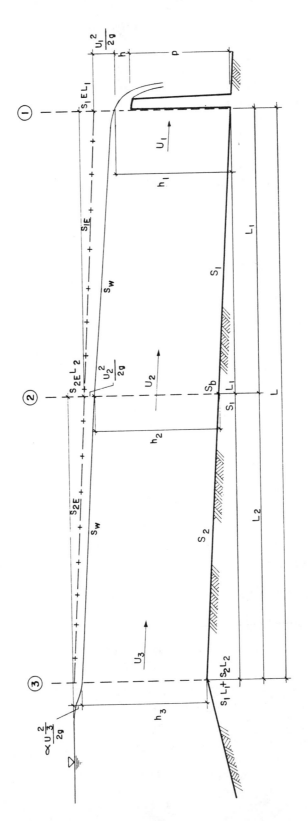

Figure 3.1. Flow in the approach channel.

in metric system, or

$$Q = \frac{1.486}{n} A R^{2/3} S^{1/2} \qquad (3.2b)$$

in English system

where   n is the Manning's coefficient of resistance

         R is the hydraulic radius in meters or feet

and   S is the slope of the energy line.

Assume the cross section of the channel rectangular; then its characteristics can be expressed in terms of the flow depth, h and the width, B:

$$R = \frac{Bh}{B + 2h}$$

is the hydraulic radius; substituting in Eq. 3.2a yields

$$Q = \frac{Bh}{n} \left(\frac{Bh}{B+2h}\right)^{2/3} S^{1/2}$$

When Q is given, an assumed value is selected for B, so h can be computed from this relationship. In general, the backwater effect is not negligible or the channel perfectly prismatical, then the flow condition is mostly gradually varied.

3.2.2 Hydraulic Computation of the Approach Channel for Gradually Varied Flow

The conditions under which the flow is gradually varied is very well known and defined. However it is difficult to find such ideal conditions in nature and therefore some approximations are always needed. These are outlined as follows:

A. The Flow is Gradually Varied When the Geometric Properties Satisfy the Following Conditions:

- The channel is almost prismatic; the cross sections are different from each other but the variation is small and any boundary effect such as return flow influences the flow

- The backwater effect is existent

- The roughness coefficient is constant throughout the channel reach under consideration

- The channel is long enough for the gradually varied flow to take place

- Abrupt changes of sections and abrupt drops of the bottom are avoided.

B. The Flow is Gradually Varied When the Hydraulic Properties Satisfy the Following Conditions:

- The slope of the energy line is small

- The velocity distribution in the cross sections is fixed and the velocity distribution coefficient constant. The mean velocity is different in each section but the differences are small.

When a channel and the flow which takes place in it conform to the conditions listed above, the computations can be conducted as shown below:

1. The channel is divided into n parts. The length of each reach is independently chosen but each reach responds perfectly to the above conditions. In the particular case shown in Fig. 3.1, $n = 2$.

2. In each reach the canal sections are perfectly constant. For the case outlined in Fig. 3.1 the cross sections are assumed rectangular.

3. If the cross section 1-1 is chosen clear of the influence region of the spillway, the energy equation can be written between sections 1-1 and 2-2 as follows

$$h_1 + \frac{U_2^2}{2g} + S_{1E}L_1 = S_bL_1 + h_2 + \frac{U_2^2}{2g} \tag{3.3}$$

where

h_1 is the flow depth in the cross section 1-1

U_1 is the mean velocity in the cross section 1-1

$$U_1 = \frac{Q}{h_1 B_1}$$

U_2 is the mean velocity in the cross section 2-2

72

$$U_2 = \frac{Q}{h_2 B_2}$$

h_2 is the flow depth in the cross section 2-2

Q is the discharge of the channel

S_{1E} is the slope of the energy line between the section 1-1 and 2-2

L_1 is the length of the reach under consideration

S_b is the slope of the bottom, and

 B_2 is the width of section 2-2; this width is assumed constant throughout the reach.

S_{1E} can be obtained by direct application of Manning's formula:

$$S_{1E} = \frac{U_o^2 n^2}{R_o^{4/3}} \qquad (3.4)$$

where

$$U_o = \frac{U_1 + U_2}{2} \text{ and } R_o = \frac{R_1 + R_2}{2} \qquad (3.5)$$

n, is constant along the channel. Substituting these values in Eq. 3.3 yields

$$h_1 + \frac{Q^2}{B_1^2 h_1^2 \, 2g} + n^2 L_1 \, \frac{Q^2}{4} \left(\frac{1}{B_1 h_1} + \frac{1}{B_2 h_2}\right)^2 \frac{1}{\left(\frac{R_1 + R_2}{2}\right)^{4/3}} =$$

$$S_b L_1 + h_2 + \frac{Q^2}{B_2^2 h_2^2} \frac{1}{2g} \qquad (3.6)$$

R_1, can be written as a function of h_1:

$$R_1 = \frac{B_1 h_1}{B_1 + 2h_1}$$

and R_2 can be written as a function of h_2:

$$R_2 = \frac{B_2 h_2}{B_2 + 2h_2}$$

Equation 3.6 is solved for the following cases:

a. Q is known but the water level in the reservoir is unknown. Q yields h_1 by using Eq. 3.1; then following the procedure outlined below, h_3 and H are obtained by the application of Eq. 3.6.

b. Q is unknown but the water level in the reservoir is known.

In this case the values of the variables in Section 1-1 are assumed; the only unknown in Section 2-2 is h_2. Eq. 3.6 can be solved for h_2. Thus, the variables of Section 2-2 being known, the energy equation can be applied between Section 2-2 and 3-3 and h_3 is solved. The hydraulic head of the flow is equal to the difference of levels between the water surface in the reservoir and the bottom of the channel at section 3-3. In other words, the hydraulic head can be computed as:

$$H = h_3 + \alpha \frac{U_3^2}{2g} \tag{3.7}$$

If the two values of H are equal, the problem is solved; if not, the assumed value of the unknown in section 1-1 is modified and the computation repeated. When the solution of Eq. 3.6 yields h_1 then the head at the spillway can be computed:

$$h = h_1 - P$$

where P is the height of the control structure.

The local head loss at the entrance to the channel is $\alpha (U_3^2/2g)$. The values of the coefficient α are given in Table 3.1.

TABLE 3.1

Values of the coefficient α at the entrance of the channel

Characteristics of the channel	α
1-Rough entrance without any stream-lined transition	1
2-Smooth entrance without any stream-lined transition	0.5
3-Smooth entrance with stream-lined transition	0.10 -0.20

Example 3.01

Figure 3.2 shows an approach channel with the following characteristics:

- Channel is infinitely large

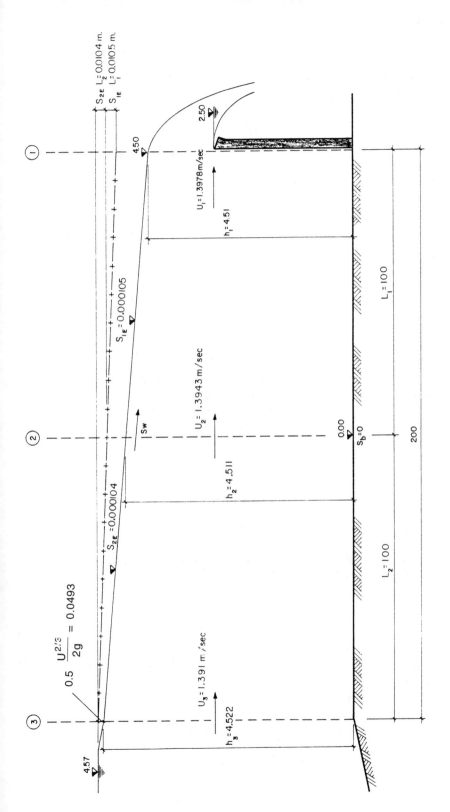

Figure 3.2. The characteristics of the approach channel in Example 3.01.

- Channel has no curvature in plan or in longitudinal profile

- Channel entrance is symmetrical, thus streamlines are parallel to each other and to the center line of the channel

- Bottom of the channel is horizontal, $S_b = 0$

- Channel is unlined with the Manning's roughness coefficient 0.02

- Reservoir water level is 4.57 m

- Channel bottom level is 0.00 m

- Coefficient of discharge of the control structure is m = 0.49

- Height of control structure is P = 2.50 m.

Determine the characteristics of the flow in the channel: $q - S_{1E} - S_{2E}$

Solution

The characteristics of the flow taking place in the approach channel are functions of the water level in the reservoir. A variation of this level causes a variation in the discharge and with it a variation in the depth of flow.

S_b being equal to "0", the flow in the channel will be gradually varied. Three sections are chosen for solving the problem; the sections are 100.00 m apart:

$$L_1 = 100.00 \text{ m}$$

and

$$L_2 = 100.00 \text{ m}$$

Also, U_1 represents the mean velocity and h_1, the depth of flow in the cross section 1-1. The linear discharge, q, is equal to

$$q = U_1 h_1$$

Assume the hydraulic head at the control structure is 2.00 m. The application of Eq. 3.6 yields the water level in the reservoir. If the value obtained is other than 4.57 m, the hydraulic head at the control structure will be modified.

Equation 3.1 yields

$$q = 0.49 \times 2.00 \left[2g \left(2 + \frac{q^2}{4.5^2} \frac{1}{2g} \right) \right]^{1/2}$$
$$q = 6.29 \text{ m}^3/\text{sec.m}$$

and

$$h_1 = 2.00 + 2.50 = 4.50 \text{ m}$$

$$U_1 = \frac{6.29}{4.50} = 1.3978 \text{ m/sec}$$

The application of Eq. 3.6 between the two sections 1-1 and 2-2 yields

$$\frac{U_1^2}{2g} = \frac{1.3978^2}{19.62} = 0.0996 \text{ m}$$

$$h_1 + \frac{q^2}{h_1^2}\frac{1}{2g} + n^2 L_1 \frac{q^2}{4}(\frac{1}{h_1}+\frac{1}{h_2})^2 \frac{1}{\left(\frac{R_1+R_2}{2}\right)^{4/3}} = S_b L_1 + h_2 + \frac{q^2}{h_2^2}\frac{1}{2g}$$

Substituting $h_1 = 4.50$ m $\quad R_1 = h_1 = 4.50$ m $\quad q = 6.29$ m³/sec.m

$$L_1 = 100.00 \text{ m and } R_2 = h_2 \quad n = 0.02$$

yields

$$4.50 + \frac{6.29^2}{4.5^2}\frac{1}{19.62} + 0.003956 (\frac{1}{4.5}+\frac{1}{h_2})^2 \frac{2.5198}{(4.5+h_2)^{4/3}} 100 =$$

$$= 0 \cdot 100 + h_2 + \frac{6.29^2}{19.62}\frac{1}{h_2^2}$$

This equation can be solved for h_2 and the characteristics of the section 2-2 be obtained:

$$h_2 = 4.5111 \text{ m} \qquad S_{1E} = 0.000105 \qquad U_2 = 1.3943 \text{ m/sec}$$

and

$$\frac{U_2^2}{2g} = 0.099086 \text{ m}$$

The application of Eq. 3.6 between the two sections 2-2 and 3-3 yields

$$h_2 = 4.5111 \text{ m} \qquad q = 6.29 \text{ m³/sec.m}$$

$$L_2 = 100.00 \text{ m} \qquad R_3 = h_3 \text{ m} \qquad \text{and} \qquad n_2 = 0.02$$

$$4.5111 + \frac{6.29^2}{4.5111^2}\frac{1}{2g} + 0.003956\,(\frac{1}{h_2} + \frac{1}{h_3})^2\,\frac{2.5198}{(4.51+h_3)^{4/3}} \cdot 100 =$$

$$S_bL_2 + h_3 + \frac{q^2}{h_3^2} \cdot \frac{1}{2g}$$

This equation can be solved for h_3 and the characteristics of the section 3-3 be obtained:

$$h_3 = 4.522\ m \qquad S_{2E} = 0.000104 \qquad U_3 = 1.391\ m/sec$$

$$\frac{U_3^2}{2g} = 0.099\ m$$

Assuming $\alpha = 0.5$, yields, $0.50(\,U_3^2/2g) = 0.0493$. The water elevation in the reservoir is then

$$h_3 + 0.5\,\frac{U_3^2}{2g} = 4.522 + 0.0493 = 4.5713 \cong 4.57$$

The water surface elevation thus obtained is equal to the reservoir water surface elevation given as data of the problem; then it is not necessary to take a new step.

3.3 HYDRAULICS OF FREE FLOW SPILLWAYS

An exact theoretical analysis of the flow taking place on the control structure is difficult due to the diversity of the involved variables. Therefore, simplifying assumptions are introduced in the theory. The first and important assumption is that the stream lines are perpendicular to the axis of the control structure, an assumption which is generally not true. Also, the approach velocity will at first be assumed to be negligible; later, the differential resulting from this assumption will be added and an adequate correction made. The hydraulic computation of a free flow spillway can be done by determining two basic variables:

1. The coefficient of discharge m in equation

$$Q = mA(2_gH_E)^{1/2} \qquad (3.8)$$

2. The negative pressures building up downstream of the control structure.

Actually, m is a function of the negative pressures. The increasing values of the pressures correspond to the increasing values of the discharge coefficient. The increase in both of these variables is not continuous, but has defined boundaries. The task of the engineer is to determine the optimum solution.

The Eq. 3.8 has a physical meaning (Scimemi, E.). It is used mostly in Europe. In United States, a simplified form of it is used

$$Q = CAh^{1/2} = CLH^{3/2} \qquad (3.9)$$

where L is the length of the crest and

$$C = m(2g)^{1/2}$$

In this case, as shown in Eq. 3.1, the velocity head is neglected. (Eq. 3.15). Both of the coefficients, m and C, will be used in this book.

3.3.1 Coefficient of Discharge m and Profile of the Free Flow Spillway

Free flow spillways have been used extensively in hydraulic engineering since the beginning of this century. Their theoretical analysis also was completed in the first half of 20th century. Two variables play a principal part in designing these structures:

a. the coefficient of discharge, and

b. the negative pressures on the profile of the spilling device including the shape of the profile defined as a function of the subpressure.

A. The Profile of a Free Flow Spillway with Zero Negative Pressure

A sharp crested spillway is shown in Fig. 3.3. A flow is taking place on it. The equation of the lower nappe can be determined by considering the mechanics of a projectile. The stability of this nappe is not altered when, for example, the empty space between the nappe and the spillway is filled with concrete. Theoretically, for any friction which exists between the nappe and the filling material, the hydraulic head at the spillway will not change the stability of the system. Actually, the assumed non-existent friction cannot be obtained, but in the engineering practice it is considered to be negligible. For smaller hydraulic heads, the nappe will adhere to the concrete and,

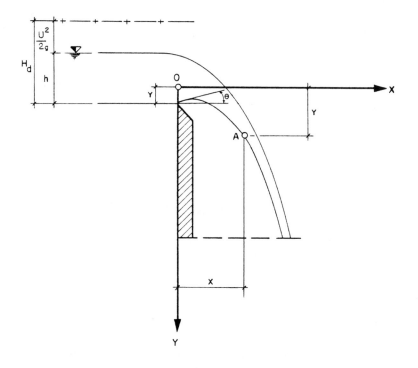

Figure 3.3. The nappe over a sharp-crested spillway.

contrarily, for greater heads the nappe will be detached from it. The hydraulic head defining the profile of the filling material is called design head, H_d, and the value of the coefficient of discharge relative to it and shown by m_0 is about 0.496. (In the Hydraulic Research Laboratory, Ankara, the value of m_0 was measured and found to be 0.4956).*

When a control structure is designed following the rule outlined above, it is generally over-sized. Figure 3.4 shows different geometrical solutions used for obtaining more economical cross sections. However, these truncated cross sections present many hydraulic inconveniences which will be discussed in the following chapters. The approach flow creates vortices with horizontal axis called helicoidal flows. These flows are projected upward from the two extremities of the spillway, disturbing the uniformity of the flow and reducing the coefficient of discharge. Costly measures must

---

*   Schimemi, E., 1946 : $m_0 = 0.4961$

        Randolph, 1937   : $m_0 = 0.4902$

        Brundenell, 1935 : $m_0 = 0.4990$

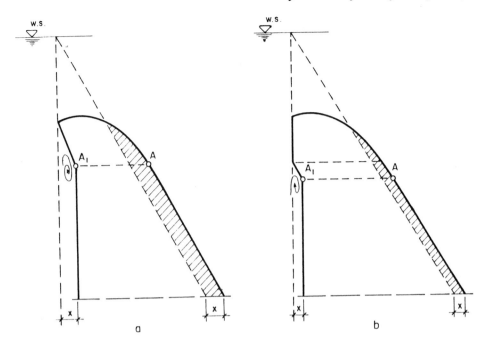

Figure 3.4. Truncated profiles.

be taken to solve this problem (side piers of Keban Dam, Turkey, Fig. 2.3). The value of x (Fig. 3.4) is generally determined by stability analysis of the gravity structure. The engineer must seriously review the final design and balance economic factor with the loss due to hydraulic turbulence.

The equation of the profile of the nappe taking place over a sharp crested spillway can be obtained by applying the principle of the projectile. Let us consider a point $A(x,y)$ on the lower nappe of the flow taking place on the sharp crested weir shown in Fig. 3.3. If the velocity of the flow at point 0 is U_0, the abscissa x of point A can be written as

$$x = U_0 t \cos\theta$$

where t is the time and θ is the angle of attack. Similarly y is

$$y = -U_0 t \sin\theta + \frac{1}{2} g t^2 + y_0$$

where y_0 is the ordinate at the origin. Eliminating the time t and dividing by H_d yields

$$\frac{y}{H_d} = \frac{g H_d}{2 U_0^2 \cos^2\theta} \left(\frac{x}{H_d}\right)^2 - \tan\theta \frac{x}{H_d} + \frac{y_0}{H_d} \qquad (3.10)$$

81

This relation is the equation of a parabola of the second degree, it is completely defined when the velocity U_0 and the angle θ are determined (Table 3.2). Raising the axis of abscissa up by a quantity y_0, yields

$$\frac{y}{H_d} = \frac{gH_d}{2\,U_0^2\,\cos^2\theta}\,(\frac{x}{H_d})^2 - \frac{x}{H_d}\,\tan\theta \tag{3.11}$$

then the origin of the coordinates is situated on the crest of the parabola.

In Fig. 3.5 an orifice is directing the flow towards the positive values of the coordinates; the application of the principle of the projectile yields

$$y = \frac{x^2}{4H_d\,\cos^2\theta} + x\,\tan\theta \tag{3.12}$$

Equation 3.12 is obtained by substituting U_0 by its value from

$$U_0^2 = 2gH_d$$

Assume $\tan\theta = b$ and $\dfrac{1}{4H_d\cos^2\theta} = a$

Substituting these values in Eq. 3.12 yields

$$y = ax^2 + bx \tag{3.13a}$$

For $q = 0$, Eq. 3 11 and 3.13a take the following forms

$$\frac{y}{H_d} = ax^2 = \frac{x^2}{4\,H_d^2} \tag{3.13b}$$

or

$$\frac{y}{H_d} = \frac{1}{4}\,(\frac{x}{H_d})^2 \tag{3.14}$$

Research has been done to determine the constant values (a) and (b) but the solutions given are only approximate. The best solution is obtained from an approximation of the control structure profile chosen without bypassing the parameters defining the nappe. The chosen profile in this case will be defined directly (Eq. 3.16a, to Eq. 3.16d and 3.17); but the general form of the relation defining it will be in the form of Eq. 3.14. This form is only true when $\theta = 0$ and $y_0 = 0$.

TABLE 3.2

Coordinates of the Lower Nappe of a Sharp Crested Spillway in Function of H_d and U_o

$\left[\dfrac{U^2/2g}{(h+y_0)+U^2/2g}\right]$	0.002	0.020	0.040	0.060	0.080	1.000	0.120	0.140	0.160	0.180	0.200	0.220
$\left[\dfrac{x}{(h+y_0)+U^2/2g}\right]$						$\left[\dfrac{y}{(h+y_0)+U^2/2g}\right]$						
0.000	0.0000	0.0000	0.0000	0.0000	0.0000	0.0000	0.0000	0.0000	0.0000	0.0000	0.0000	0.0000
0.040	0.0280	0.0265	0.0250	0.0235	0.0220	0.0210	0.0200	0.0195	0.0190	0.0175	0.0160	0.0150
0.020	0.0190	0.0460	0.0440	0.0400	0.0380	0.0355	0.0340	0.0320	0.0310	0.0295	0.0280	0.0265
0.060	0.0650	0.0605	0.0570	0.0530	0.0500	0.0465	0.0440	0.0420	0.0405	0.0490	0.0365	0.0355
0.080	0.0860	0.0720	0.0670	0.0630	0.0585	0.0555	0.0520	0.0500	0.0480	0.0455	0.0430	0.0415
0.100	0.0860	0.0810	0.0760	0.0710	0.0655	0.0615	0.0580	0.0550	0.0525	0.0500	0.0480	0.0450
0.120	0.0910	0.0880	0.0825	0.0770	0.0715	0.0665	0.0625	0.0595	0.0560	0.0535	0.0505	0.0475
0.140	0.1000	0.0940	0.0880	0.0820	0.0775	0.0700	0.0655	0.0620	0.0580	0.0555	0.0515	0.0485
0.160	0.1015	0.0980	0.0920	0.0845	0.0780	0.0725	0.0675	0.0635	0.0595	0.0560	0.0525	0.0490
0.180	0.1080	0.1010	0.0945	0.0870	0.0800	0.0740	0.0685	0.0640	0.0600	0.0560	0.0530	0.0485
0.200	0.1105	0.1030	0.0965	0.0880	0.0815	0.0750	0.0695	0.0645	0.0595	0.0560	0.0520	0.0480
0.220	0.1120	0.1040	0.0975	0.0885	0.0820	0.0750	0.0690	0.0640	0.0590	0.0550	0.0510	0.0465
0.240	0.1120	0.1050	0.0980	0.0880	0.0815	0.0745	0.0680	0.0630	0.0575	0.0535	0.0480	0.0450
0.260	0.1120	0.1045	0.0975	0.0875	0.0800	0.0730	0.0665	0.0610	0.0560	0.0510	0.0465	0.0425
0.280	0.1115	0.1040	0.0960	0.0860	0.0780	0.0710	0.0645	0.0590	0.0535	0.0490	0.0445	0.0400
0.300	0.1105	0.1020	0.0940	0.0845	0.0760	0.0690	0.0620	0.0560	0.0505	0.0460	0.0415	0.0365
0.320	0.1090	0.1005	0.0920	0.0820	0.0735	0.0660	0.0590	0.0525	0.0480	0.0430	0.0380	0.0330
0.340	0.1070	0.0980	0.0885	0.0785	0.0700	0.0630	0.0555	0.0490	0.0440	0.0390	0.0340	0.0290
0.360	0.1040	0.0950	0.0855	0.0750	0.0670	0.0590	0.0515	0.0450	0.0400	0.0350	0.0300	0.0250
0.380	0.1010	0.0910	0.0815	0.0710	0.0625	0.0550	0.0475	0.0415	0.0360	0.0305	0.0255	0.0200
0.400	0.0970	0.0870	0.0770	0.0670	0.0580	0.0500	0.0425	0.0360	0.0310	0.0260	0.0200	0.0150
0.450	0.0815	0.0745	0.0650	0.0540	0.0450	0.0365	0.0290	0.0200	0.0165	0.0110	0.6000	0.000
0.500	0.70	0.060	0.050	0.030	0.030	0.020	0.013	0.006	0.000	0.006	0.012	0.018
0.600	0.022	0.022	0.012	0.000	0.010	0.020	0.028	0.035	0.042	0.018	0.054	0.059
0.700	0.016	0.028	0.038	0.019	0.059	0.068	0.076	0.084	0.091	0.098	0.104	0.108
0.800	0.071	0.084	0.095	0.108	0.117	0.126	0.134	0.142	0.149	0.156	0.160	0.164
0.900	0.138	0.150	0.162	0.173	0.183	0.192	0.201	0.208	0.215	0.222	0.226	0.228
1.00	0.214	0.224	0.236	0.247	0.258	0.266	0.276	0.282	0.290	0.297	0.300	0.298
1.20	0.393	0.402	0.412	0.422	0.432	0.440	0.448	0.455	0.463	0.470	0.473	0.466
1.40	0.606	0.614	0.623	0.632	0.641	0.650	0.657	0.664	0.670	0.672	0.670	0.661
1.60	0.850	0.860	0.867	0.874	0.883	0.890	0.896	0.904	0.905	0.900	0.893	0.886
1.80	1.132	1.140	1.147	1.156	1.164	1.170	1.175	1.180	1.173	1.166	1.156	1.146
2.00	1.451	1.460	1.467	1.476	1..485	1.490	1.494	1.496	1.476	1.464	1.454	1.440
2.20	1.798	1.807	1.816	1.825	1.834	1.839	1.841	1.834	1.818	1.801	1.785	1.767
2.40	2.179	2.188	2.198	2.198	2.212	2.213	2.212	2.198	2.179	2.164	2.147	2.125
2.60	2.602	2.611	2.622	2.621	2.620	2.618	2.604	2.591	2.568	2.548	2.326	2.504
2.80	3.048	3.060	3.068	3.052	3.055	3.036	3.011	3.000	2.976	2.953	2.916	2.893

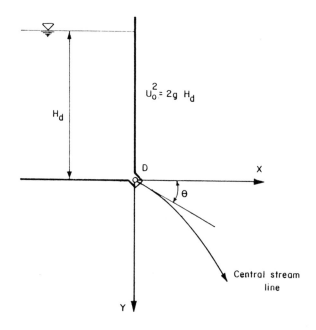

Figure 3.5. The downward directed flow from an orifice.

When $\theta = 0$, this means that the direction of the flow on the crest is horizontal. It is always possible to find a section on the spillway where the direction of the flow is horizontal. The condition which is most difficult to satisfy is to place the origin of the coordinates in a section which is not constant and varies with the flow characteristics. The engineer must be aware of these uncertainties and solve his problem accordingly.

For H_d given, Eq. 3.14 is known and theoretically no friction exists between the nappe and the surface of the control device. For smaller values of H_d the nappe is supported by the control structure and exerts a pressure on it. For greater values of H_d, the nappe is projected far from the control structure and will exert a suction on it by reducing the pressure. This way of functioning is similar to a siphon and the discharge of the control structure is increased. Equation 3.14 is generalized by writing*

$$\frac{y}{H_d} = a \left(\frac{x}{H_d}\right)^n \tag{3.15}$$

---

* In this book capital Y and X represent the coordinates of the profile and x and y the coordinates of the nappe.

Many attempts have been made to determine the value of a and n. Some of them are reproduced here (Maître, R., 1960; Scimemi, E., 1937; Creager, Justin, Hinds, 1945; etc.)

$$\frac{Y}{H_d} = 0.5 \left(\frac{X}{H_d}\right)^{1.85} \qquad \text{Creager-Scimemi} \qquad (3.16a)$$

$$\frac{Y}{H_d} = 0.47 \left(\frac{X}{H_d}\right)^{1.80} \qquad \text{Creager-Scimemi} \qquad (3.16b)$$

$$\frac{Y}{H_d} = 0.461 \left(\frac{X}{H_d}\right)^{1.85} \qquad \text{Smetana} \qquad (3.16c)$$

$$\frac{Y}{H_d} = 0.556 \left(\frac{X}{H_d}\right)^{2} \qquad \text{de Marchi} \qquad (3.16d)$$

WES, "Waterways Experiment Station," USA, performed a series of experiments on special forms of control structures and suggested a general formula of the following type

$$\frac{Y}{H_d} = \frac{1}{k} \left(\frac{X}{H_d}\right)^{n} \qquad (3.17)$$

The values of k and n are given as a function of the slope of the upstream face of the control structure, which means that the influence of θ is taken into consideration; in fact, the variation of $X \tan\theta$ is represented by k in this type of formula (Chow, V.T., 1959), (Table 3.3).

TABLE 3.3

Variation of Parameter k and n in WES Profile

Slope of the upstream face	k	n
Vertical	2.000	1.850
3 on 1	1.936	1.836
3 on 2	1.939	1.810
3 on 3	1.873	1.776

Figure 3.6 represents the WES-Standard Spillway shapes.

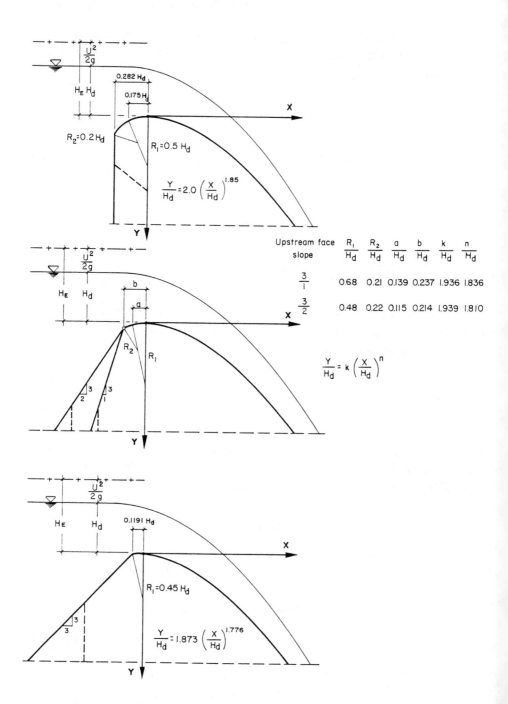

Figure 3.6. The WES - Standard Spillway shapes (U.S. Army Engineers Waterways-Experiment Station. Smother and Fimer).

Figures 3.7 and 3.8 define the lower and the upper nappes of the Creager Profile for $H_d = 1.00$. The definition of different ogee type spillways will be given in this chapter, but the basic type is Creager's profile having zero pressure on its downstream face. In the Creager Profile and in the WES Profile, the pressures exerted are nil if the profile is designed for design head. As stated earlier for $H < H_d$, an increase of pressure on the profile will be observed. (When this increase is taken into consideration in the design, the structure will always be secure.)

Practical rules are suggested for defining different ogee profiles relative to appreciable velocity heads or zero velocity heads. Figure 3.9 shows an ogee profile prepared by using USBR's data corresponding to the relationship

$$\frac{y}{H_d} = K \left(\frac{X}{H_d}\right)^n$$

Negative pressures are measured on this profile for $P \geq H_d/2$, where P represents the height of the control structure (USBR, 1957). It is recommended for higher values of velocity head. If the velocity head is negligible, USBR recommends the profile defined in Fig. 3.10. The practical application of this profile is easier due to the existence of simple geometrical elements incorporated in it (the circles with radius R_1 and R_2); then the form work will be economical.

The USBR profiles given in Figs. 3.9 and 3.10 are similar to each other. However, the first is particularly applicable for a high approach velocity. It is recommended for use in designing weirs or spillways where there is a high approach velocity and the spillway design flood discharges into a narrow downstream channel.

The ogee spillways shown in Figs. 3.7 and 3.8 are recommended by Creager. The upper and lower nappe projected from a sharp crested spillway are given point by point as a result of a multitude of observations. Figure 3.7 shows a control device with a vertical upstream face and Fig. 3.8 with 45° inclined upstream face. These two figures are classic and the profiles they define have been applied for a range of conditions. Where approach velocities and hydraulic heads are greater than 15.00 m, some instability occurs on the crest. The profile defined in Fig. 2.11 is recommended for practical application because of its simplicity and economy of the work form. The inconvenience of this profile is that the coefficient of discharge varies with each particular case and the stability of the nappe is not completely assured. The theoretical computation of the negative pressures is not possible because each diameter change induces a different air entry which, in turn, generates an extra loss of hydraulic head or energy. If the energy is dissipated by means of a ski jump terminal structure, the efficiency of the jump will be dangerously reduced. It is recommended in such cases, to use stilling basins for the dissipation of energy.

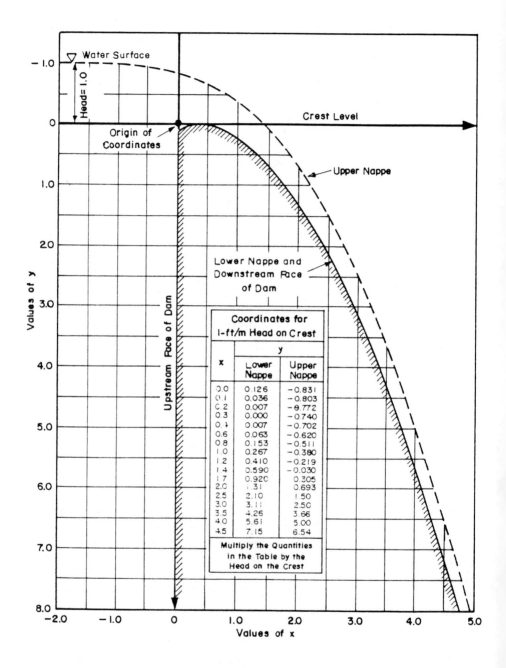

Figure 3.7. Coordinates of the nappes of the flow taking place on a sharp crested weir with vertical upstream face.

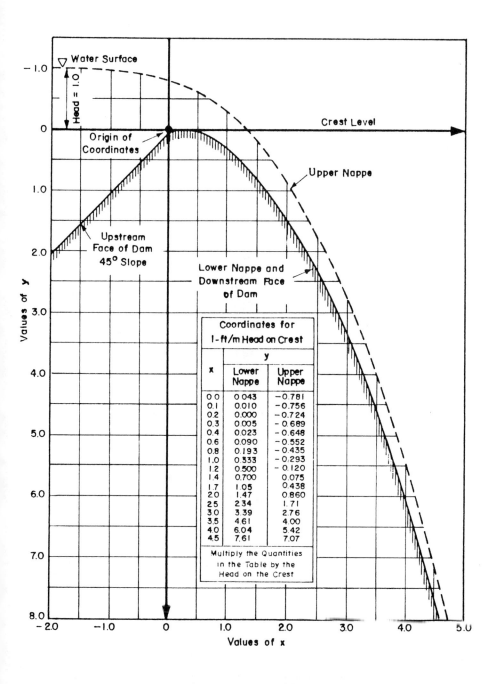

Figure 3.8.   Coordinates of the nappes of the flow taking place on a
sharp crested weir with inclined upstream face.

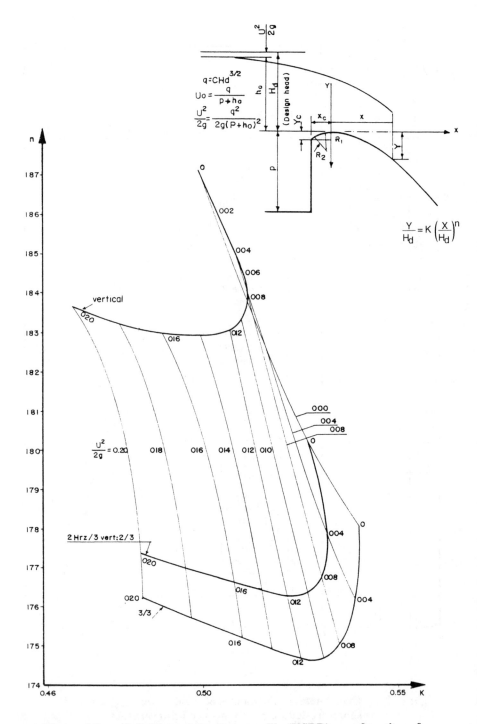

Figure 3.9. Definition of the Ogee profile (USBR) as a function of $(U^2/2g)$.

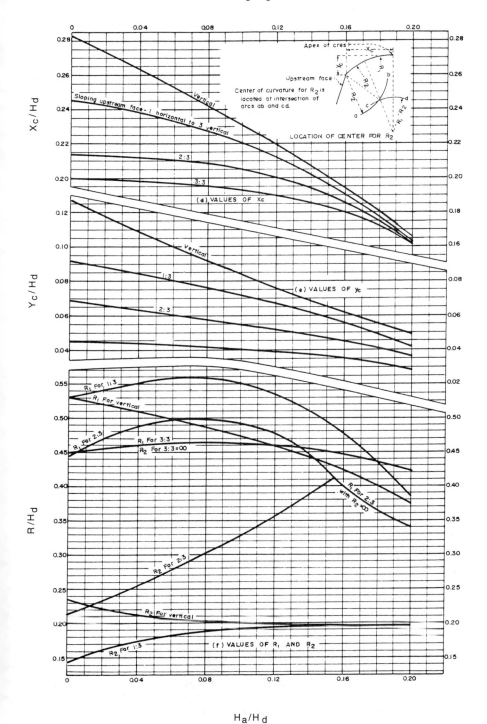

H_0/H_o

Figure 3.10. Definition of the Ogee profile for $U^2/2g = 0$ after USBR (book "Design of small dams").

F. Sentürk

If the free flow spillway is used for discharging large floods under high hydraulic heads it is recommended that an ogee where n < 2 be used. In case of a gated spillway, the exponent n can be chosen greater than or equal to 2. The hydraulic efficiency of the system is then under perfect control. For this reason the ogee of Keban Dam is a second degree parabola. Many other examples can be given for illustrating this case. Table 3.4 summarizes some of them.

TABLE 3.4

Spillway Profiles Used in Large Dams

Name of the Dam	Eq. of the Profile	n	m (*)	Δ %
The Canyon Ferry Dam Spillway (USA)	$Y = \dfrac{X^2}{127} + \dfrac{X}{6}$	2	0.461	7.1
The Hoover Dam Spillway	$Y = \dfrac{X^2}{75} + \dfrac{X}{5} + 1.39$	2	0.459	7.5
The Angostura Dam Spillway (Mexico)	$Y = \dfrac{X^2}{120}$	2	0.483	2.6
The Keban Dam Spillway (Turkey)	$Y = \dfrac{1}{39.3}X^2$	2	0.44	11.3
Remark:	$\Delta = \dfrac{0.496 - m}{0.496}$	(*) Measured on model		

Each one of the profiles was chosen to spill large floods. The scale model investigation is recommended. Then it becomes possible to compare computed theoretical results using relations similar to the given formula in subsection 3.3.2, with those obtained from model studies (Examples 3.05, 3.06, and 3.07). Particularly in the United States, negative pressures are avoided for a complete security (5% - 7.5% in excess, Table 3.4). The ogee, if required, is flattened, as in the case of the Canyon Ferry and the Hoover Dams Spillway (Bradley, J. N., 1952; Şentürk, F, 1957).

B. Variation of the Coefficient of Discharge of a Creager Profile

a. Variation of the coefficient of discharge as a function of h/H_dc

The variation of the coefficient of discharge is studied by many researchers in hydraulics research laboratories throughout the world. The work of Brudenell, 1935; Randolph, R.R., 1937; USBR; Rouse, H.; Creager; Scimemi, E., 1946; and recently the research work undertaken in DSI Research Laboratories, Ankara, have shed some light on this hydraulic phenomenon. The relations describing this variation are reproduced here:

$$\text{Randolph, 1937} \quad m = m_o \left(\frac{h}{H_d}\right)^{0.17} = 0.4902 \left(\frac{h}{H_d}\right)^{0.17} \quad (3.18)$$

$$\text{Brudenell, 1935} \quad m = m_o \left(\frac{h}{H_d}\right)^{0.12} = 0.4990 \left(\frac{h}{H_d}\right)^{0.12} \quad (3.19)$$

$$\text{DSI, 1977} \quad m = m_o \left(\frac{h}{H_d}\right)^{0.16} = 0.4956 \left(\frac{h}{H_d}\right)^{0.16} \quad (3.20)$$

For appreciable velocity heads, h is replaced by

$$H_E = h + \frac{U^2}{2g}$$

Different formulas suggested for determining the variation of m as a function of h are compared in Fig. 3.11. This comparison is observed in a laboratory flume under the following conditions:

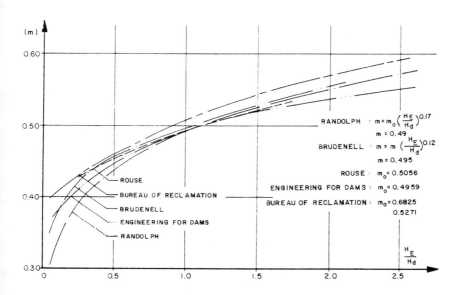

Figure 3.11. Comparison of different formulas defining the variation of m as a function of h/H_d.

1. Stream lines are parallel to the axis of the canal and perpendicular to the axis of the spillway

2. Velocity head is negligible

3. Side contraction in the canal where the experiments are conducted is negligible.

It is easy to see from the representative curves that discrepancies are at a minimum value for:

$$m = 0.8 < \frac{H_E}{H_d} < 1.4$$

USBR suggests using the diagram shown in Fig. 3.12 for determination of the variation of coefficient m, according to different geometrical properties of the spillway and different hydraulic characteristics of the flow. Figure 3.13, also suggested by USBR, shows the variation of m as a function of the height P of the control structure when the upstream face is vertical. For inclined upstream faces this variation is shown in Fig. 3.14.

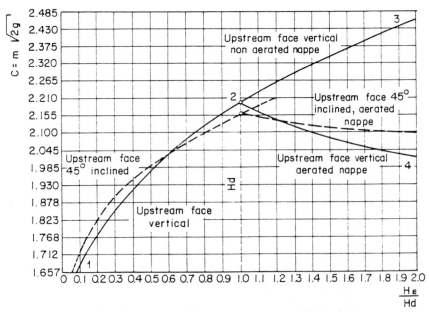

Figure 3.12. Variation of C as a function of H_E/H_d.

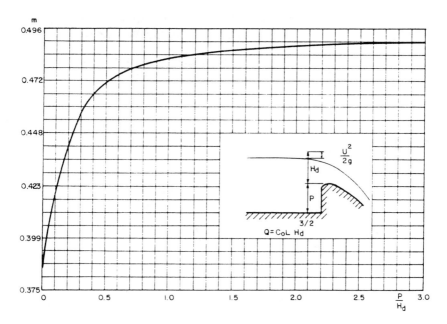

Figure 3.13. Variation of m as a function of P/H_d in case of vertical upstream face (book "Design of Small Dams", USBR).

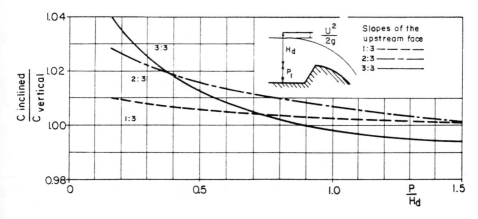

Figure 3.14. Variation of C as a function of P/H_d in case of sloping upstream face (book "Design of Small Dams", USBR).

b. Variation of the coefficient of discharge as a function of the form of the upstream face

The upstream face of a spillway can be vertical, inclined or truncated as shown in Fig. 3.4. The influence of such irregular faces has been studied in

95

the DSI Research Laboratories, Ankara, with Fig. 3.15 showing different profiles used in research work, using the following considerations:

1. Value of X is chosen arbitrarily, without taking stability considerations into account.

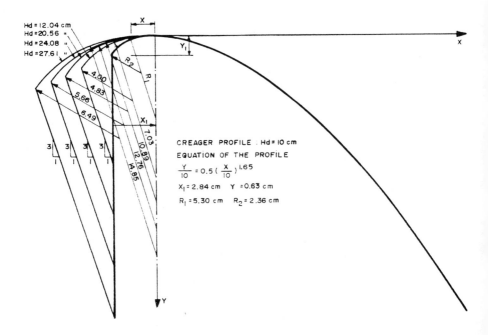

Figure 3.15. **Different profiles used in DSI laboratories to show the influence of upstream irregularities.**

2. Upstream extension is added to the Creager profile directly but the design heads differ from each other.

3. Variation of the discharge coefficient m and the build-up of the negative pressures are investigated as a function of H_d. A relation corresponding to the general form suggested by Randolph, Eq. 3.18, is used.

$$m = m_o \left(\frac{H_E}{H_d}\right)^n \qquad (3.21)$$

This preliminary research shows that m is influenced by the addition of extensions which create extra head losses and a reduction in the value of m. What is gained by reducing some of the concrete volume is offset by the increase in the head loss. The designer must evaluate a range of shapes to find the optimum solution. In the Keban Dam Spillway the introduction of such turbulence, reduced m to a small value, 0.44. A detailed analysis of the flow on this spillway taking negative pressures into account (approximately -4.0 kg/cm^2) has shown that the boundary 0.52 can easily be reached. This problem will be discussed in the following sections.

c. Variation of the coefficient of discharge as a function of the geometrical form of the pierso

The influence of the angle of attack of the piers upon the coefficient of discharge has been studied in Europe (Maître, R., 1958) and in USA (USBR and U.S. Army Corps of Engineers). More details are given in Chapter 4.

C. Profile of a Free Flow Spillway With Negative Pressures

For added security, as stated in earlier sections, negative pressure was avoided in the design of spillways during the first half of this century. Even today many designers prefer the security of positive pressures. It is interesting to know exactly what happens if negative pressure is built up on the downstream face of a spillway. Different questions can be asked and answers given.

Question No. 1: What happens when negative pressures are accepted in the design of spillways?

When negative pressures are accepted in the design, two factors must be taken into consideration:

1. Pulsating flow along the downstream face of the spillway

2. Buildup of the negative pressures (the construction material, generally concrete, will be subject to traction).

Question No. 2: When do negative pressures occur?

The negative pressures occur only when the hydraulic head exceeds the design head. If the design head H_d is chosen slightly smaller than the hydraulic head corresponding to the maximum flood discharge, the negative pressures likely will never occur.

Question No. 3: What is the utility of negative pressures?

Negative pressures increase the coefficient of discharge, m. Then the dimensions are reduced. Particularly in case of large dam spillways, discharging more than 2,000 m^3/sec, economy is important. It is interesting to discuss the introduction of negative pressures in the design of spillways. This method has been used since the second half of this century in Europe. In Turkey, the Karakaya, the Atatürk and the Aslantas, etc., spillways have been studied according to this principle. A detailed study of the design of spillways where negative pressures occur, follows.

D. Variation of the Coefficient of Discharge as a Function of Pressure

The variation of the coefficient of discharge m is defined by Eq. 3.20

$$m = 0.4956 \left(\frac{H_E}{H_d}\right)^{0.16} \qquad 3.20$$

where

$$H_E = h + \frac{U^2}{2g}$$

h is the hydraulic depth on the crest of the spillway and U is the mean velocity in the approach channel. For $H_E = H_d$, $m = m_o = 0.4956$; for $H_E > H_d$, $m > 0.4956$. This is only possible when negative pressures are built up on the downstream face of the control structure. Figure 3.16 shows such a case. Increasing values of m correspond to increasing values of subpressures. The existing pressure under the lower nappe decreases due to the separation of it from the face of the structure. This separation of the nappe from the downstream face of the control structure is a consequence of the increased curvature of the nappe. The system functions like a siphon.

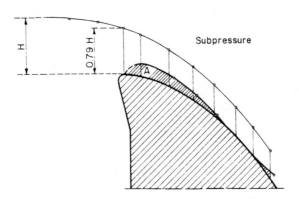

Figure 3.16. Subpressure buildup on a spillway.

The region showed by A is subject to a pressure smaller than the atmospheric pressure. Its value is around -4.0 m when m $\cong$ 0.52. At this point the external atmospheric pressure breaks the nappe and air enters region A. The coefficient m reduces to 0.4956, h decreases suddenly but subatmospheric pressure continually builds up on the crest and the hydraulic phenomenon repeats itself. This is called pulsating flow. Escande, L., 1952 has shown in the laboratory that it was possible to increase the value of m up to 0.54 for h = 1.65H$_d$; and by eliminating the boundary layer by a particular device, he has obtained values as high as 0.70 for m. Naturally the negative pressures relative to these higher values of m are higher than -4.00 m.

The critical value -0.4 kg/cm^2 is obtained as a result of research conducted in different countries and it may differ from one country to another. The experimental results have shown that the concrete material is resistant to a traction having a mean value of 29 kg/cm^2. For example, in Turkey, a coefficient of security equal to 5 is adopted for determining the maximum value of the subatmospheric pressure to be used for designing the spillway control structures.

It is suggested that the following values be used for economical spillway design:

$$p < -0.4 \text{ kg/cm}^2$$

$$0.4956 < m < 0.515$$

Example 3.02

The design discharge of the Keban Dam in Turkey is 17,000 m^3/sec. The observed maximum flood discharge is only 7,000 m^3/sec. This flood has a return period of 45 years. The length of the spillway is 96.00 m and the hydraulic head is 18.00 m. Subatmospheric pressures are not accepted in the design. Analyze the design for negative pressures.

Solution

Assume the flow is bi-dimensional, then

$$q_{max} = \frac{Q}{96} = \frac{17,000}{96} = 177 \text{ m}^3/\text{sec.m}$$

and

$$q = \frac{7,000}{96} = 73 \text{ m}^3/\text{sec.m}$$

F. Sentürk

Furthermore assume that H_d is chosen corresponding to $q = 73$ m³/sec.m; in this case $m_o = 0.4956$, and

$$73 = 0.4956\, H_d\, (2gH_d)^{1/2}$$

$$H_d = 10.34\ m$$

Then, if H_d is chosen equal to 10.34 m, a flood up to 45-year return period can be evacuated by this spillway without any subpressure building up on it. However, if q_{max} occurs, then the coefficient of discharge will take the following value

$$q = m\, 4.43\, h^{3/2} = 177\ m^3/sec.m$$

$$m = 0.4956\, (\frac{h}{10.34})^{0.16}$$

$$q = 0.4956\, (\frac{h}{10.34})^{0.16}\, 4.43\, h^{3/2} = 177\ m^3/sec.m$$

$$h = 17.629 \cong 17.63\ m \qquad m = 0.54$$

This spillway is economical but it cannot be chosen due to $m = 0.54 > 0.52$; the magnitude of the subpressure is likely higher than -0.4kg/cm²; then H_d must be increased. A solution to the problem can be

$$H_d = 13.50\ m \qquad h = 18.089 \cong 18.09\ m$$

and

$$m = 0.519$$

The so-defined spillway must be checked for subpressures (see subsection 3.3.2).

3.3.2 Buildup of Pressure on Downstream Face of Free Flow Spillways

The buildup of pressure on steep channels with curvature in the vertical plane, studied by Läuffer in 1936 and Maître, R., 1960, was also solved by Dressler and Yevjevich in 1984. Flow taking place in a curved channel is visualized in Fig. 3.17. *The depth of this flow is assumed small relative to the radius of curvature of the channel.* If this depth is appreciable, the following analysis is inadequate.

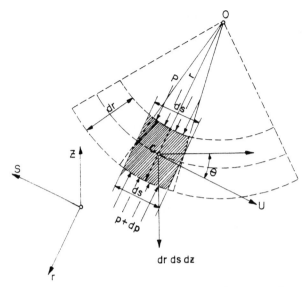

Figure 3.17. Forces acting on flow in a curved rectangular canal.

Consider a volume element of flow with length ds and depth dr. If the depth of flow is small compared to the radius of curvature, this volume element can be converted to a prism. The third dimension of this prism is shown by dz. This volume of water is under the action of three forces:

1. The weight of the water contained in the prism γ dr.ds.dz

2. The force due to the pressure acting on the prism - (p + dp) ds.dz + pds.dz

3. The centrifugal force acting on the prism (U^2/r) (γ/g) dr.ds.dz

where

- γ is the density of water

- dr, ds, dz are the dimensions of the prism

- p is the pressure acting on the prism in the positive direction of the axis of pressure

- U is the local velocity

- r is the radius of curvature at the center-line. and

- g is the gravitational acceleration.

Consider a system of orthogonal axes. The Oz axis is perpendicular to the plane defined by the two axes Or and Os. (Or axis is parallel to the radius of curvature, Os axis is perpendicular to Or axis).

The projection of these three forces on the Or axis is assumed to be in equilibrium for the uniform flow taking place at the curved surface

$$\gamma dr\, ds\, dz\, \cos\theta + \frac{U^2}{r}\frac{\gamma}{g}\, dr\, dz\, ds + pds\, dz - pds\, dz - dp\, ds\, dz = 0$$

and

$$\cos\theta + \frac{U^2}{gr} = \frac{d(p/\gamma)}{dr} \tag{3.22a}$$

The two unknowns of Eq. 3.22a are:

1. the local velocity U

2. the pressure   p

A second equation is necessary to solve Eq. 3.22a; the energy equation can be used for obtaining the unknowns

$$H = z + \frac{p}{\gamma} + \frac{U^2}{2g}$$

and

$$dH = dz + \frac{dp}{\gamma} + \frac{2UdU}{2g} = 0$$

and

$$\frac{dp}{\rho} + gdz + UdU = 0 \tag{3.23a}$$

where

$$\gamma = \rho g$$

Both Eqs. 3.22a and 3.23a are written assuming that the forces are applied to point C in Fig. 3.17. Then

$$\cos\theta + \frac{U_c^2}{gr_c} = (\frac{dp_c}{\gamma}/dr_c) \tag{3.22b}$$

$$\frac{dp_c}{\rho} + gdz_c + U_c dU_c = 0 \qquad (3.23b)$$

are two equations with two unknowns; they can be solved for p and U. Equation 3.23a can be integrated if H_c is assumed constant; then

$$d\left(\frac{p}{\gamma}\right) + dz + U\left(\frac{dU}{g}\right) = 0$$

and from Fig. 3.18

$$d\left(\frac{p}{\gamma}\right) = dr\cos\theta - U\left(\frac{dU}{g}\right)$$

Eliminating $d(p/\gamma)$ between Eqs. 3.22b and 3.23b yields:

$$r_c dU_c + U_c dr_c = 0 \qquad (3.24)$$

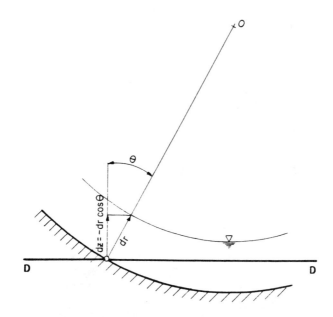

Figure 3.18. Flow in a concave rectangular channel.

Then Eqs. 3.22b and 3.23b are reduced to

$$U_c r_c = U_y r_y$$

$$U_c r_c = Ct \qquad (3.25a)$$

103

$$\frac{p_c}{\gamma} + z_c + \frac{U_c^2}{2g} = Ct \qquad (3.25b)$$

Two constant values exist in this system of equations.

If they can be determined, the problem is solved. Let's consider Fig. 3.19. Assume U_y as the velocity of the flow at the surface at a point Y. At point C, chosen on a perpendicular line to the bottom of the channel along a radial through Y, the velocity is U_c. Applying Eq. 3.25a between points Y and C yields

$$U_c \, r_c = U_y \, r_y$$

or

$$U_c \, (r - m) = U_y \, (r-d)$$

Applying Eq. 3.25b to the flow at point C yields

$$\frac{p_c}{\gamma} + m \cos\theta + \frac{U_c^2}{2g} = H_c$$

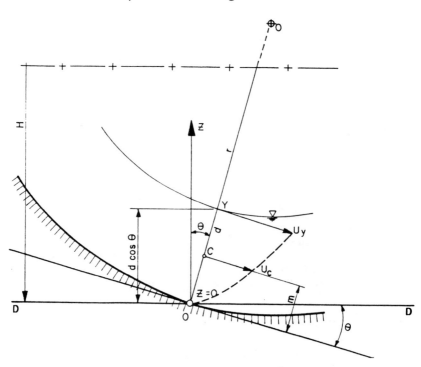

Figure 3.19. Definition of symbols.

and

$$\frac{p_c}{\gamma} = H_c - m\cos\theta - \frac{U_c^2}{2g} = H_c - m\cos\theta - \frac{U_y^2}{2g}(\frac{r-d}{r-m})^2$$

The pressure at the surface of the flow being assumed equal to zero $(p_y/g)=0$; then

$$d\cos\theta + \frac{U_y^2}{2g} = H_y \cong H_c$$

The difference $(H_y - H_c)$ is assumed negligible; if this assumption is not valid, the result of the computation will not be representative of the hydraulic phenomenon.

Substituting this value of $(U_y^2/2g)$ into the above relationship yields

$$\frac{p_c}{\gamma} = H_c - m\cos\theta - (H_c - d\cos\theta)\frac{(r-d)^2}{(r-m)^2} \qquad (3.26a)$$

and

$$U_c = \sqrt{2g(H_c - d\cos\theta)}\,\frac{r-d}{r-m} \qquad (3.26b)$$

Applied to point O at the bottom of the channel, Eqs. 3.26a and 3.26b take the following forms

$$\frac{p_o}{\gamma} = H - (H - d\cos\theta)\frac{(r-d)^2}{r^2} \qquad (3.27a)$$

$$U_o = \sqrt{2g(H_c - d\cos\theta)}\,\frac{(r-d)}{r} \qquad (3.27b)$$

Equations 3.26a - 3.26b and Eqs. 3.27a - 3.27b were suggested by Läuffer, 1936 and Maître, R., 1960. They are applicable only if the following two conditions are satisfied:

1. the depth of flow is small compared to the radius of curvature.

2. $H_y \cong H_c$.

F. Sentürk

The boundary condition imposed by these equations is also important and an analysis is given for a better understanding.
Let us divide Eq. 3.27a by H

$$\frac{p/\gamma}{H} = 1 - [1 - (\frac{d\cos\theta}{H})] [1 - \frac{d\cos\theta}{H} \cdot \frac{H}{r\cos\theta}]^2$$

Furthermore, let us proceed with a change of variable

$$z = \frac{H}{r\cos\theta} \qquad x = \frac{d\cos\theta}{H}$$

$$y = (p/\gamma)/H$$

Substituting in the above equation yields

$$y = 1 - (1 - x)(1 - zx)^2$$

or

$$y = (1 + 2z)x - z(2 + z)x^2 + z^2x^3 \qquad (3.28)$$

a. The curve represented by Eq. 3.28 has a maximum and a minimum value for

$$x_n = \frac{1}{z} \qquad x_m = \frac{2}{3} + \frac{1}{3z} \qquad (3.29a)$$

$$y_n = 1 \qquad y_m = \frac{1}{27}(1 + 2z)^2 (\frac{4}{z} - 1) \qquad (3.29b)$$

(Problem 3.11)

b. The following expression of the hydraulic head corresponds to a point on the water surface

$$H = d\cos\theta + \frac{U^2}{2g}$$

$\frac{U^2}{2g}$ being positive $\qquad H > d\cos\theta; \quad$ then $0 < x < 1$

This condition is very important; for $z < 1$, the equation has no physical meaning

$$x_n = \frac{1}{z} \qquad z < 1 \qquad \text{then } n_n > 1$$

This cannot be true. Negative pressures correspond to $z < 0$ (Fig. 3.20). The limiting value for which negative pressures are observed is $z = -1/2$. For $z < 0$, $r < 0$, the center of curvature is located at the lower region of the profile which is convex.

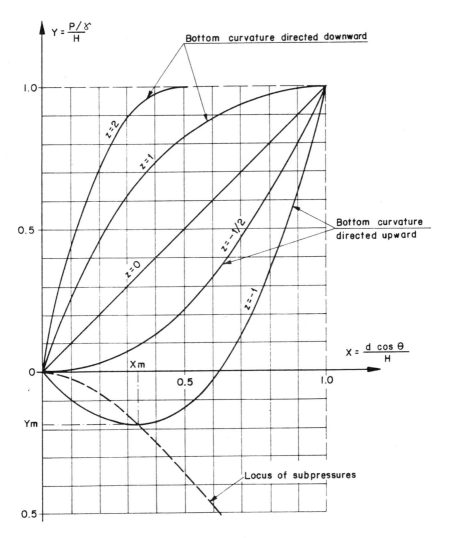

Figure 3.20. Diagram showing the buildup of pressures on the downstream face of a spillway.

c. For $z < -1/2$, the negative pressures are built up on the profile. These pressures increase up to x_m and decrease for $x > x_m$ up to 0; then, changing their sign, they continue to increase in the region of positive pressures. The coordinates of the nadir are shown by x_m and y_m in Fig. 3.20.

d. The variation of pressures is defined by Eq. 3.28 and shown in Fig. 3.20. Their position on the profile can be determined using either Eq. 3.28 or Fig. 3.20. Then d and r can be used for computing p/γ.

e. Subpressures observed on the profile of a control structure increase with the hydraulic head and decrease for increasing values of the radius of curvature. Equation 3.28 can be used for showing this property as observed on hydraulic scale models and measured on prototypes. The magnitude of negative pressures decreases towards the downstream end of the hydraulic structure due to the increased radius of curvature. Then there is a change and positive pressures occur at the upstream start of the ski jump which is a concave curve.

When H is constant and r decreases, $-p/\gamma$ increases. Increasing values of negative pressures correspond to an increase in the coefficient of discharge. If the depth of flow increases at the crest of the control structure, then the first condition is not fulfilled and the proposed relations cannot be used to compute the pressures.

Variation of Coefficient of Discharge m as a Function of Subpressures

The discharge of the flood to be evacuated by a spillway can be computed when the coefficient of discharge is known

$$Q = mLh\,(2gH)^{1/2}$$

The variation in the value of the coefficient m is defined by Eq. 3.20, or

$$m = 0.4956\,(\frac{H}{H_d})^{0.16} \qquad\qquad 3.20$$

Substituting m by its equivalent in the formula which defines the discharge yields

$$Q = 0.4956\,(\frac{H}{H_d})^{0.16}\,Lh\,(2gH)^{1/2}$$

L and H_d are data which depend on the problem. Then values of H result from the given values of Q.
Substituting

$$z = \frac{H}{r\cos\theta} \qquad \text{and} \qquad y = \frac{p}{\gamma}$$

in Eq. 3.29b yields

$$\left(\frac{p}{\gamma}\right)_{max} = \frac{1}{27}(1+\frac{2H}{r\cos\theta})^2\,(\frac{4r\cos\theta}{H}-1) \qquad (3.30)$$

Values of r and θ correspond to a given profile and H being known, $(p/\gamma)_{max}$ can be computed from Eq. 3.30. If the obtained value is greater than the allowable value, -4.00 m, H or H_d must be changed and the computation repeated.

Values of r and q can be computed by applying the following expressions:

$$r = \frac{(1-f')^{3/2}}{f''}$$

$$\tan\theta = f'$$

where f' and f'' are the first and the second derivatives of the equation which defines the profile chosen for the design of the control structure.

Example 3.03

The control structure of a spillway is defined by:

$$h = 18.00\text{ m}, \qquad H_d = 12.00\text{ m}, \qquad m = 0.52$$

Compute the magnitude of subpressures building up on the profile.

Data

Equation of the profile: $Y = 0.50\,\dfrac{X^{1.85}}{H_d^{0.85}}$

$$q = 73.00\text{ m}^3/\text{sec.m}$$

Solution

The magnitude of the subpressure is given by Eq. 3.30, or

$$\left(\frac{p}{\gamma}\right)_{max} = \frac{1}{27}(1+\frac{2H}{r\cos\theta})^2\,(\frac{4r\cos\theta}{H}-1) \qquad 3.30$$

$$\cos\theta = \frac{1}{(1+\tan^2\theta)^{1/2}} \qquad \tan\theta = Y'$$

$$Y = 0.50\,\frac{X^{1.85}}{H_d^{0.85}} \qquad r = \frac{(1+Y')^{3/2}}{Y''}$$

Then

$$Y' = \frac{1.85}{2}\,\frac{X^{0.85}}{H_d^{0.85}} = 0.93\,\frac{X^{0.85}}{H_d^{0.85}}$$

$$Y'' = 0.93 \times 0.85\,\frac{X^{-0.15}}{H_d^{0.85}} = \frac{0.79}{H_d^{0.85}\,X^{0.15}}$$

$$r = \frac{(H_d^{1.70} + 0.86\,X^{1.70})^{3/2}}{0.79\,H_d^{1.70}}\,X^{0.15}$$

Substituting H_d by its value r, is computed as

$$r = \frac{(12^{1.7} + 0.86 X^{1.7})^{3/2}\,X^{0.15}}{0.79 \cdot 12^{1.70}} = \frac{(68.33 + 0.86 X^{1.7})^{3/2}}{53.98}\,X^{0.15}$$

The values of (r) as a function of X are listed in Table 3.5a.

TABLE 3.5a

p/γ and (r) values as a function of X

X	Y	H	cosθ	r	p/γ	p
m	m	m		m	kg/cm²	m
0.50	0.02	18.02	0.9985	-9.49	-0.90	-9.00
1.00	0.06	18.06	0.9942	-10.66	-0.72	-7.20
2.00	0.22	18.22	0.9806	-12.33	-0.55	-5.50
10.00	4.28	22.28	0.7834	-30.78	-0.14	-1.40

Since the center of curvature is at the left side of the profile, (r) is negative and (p/γ) is also negative; p_{max} being greater by the absolute value than -4.00 m, this profile cannot be used. This inconvenience is corrected by increasing the value of H_d.

H_d = 15.00 m is tried and the results are listed in Table 3.5b.

TABLE 3.5b

H_d = 15.00 m

X	Y	H	$\cos\theta$	r	p/γ	p
m	m	m		m	kg/cm^2	m
0.50	0.01	18.01	0.9987	-11.4451	-0.6066	-6.066
1.00	0.05	18.05	0.9957	-12.8126	-0.4745	-4.745
2.00	0.18	18.18	0.9863	-14.6280	-0.3573	-3.573
5.00	0.98	18.98	0.9395	-19.4160	-0.2097	-2.097
10.00	3.54	21.54	0.8358	-30.6063	-0.0997	-0.997

If the results of computation in the vicinity of the crest are not valid; then $p_{max} \cong$ - 4.00 m.

Example 3.04

Compute the evolution of pressures on the control structure of the Keban Dam Spillway, Turkey.

Data

Equation of the profile $\quad Y = \dfrac{1}{39.3} X^2$

Solution

The pressures and the velocity on the profile are computed using

$$\frac{p}{\gamma} = H - (H - d\cos\theta) \left(\frac{r-d}{r}\right)^2 \qquad 3.27a$$

$$U_o = \sqrt{2g\,(H - d\cos\theta)}\ \left(\frac{r-d}{r}\right) \qquad 3.27b$$

where   H is the hydraulic head where the pressure is computed

d is the depth of flow in the cross section considered

r is the radius of curvature, and

q is the angle of tangent to the curve with the horizontal line where the pressure is computed.

The equation of the profile of the Keban Dam Spillway is written for a Cartesian system of axes, the origin of which is placed at the spillway crest. Table 3.6 summarizes the results of computations.

TABLE 3.6

Pressures building up on the profile of the Keban Dam Spillway

Station	X	Y	r	$\cos\theta$	d	H	p/γ
	m	m	m		m	m	m
1	2	3	4	5	6	7	8
1	0.00	0.00	-19.650	1.0000	15.20	20.00	4.900
2	4.06	0.42	-20.922	0.9793	12.80	20.42	-0.064
3	10.30	2.70	-28.282	0.8857	9.70	22.70	-2.750
4	16.83	7.21	-44.851	0.7595	8.00	27.21	-2.140
5	27.54	19.11	∞	0.6199	6.60	39.11	4.090
6	35.99	29.81	∞	0.6199	6.20	49.81	3.840

Column 1:   Stations chosen on the profile where the pressures is computed

Column 2:   Abscissa of the stations shown in Column 1

Column 3:   Ordinates of the stations shown in Column 1

Substituting the value of X from Column 2 and introducing it in the Keban Dam Spillway's formula gives the value of Y

$$Y = \frac{1}{39.3}X^2$$

The point of tangent to the spillway curve profile occurs at $X = 24.85$, with $\tan\theta = 1/0.79$. This factor is also included in the example as follows:

Station 1: $X_1 = 0$ $\qquad$ $Y_1 = 0$

Station 2: $X_2 = 4.06$ m $\quad$ $Y_2 = \dfrac{1}{39.3} \times 4.06^2 = 0.42$ m

Station 3: $X_3 = 10.30$ m $\quad$ $Y_3 = \dfrac{1}{39.3} \times 10.3^2 = 2.70$ m

Station 4: 16.83 m $\qquad$ $Y_4 = \dfrac{1}{39.3} \times 16.83^2 = 7.21$ m

Station 5: 27.54 m $\qquad$ $Y_5 = \dfrac{1}{39.3} \times 24.89^2 +$

$$\dfrac{1}{0.79} \times 2.65 = 19.11 \text{ m}$$

Station 6: 35.99 m $\qquad$ $Y_6 = \dfrac{1}{39.3} \times 24.89^2 +$

$$\dfrac{1}{0.79} \times 11.10 = 29.81 \text{ m}$$

Column 4 $\quad$ Radius of curvature of the points shown in column 1

$$r = \frac{[1 + (dY/dX)^2]^{3/2}}{d^2Y/dX^2} \qquad (3.31)$$

where

$$Y = (1/39.3)X^2 \qquad \frac{dY}{dX} = (2/39.3)X$$

$$\frac{d^2Y}{dX^2} = (2/39.3)$$

Radius of curvature of station 1, $\quad X_1 = 0$

$$\frac{dY}{dX} = 0$$

$$\frac{d^2Y}{dX^2} = \frac{2}{39.3}$$

113

$$r_1 = \frac{(1+0)^{3/2}}{2/39.3} = 19.650 \text{ m}$$

Radius of curvature of point 2, $\quad X_2 \quad = \quad 4.06 \text{ m}$

$$\frac{dY}{dX} = \frac{2}{39.3} 4.06$$
$$= 0.2066$$

$$\frac{d^2Y}{dX^2} = \frac{2}{39.3}$$

$$r_2 = \frac{(1+0.2066^2)^{3/2}}{2/39.3} = 20.922 \text{ m}$$

Radius of curvature of station 3, $\quad X_3 \quad = \quad 10.30 \text{ m}$

$$\frac{dY}{dX} = \frac{2}{39.3} 10.3$$
$$= 0.5242$$

$$\frac{d^2Y}{dX^2} = \frac{2}{39.3}$$

$$r_3 = \frac{(1+0.5242^2)^{3/2}}{2/39.3} = 28.282 \text{ m}$$

Radius of curvature of station 4, $\quad X_4 \quad = \quad 16.83 \text{ m}$

$$\frac{dY}{dX} = \frac{2}{39.3} = 16.83$$
$$= 0.8565$$

$$\frac{d^2Y}{dX^2} = \frac{2}{39.3}$$

$$r_4 = \frac{(1+0.8565^2)^{3/2}}{2/39.3} = 44.851 \text{ m}$$

Stations 5 and 6, being on the straight line section of the spillway, the corresponding radius of curvatures are equal to infinity. Since the centers of curvature are at the left side of the profile, the radii have a negative sign.

Column 5:   Computations of $\cos\theta$

Station 1: $X_1 = 0$ $\tan\theta_1 = Y' = 0$

$$\cos\theta_1 = \frac{1}{\sqrt{1+\tan^2\theta_1}} = \frac{1}{\sqrt{1+0}} = 1$$

Station 2: $X_2 = 4.06 \quad \tan\theta_2 = 0.2066$

$$\cos\theta_2 = \frac{1}{\sqrt{1+0.2066^2}} = 0.9793$$

Station 3: $X_3 = 10.30 \quad\quad \tan\theta_3 = 0.5242$

$$\cos\theta_3 = \frac{1}{\sqrt{1+0.5242^2}} = 0.8857$$

Station 4: $X_4 = 16.83 \quad\quad \tan\theta_4 = 0.8565$

$$\cos\theta_4 = \frac{1}{\sqrt{1+0.8565^2}} = 0.7595$$

Stations 5 and 6 are located on the straight line reach of the profile then $\tan\theta_5 = \tan\theta_6 = 0.2658$ and $\cos\theta_5 = \cos\theta_6 = 0.6199$.

Column 6:   The depth d of the cross section

In this particular case, in order to be able to compare the results obtained from the physical model with the mathematical model, the depth of the cross section is measured directly on the model of the Keban Dam Spillway. Equation 3.27b can also be used to obtain these values. In this case the two basic equations given in subsection 3.3.2 must be satisfied.

These conditions are:

1. The depth of flow must be small compared to the radius of curvature

2. $H_y \cong H_c$

Examinations of Table 3.6 shows that

- at points 1, 2, and 3, the radius of curvature is comparable to the depth of flow

- $H_y \neq H_c$

Then d can only be computed at point 4 as follows:

The linear discharge of the Keban Dam Spillway is

$$q = \frac{17,000}{L}$$

where

$$L = L_o - KnH \qquad\qquad 4.16$$

$$L_o = 96.00 \text{ m} \qquad K = 0.04 \qquad n = 12 \qquad H = 27.21 \text{ m}$$

$$L = 96 - 0.04 \cdot 12 \cdot 27.21 = 82.939 \text{ m}$$

$$q = \frac{17,000}{82.939} = 204.97 \text{ m}^3/\text{sec.m}$$

$$U = \frac{q}{d} = \frac{204.97}{d} \text{ m/sec}$$

$$(204.97/d)^2 = 2g (H - d \cos\theta) \left(\frac{r-d}{r}\right)^2$$

At station 4, $H = 27.21$ m, $r = -44.851$ m, $\cos\theta = 0.7595$

Then

$d \cong 8.54$ m. The measured value being 8.00 m, the error involved is 6.3%.

Column 7: Computation of the hydraulic head

Station 1: $X_1 = 0$ $H_1 = 20.00$ m

Station 2: $X_2 = 4.06$ $H_2 = 20.00 + 0.42 = 20.42$ m

Station 3: $X_3 = 10.30$ $H_3 = 20.00 + 2.70 = 22.70$ m

Station 4: $X_4 = 16.83$ $H_4 = 20.00 + 7.21 = 27.21$ m

Station 5: $X_5 = 27.54$ $H_5 = 20.00 + 19.11 = 39.11$ m

Station 6: $X_6 = 35.99$ $H_6 = 20.00 + 29.81 = 49.81$ m

Column 8: Computation of pressures

Station 1: $X_1 = 0.00$ m

$$\frac{P_1}{\gamma} = 20.00 - (20.00 - 15.20 \bullet 1.00) \left(\frac{-19.65 - 15.20}{-19.65}\right)^2 = 4.9019 \text{ m}$$

Station 2: $X_2 = 4.06$ m

$$\frac{P_2}{\gamma} = 20.42 - (20.42 - 12.80 \bullet 0.9793) \left(\frac{-20.922 - 12.80}{-20.922}\right)^2 = -0.0642 \text{ m}$$

Station 3: $X_3 = 10.30$ m

$$\frac{P_3}{\gamma} = 22.70 - (22.70 - 9.70 \bullet 0.8857) \left(\frac{-28.282 - 9.70}{-28.282}\right)^2 = -2.7462 \text{ m}$$

Station 4: $X_4 = 16.83$ m

$$\frac{P_4}{\gamma} = 27.21 - (27.21 - 8.00 \bullet 0.7595) \left(\frac{-44.851 - 8.00}{-44.851}\right)^2 = -2.1357 \text{ m}$$

Station 5: $X_5 = 27.54$ m

$$\frac{P_5}{\gamma} = d_5 \cos\theta_5 = 6.60 \bullet 0.6199 = 4.0913 \text{ m}$$

Station 6: $X_6 = 35.99$ m

$$\frac{P_6}{\gamma} = d_6 \cos\theta_6 = 6.20 \bullet 0.6199 = 3.8434 \text{ m}$$

The computed and measured pressures on the Keban Dam Spillway are shown in Fig. 3.21. They are comparable if the boundary conditions can be defined adequately; then a mathematical analysis based upon the Creager profile will be sufficient for determining the dimensions of the control structure.

Example 3.05

Compute the pressures on the control structure of the Oymapinar Dam Spillway, Turkey.

Data

• The nappe is sustained

117

F. Sentürk

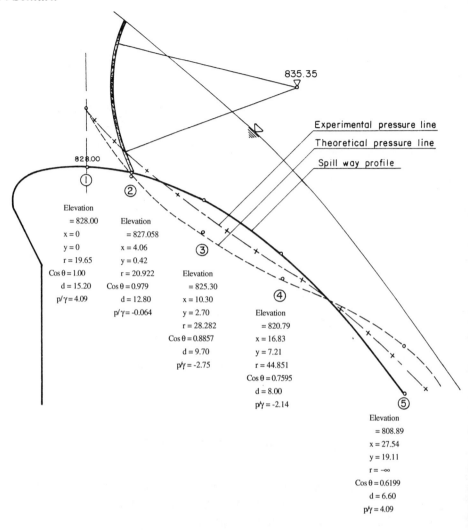

Figure 3.21. Comparison of computed and measured pressures on the Keban Dam Spillway.

- The equation of the spillway profile relative to a coordinate system with the origin on the crest is

$$\frac{Y}{13.30} = 0.50 \left(\frac{X}{13.30}\right)^{1.85}$$

- H_d = 13.30 m $\qquad$ H= 15.00 m $\qquad$ L = 26.80 m

- Q = 3,100 m³/sec

118

Solution

The solution of the problem is summarized in Table 3.7.

TABLE 3.7

Pressures built up on the profile of the Oymapinar Dam Spillway

Stations	X m	Y m	r m	$\cos\theta$	d m	H m	p/γ m
1	0	0.000	0.000	1.0000	12.00	15.000	∞
2	1	0.055	-11.658	0.9946	11.10	15.055	-0.246
3	3	0.423	-14.938	0.9673	9.95	15.423	-0.672
4	5	1.089	-18.304	0.92735	9.00	16.089	-1.140
5	7	2.029	-22.443	0.8813	8.25	17.027	-1.220
6	9	3.230	-27.580	0.8832	7.90	18.227	-1.044

Column 1:   Stations chosen on the profile of the control structure

Column 2:   Abscissa of the stations shown in column 1

Column 3:   Ordinates of the stations shown in column 2 (using the equation of the spillway profile)

Column 4:   Computation of the radius of curvature

$$Y = 0.0554\ X^{1.85} \qquad\qquad Y' = 0.1025\ X^{0.85}$$

$$Y'' = 0.087\ X^{-0.15}$$

Station 1: $X_1 = 0$ m $Y' = 0$ $Y'' = \infty$ $r_1 = \dfrac{(1+0)^{1.5}}{\infty} = 0$

Station 2: $X_2 = 1$ m $Y' = 0.10251$ $Y'' - 0.08712$ $r_2 = 11.658$ m

Station 3: $X_3 = 3$ m $Y' = 0.2608$ $Y'' = 0.07389$ $r_3 = 14.9376$ m

Station 4: $X_4 = 5$ m $Y' = 0.4026$ $Y'' = 0.06844$ $r_4 = 18.3040$ m

Station 5: $X_5 = 7$ m $Y' = 0.5359$ $Y'' = 0.06507$ $r_5 = 22.4431$ m

Station 6: $X_6 = 9$ m $Y' = 0.6635$ $Y'' = 0.06267$ $r_6 = 27.5797$ m

The profile being convex r is negative.

Column 5:   Computation of $\cos\theta$

$$\tan\theta = Y' \qquad\qquad \cos\theta = \frac{1}{\sqrt{1 + Y'^2}}$$

Station 1: $X_1 = 0$ m $\tan\theta_1 = Y' = 0$ $\cos\theta_1 = 1$

Station 2: $X_2 = 1$ m $\tan\theta_2 = 0.1025$ $\cos\theta_2 = 0.99479$

Station 3: $X_3 = 3$ m $\tan\theta_3 = 0.2608$ $\cos\theta_3 = 0.96763$

Station 4: $X_4 = 5$ m $\tan\theta_4 = 0.4026$ $\cos\theta_4 = 0.92735$

Station 5: $X_5 = 7$ m $\tan\theta_5 = 0.5359$ $\cos\theta_5 = 0.88131$

Station 6: $X_6 = 9$ m $\tan\theta_6 = 0.6635$ $\cos\theta_6 = 0.83327$

Column 6:   Depth of cross section

The depth of cross section is measured on the scale model of the Oymapinar Dam Spillway; Eq. 3.27b can also be used to obtain these values. In this case the two basic conditions are:

1. the depth of flow must be small compared with the radius of curvature

2. $H_y \cong H_c$

In this particular case, the depth at point 6 is 7.90 m, and the radius of curvature is (-27.58 m). The depth is not small compared with the radius of curvature. Then Eq. 3.27b cannot be used satisfactorily.

Column 7:   Computation of hydraulic head

Station 1: $X_1 = 0$ $H_1 = 15.00$ m

Station 2: $X_2 = 1$ m $H_2 = 15.00 + 0.055 = 15.055$ m

Station 3: $X_3 = 3$ m $H_3 = 15.00 + 0.423 = 15.423$ m

Station 4: $X_4 = 5$ m $H_4 = 15.00 + 1.088 = 16.088$ m

Station 5: $X_5 = 7$ m $H_5 = 15.00 + 2.027 = 17.027$ m

Station 6: $X_6 = 9$ m $H_6 = 15.00 + 3.227 = 18.227$ m

Column 8:   Computation of pressures

Station 1: $X_1 = 0$ m $\dfrac{P_1}{\gamma} = 15.00 - (15.00 - 12.00 \bullet$

$$1.00) \, (\frac{0 - 12.00}{0})^2$$

$$\frac{P_1}{\gamma} = \infty$$

Station 2: $X_2 = 1$ m $\dfrac{P_2}{\gamma} = -0.246$ m

Station 3: $X_3 = 3$ m $\dfrac{P_3}{\gamma} = -0.672$ m

Station 4: $X_4 = 5$ m $\dfrac{P_4}{\gamma} = -1.140$ m

Station 5: $X_5 = 7$ m $\dfrac{P_5}{\gamma} = -1.225$ m

Station 6: $X_6 = 9$ m $\dfrac{P_6}{\gamma} = -1.046$ m

The computed and measured pressures on the Oymapinar Dam Spillway are shown in Fig. 3.22. Subpressures are measured on the profile for partial gate opening. Subpressure does not exist when the gates are fully open. This is due to the form of the nappe and the concave shape of the profile which occurs immediately downstream of the Creager profile. The nappe does not follow the Creager profile and it is sustained by the concave shape of the spillway.

Equation 3.20 being applicable only for $p/H_d > 5$, Fig. 3.13 can be used for computing the coefficient of discharge.

$$H_d = 13.30 \text{ m} \qquad\qquad P = 4.00 \text{ m}$$

then, $m = 0.454$

m is smaller than the limiting value 0.52.

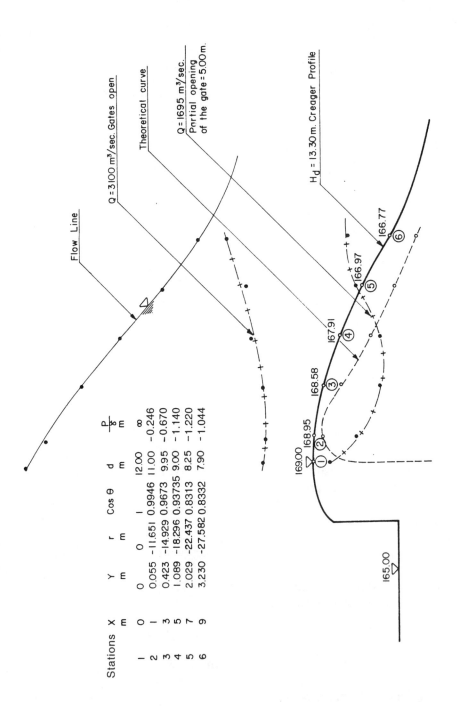

Figure 3.22. Comparison of computed and measured pressures on the Oymapinar Dam Spillway.

The theoretical variation of pressures must be corrected taking the properties of the nappe into account. This nappe is not a free nappe; it is a sustained nappe. The measured pressures are positive. This is also due to the sustained characteristics of the nappe.

Example 3.06

Compute the pressures on the control structure of the Aslantaş Dam Spillway, Turkey.

Data

- The equation of the profile relative to a coordinate system with the origin on the crest is:

$$\frac{Y}{16.30} = \frac{1}{1.791} \left(\frac{X}{16.30}\right)^{1.776}$$

- $H_d = 16.30$ m

- $P = 5.00$ m

- Upstream face slope (1/1)

Solution

The hydraulic computations are conducted following the routine shown in Examples 3.04 and 3.05. The results are summarized in Fig. 3.23 and Table 3.8.

This theory is not applicable for Stations 1 and 2, when d is comparable to r. The discharge coefficient can be computed by direct application of Fig. 3.14.

$$P = 5.00 \text{ m} \qquad H_d = 16.30 \text{ m} \qquad \frac{P}{H_d} = 0.31 \qquad \frac{m}{m_o} = 1.08$$

m_o can be obtained from Fig. 3.13 as $m_o = 0.450$, then

$$m = 1.08 \cdot 0.4570 = 0.4936 < 0.52$$

123

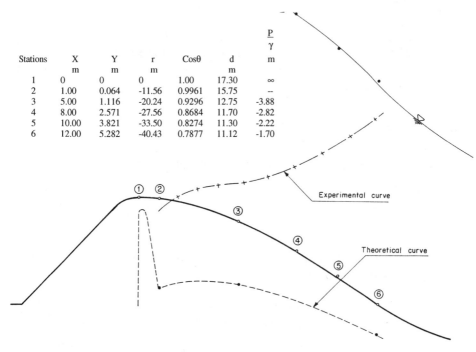

Stations	X m	Y m	r m	Cosθ	d m	$\frac{P}{\gamma}$ m
1	0	0	0	1.00	17.30	∞
2	1.00	0.064	-11.56	0.9961	15.75	--
3	5.00	1.116	-20.24	0.9296	12.75	-3.88
4	8.00	2.571	-27.56	0.8684	11.70	-2.82
5	10.00	3.821	-33.50	0.8274	11.30	-2.22
6	12.00	5.282	-40.43	0.7877	11.12	-1.70

Figure 3.23. Comparison of computed and measured pressures on the Aslantaş Dam Spillway.

TABLE 3.8

Pressure buildup on the profile of the Aslantaş Dam Spillway

Station	X m	Y m	r m	cos θ	d m	H m	p/γ m
1	2	3	4	5	6	7	8
1	0.00	0.000	0.0000	1.0000	17.30	20.20	
2	1.00	0.064	-11.5584	0.9961	15.75	20.26	-
3	5.00	1.116	-20.2395	0.9296	12.75	21.32	-3.83
4	8.00	2.571	-27.5594	0.8684	11.70	22.77	-2.82
5	10.00	3.822	-33.5013	0.8274	11.30	24.00	-2.22
6	12.00	5.282	-40.4301	0.7877	11.12	25.48	-1.70

Example 3.07

Compute the maximum value of the subpressure on the Saint Etienne Cantalès and the Aigle Dam Spillway, France.

Data	St. Etienne Dam	Aigle Dam
Equation of profile (origin on the crest)	$Y = \frac{1}{17.68} X^2$	$Y = \frac{1}{30} X^2$
H_d	4.42 m	7.70 m
Hydraulic head	7.70 m	12.00 m
Radius of curvature at the crest	8.84 m	15.00 m

Solution

Figures 3.24 and 3.25 illustrate the subpressures near the crest of the St. Etienne Cantalès Dam Spillway and the Aigle Dam Spillway.

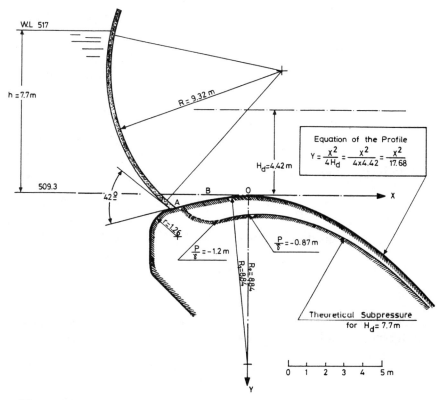

Figure 3.24. The Saint Etienne Cantalès Dam Spillway subpressures.

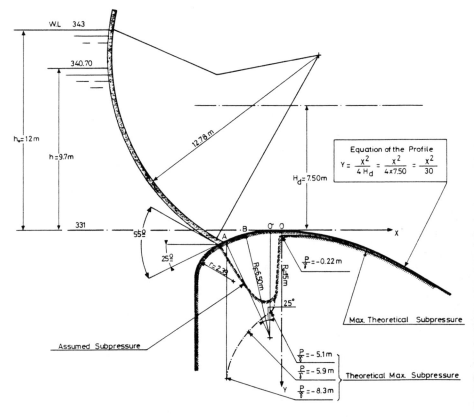

Figure 3.25. The Aigle Dam Spillway subpressures.

Equation 3.29b is used to compute $(p/\gamma)_{max}$. This equation has a minimum value corresponding to

$$z = -0.50\,H_m$$

(see Fig. 3.20). Substituting this value in Eq. 3.29b yields

$$\frac{(p/\gamma)}{H_m} = -\frac{1}{27}\left(\frac{H}{H_m} - 1\right)^2\left(8 + \frac{H}{H_m}\right) \qquad 4.9$$

where H_m is the head over a reference point M. If the crest is considered as this reference point $H_m = H_d$ and

$$\frac{(p/\gamma)}{H_d} = -\frac{1}{27}\left(\frac{H}{H_d} - 1\right)^2\left(8 + \frac{H}{H_d}\right)$$

where H becomes the hydraulic head of the spillway (see Chapter 4).

- **The St. Etienne Cantalès Dam Spillway**

$$\frac{H}{H_d} = \frac{7.70}{4.42} = 1.74 \text{ and } (\frac{p}{\gamma})_{max} = -4.42 \cdot 0.2 = -0.87 \text{ m}$$

The subpressure appears for H = 5.20 m on the prototype then $(p/\gamma)_{max} = -0.05$ m. For this reason the nappe adheres smoothly to the spillway profile.

- **The Aigle Dam Spillway**

$$\frac{H}{H_d} = \frac{12.00}{7.50} = 1.60 \text{ and } (\frac{p}{\gamma})_{max} = -0.13 \cdot 7.5 = -0.98 \text{ m}$$

The subpressure begins for H = 9.70 m, then

$$\frac{H}{H_d} = \frac{9.70}{7.50} = 1.29 \qquad\qquad (\frac{p}{\gamma})_{max} = -0.029 \cdot 7.5 = -0.22 \text{ m}$$

3.4   PROBLEMS

Problem 3.01

a. Assuming q = 6.29 m³/sec.m, determine the water surface elevation in the reservoir in Example 3.01.

b. If q increases to 8.74 m³/sec.m, n to 0.022, P to 4.00 m, check the variation of water level.

Data

$$n = 0.020 \quad P = 2.50 \text{ m} \quad m = 0.49 \quad L = 200.00 \text{ m} \quad \alpha = 0.50$$

where n is the Manning's coefficient of resistance

P is the dam height

h is the hydraulic head of the control structure at the downstream end of the canal

m is the discharge coefficient

α is the coefficient of the energy loss at the entrance, and

L is the length of the approach canal

Problem 3.02

a. The control structure of a spillway presents the following characteristics:

$$m = 0.444 \qquad P = 15.00 \text{ m} \qquad q = 45.00 \text{ m}^3/\text{sec.m}$$

The bottom slope of the approach channel is horizontal

$$n = 0.024 \qquad L = 1200.00 \text{ m, and} \qquad \alpha = 0.45$$

where

m is the coefficient of discharge

P is the height of the control structure

q is the linear discharge taking place from the control structure

n is the Manning's coefficient of resistance

L is the length of the approach channel, and

α is the coefficient of the energy loss at the entrance.

Assuming that $q = 45.00 \text{ m}^3/\text{sec.m}$ during a catastrophic flood, determine the variation in the water surface elevation in the reservoir and control section.

b. Check the results assuming that $P = 5.00$ m and the hydraulic head $h = 8.00$ m, $m = 0.49$ and $q = 52.00 \text{ m}^3/\text{sec. m}$

Problem 3.03

Compare mathematically and graphically WES and Creager-Scimemi profiles.

Data

$$H_d = 13.50 \text{ m}$$

Upstream face vertical

Problem 3.04

Compare the Creager profile defined in Figure 3.10 with the profile defined in Problem 3.03.

Problem 3.05

Design a spillway profile that will require a minimum volume of concrete.

Data

$$H_d = 11.00 \text{ m} \qquad\qquad P = 30.00 \text{ m}$$

$h + (U^2/2g) = 4.00$ m and the downstream face slope will be chosen by the designer. Use a profile of the type,

$$\frac{Y}{H_d} = 0.50 \left(\frac{X}{H_d}\right)^{1.85}$$

Problem 3.06

Design the ogee of a spillway using Fig. 3.7.

Data

Hydraulic head = 10.00 m

$$P = 20.00 \text{ m}$$

Upstream face vertical.

Problem 3.07

Compare the profile of the Keban Dam Spillway with the profile of the Angostura Dam Spillway, USA. Make a literature search for finding the equation of the Angostura Dam Spillway.

The equation of the ogee of the Keban Dam Spillway is given in the text (see Example 3.04; page 111).

Problem 3.08

Design a spillway by using Fig. 3.9.

Data

$$H_d = 10.75 \text{ m} \qquad\qquad \frac{U^2}{2g} = 0.20 \text{ m}$$

Upstream face vertical.

Problem 3.09

Show the variation of discharge of a spillway with the Creager profile as a function of h on a diagram responding to the following conditions:

$$10 < q < 100 \text{ m}^3/\text{sec.m}$$

$$H_d = 10.00 \text{ m}$$

$$P = 10.00 \text{ m}$$

$$q = mh \sqrt{2g \left(h + \frac{U^2}{2g}\right)}$$

where P is the height of spillway. Use Eqs. 3.18, 3.19 and 3.20 for determining m. Compare the obtained results on a graph.

Problem 3.10

Determine the maximum negative pressure at the abscissa $X = 10.00$ m building up on the control structure of a spillway the characteristics of which are given.

Data

1. Equation of the selected profile:

$$\frac{Y}{H_d} = 0.50 \left(\frac{X}{H_d}\right)^{1.85}$$

2. $H_d = 14.00$ m

3. $L = 90.00$ m

4. $m = 0.4956 \left(\frac{H}{H_d}\right)^{0.16}$

5. $U = 1.50$ m/sec, mean velocity in the approach channel

6. $q = 120$ m^3/sec.m

7. $P = 80.00$ m

Problem 3.11

Compute the coordinates of the maximum and minimum turning points of Eq. 3.28.

Chapter 4

HYDRAULICS OF GATED SPILLWAYS

4.1 INTRODUCTION

A control structure equipped with gates is called a "gated spillway". This type of spillway is generally used when the reservoir is to be maintained at a constant level to stabilize production at a hydroelectric power plant or to store a fixed volume of water for irrigational purposes. The policy to be used for reservoir operation constitutes separate aspects not discussed here.

As outlined in previous chapters, the sizing of spillways is done taking the catastrophic discharges into consideration. If the gates are hoisted accidentally when the reservoir is full, an artificial catastrophic flood may be generated downstream of the dam. This possibility has been discussed at international conventions. Nevertheless, the gated spillways are constructed and are in construction throughout the world.

4.1.1 Objectives

The aim of this chapter is the study of:

- the buildup of negative pressure on the control structure for partial opening of the gates

- the determination of dimensions of the gates in respect to hydraulic behavior of the spillway

- the variation of discharge for partial opening of the gates

- the determination of flow characteristics on the control structure and particularly around the piers.

4.1.2 Classifications of Gates

Different types of gates are used on weirs and on large dam spillways. Where large discharges are to be evacuated, tainter gates are most often chosen.

F. Sentürk

Gates can be classified as follows (Creager, W.P., Justin, J.D., 1950):

Crest control

- Flash boards
- Tilting gates
- Bear-trap crests
- Drum crests.

Crest gates

- Sliding gates
- Tainter gates (radial gates)
- Wheeled gates
- Stony gates
- Caterpillar gates
- Rolling gates
- Stop logs or Gate: needles.

4.2   HYDRAULIC PROPERTIES OF GATED SPILLWAYS

4.2.1   Hydraulic Problems of Gated Spillways

Various problems must be solved in designing the gated spillways. They may be grouped into two major categories:

- Problems which must be solved at the design level
- Problems which must be solved at the operational level.

The first group can be divided into:

1. Problems related to geometry of end piers
2. Problems related to geometry of intermediate piers
3. Problems related to choice of type of gates
4. Problems related to position of gate at the profile
5. Problems related to buildup of pressures at the profile.

The most important problems related to the operational phase is the uncontrolled hoisting of gates.

4.2.2 Hydraulic Computation for Gated Spillways

Subpressures are built up at a given point on the profile for certain gate openings. They increase with the increasing opening, reach a maximum value at some partial opening, then decrease to the free flow value at full opening of the gate.

In general, as stated earlier, spillways of large dams are equipped with radial gates. The position of the seat of the gate at the spillway profile influences the coefficient of discharge.

The characteristics of gates must be chosen in such a way that all kinds of vibration, the formation of a vortex in the upstream reservoir and the disturbance of the nappe be avoided.

A. Location of Gates in Relation to Spillway Axis

The location of the gate relative to the axis of the control structure influences the buildup of negative pressures at the profile. An analytical procedure of computation of these pressures is given in this section.

Empirical rules are usually suggested for determination of the position of the gates in respect to the axis of the control structure. In the US Corps of Engineers Manual EM 1110-2-1603, the use of a statistical study is suggested. This study has been completed by adding the Turkish experience on large dam spillways and reproduced in Table 4.1.

Research was conducted all over the world to determine the buildup of subpressure at the profiles of control structures (Rohne, T.J., 1959; Maître, R., 1960). The problem can be summarized as follows:

1. Determination of the evolution of subpressures as a function of the variation of the design head of control structure, assuming a constant position of the gate in respect to the axis of spillway and also constant gate characteristics.

2. Determination of the evolution of subpressures as a function of the position of the gate seat on the crest of the control structure.

3. Determination of discharge in respect to the characteristics of the gate.

These questions should be answered in order to prepare a good design for the gated spillway.

TABLE 4.1

Characteristics of Radial Gates

Spillway		H_d	X_E/H_d	Y_E/H_d	X_M/H_d	Y_M/H_d	h_k/H_d	L/H_d	R/H_d
High Sills									
Pine Flat,	USA	11.89	0.20	0.03	1.13	0.35	0.97	1.08	1.00
Detroid	"	9.27	0.38	0.08	1.09	0.34	1.00	1.38	0.89
Folsom	"	15.24	0.16	0.02	1.04	0.30	1.02	0.84	0.94
Chief Josef	"	12.68	0.24	0.04	1.15	0.44	1.11	0.96	1.01
Sutton	"	11.67	0.19	0.02	1.00	0.29	0.81	1.04	0.85
Table Rock	"	12.50	0.17	0.02	1.01	0.32	0.92	1.15	0.90
Center Hill	"	13.11	0.12	0.01	0.85	0.30	0.87	1.16	0.79
Eufaula	"	9.14	0.20	0.03	1.24	0.30	1.07	1.33	1.10
Whitney	"	12.14	0.27	0.04	1.16	0.28	0.99	1.00	0.95
Osceola	"	10.79	0.17	0.02	1.01	0.28	0.87	1.13	0.90
Mean values		11.84	0.21	0.03	1.07	0.32	0.96	1.11	0.93
Kiralkizi Türkiye		8.75	0.23	0.03	1.35	0.61	1.15	0.94	1.26
Karakaya	"	13.00	0.36	0.07	1.33	0.38	1.06	1.08	1.08
Karakuz	"	5.00	0.30	0.06	1.48	0.24	1.17	1.16	1.20
Keban	"	9.83	0.31	0.03	1.55	0.70	1.55	1.63	1.42
H. Uğurlu	"	15.00	0.10	0.01	1.08	0.47	1.01	0.89	1.17
Mean values		10.32	0.26	0.04	1.36	0.48	1.19	1.14	1.23
Low Sills									
Garrison,	USA	9.20	0.13	0.01	1.14	0.31	0.97	1.32	1.06
Gavins Point	"	12.50	0.18	0.02	0.84	0.29	0.76	0.098	0.73
Ft. Randall	"	10.24	0.15	0.01	0.99	0.88	0.88	1.19	0.83
Mean values		10.65	0.15	0.01	0.99	0.29	0.87	1.16	0.89

Location of the gate lip: X_E: abscissa Y_E: ordinate

Location of the trunnion: X_M: abscissa Y_M: ordinate

Height of the gate: h_k Width of the gate: L

Radius of the gate: R

It can be mathematically shown that the magnitude of the negative pressures decreases when the gate lip is located downstream of the control structure. Such a solution increases the dimension of the gates and increases the cost of the structure. It can also be mathematically shown that with a constant effective hydraulic head, the increase of the design head, H_d, decreases the magnitude of the subpressures. Many attempts have been made to find a correct answer to this problem. Maître, R., (1960) suggested the inequality relationship of Eq. 4.1 for determining the location of the gate for profiles with equation

$$\frac{Y}{H_d} = 0.47 \left(\frac{X}{H_d}\right)^{1.80}$$

$$0 < \frac{X}{H_d} < 1.33 \tag{4.1}$$

and the inequality, Eq. 4.2, for profiles with equation

$$\frac{Y}{H_d} = 0.50 \left(\frac{X}{H_d}\right)^{1.85}$$

$$0 < \frac{X}{H_d} < 1.17 \tag{4.2}$$

where X is the abscissa of the gate seat on the control structure. The application of Eqs. 4.1 and 4.2 yields rather large limits for X, but can be used as a first approximation.

The highest value of subpressure occurs immediately downstream of the gate for partial openings. An optimum solution must be determined for solving the problem as outlined above, recognizing that:

- Increasing values of X occur for decreasing values of subpressures,

- Increasing values of H_d correspond to decreasing values of subpressures.

In the first case, the dimension of the gate increases; then an optimum solution is necessary which satisfy both conditions, namely, the decrease in cost of the structure and the decrease in the magnitude of subpressures.

In second case, the increase of H_d yields a heavier cross section of the control structure, then an optimum solution is necessary which satisfy both conditions, namely the increase of H_d for decreasing the subpressures, and the resulting increase in the cost of the control structure.

In gated spillways, the buildup of negative pressures is a function of the opening of the gate. It is then possible to define a particular opening at which the magnitude of pressures attains a maximum value. If this opening can be mathematically determined, the designer's objective will be completely satisfied. In the following analysis this particular value will be computed and its variation as a function of H_d and X studied.

A partially open gate constitutes an orifice. Considering a flow taking place from it, it is possible to say that the profile of the nappe immediately downstream of the orifice is different from the profile of the control structure. At this specific reach, the mathematical analysis of the flow is difficult to perform and only the results of laboratory or prototype experiments shed some light on this phenomenon (Figs. 3.21, 3.22, and 3.23). Particularly, in the vicinity of the crest, discrepancies are very important due to the following reasons:

a. Immediately downstream of the gate the flow jet shows a contraction. The nappe is cut off abruptly from the supporting body causing an increase in subpressures.

b. The radius of curvature r of the Creager profile is equal to zero at the crest. The theory is only applicable for depths of flow much smaller than r and it cannot be applied around the crest.

Figure 4.1 shows the flow taking place through an orifice defined by the control structure and the gate. The variation of observed pressures as a function of the angle α between the tangent to the gate and the ogee, and the theoretical variation of pressures are shown in the figure. The influence of h/H_d is also shown on the same figure (Escande, L.).

It is interesting to observe that at a given point the maximum values of pressures in both cases, namely the mathematically determined and measured ones, coincide with each other. The engineer has to compute the pressure up to this point and use the obtained value in the design. Upstream of this point the mathematical analysis shows a rapid increase which has no practical meaning.

It has been stated earlier that location of the gate downstream increases the size of the gate. Figure 4.2 illustrates this case. A solution to this disadvantage is shown in Fig. 4.3. The main characteristics of this very elegant solution are: *

• The dimensions of the gate are a function of the discharge, which varies as a function of the hydraulic head. The designer can choose the hydraulic head according to the geometric properties of the control structure. The crest of the spillway can be lowered without changing the dimensions of the gate, the gate opening will be constant but the hydraulic head is increased; then the discharge increases accordingly (Fig. 4.3).

---

* See also Fig. 2.6a

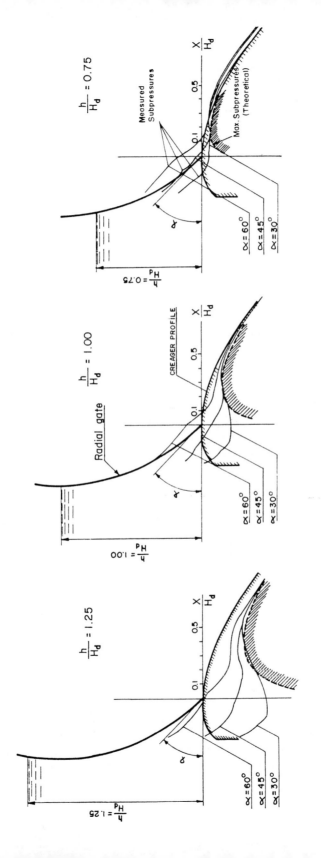

Figure 4.1. Variation of the negative pressure as a function of α and h/Hd.

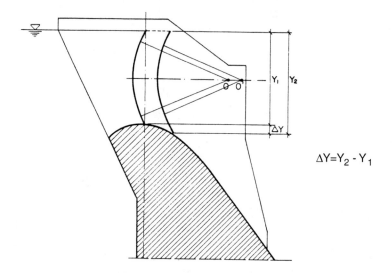

$$\Delta Y = Y_2 - Y_1$$

Figure 4.2. The increase in the dimensions of a gate due to a downstream location.

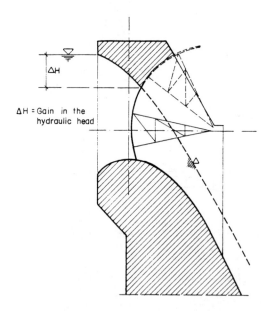

ΔH = Gain in the hydraulic head

Figure 4.3. Orifice spillway.

- The spillway will work as an orifice for partial and full opening of the gate. The velocity is increased, but the nappe follows the profile of the control structure rather satisfactorily. The negative pressures can be controlled easily.

138

The inconvenience of this solution is that sealing is difficult; this means that the prevention of leakage through the seats of the gate becomes a problem. This solution cannot be used for small reservoirs, (Fig. 2.6b).

For the Chastang Spillway, France, this solution was applied and it was noted that the operation of the gate was difficult and loss of water by leakage due to the high hydraulic head was significant. Nowadays, the techniques of sealing have been improved; similar solutions can be selected and applied satisfactorily (EDF, 1951 and 1952).

B. Influence of Geometry of Profile of Control Structure on Subpressures Building up Directly Downstream of the Gate

The nappe is continuous in free flow spillways. It is shown by dotted line in Fig. 4.4 for full opening of the gate. A partially open gate causes a discontinuity in gated spillways.

The hydraulic head at a point M is H in free flow spillways (the spillway is assumed to be a short structure) and in a gated spillway it is H_m, with $H > H_m$. These are the major hydraulic differences between a free flow and a gated spillway.

Subpressure does not occur in a free flow spillway for $H = h+Y$, and in a gated spillway $H_m = h_m+Y$ with $H > H_m$. Then in a free flow spillway, if $h < H_d$ for a hydraulic head, H, the subpressure does not buildup for the same H, which is observed in a gated spillway. The subpressure can be computed using Eq 3.27, as shown in Chapter 3. Considering the hydraulic head, H, constant, and varying water depth, d, the value of subpressure varies. This variation shows a maximum value.

H can be taken as a constant in gated spillways, and the partial opening of the gate acts upon the depth d. There is a value for the partial opening at which the subpressure attains its maximum value. For the designer, this value is of paramount importance. Equation 3.29b defines this value:

$$x = (d \cos\theta / H)$$

$$\frac{p/\gamma}{H} = \frac{1}{27} (1 + 2z)^2 (\frac{4}{z} - 1)$$ 3.29b

where

$$z = (H/r \cos\theta)$$

The subpressure attains its maximum value immediately downstream of the gate. These general considerations are explained in full detail in the following subchapters.

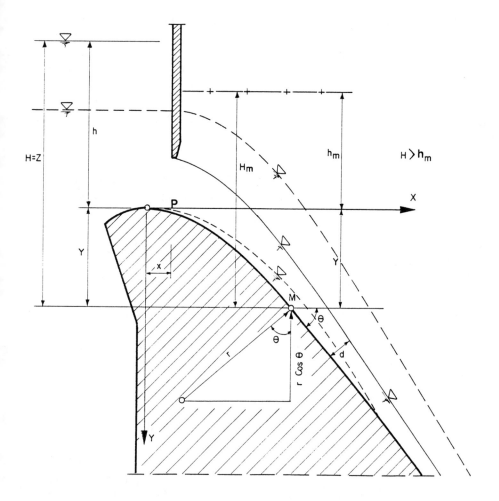

Figure 4.4. Maximum subpressures expected at X_P, the abscissa of the lip of the gate.

a. Creager profile

The design of a spillway requires a joint effort of civil, hydraulic and mechanical engineers. The first step to be taken is the determination of the pressures buildup at the control structure. Maître, R. (1960) suggested the diagram of Figs. 4.5 and 4.6 for a first approximation of the negative pressures expected downstream of the gate (Eq 3.27a). The maximum value of $(p/\gamma)/H_d$ is given as a function of (h/H_d) and X/H_d, where p is the

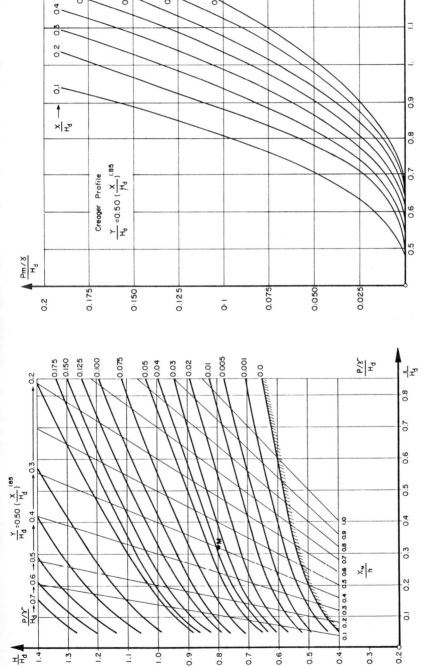

Figure 4.5. Maximum negative pressures expected immediately downstream of the gate of a gated spillway.

Figure 4.6. Build up of subpressure on a gated spillway.

pressure, h the hydraulic head*, X the abscissa at which the maximum negative pressure occurs and H_d the design head. In case of a free flow spillway for

$$\frac{h}{H_d} < 1$$

negative pressure does not occur, but in case of a gated spillway even for $(h/H_d) < 1$ negative pressures can be observed at the profile.

The maximum negative pressure occurs at X_p which is the abscissa of the seat of the gate at the control structure as shown in Fig. 4.4. The graph given in Fig. 4.6 is directly derived from the graph of Fig. 4.5.

In fact, if the expression

$$\frac{X_M/H_d}{h/H_d} = \frac{X_M}{h}$$

is chosen by the designer, then X_M/H_d can be computed for a given point H.

The example given by Maître, R. (1960), for illustrating the use of these graphs, is reproduced in the following figures.

Example 4.01

Determine the maximum negative pressure expected at a spillway with the Creager profile.

Data

$$H = h = 10.00 \text{ m} \qquad X_M = 4.00 \text{ m (**)} \qquad H_d = 12.50 \text{ m}$$

Solution

H_d, X_M and h being known, $\dfrac{X_M}{h}$ and $\dfrac{h}{H_d}$ can be computed

$$\frac{X_M}{h} = \frac{4.00}{10.00} = 0.40 \qquad\qquad \frac{h}{H_d} = \frac{10.00}{12.50} = 0.80$$

---

*   When the approach velocity is appreciable, h is substituted by H.

**   If x_m is the abscissa of the seat of the gate then the subpressure is the max subpressure which can occur on the profile.

Entering the graph in Fig. 4.5 with these values yields a value for Point M, and

$$\frac{P_M}{\gamma} = 0.04 \text{ m}$$

b. Second degree parabola profile: Partially opened gates

In gated spillways and for partial opening of the gate, the vector velocity, U_o, is directed downward towards a point located below the horizontal. It is possible to determine the angle of the vector velocity, U_o, with the horizontal line as a function of the geometry of the control structure and the position of the gate. The equation defining the flow issuing from the opening of the gate is

$$y = \frac{x^2}{4H_d \cos^2\theta} + x\tan\theta \qquad\qquad 3.12$$

where x is the abscissa of a point on the flow

y is the ordinate of the same point

H_d is the design head

and θ is the angle of the vector velocity with the horizontal at the origin.

The equation of the second degree parabola is

$$Y = aX^2 + bX \qquad\qquad (4.3)$$

Comparing Eq. 3.12 with Eq. 4.3 yields

$$H_d = \frac{1+b^2}{4a} \qquad a \neq 1.00 \qquad a > 1.00 \qquad (4.4)$$

$$\tan\theta = b \qquad\qquad (4.5)$$

and for

$$\theta = 0$$

$$\frac{Y}{H_d} = \frac{1}{4}\left(\frac{X}{H_d}\right)^2$$

Considering the profile of a spillway defined by an equation of the form of Eq. 3.12, it is possible to compute the subpressure buildup at the point M on it, (see Fig. 4.6).

Assume H_m the real hydraulic head at point M and H_d, the design head. The hydraulic head of the spillway corresponding to H_m is shown by h_m. Using these variables, and Eq. 4.6 yields:

Computation of the parameter, z:

$$z = \frac{H}{r \cos\theta} \qquad r \cos\theta = -\frac{1 + y'^2}{y''} \;(*) \qquad\qquad (4.6)$$

The first derivative of Eq. 3.12 yields

$$y' = \frac{x}{2H_d \cos^2\theta} + \tan\theta$$

and the second derivative yields

$$y'' = \frac{1}{2H_d \cos^2\theta}$$

Substituting these two values in Eq. 4.6 yields

$$r \cos\theta = -2(h_m + y) = -2H_m$$

and finally **

$$z = -\frac{1}{2}\frac{H}{H_m} \qquad\qquad (4.6a)$$

The subpressure builds up for $z \leq -(1/2)$ (Fig. 3.20 and subsection 3.3.2), then $H > H_m$, which has already been chosen arbitrarily. When the subpressure occurs at a given point at the profile of a partially opened gate, it continues for the entire profile. However, in a free flow spillway, or when the gated spillway is fully open, subpressures may occur near the crest and positive pressures may still take place below the crest. Figure 3.21 illustrates this phenomenon.

When point M is located further downstream, the value of z approaches $(-1/2)$ asymptotically, then subpressures are equal to zero at the boundary.

---

*   The sign (-) is a consequence of the direction of y axis (See Fig. 4.4).

**   See Problem 4.02

The maximum value of subpressure as a function of the parameter z was calculated in Chapter 3 and found equal to

$$\frac{p/\gamma}{H_d} = \frac{1}{27}(1 + 2z)^2 \left(\frac{4}{z} - 1\right) \tag{4.7}$$

$$x = \frac{d \cos\theta}{H}$$

(see Chapter 3) and

$$z = \frac{H}{r \cos\theta}$$

The variation of $(p/\gamma)/H$ as a function of x is shown in Fig. 3.20. The representative curve presents two singular points

$$\left(\frac{d \cos\theta}{H}\right)_n = \frac{1}{z} \qquad \left(\frac{d \cos\theta}{H}\right)_m = \frac{2}{3} + \frac{1}{3z} \qquad \text{3.29a}$$

$$\left(\frac{p/\gamma}{H}\right)_n = 1 \qquad \left(\frac{p/\gamma}{H}\right)_m = \frac{1}{27}(1 + 2z)^2 \left(\frac{4}{z} - 1\right) \qquad \text{3.29b}$$

Equation 3.29a yields

$$\frac{d \cos\theta}{H} = \frac{1 + 2z}{3z} \tag{4.8}$$

and substituting

$$z = -\frac{1}{2}\frac{H}{H_m}$$

in Eq. 3.29b yields

$$\frac{p/\gamma}{H_m} = -\frac{1}{27}\left(\frac{H}{H_m} - 1\right)^2 \left(8 + \frac{H}{H_{\text{111}}}\right) \tag{4.9}$$

and

$$d \cos\theta = \frac{2}{3}(H - H_m) \tag{4.10}$$

which gives the flow depth.

Equations 4.7 and 4.8 can be used for computing the water depth d and the subpressure p/γ at a point M of a flow taking place under a gated spillway. When H and H_m are given, p/γ is determined accordingly (Fig. 4.4).

Conclusions:

1. The magnitude of subpressure decreases when the gate is located downstream of the crest.

2. In parabolic profiles the subpressure continues for the entire profile if it is built up at one point (Fig. 4.7).

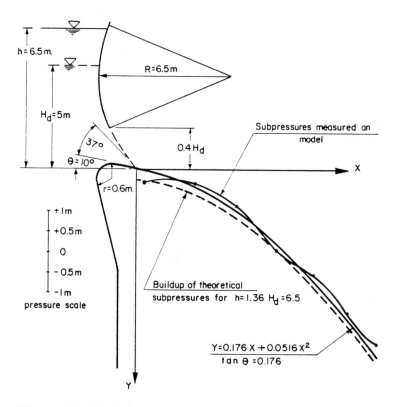

Figure 4.7. Variation of subpressures on the profile of a gated spillway.

3. The position of the gate in relation to the spillway axis varies as a function of the geometry of the gate and of the spillway. A suitable study of this variation leads to a good design.

C. Variation of Critical Head at which Subpressures are Built up in Case of a Second Degree Parabola Profile: Gates Fully Open

Figure 4.8 illustrates a spillway with a second degree parabola profile. The equation of the flow issuing from the gate is

$$y = \frac{x^2}{4H\cos^2\theta} + x\tan\theta \qquad\qquad 3.12$$

where H is the hydraulic head at a chosen point. It can also be taken equal to H_d.

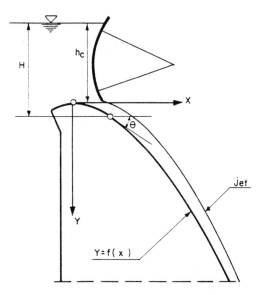

Figure 4.8. Flow on a second degree parabola spillway.

When the hydraulic radius of the flow at a point M is equal to the hydraulic radius of the profile, subpressure is built up and for $r_{profile} < r_{nappe}$ it increases. The boundary condition is:

$$y' = Y' \qquad \text{and} \qquad y'' = Y''$$

where y' and y" are the first and the second derivatives of the flow equation and Y' and Y" the first and the second derivatives of the equation defining the profile. Then

$$|r| = \frac{(1 + Y')^{3/2}}{Y''} = \frac{(1 + \tan^2\theta)^{3/2}}{Y''} = \frac{1}{Y''\cos^3\theta}$$

The second derivative of Eq. 3.12 yields

$$Y'' = \frac{1}{2H\cos^3\theta}$$

and

$$\frac{1}{2H''\cos^2\theta} = \frac{1}{|r|\cos^3\theta}$$

$$2H = |r|\cos\theta \qquad \frac{H}{|r|\cos\theta} = \frac{1}{2} = -z$$

It is clear that for $H \leq H_d$, subpressures are eliminated because it was previously shown that the buildup of subpressures begins from the basic condition $z < -1/2$.

The radius of curvature of the nappe near the gate is larger than the radius of curvature of the profile. It decreases rapidly until the two radii become equal to each other. Downstream of this point, the curvature of the nappe continues to decrease and it becomes smaller than the hydraulic radius of the profile and the subpressures are eliminated

$$y'' = \frac{1+\tan^2\theta}{2Hd} = \frac{1+y'^2}{2H} = \frac{1+y'^2}{2(h+y)}$$

where $h = h_m$ (Fig. 4.4).

At the point where the two radii of curvature are equal, $Y' = y' = \tan\theta$ and $Y'' = y''$. Then

$$\frac{1+y'^2}{2(h+y)} = Y'' = \frac{1+Y'^2}{2(h+Y)} = y''$$

and

$$h = \frac{1-2yy''+y'^2}{2y''} = \frac{1-2YY''+Y'^2}{2Y''} \qquad (4.11)$$

This relation defines the depth of flow at the beginning of the buildup of subpressures. Considering a profile defined by

$$\frac{Y}{H_d} = a\left(\frac{X}{H_d}\right)^n$$

the above given relation yields

$$\frac{h}{H_d} = \frac{1-2YY''+Y'^2}{2Y''H_d}$$

and

$$Y = \frac{aX^n}{H_d^{n-1}} \qquad Y' = an\frac{X^{n-1}}{H_d^{n-1}} \qquad Y'' = an(n-1)\frac{X^{n-2}}{H_d^{n-1}}$$

Substituting the value of Y' and Y" in Eq. 4.11 yields

$$\frac{h}{H_d} = \frac{1 - a^2 n (n-2) \left(\dfrac{X}{H_d}\right)^{2(n-1)}}{2an(n-1)\left(\dfrac{X}{H_d}\right)^{n-2}} \qquad (4.12)$$

In case of a second degree profile n = 2, then

$$\frac{h}{H_d} = \frac{1}{4a}$$

and the value of h is independent of x.

Using Eq. 4.12 and for given values of hydraulic characteristics of a flow, the variation of h can be shown on a graph. This equation is valid for H = h + Y. Then the value of h defines the position of the point at which the subpressures appear.

Example 4.02

Determine the limiting head at which subpressures appear on a gated spillway where the abscissa of the gate seat is X.

Data

- Equation of the profile: $\dfrac{Y}{H_d} = 0.50 \left(\dfrac{X}{H_d}\right)^{1.85}$

- $H_d = 13.30$ m

- $a = 0.50$

- $n = 1.85$

Solution

- For fully open gate

The problem is reduced to a free flow spillway on which the flow is taking place (see Chapter 3).

- For partially open gate

F. Sentürk

It has previously been shown that subpressures can appear in the following conditions:

$$y' = Y' \qquad \text{and} \qquad Y'' = y''$$

Equation 4.12 defines this condition:

$$\frac{h}{H_d} = \frac{1-a^2n\,(n-2)\,\left(\dfrac{X}{H_d}\right)^{2(n-1)}}{2an\,(n-1)\,\left(\dfrac{X}{H_d}\right)^{n-2}} \qquad\qquad 4.12$$

where   h is the hydraulic head of the spillway

          H_d is the design head

          X is the abscissa of the gate seat

          a = 0.50 and

          n = 1.85

Substituting these values in Eq 4.12 yields

$$h = 5.74\,(1 + 0.0009X^{1.7})\,X^{0.15}$$

The solution of this equation is given in Table 4.2, and shown in Fig. 4.9. If, for example, the abscissa of the gate seat is 3.00 m, which is a value generally encountered in practice, the subpressure appears for h = 6.80 m.

TABLE 4.2

Variation of the boundary value h at which the subpressure appears

X	Y	h
m	m	m
0	0	0
1	0.055	5.74
3	0.423	6.80
5	1.089	7.40
7	2.029	7.87
9	3.230	8.27

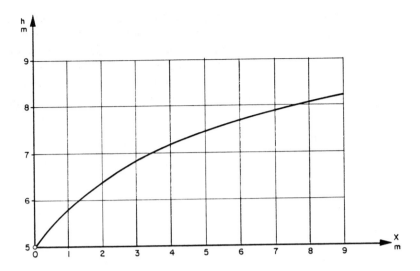

Figure 4.9. Variation of the boundary value (h) at which the subpressure appears as a function of X.

Conclusions:

1. Consider Eq. 3.29b

$$\frac{p/\gamma}{H} = \frac{1}{27} (1 + 2z)^2 \left(\frac{4}{z} - 1\right)$$

$$z = \frac{H}{r \cos \theta}$$

It is possible to say that at a given point, an increase of H corresponds to a decrease of the pressures (Problem 4.03).

2. At a given point, (p/γ) decreases for increasing values of $r \cos\theta$ (Problem 4.04). Then for the profiles of the form

$$\frac{Y}{H_d} = a \left(\frac{X}{H_d}\right)^n$$

an increase of the design head with a constant hydraulic head yields a decrease of the pressures at all points M corresponding to

$$X < \frac{H_d}{(an)^{n-1}}$$

151

3. For profiles of the form given above and for $n < 2$, (p/γ) and (z) vary in the same direction. This variation is shown in Fig. 4.10, prepared for a profile of the Creager type. The value, z, decreases rapidly for the increasing values of X/H_d and for $h/H_d = Ct$ and $(p/\gamma) = 0$ for $z = 0.50$ as shown previously. Then for a gated spillway, the maximum subpressure occurs immediately downstream of the gate.

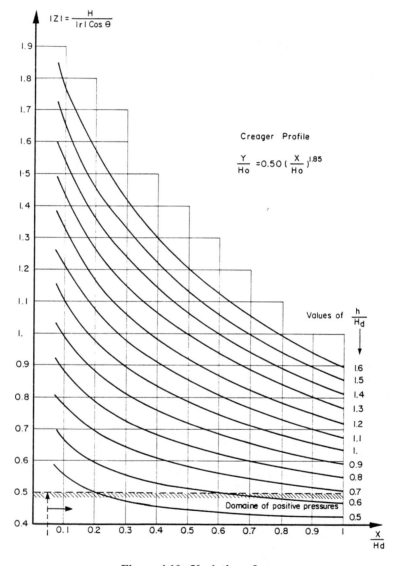

Figure 4.10. Variation of z.

D. Influence of the Geometry of Gates on Discharge Coefficient and on Subpressures

a. Criteria

The term geometry includes:

- The horizontal distance from the crest axis to the gate seat

- The trunnion location

- The radius of the gate

When these characteristics are known, the angle, θ, between the tangents at the gate and at the spillway profile and gate seat, is also known.
The design of a gated spillway should answer the following questions:

- Are the subpressures built up downstream of the gate smaller than the boundary value which is equal to - 4 m?

- Are the dimensions of the gate adequately sized to discharge the design flood?

A proper choice of the geometry of the gate can adequately resolve both of these questions.

b. Influence of geometry of the gate on subpressures

The buildup of subpressure is studied in subsection 4.2.2, B; but this study does not take into account the influence of geometry on the subpressure. Rohne, T.J. (1959) has given two graphs showing this influence (Figs. 4.11 and 4.12). It has been previously shown that a location of the gate downstream of the crest causes a decrease of subpressures. The example of the Altus and the Ross Dams, USA, constitutes an example observed in nature. (Urrutia, C.M., 1955, "The effect of Radial Gates upon Pressure Distribution on Overfall Spillways," Tests performed in USBR laboratory).
Escande, L., performed complete model studies showing the influence of the angle θ on subpressures (Fig. 4.1).* According to this research

- For $\theta = 30°$ the theoretical subpressures coincide in general with the measured ones on the scale model

- For $\theta = 60°$ the theoretical subpressures are greater than the measured subpressures on the scale model.

---

* θ is substituted by α in this figure

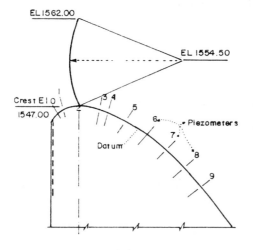

CREST PROFILE

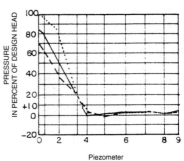

(2) GATE SEAT AT PIEZOMETER 3.6" BELLOW CREST

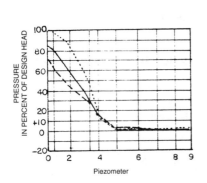

(1) GATE SEAT AT PIEZOMETER 12" BELLOW CREST

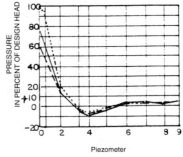

(3) GATE SEAT AT PIEZOMETER 2" ON CREST

NOTES

Head and openings measured vertically above crest

------ Gate open 13.9% of design head
————— Gate open 33.3% of design head
— — — Gate open 46.7% of design head

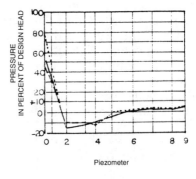

(4) GATE SEAT AT PIEZOMETER 1.6" BELOW CREST

Figure 4.11. Influence of gate position on pressure distribution, the Altus Dam, USA.

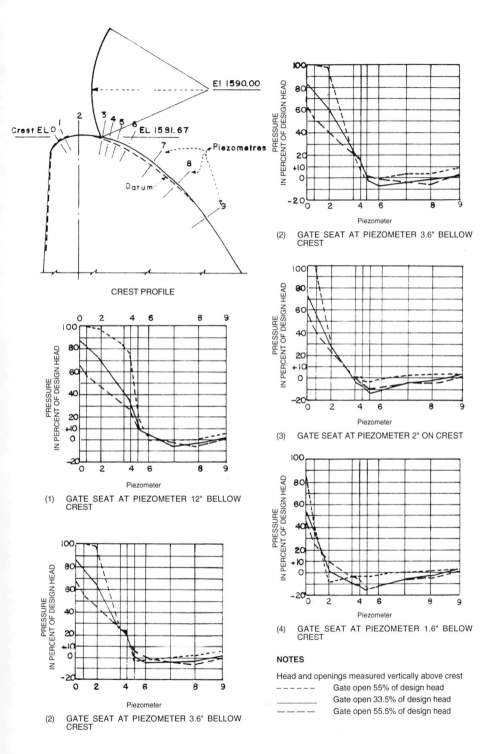

Figure 4.12. Influence of the gate position on pressure distribution, the Ross Dam, USA.

c. Influence of geometry on discharge coefficient

Rohne, T.J. (1959) and Corps of Engineers have defined this influence. Figures 4.13, 4.14, 4.15, and 4.16 show the variation of the coefficient of discharge as a function of the geometry of the gate.

Example 4.03

Determine the coefficient of discharge of a tainter gate using the diagram given in Fig. 4.13.

Data

$$h = H_d$$
$$H_d = 5.00 \text{ m}$$
$$D = 1.00 \text{ m, and}$$
$$\theta = 70°$$

Solution

For $h = H_d$, $m = 0.49$

For $\dfrac{D}{h} = \dfrac{1.00}{5.00} = 0.20$ and for $\theta = 70°$, Fig. 4.13 yields

$\dfrac{m}{m_o} = 1.06$, then $m = 1.06 \bullet 0.49 = 0.52$

The coefficient C is given as $C = m (2g)^{1/2}$ then

$$\frac{C}{C_*} = \frac{m(2g)^{1/2}}{m_o(2g)^{1/2}} = \frac{m}{m_o}$$

The discharge is defined by

$$Q = CL (h_1^{3/2} - h_2^{3/2}) \tag{4.13}$$

where

Q is the discharge under the radial gate

C is the coefficient of discharge

L is the width of the gate, and

h_1 and h_2 are shown in Fig. 4.13.

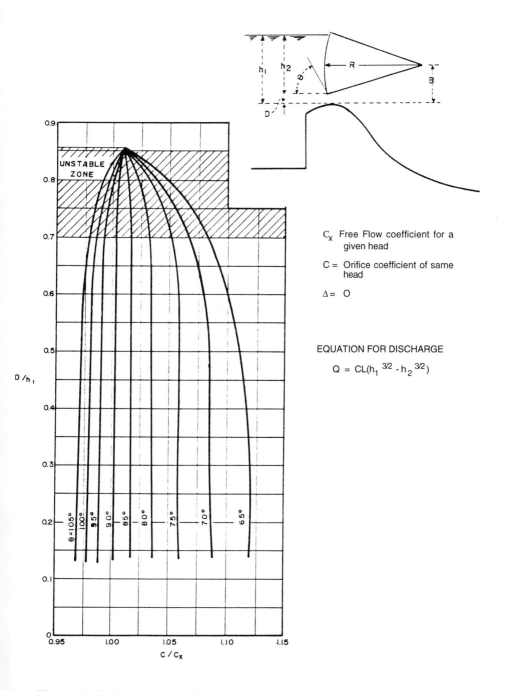

Figure 4.13. Discharge coefficient for radial gates (Rohne, T.J., 1959).

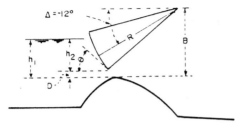

EQUATION FOR DISCHARGE

$$Q = CL(h_1{}^{3/2} - h_2{}^{3/2})$$

C_x Free Flow coefficient for a given head

$C =$ Orifice coefficient of same head

$\Delta = -12°$

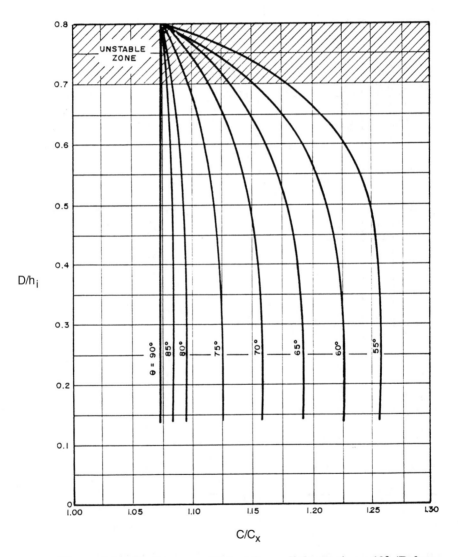

Figure 4.14. Discharge coefficient for radial gate $\Delta = -12°$ (Rohne, T.J., 1959).

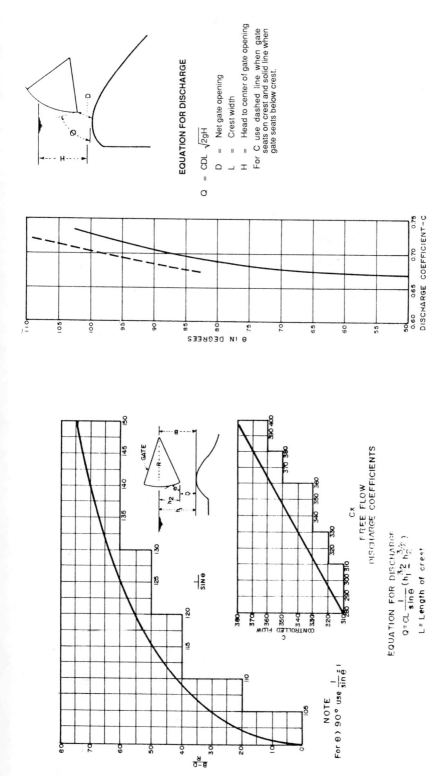

EQUATION FOR DISCHARGE

$$Q = CDL\sqrt{2gH}$$

D = Net gate opening
L = Crest width
H = Head to center of gate opening

For C use dashed line when gate seats on crest and solid line when gate seats below crest.

Figure 4.16. Discharge under a radial gate (Rohne, T.J., 1959).

GATE

$$\frac{1}{\sin\theta}$$

EQUATION FOR DISCHARGE

$$Q = CL\frac{1}{\sin\theta}(h_1^{3/2} h_2^{3/2})$$

L = Length of crest

NOTE
For θ > 90° use $\frac{1}{\sin\theta}$ = 1

CONTROLLED FLOW C   FREE FLOW Cx
DISCHARGE COEFFICIENTS

Figure 4.15. Discharge under a radial gate (Rohne, T.J., 1959).

F. Sentürk

The discharge under the radial gate shown in Fig. 4.15 is given as

$$Q = CL \frac{1}{\sin\theta} (h_1^{3/2} - h_2^{3/2}) \tag{4.14}$$

In Fig. 4.16 the definition of symbols are different from the prevailing definition in Figs. 4.13 and 4.15

$$Q = CDL (2gH)^{1/2} \tag{4.15}$$

where

D is the net gate opening, in feet

L is the crest width, in feet, and

H is the hydraulic head measured from the center of gate opening.

(The unit of Q is cfs).

E. Influence of Geometry of Piers on Discharge Coefficient;

a. Criteria

The shape and position of piers have an appreciable influence on the discharge coefficient. Three questions must be answered for a better understanding of the problem:

- How to evaluate the decrease in discharge due to contraction of the flow lines between the piers?

- How to evaluate the influence of the upstream extension of the pier on the discharge coefficient and on subpressures?

- How to evaluate the influence of side piers on the flow?

b. Contraction of flow by piers

The streamlines in turbulent flow cannot follow perfectly the rigid boundaries. The boundary layer causes a contraction of the flow which decreases the discharge. Figure 13.25 shows this phenomenon. A complete theoretical analysis is not necessary for solving the engineering problem. The accepted solution foresees a decrease of the clear length of the discharging sill expressed by Eq. 4.16

$$L = L_o - KHn \tag{4.16}$$

160

where

L   is the effective length of the spillway

L_o is the geometric length of the spillway

K   is the contraction coefficient, and

n   is the number of side contractions, equal to 2 for each gate bay.

In case of a spillway with two gates and one pier

$$n = 4$$

Two of the side contractions derive from the center pier and the remaining two from the side piers.

H is the hydraulic head and is equal to $h + (U^2/2g)$

h is the head on the spillway, and

U is the mean velocity in the approach channel.

The variation of K is defined differently in different countries. Corps of Engineers definition is used in this book (Figs. 4.17, 4.18 and 4.19).

c. Influence of upstream extension of piers on discharge coefficient and on subpressures

The extension of the pier upstream increases the discharge coefficient. This increase has been measured on scale models and observed on prototypes. For this reason many piers have a shape similar to the prow of a ship (Fig. 2.3). Increasing the extension of the pier upstream flattens the nappe causing an increase of the discharge coefficient. Figure 4.17 shows an increase of up to 0.10 in K parameter, for type-2 pier. A theoretical approach for calculating the angle of attack at the nose of the pier is difficult to attain. The only acceptable solution can be obtained from scale model investigations. Similarly, the increase in the discharge coefficient and in the subpressure can only be estimated from model studies. Figures 4.17, 4.18, 4.19 and 4.20 can also be used for the first approximation.

F. Influence of Side Piers on Flow

A theoretical approach for solving this problem is difficult. Model investigation is used in engineering practice.

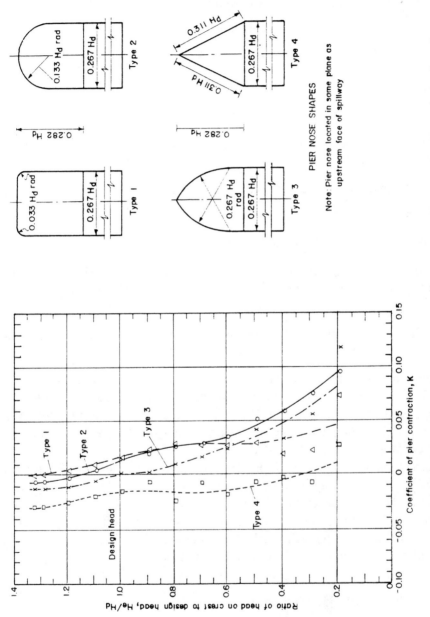

Figure 4.17. Coefficient of contraction for round-nose pier in high dams. Effect of pier length, US Army Corps of Engineers.

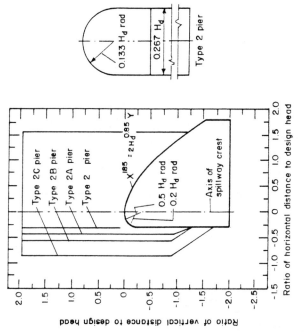

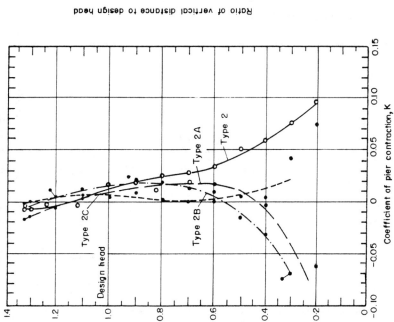

Figure 4.18. Coefficient of contraction for piers of various nose shapes in high dams with the nose located in the same vertical plane as the upstream face of the WES spillway (US Army Eng. WES, Hydraulic Design Chart 111-5, WES 4.1.53). Effect of nose shape.

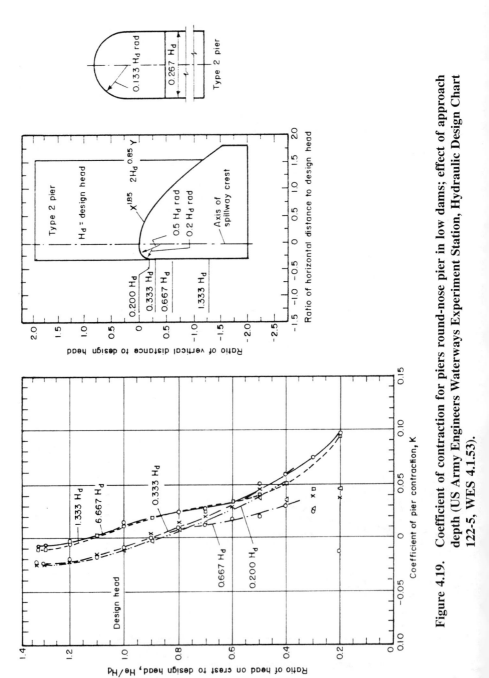

Figure 4.19. Coefficient of contraction for piers round-nose pier in low dams; effect of approach depth (US Army Engineers Waterways Experiment Station, Hydraulic Design Chart 122-5, WES 4.1.53).

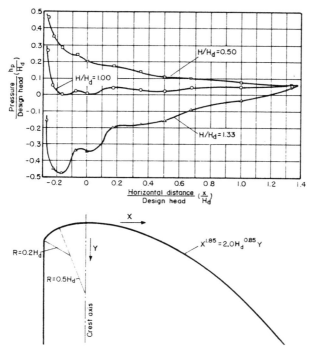

Figure 4.20. Crest pressures on WES high overflow spillways. No piers. (US Army Engineers Waterways Experiment Station, Hydraulic Design Chart 111-16 WES 9-54).

Example 4.04

Compare the subpressure buildup along the center line of pier bay and along piers.

Data

$L_{effective} = 10.00$ m $\quad\quad$ $H_d = 9.00$ m $\quad\quad$ $H = 12.00$ m

Types of pier: WES $\quad\quad$ $h = 12.00$ m $\quad\quad$ $X = 5.00$ m

Solution

Figure 4.21 gives the crest pressure along the center line of pier bay:*

$$\frac{X}{H_d} = \frac{5.00}{9.00} = 0.56$$

---

* Figure 4.20 can only be used for spillways with no piers.

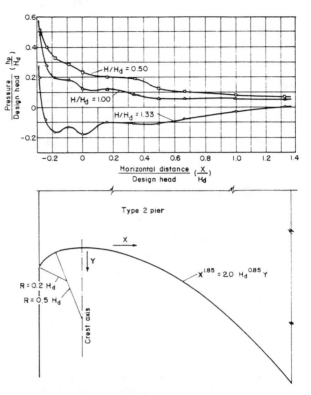

Figure 4.21. Crest pressures on WES high overflow spillways: b-along center line of pier bay (US Army Engineers WES, Hydraulic Design Chart 111-16/1 WES, 3.55).

$$\frac{H}{H_d} = \frac{12.00}{9.00} = 1.33$$

then

$$\frac{p/\gamma}{H_d} = -0.1 \text{ and } \frac{p}{\gamma} = -0.1 \bullet 9.00 = -0.9 < -4.00 \text{ m}$$

Figure 4.22 gives the crest pressure along the piers

$$\frac{X}{H_d} = \frac{0}{9.00} = 0 \qquad \frac{p/\gamma}{H_d} = -0.3 \qquad \frac{p}{\gamma} = -0.3 \bullet 9.00 = -2.7 \text{ m}$$

The pressure is larger upstream of the center line

$$X = 0.2 \bullet 9.00 = 1.80 \text{ m}$$

$$\frac{p/\gamma}{H_d} = -0.62 \qquad \frac{p}{\gamma} = -0.62 \bullet 9.00 = -5.58 > -4.00 \text{ m}$$

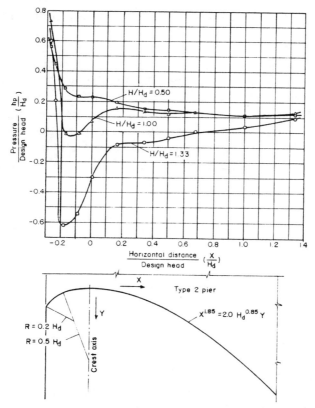

Figure 4.22. Crest pressures on WES high overflow spillways along piers; c-along of piers (US Army Engineers WES, Hydraulic Design Chart 111-16/2 WES, 3.55).

This example shows that when the subpressure along the center line of the pier bay is smaller than the allowable value, - 4.00 m, a subpressure larger than the allowable value occurs along the piers. The best solution will be to use reinforced concrete near the piers capable of absorbing the extra subpressures.

4.3 AERATION PHENOMENA IN SPILLWAY FLOW

Aeration of the flow is an important phenomenon in spillway flow. The following problems must be solved in an acceptable spillway design:

• Increase of flow depth due to air entrainment

• Aeration of lower nappe to prevent pulsating flow

• Aeration of lower nappe in the downstream channel to prevent cavitation.

167

4.3.1 Increase of Depth of Flow due to Air Entrainment

The hypercritical flow taking place in the downstream channel of a spillway causes various problems. Air entrainment decreases the density of air-water mixture by increasing the velocity of flow. At the same time, the excess velocity of flow increases the air friction, which decreases the velocity of flow. An equilibrium is established if the downstream channel is long enough. The depth of this flow is greater than the depth of a supercritical flow carrying the same discharge. Empirical formulas have been suggested for solving this problem. The Douma formula is given, as an example:

$$u = 10 \left(\frac{0.2U^2}{gR} - 1 \right)^{1/2} \qquad (4.17)$$

where

u is the percent of entrained air by volume

U is the mean velocity of the flow, and

R is the hydraulic radius.

The dimensionless square of the Froude number is (U^2/gR). Thus the formula can be used equally in metric and in English units.

The water surface line obtained from this expression is not an exact representation of the water surface in nature, but it does supply a close approximation.

The depth of flow in downstream channel forms the basis for calculation of heights of the side walls and divider walls. If the walls are designed to insufficient heights, overtopping may occur, causing dangerous erosion on the outside of the side channel walls.

Example 4.05

Determine the depth of flow in the downstream channel of the Karakuz Dam Spillway using the Douma formula. The analysis of flow using gradually varied flow procedure follows. (Table 4.3).

Solution

Entrained air increases the depth of flow. The Douma formula is used for computing this increase. The computation is summarized in Tables 4.3 and 4.4. Figure 4.23 shows the new water surface line.

TABLE 4.3

Station	h	R	U	Width of Channel	Q	L
	m	m	m/sec	m	m³/sec	m
1	4.44	3.38	21.53	28.25	27.00	19.64
2	3.69	2.93	25.90	28.25	27.00	40.00
3	3.38	2.71	29.13	27.42	27.00	60.00
4	3.57	2.82	28.40	26.63	27.00	75.00
5	3.44	2.72	30.36	25.85	27.00	90.00
6	3.39	2.67	31.46	25.32	27.00	100.00
7	3.35	2.64	32.49	24.80	27.00	110.00
8	3.32	2.60	33.81	24.01	27.00	125.00
9	3.31	2.56	35.92	22.71	27.00	150.00

TABLE 4.4

Station	h	R	U	B	u	h(*)	L
	m	m	m/sec	m	%	m	m
1	4.44	3.38	21.53	28.25	13.40	5.03	19.64
2	3.69	2.93	25.90	28.25	19.15	4.40	40.00
3	3.38	2.71	29.13	27.42	23.20	4.16	60.00
4	3.57	2.82	28.40	26.63	21.98	4.35	75.00
5	3.44	2.72	30.36	25.85	24.31	4.28	90.00
6	3.39	2.67	31.46	25.32	25.61	4.26	100.00
7	3.35	2.64	32.49	24.80	26.74	4.25	110.00
8	3.32	2.60	33.81	24.01	28.21	4.26	125.00
9	3.31	2.56	35.92	22.71	30.45	4.32	150.00

B: Width of channel

---

*   h corrected

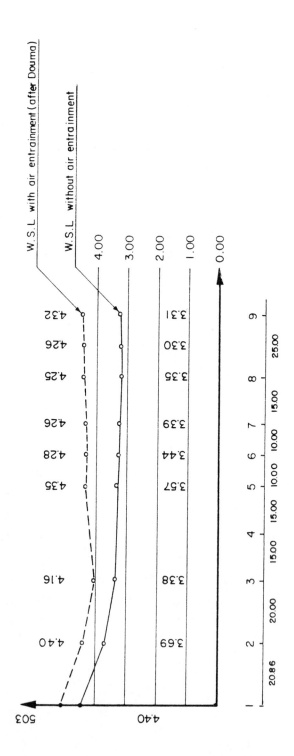

Figure 4.23. Flow line, in Example 4.05.

4.3.2 Aeration of Lower Nappe to Prevent Pulsating Flow and Cavitation

Figure 4.24 shows a gate working simultaneously as a sluice gate and an overflow gate. The lower nappe of this flow must be aerated for a stable operation. If not, pulsating flow occurs. If pulsating flow creates vibrations severe enough to affect the structure, dam failure could take place. Pulsating flow is a result of insufficient aeration of the lower nappe. To prevent this phenomenon air is injected into the confined area.

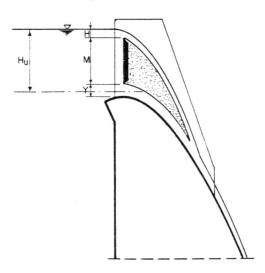

Figure 4.24. **Simultaneous action of a gate as a sluice gate and an overflow gate.**

The flow conditions cause the nappe to intermittently entrain air and at other times to so compress water as to prevent air from entering.

The computation of the needed air inflow is difficult. Escande, L., and Hickox (1944) have done extensive research on it and Hickox has given an empirical formula for computing the air demand:

$$q_a = \frac{568\,(CH)^{3.64}}{p^{1.14}} \qquad (4.18)$$

where

q_a is air demand (ft^3/sec/ft)

H is hydraulic head of the spillway (ft)

p   is pressure difference between higher and lower nappe (ft),

and

C is a parameter defined by

$$d = \frac{y\,(H_u)^{1/2}}{H^{1.50}}$$

d	0	0.500	1.000	1.500	2.000	2.500
C	0.077	0.135	0.175	0.202	0.220	0.225

H and H_u are defined in Fig. 4.24

In the case of free flow spillway $\delta = 0$

The results obtained by the application of this formula must be analyzed seriously. To prevent pulsating flow, Escande, L. (1951) suggested the following design criteria:

$$M \geq 0.60\,H_u \tag{4.19}$$

and

$$\frac{H}{M} \geq 0.45 \qquad \frac{Y}{M} \geq 0.20 \tag{4.20}$$

Dubs (1947) suggested Eq. 4.21 for computing the cross section of the aeration conduit

$$A = \frac{5}{1,000}\,Lh \tag{4.21}$$

where   A is cross sectional area, m^2

L is width of the spilling section

and   h is thickness of the nappe.

The How's (1955) suggestion for the computation of the needed air is given in Fig. 4.25.

Non-aerated flow also causes cavitation on the downstream face of spillways and generates heavy erosion. Figure 4.26 shows such an eroded surface observed on the Keban Dam Spillway. Aeration remedies this inconvenience.

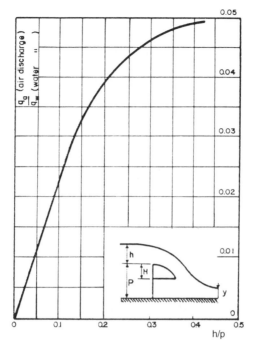

Civil Engineering, May 1955

Figure 4.25. Air need for a spillway.

Figure 4.26. Erosion due to cavitation on the Keban Dam Spillway.

Example 4.06

Design the gated spillway, the characteristics of which are given as

$$Q_{max} \text{ outflow } = 3550 \text{ m}^3/\text{sec}$$
$$Q_{100} = 2511 \text{ m}^3/\text{sec}$$

The design hydrograph is given in Fig. 4.27.

- The number of gates is determined by an optimization analysis and found to be equal to 4.

- Number of piers: 3

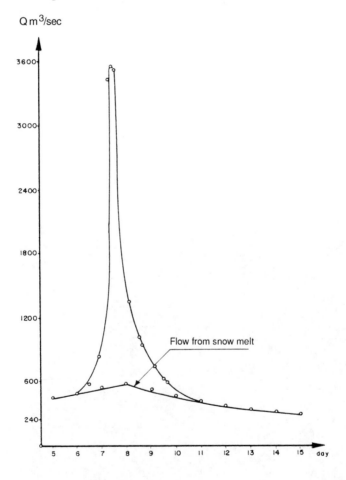

Figure 4.27. Design hydrograph, in Example 4.06.

- Maximum elevation of the upstream reservoir 815.75

- Maximum elevation of the downstream reservoir 715.00

- A power plant is supplied by the reservoir

- $P = 5.00$ m (height of the spillway above the upstream channel bottom)

- The spillway is located in an approach channel

It is required to design the spillway taking into account hydraulic characteristics given by the data.

Solution

A complete example of hydraulic design of the gated spillway includes the examination and determination of the length of the spillway crest, the design head, the true hydraulic head on the spillway while discharging the design flood, the spillway profile, the coordinates of the trunnion and the radius of the gates (if the spillway is equipped with radial gates), and the evaluation of the pressure buildup on the spillway body, and the evaluation of the forces acting on the spillway.

The gate number may be determined by an optimization. The type of spillway profile may be determined by research.

In general, when the magnitude of the design flood is determined, the geometric characteristics of hydraulic structure are computed according to those data. In this particular case, it is assumed that hydrologic data obtained by direct measurements cover a very short period of time and that the flood hydrograph given in Fig. 4.27, rests on theoretical computations. The value $Q_{max} = 3,550$ m^3/sec may include a large safety margin. To solve this problem, the following rules may be applied:

- The spillway is sized for a smaller discharge

- The design flood will be evacuated utilizing some of the freeboard reserve height, which is determined by flood routing.

The determination of geometric properties of the spillway structure is done by steps:

Step 1: Choice of the type of spillway

The existence of a power plant implies a controlled reservoir level. Then a gated spillway is chosen. This choice is based upon Q_{max}, which is greater than 2,000 m^3/sec. The value of 2,000 m^3/sec is arbitrary, however the economic analysis has shown that for discharges greater than this value, a gated spillway may be more economical than a free flow spillway.* In any event, the existence of the power plant requires a controlled crest spillway.

Step 2: Three types of spillway profiles with controlled crests will be discussed:

- second degree parabola profile
- WES profile
- Creager's profile.

A second degree parabola profile is generally used for high spillways, directly in contact with the reservoir, which is not the case in this problem. Therefore, either the WES or the Creager profile may be preferred. In general, equation of the profile in each case is similar and can be written as

$$X^a = z \, H_d^{(a-1)} \, Y$$

where

H_d is design head

X and Y are coordinates of the profile,

and a and z are parameters to be defined later.

- Hydraulic computation of the spillway: determination of the design flood.

The design is done in steps as:

Step 1: A profile without subpressure

$$m = 0.4956$$

$$\frac{p}{\gamma} > 0$$

(where m is discharge coefficient, and p is pressure) is chosen (see Chapter 3). If these limiting values are exceeded, subpressure will appear at the profile. If Q is taken equal to Q_{max}, (m) will be equal

---

* See also Chapter 2.

to 0.4956 and pressures observed on the profile will probably be very large. For solving this problem, truncated profiles as shown in Fig. 3.4 may be chosen.

Step 2: Profile with subpressure,

In this case

$$0.4956 < m < 0.52$$

$$-4 < \frac{p}{\gamma} < 0$$

The choice of Q_d will determine the real value of (m) and (p/γ).

• Procedure used for hydraulic computation:

Step 1: The variation of discharge coefficient is determined using Eq. 3.20

$$m = 0.4956 \left(\frac{H}{H_d}\right)^{0.16} \qquad\qquad 3.20$$

$$H = h + \frac{U^2}{2g}$$

where   h is flow depth in the approach channel

U is mean velocity in the approach channel

H is hydraulic head of the spillway

Step 2: If discharge of the central bay is Q, assume 0.80 Q to be the discharge from side bay. This assumption can also be expressed so that the length of spillway crest in the side bay is equal to 0.80 of the crest length of the central bay. Since there are four gates, the total effective crest length of the spillway will be

$$L = 0.80\ell + \ell + \ell + 0.80\ell = 3.6\ell$$

where   (ℓ) represents the crest length in one bay (Fig. 4.28).

Step 3: As a first approximation, Q_d is assumed equal to 2,750 m³/sec > Q_{100}.

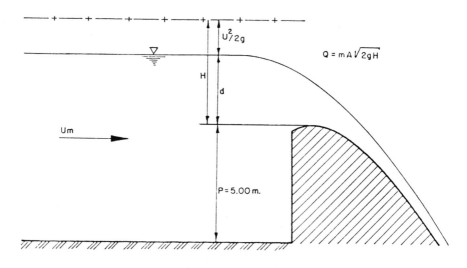

Figure 4.28. Definition of symbols.

Step 4: (d/ℓ) is assumed equal to 1.17

$$\frac{d}{\ell} = 1.17$$

d is the flow depth in one bay.

This ratio also gives the dimension of the gate. If (ℓ) is computed, the gate height (d) is obtained as,

$$d = 1.17\,\ell$$

With $Q_{o\,max} = 3,550\ m^3/sec$, the extra discharge to be evacuated is

$$Q = 3,550 - 2,750 = 800\ m^3/sec$$

This quantity may be discharged by an additional increment of the hydraulic head. This increment should, in all cases, less than the freeboard. Figure 4.28 gives the definition of symbols used in this analysis.

- First approximation: the Creager profile.

Figure 4.28 illustrates the flow taking place on the spillway. The mathematical expression of this flow is

$$Q = mA\,(2gH)^{1/2} = mA\,(2g)^{1/2}H^{1/2} \qquad 3.1$$

$$U = \frac{Q}{(d+P)(4\ell+3e)} \tag{4.21}$$

$$\frac{U^2}{2g} = \left[\frac{Q}{(d+5.0)(4\ell+9)}\right]^2 \frac{1}{2g}$$

$$m = 0.4956 \left[\frac{d+\left(\frac{Q}{(d+5)(4\ell+9)}\right)^2 \frac{1}{2g}}{H_d}\right]^{0.16} \qquad 3.20$$

$$A(2g)^{1/2} = d \bullet 3.6\,\ell\,(2g)^{1/2} = 1.17\,\ell \bullet 3.6\,\ell \bullet 4.43 = 18.66\,\ell^2$$

Then

$$Q = 0.4956 \left[\frac{d+\left(\frac{Q}{(d+5)(4\ell+9)}\right)^2 \frac{1}{2g}}{H_d}\right]^{0.16} \bullet 18.66\ell^2 \left[d+\frac{\left(\frac{Q}{(d+5)(4\ell+9)}\right)^2}{2g}\right]^{1/2} \tag{4.22}$$

where

> m is coefficient of discharge
>
> A is wetted area of the flow taking place on the spillway (m^2)
>
> g is gravitational acceleration (m/sec^2)
>
> d is flow depth on the spillway crest (m)
>
> P is height of the control structure (m)
>
> ℓ is crest length of the central bay (m)
>
> e is thickness of a pier (m)
>
> H$_d$ is design head (m)
>
> Q is spillway discharge (m^3/sec)
>
> U is mean velocity in the approach channel (m/sec)
>
> B is width of the approach channel (m).

This relation can be used to study and define H$_d$ as a function of remaining variables. Its solution is summarized in Table 4.5.

The parameters in this study are:

1. Q, outflow discharge

TABLE 4.5

First step investigation in design the Creager profile

Q m³/sec	H_d m	ℓ m	d m	H m	X/H	m	B m	H/H_d	p/γ m
1	2	3	4	5	6	7	8	9	10
2,750	11.50	10.00	8.55	9.42	0.32	0.480	49.00	0.819	-0.633
3,550	11.50	10.00	13.05	13.87	0.22	0.511	49.00	1.206	-3.565
2,750	11.50	9.45	10.45	11.19	0.27	0.493	46.80	0.973	-1.725
3,550	11.50	9.45	15.80	16.48	0.18	0.525	46.80	-	>-4.000
2,750	10.85	9.45	10.28	11.03	0.27	0.497	46.80	1.017	-1.649
3,550	10.85	9.45	15.56	16.25	0.19	0.529	46.80	-	>-4.000
2,750	10.00	9.40	10.22	10.99	0.27	0.503	46.60	1.099	-2.000
3,550	10.00	9.40	-	-	-	0.535	-	-	>-4.000
2,750	9.50	9.37	10.19	10.96	0.27	0.507	46.48	1.154	-2.375
2,750	9.00	9.34	10.14	10.92	0.27	0.510	46.36	1.217	-2.610
2,750	8.50	9.30	10.12	10.91	0.27	0.516	46.20	1.284	-2.390

Two values of Q are investigated:

$$Q_{max} = 3,550 \text{ m}^3/\text{sec} \quad \text{and} \quad Q = 2,750 \text{ m}^3/\text{sec}$$

2. H_d, design head

Six values of H_d are considered. One of them corresponding to the limiting condition will be chosen.

3. ℓ

Six values of ℓ are considered

4. X, abscissa of the sill of the gate with respect to the spillway axis.

The variables which depend on parameters are: d, H, m, (p/γ), and the pressure buildup on the profile.

In order to determine H_d, which is the governing factor in spillway design, the acceptable values of discharge coefficient, m, vary between 0.4956 and 0.5150. The crest length of one bay ℓ is computed corresponding to the two values of Q determined previously; then H_d is defined accordingly.

The columns of Table 4.5 are defined as follows:

Column 1: The variation of pressure buildup on the spillway profile as a function of H_d and Q is the subject of this analysis. The outflow discharge corresponding to inflow discharge is smaller than the inflow discharge. But, in a gated spillway, an accidental opening of the gates can be the cause of an outflow equal to an inflow, $Q_i = Q_o$, which is dangerous. For this reason $Q_{max} = 3,550 \text{ m}^3/\text{sec}$ is investigated and it is shown in Table 4.5, Column 1.

Column 2: Different possible values of H_d are introduced into the analysis. They are chosen by designer.

Column 3: Possible values for ℓ are chosen by designer

Column 4: (d) is computed using Eq. 4.23

$$Q = 0.4956 \left[\frac{d + \left(\dfrac{Q}{(d+5)(4\ell+3e)} \right)^2 \dfrac{1}{2g}}{H_d} \right]^{0.16} \bullet 18.66\ell^2 \left[d + \frac{\left(\dfrac{Q}{(d+5)(4\ell+3e)} \right)^2}{2g} \right]^{1/2} \quad (4.23)$$

$e = 3.00 \text{ m}$ is the thickness of the pier.

Column 5: $H + \dfrac{U^2}{2g}$

where $\qquad U = \dfrac{Q}{(d+P)\,(4\,\ell+3e)} \qquad\qquad$ 4.21

$\qquad\qquad\quad P = 5.00 \text{ m}$

$\qquad\qquad\quad \ell$, from Column 3

$\qquad\qquad\quad$ 3, is the number of piers, and

$\qquad\qquad\quad$ e is the chosen thickness of the pier.

Column 6: $\dfrac{X}{H}$; X is chosen equal to 3.00 m

Column 7: $\quad m = 0.4956 \left[\dfrac{d + \left(\dfrac{Q}{(d+5)(4\ell+3e)} \right)^2 \dfrac{1}{2g}}{H_d} \right]^{0.16} \qquad$ 3.20

The computed value of m in this column must be smaller than 0.515, as an acceptable value. If m = 0.52, the chosen value of H_d and ℓ in Column 2 and 3 must be changed.

Column 8:   B = 4 ℓ + 3e

Column 9: $\dfrac{H}{H_d}$ this value is selected between 0.819 and 1.289; it influences directly the pressures

Column 10: (p/γ); this value is computed using Fig. 4.5, as a first approximation.

In the final computation, Eq. 3.27a will be used. If (p/γ) is greater than (-4.00 m), the previously selected value of parameters must be changed so that

$$\left(\frac{p}{\gamma}\right) \leq -4.00 \text{ m}$$

is obtained.
Table 4.6 shows that for

$$8.50 < H_d < 11.50 \text{ m}$$

and for

$$Q = 2{,}750 \text{ m}^3/\text{sec}$$

the subpressures are acceptable. Under these conditions the above analysis can be repeated for smaller values of Q.

TABLE 4.6

Determination of Q_{min} for P = 5.00 m

Q m³/sec	H_d m	d m	H/H_d	H m	ℓ m	m	B m	X/H	p/γ m
1	2	3	4	5	6	7	8	9	10
2,250	8.50	8.835	1.121	9.525	8.80	0.505	44.20	0.315	-1.570
2,750	8.50	12.25	1.518	12.913	8.80	0.530	44.20	0.232	-4.000
2,500	8.50	9.78	1.235	10.501	8.99	0.513	44.92	0.286	-2.550
2,500	8.75	9.87	1.210	10.583	8.99	0.511	44.96	0.283	-2.45
2,750	8.75	11.52	1.397	12.219	8.99	0.523	44.96	0.245	-3.98
2,500	9.00	9.73	1.160	10.439	9.13	0.508	45.52	0.287	-2.61
2,750	9.00	10.97	1.305	11.699	9.13	0.517	45.52	0.256	-3.60

Let us consider, for example, a profile corresponding to:

$$H_d = 11.50 \text{ m} \qquad \text{and} \qquad \ell = 10.00 \text{ m}$$

Using Table 4.5, it can be seen that this profile is subject to a subpressure equal to (-3.565) m, when discharging $Q_{max} = 3,550$ m³/sec. The subpressure is smaller than (- 4.00) meter. But for Q=3.550 m³/sec, H_d=11.50 m, ℓ =9.45 m or for H_d = 10.00 m, and ℓ = 9.40 m, $(p/\gamma) > -4.00$ m. These values are not acceptable.

Smaller values of Q_0 are investigated in Table 4.6

$$2,250 < 2,600 \text{ m}^3/\text{sec}$$

Values of H_d, ℓ and X are determined according to this new range of outflow discharges. Examination of Table 4.6 shows that for Q = 2,500 m³/sec or for Q = 2,600 m³/sec the values of subpressures are in the range of acceptable values. Then the previously determined value of the outflow discharge Q = 2,750 m³/sec, should be decreased to 2,500 - 2,600 m³/sec. This opportunity will be studied in the final computation.

Results obtained from the first approximation

Q = 2,500 m³/sec H_d = 8.75 m   P = 5.00 m   H = 10.583 m

m = 0.511 $\dfrac{X}{H} = 0.28$ ℓ = 8.99 m $\dfrac{d}{\ell} = 1.10$

$\dfrac{p}{\gamma}$ = -2.45 m

If the profile is prepared according to data

$$Q = 2,750 \text{ m}^3/\text{sec} \qquad H = 8.75 \text{ m}$$

and p/γ = -3.98 < -4.00 m, it cannot discharge Q_{max} = 3,550 m³/sec with safety due to the fact that for this value of Q, p/γ > - 4.00 m.

Second approximation: WES Profile (H/V : 2/3)

The analysis for WES Profile is conducted considering the results of the first approximation.

The inclined upstream face of the chosen WES profile has a slope of 2/3. Data pertaining to this type, are:

Slope of the upstream face	$\frac{2}{3}$ (H/V)
H_d	8.75 m
P	5.00 m
m_o	0.4956
ℓ	8.75
$\dfrac{P}{H_d}$	0.571
X	3.00 m
$\dfrac{m}{m_o}$	1.009
m	0.50
e	2.50 m

The computation of m is done according to Fig. 3.14; the obtained result is:

$$Q = 0.50 \bullet 3.60 \bullet 8.75 \bullet 10.78 \bullet 4.43 \left[d + \frac{Q^2}{(4\ell + 3e)^2 (P+d)^2} \frac{1}{2g} \right]^{1/2}$$

For Q = 2,500 m³/sec obtained as a result of the first approach, d = h = 10.78 m, and p/γ ≅ -2.80 m (Fig. 4.5).

Final approach

A WES Profile with an upstream slope of 2/3 is chosen, for the following reasons:

- P < 10.00 m
- the Creager Profile conforms perfectly only to theoretical considerations of a high profile only with a vertical upstream face.

The parameters already accepted and checked in the previous section were:

$$H_d = 8.75 \text{ m} \qquad P = 5.00 \text{ m} \qquad m_o = 0.4956$$

$$\frac{P}{H_d} = 0.571 \qquad \frac{m}{m_o} = 1.009 \qquad m = 0.50$$

This value of m varies according to boundary conditions and it should be checked on a scale hydraulic model.

Complementary assumptions are:

1. e = 2.50 m (the piers are thinner)

2. ℓ = 8.20 m

    Then L = 4 • 8.20 + 3 • 2.50 = 40.30 < 44.96 m

3. Spillway crest elevation = 806.16

4. Q = 2,000 m^3/sec

A smaller value of Q_o is considered in this case. The utility of this assumption is:

- in case of an accidental opening of gates, the downstream flood is less hazardous,

- hydraulic structure is less costly.

The disadvantages of this assumption are:

- overtopping possibilities are increased

- the possibility of extra subpressure (greater than the limiting value) are increased

- the possibility of gate vibration must also be checked (this aspect should be investigated on a special scale model).

Hydraulic computation of the spillway is:

$$Q = 0.50 \bullet 3.60 \bullet 8.20 \bullet 4.43d \left[d + \frac{Q^2}{(4\ell+3e)^2 (P+d)^2} \frac{1}{2g} \right]^{1/2}$$

For $\quad Q = 2,000 \text{ m}^3/\text{sec} \quad d = 9.53 \text{ m} \qquad \dfrac{d}{\ell} = 1.16$

The ratio $d/\ell = 1.16$ is used for gate sizing. The rule for gate operation is as follows:

1. When Q reaches 2,000 m³/sec, gates should be partially open so that the reservoir water surface elevation stays constant (the reservoir water surface elevation is assumed to be at the normal operating level).

2. The gate opening will be increased with an increase of inflow and a rise of reservoir water surface elevation. For $Q_{max} = 3,550$ m³/sec, gates will be fully open. The outflow discharge corresponding to this inflow may be determined by flood routing. The flood is routed taking into consideration the given conditions, and the outflow is determined as:

$$Q_o = 2,318 \text{ m}^3/\text{sec}$$

Then

$$h \quad = \quad 10.56 \text{ m}$$

$$H \quad = \quad h + (U^2/2g) = 10.56 + 0.70 = 11.26 \text{ m}$$

$$\frac{H}{H_d} \quad = \quad \frac{11.26}{8.75} = 1.29$$

$$\frac{X}{H} \quad = \quad 0.27$$

Maximum subpressure computed on the spillway crest for fully open gates:

$$\frac{p}{\gamma} = -2.98 \text{ m}$$

Maximum subpressure along the pier interface is

$$\frac{p}{\gamma} = -5.08 \text{ m}$$

This section of the spillway should be reinforced. Subpressures are determined using Figs. 4.5 and 4.20. The subpressure, -2.98 m, must be checked mathematically.

Computation of subpressure for fully open gates

The following equations are used for computing the subpressure and the velocity on the profile:

$$\frac{p}{\gamma} = H - (H - d\cos\theta)\left(\frac{r-d}{r}\right)^2 \qquad 3.27a$$

$$U = [2g(H - d\cos\theta)]^{1/2}\left(\frac{r-d}{r}\right) \qquad 3.27b$$

where

$\frac{p}{\gamma}$ is subpressure (m)

H is head in the section under consideration, (m)

d is flow depth at the section, (m)

θ is angle of the tangent to the profile with the horizontal line at the section under consideration, and

r is radius of curvature of the profile at the section under consideration (m)

The equation of the profile selected for final design is (Fig. 4.29)

$$X^{1.81} = 1.939\, H_d^{0.81}\, Y$$

or

$$Y = \frac{1}{11.236}\, X^{1.81}$$

$$r = \frac{(1 + Y'^2)^{3/2}}{Y''}$$

$$r = \frac{[1+(0.161\, X^{0.81})^2]^{3/2}}{0.13\, X^{-0.19}}$$

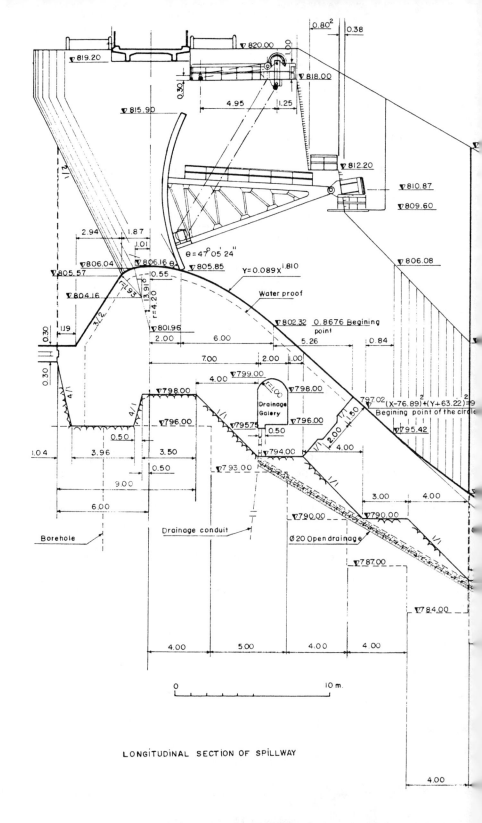

Figure 4.29. Longitudinal section of spillway.

COORDINATES OF THE SPILLWAY PROFILE

X	Y
0	806.16
1	806.07
2	805.85
3	805.51
4	805.07
5	804.52
6	803.88
7	803.15
8	802.32

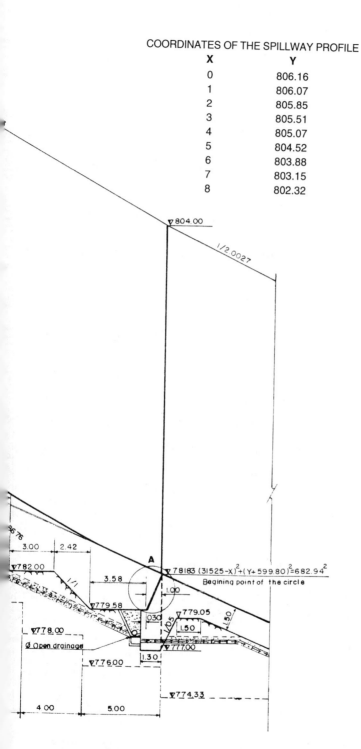

Figure 4.29. Continued.

$$\cos\theta = \frac{1}{(1+Y'^2)^{1/2}}$$

$$\tan\theta = Y'$$

$$\cos\theta = \frac{1}{[1+(0.161\ X^{0.81})^2]^{1/2}}$$

As a first approximation, the flow depth can be computed using Eq. 3.27b. The computations are summarized in Table 4.7.

TABLE 4.7

Computation of subpressure for free flow on the spillway profile [*]

$Y = 0.089\ X^{1.81}$ $Q = 2318\ m^3/sec$ $h = 9.53\ m$
$L = 40.30\ m$

X	Y	r	$\cos\theta$	H	d	p/γ
m	m	m		m	m	m
1	2	3	4	5	6	7
3	0.65	-11.7440	0.931	10.98	3.600	-2.04
5	1.64	-16.4100	0.860	11.97	3.575	-1.22
10	5.75	-35.7540	0.693	16.08	3.201	-0.37
15	11.97	-69.6960	0.569	22.30	2.744	-0.10
20	20.15	-122.092	0.481	30.48	2.352	-0.01

The columns of Table 4.7 are defined as follows:

Column 1:   Abscissa is respect to the cartesian axis passing through the crest of the spillway (Fig. 4.30).

Column 2: $Y = 0.089\ X^{1.81}$

---

[*]   The definition of symbols are given in Fig. 4.30.

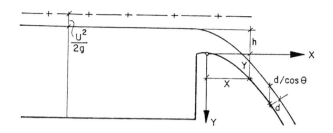

Figure 4.30. Definition of symbols in Table 4.7.

Column 3: $\quad r = \dfrac{[1+(0.161\,X^{0.81})^2]^{3/2}}{0.13\,X^{-0.19}}$

Column 4: $\quad \cos\theta = \dfrac{1}{[1+(0.161\,X^{0.81})^2]^{1/2}}$

Column 5: $\quad H = Y+h+\dfrac{U^2}{2g} \qquad U = \dfrac{Q}{L(5+h)} \qquad L = 40.30$

$\qquad\qquad\qquad\qquad\qquad\qquad\qquad\qquad\qquad\qquad h = 9.53\ \text{m}$

Column 6: $\quad Q = Bd\,[2g(H-d\cos\theta)]^{1/2}\,(\tfrac{r-d}{r}) \qquad B = L$

Column 7: $\quad \dfrac{p}{\gamma} = H - (H-d\cos\theta)\,(\tfrac{r-d}{r})^2$

The spillway profile is shown in Figs. 4.31 and 4.32. The maximum subpressure building up on it is equal to -1.23 m, which is smaller than the limiting value of -4.00 m.

In this table, the most interesting value for subpressure buildup is not $p/\gamma = -2.06$ m for fully open gates, because the subpressure is higher at the same section for partially open gates. The subpressure at the section $X = 5$ m, for example, is more interesting although the subpressure is 1.23 m. This subpressure is smaller than -2.06 m, which occurs at the gate section indicating that the subpressure at a partially open gate is indeed the largest one.

Table 4.8 summarizes the characteristics of the spillway profile. Figures 4.29, 4.31, and 4.32 show the general spillway drawings.

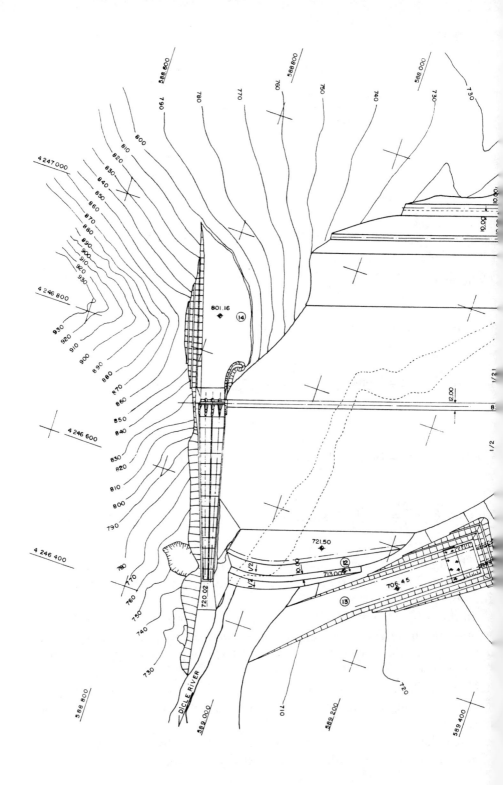

Figure 4.31. Spillway layout.

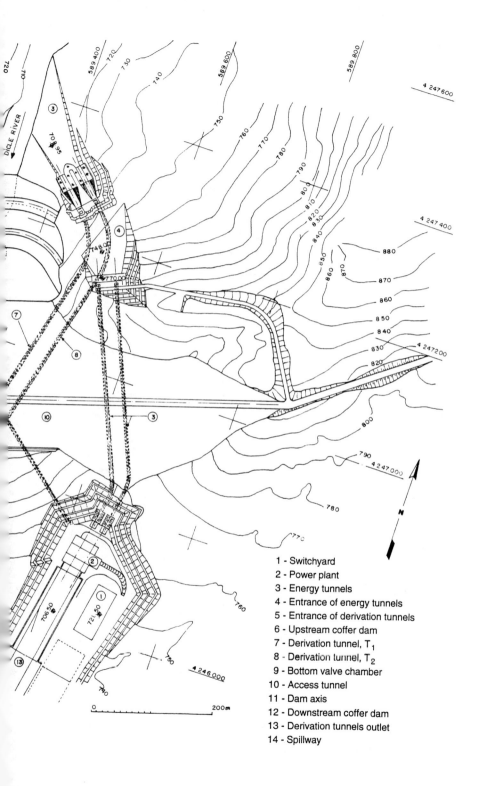

1 - Switchyard
2 - Power plant
3 - Energy tunnels
4 - Entrance of energy tunnels
5 - Entrance of derivation tunnels
6 - Upstream coffer dam
7 - Derivation tunnel, T_1
8 - Derivation tunnel, T_2
9 - Bottom valve chamber
10 - Access tunnel
11 - Dam axis
12 - Downstream coffer dam
13 - Derivation tunnels outlet
14 - Spillway

Figure 4.31. Continued.

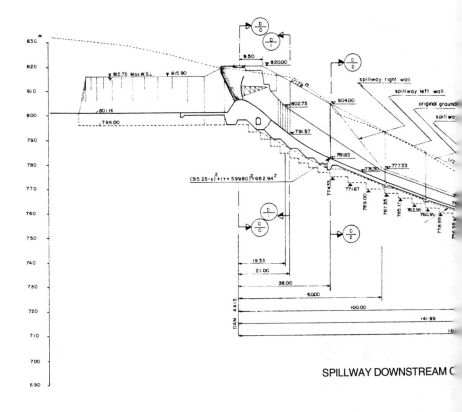

SPILLWAY DOWNSTREAM C

Intermediate distances		42.60		9.89	14.61	21.00		17.00	23.00		17.00	17.00
Distance from spillway axis	67.20			24.50	14.61	0.00		21.00	38.00		61.00	78.00
Bottom elevation	801.16			801.16	801.16	806.16		791.57	781.83		772.11	765.55

Figure 4.32. Spillway downstream channel longitudinal section.

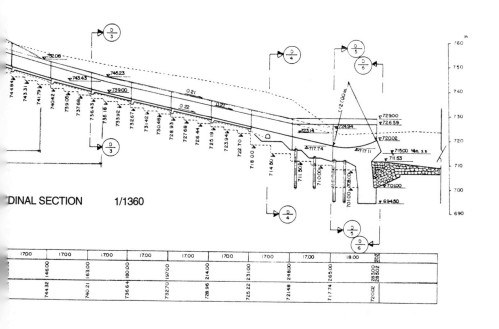

Figure 4.32. Continued.

TABLE 4.8

Characteristics of the spillway under consideration

Q	2,000 m³/sec
H_d	8.75 m
d	9.53 m
Type chosen: WES with an inclined upstream face; the slope is 2/3	

Equation profile	$Y = 0.089 \, X^{1.81}$
Flow depth on the spillway crest for Q_{max}	10.56 m
Hydraulic head for Q_{max}	11.26 m
Extra head	1.03 m
Final surface elevation	816.72
Final freeboard	3.28 m
X	3.00 m
Max subpressure at the gate section	-2.98 m
Max subpressure for free overfall	-2.06 m
Number of gates	4
Crest length in one bay	8.20 m
e	2.50 m
Total spillway length	40.30 m
Gate dimensions: Y	10.24
G(gate width)	8.20 m
Y/G	1.24

4.4   PROBLEMS

Problem 4.01

The equation of a second degree parabola is

$$Y = \frac{1}{40}X^2$$

Assuming that this relationship defines the profile of a spillway, determine the hydraulic head of this spillway for which the subpressure begins and the coordinates of the section where this subpressure builds up.

Problem 4.02

Verify Eq. 4.6a

$$z = -\frac{1}{2}\frac{H}{H_m}$$

Problem 4.03

Show the variation of the positive and negative pressures buildup on the profile of a spillway as a function of flow characteristics, the flow depth and the hydraulic head.

Problem 4.04

Show that at a given point the increase of H decreases the pressure following the variation of d/H.

Problem 4.05

A high head overflow spillway has an 18 m (60 ft) radius flip bucket at its downstream end. The bucket is not submerged, but acts to change the direction of flow from the slope of the spillway face to the horizontal line and to discharge the flow into the air between vertical training walls ~25.00 m (82 ft) apart. At a discharge of 1,500 m³/sec ~(53,000 cfs), the water surface at the vertical section OB is at elevation 0. Determine the water surface line along the downstream channel based upon following assumptions:

1. The velocity is uniformly distributed across the section.

2. The flow is gradually varied and n = 0.028 due to high velocity of the flow.

3. The flow entrains the air, and the density of the air-water mixture is estimated by Eq. 4.17 (Douma formula).

4. The coefficient of discharge is assumed equal to 0.49, and the velocity head in the approach channel is negligible.

Problem 4.06

Design the pier of a spillway corresponding to:

$$h = 13.20 \text{ m} \quad H_d = 11.00 \text{ m}$$

The equation of the spillway profile is:

$$\frac{Y}{H_d} = 0.50 \left(\frac{X}{H_d}\right)^{1.85}$$

Determine the subpressures along the pier.

Problem 4.07

Check M, Y and H_u, to prevent the pulsating flow at the double-action gate shown in Fig. 1, Problem 4.07, for case A and case B defined as data of the problem.

Data

A:			B:		
M	=	9.00 m	M	=	6.00 m
Y	=	1.00 m	Y	=	1.20 m
H_u	=	11.00 m	H_u	=	9.00 m
H	=	1.50 m			

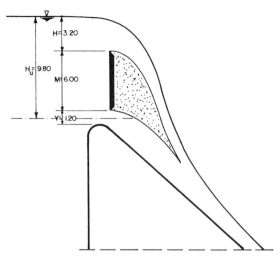

Fig. 1, For Problem 4.07.

Problem 4.08

Prepare a chart to show the discharge under gate for partial openings.

Data

The characteristics of the gate are shown in Fig. 1, Problem 4.08

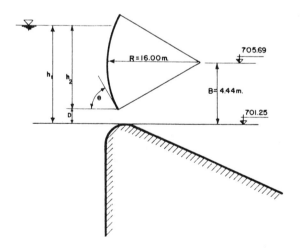

Fig. 1, Problem 4.08.

- The gate geometry corresponds to the case shown in Fig. 4.13, where $\Delta = 0$

The discharge formula is:

$$Q = CL \, (h_1^{3/2} - h_2^{3/2}) \, m^3/\text{sec}$$

where C is the discharge coefficient corresponding to the given hydraulic head

- L is the gate width = 31.85 m

- H_d = 13.75 m

Problem 4.09

Verify Eq. 4.6

$$r \cos\theta = -\frac{1 + y'^2}{y''} \qquad\qquad 4.6$$

Problem 4.10

Determine the airflow needed for a spillway, the characteristics of which are shown below, and compute the cross sectional area of the air conduit.

L = 15.00 m   h = 2.00 m   P = 5.00 m H_d = 2.00 m

Spillway profile: Creager.

Chapter 5

HYDRAULICS OF SPECIAL TYPES OF SPILLWAYS

5.1 INTRODUCTION

Previous chapters have described spillways which are most often used in engineering practice. This chapter discusses the hydraulics of special types of spillways.

These are:

- Broad-crested weir spillways, often used in turnouts,

- Side spillways or lateral spillways, often used where a long overflow crest is desired and the discharge channel is narrow,

- Shaft spillways, often used at the narrow dam sited where abutments rise steeply,

- Siphon spillway, often used for discharging large quantities of water within the narrow headwater differences, and

- Compound spillways, often used on large dams for solving specific problems.

5.2 HYDRAULICS OF BROAD-CRESTED WEIR SPILL-WAYS

The broad-crested weirs are a form of overflow spillway. Figure 5.1 shows the cross section of a broad-crested weir. When the gates are fully opened, the discharge flows between the piers and the system works as a broad-crested weir spillway.

Flow over a broad-crested weir spillway is shown on Fig. 5.2. Many attempts have been made to solve hydraulic problems arising from the flow taking place over a broad-crested weir (Doeringfeld-Barker, 1941; King and Brater, 1953; Rouse, H., 1956; etc.). Generally speaking, a broad-crested

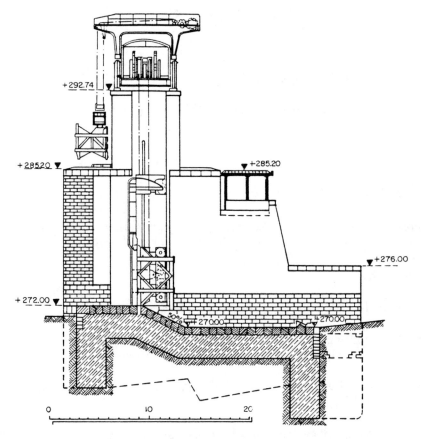

Figure 5.1. Longitudinal section of the Rybourg-Schwörstad weir (Switzerland).

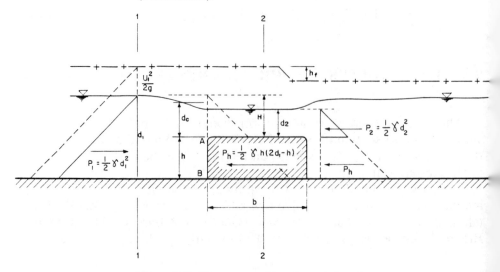

Figure 5.2. Flow over a broad-crested weir.

weir is an overflow structure with a horizontal or slightly inclined crest above which the deviation from hydrostatic pressure distribution may be neglected. Thus, the streamlines are practically parallel to each other. It was primarily believed that flow over a broad-crested weir would occur at the critical depth. Woodburn's (1932) measurements of flow over broad-crested weirs showed that the flow passes throughout the critical stage at some section at the crest and the location of this section varies with the hydrostatic head and weir dimensions. In order to obtain straight and parallel stream lines over the weir the breath b of the weir should be related to the total energy head over the crest

$$0.08 \leq \frac{H}{b} \leq 0.50$$

Under these conditions the flow characteristics are functions of the downstream water depth (Fig. 5.3).

It is evident that the use of broad-crested weirs as a measuring device is difficult and very complicated. However, their use as a control structure is worldwide. A simplified analysis of flow is given here.

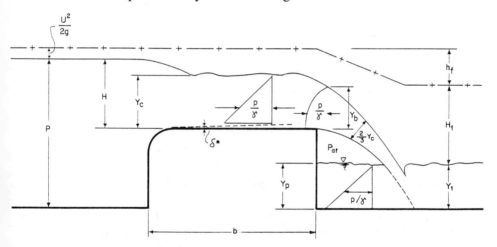

Figure 5.3. **Characteristics of flow over broad-crested weirs (Ippen, A.T.).**

Applying the momentum principle between section 1-1 and 2-2 of a flow taking place on a broad-crested weir (Fig. 5.2) yields:
Change of momentum:

$$\Delta M = \frac{\gamma}{g} q \left(\frac{q}{d_2} - \frac{q}{d_1}\right) \tag{5.1}$$

Hydrostatic pressure at section 1-1:

$$\frac{1}{2}\gamma\, d_1^2 \tag{5.2}$$

Hydrostatic pressure at section A-B:

$$\frac{1}{2}\gamma\, h[d_1+(d_1-h)] = \frac{1}{2}\gamma\, h\,(2d_1-h) \tag{5.3}$$

Hydrostatic pressure at section 2-2:

$$\frac{1}{2}\gamma\, d_2^2 \tag{5.4}$$

The forces due to friction are assumed negligible. Equating the change of momentum to the resultant of the external forces yields:

$$\frac{\gamma}{g}\, q\,(\frac{q}{d_2} - \frac{q}{d_1}) = \frac{1}{2}\gamma\, d_1^2 - \frac{1}{2}\gamma\, h\,(2d_1-h) - \frac{1}{2}\gamma\, d_2^2 \tag{5.5}$$

where

γ is the specific weight of water (T/m^3, kg/m^3),

g   is the gravitational acceleration (m/sec^2),

q   is the linear discharge (m/sec^3.m),

d_2 is the depth of flow on the crest of spillway (m), and

d_1 is the depth of flow in the approach channel (m);

q, d_1, and d_2 must be known to apply this equation. Then, the sill height can be computed. In engineering practice, d_1 and q are known and h is chosen by the designer. The problem reduces to the computation of d_2. As a result of their experimental work, Doeringfeld and Barker (1941) have shown that

$$d_1-h \cong 2d_2$$

then Eq. 5.5 can be simplified and solved for q:

$$q = 0.433\ \sqrt{2g}\ (\frac{d_1}{d_1+h})^{1/2} H^{3/2} \tag{5.6}$$

The discharge coefficient m is found equal to *

$$m = 0.433 \tag{5.7}$$

In practice this value cannot be reached and

$$m = 0.385 \tag{5.8}$$

is a reasonable assumption.

If h varies between 0 and ∞, then

$$q = 0.433 \sqrt{2g} \ H^{3/2}$$

and

$$q = 0.307 \sqrt{2g} \ H^{3/2}$$

These theoretical values are reduced to

$$0.333 < m < 0.380 \tag{5.9}$$

in practice.

King and Brater (1953) computed the variation of the discharge coefficient $C = m\sqrt{2g}$ for broad-crested weirs and presented the results in Table 5.1 and 5.1a.

For rounded corners the discharge coefficients are higher. They are summarized in Table 5.2 in the metric system and in Table 5.2a in the English system.

The discharge of broad-crested weir is expressed in Eq. 5.10, as the best approximation

$$Q = 1.705 \ LH^{3/2} = 0.385 \ L \ (2g \ H^3)^{1/2} \tag{5.10}$$

For h = 0 the broad-crested weir is reduced to a drop as shown in Fig. 5.4. The flow over a drop has been investigated intensely by Craya, A. (1948), Jäger, C. (1948), Roy (1949), Rouse, H. (1956), etc. The result reached by Rouse shows that the depth of flow at point A is $0.715d_{cr}$ and that d_{cr} happens some $(3 \cong 4)d_{cr}$ upstream of point A. Then $d_o < d_{cr}$, and it also corresponds to the minimum value of the hydraulic head. Similar hydraulic structures have been used and are still in use in irrigation systems.

---

* These values of m are computed in English system of units.

TABLE 5.1

Value of C in the formula $Q = CLH^{3/2}$ for broad-crested weirs (metric system)

Measured head	Breadth of crest								
	0.15	0.225	0.300	0.450	0.600	0.750	0.900	1.200	1.500
0.06	1.546	1.518	1.485	1.446	1.402	1.369	1.347	1.314	1.292
0.12	1.612	1.546	1.501	1.457	1.441	1.435	1.424	1.402	1.380
0.18	1.700	1.595	1.518	1.457	1.441	1.435	1.479	1.485	1.490
0.24	1.822	1.678	1.573	1.479	1.435	1.435	1.474	1.479	1.479
0.30	1.833	1.733	1.645	1.518	1.468	1.457	1.463	1.474	1.479
0.36	1.833	1.766	1.700	1.579	1.490	1.463	1.457	1.474	1.468
0.42	1.833	1.799	1.766	1.612	1.529	1.479	1.457	1.463	1.463
0.48	1.833	1.816	1.811	1.695	1.595	1.518	1.479	1.468	1.463
0.54	1.833	1.833	1.827	1.695	1.590	1.512	1.479	1.468	1.463
0.60	1.833	1.827	1.822	1.673	1.573	1.524	1.501	1.479	1.463
0.75	1.833	1.833	1.827	1.811	1.695	1.595	1.551	1.501	1.474
0.90	1.833	1.833	1.833	1.833	1.766	1.684	1.612	1.507	1.468
1.05	1.833	1.833	1.833	1.833	1.833	1.761	1.639	1.524	1.479
1.20	1.833	1.833	1.833	1.833	1.833	1.833	1.695	1.540	1.490
1.35	1.833	1.833	1.833	1.833	1.833	1.833	1.833	1.590	1.512
1.50	1.833	1.833	1.833	1.833	1.833	1.833	1.833	1.695	1.540
1.65	1.833	1.833	1.833	1.833	1.833	1.833	1.833	1.833	1.590

TABLE 5.1a

Value of C in the formula $Q = CLH^{3/2}$ for broad-crested weirs (metric system)

Measured head in feet. H	Breadth of crest of weir in feet										
	0.50	0.75	1.00	1.50	2.00	2.50	3.00	4.00	5.00	10.00	15.00
0.2	2.80	2.75	2.69	2.62	2.54	2.48	2.44	2.38	2.34	2.49	2.68
0.4	2.92	2.80	2.72	2.64	2.61	2.60	2.58	2.54	2.50	2.56	2.70
0.6	3.08	2.89	2.75	2.64	2.61	2.60	2.68	2.69	2.70	2.70	2.70
0.8	3.30	3.04	2.85	2.68	2.60	2.60	2.67	2.68	2.68	2.69	2.64
1.0	3.32	3.14	2.98	2.75	2.66	2.64	2.65	2.67	2.68	2.68	2.63
1.2	3.32	3.20	3.08	2.86	2.70	2.65	2.64	2.67	2.66	2.69	2.64
1.4	3.32	3.26	3.20	2.92	2.77	2.68	2.64	2.65	2.65	2.67	2.64
1.6	3.32	3.29	3.28	3.07	2.89	2.75	2.68	2.66	2.65	2.64	2.63
1.8	3.32	3.32	3.31	3.07	2.88	2.74	2.68	2.66	2.65	2.64	2.63
2.0	3.32	3.31	3.30	3.03	2.85	2.76	2.72	2.68	2.65	2.64	2.63
2.5	3.32	3.32	3.31	3.28	3.07	2.89	2.81	2.72	2.67	2.64	2.63
3.0	3.32	3.32	3.32	3.32	3.20	3.05	2.92	2.73	2.66	2.64	2.63
3.5	3.32	3.32	3.32	3.32	3.32	3.19	2.97	2.76	2.68	2.64	2.63
4.0	3.32	3.32	3.32	3.32	3.32	3.32	3.07	2.79	2.70	2.64	2.63
4.5	3.32	3.32	3.32	3.32	3.32	3.32	3.32	2.88	2.74	2.64	2.63
5.0	3.32	3.32	3.32	3.32	3.32	3.32	3.32	3.07	2.79	2.64	2.63
5.5	3.32	3.32	3.32	3.32	3.32	3.32	3.32	3.32	2.88	2.64	2.63

TABLE 5.2

Values of C in the formula $Q = CLH^{3/2}$ for broad-crested weirs (Metric system)

Name of	Radius	Breadth	Height	Head (m)				
Experimenter	(m)	(m)	(m)	0.12	0.18	0.24	0.30	0.45
Bazin	0.10	0.79	0.74	1.62	1.64	1.64	1.66	1.68
Bazin	0.10	1.97	0.74	1.49	1.56	1.58	1.60	1.61
U.S. Deep Waterways	0.10	0.79	1.37	-	1.53	1.55	1.56	1.61
U.S. Deep Waterways	0.10	1.97	1.37	-	-	1.56	1.56	1.56

TABLE 5.2a

Values of C in the formula $Q = CLH^{3/2}$ for models of broad-crested weirs with rounded upstream corner (Metric system)

Name of experimenter	Radius of curve in feet	Height of weir in feet, P	Breadth of weir in feet, B	Head in feet, H									
				0.4	0.6	0.8	1.0	1.5	2.0	2.5	3.0	4.0	5.0
Bazin	0.33	2.62	2.46	2.93	2.97	2.98	3.01	3.04					
Bazin	0.33	6.56	2.46	2.70	2.82	2.87	2.89	2.92					
U.S. Deep Waterways	0.33	2.62	4.57	....	2.77	2.80	2.83	2.92	3.00	3.08	3.17	3.34	3.50
U.S. Deep Waterways	0.33	6.56	4.56	....	....	2.83	2.83	2.83	2.82	2.82	2.82	2.82	2.81

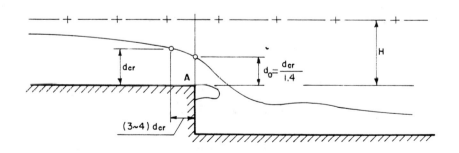

Figure 5.4. Flow over a drop.

Example 5.01

The sill of a turnout will be designed according to the rules outlined in section 5.2. The cross section of the sill will be a broad-crested weir. The streamlines are perpendicular to the axis of the structure. The water course is carrying sediment. Therefore the height of the sill will be greater than 1.50 meters. The breadth of the crest is chosen equal to B = 3.12 m. Determine the characteristics of the flow taking place at the structure.

Data

$$q = 3.00 \text{ m}^3/\text{sec.m}$$

$$d_1 = 3.20 \text{ m}$$

$$g = 9.81 \text{ m/sec}^2$$

Solution

An approximate approach follows:

1. Choose h = 1.64 > 1.50 m

2. Compute HH = 3.20 - 1.64 = 1.56 m

3. Choose C using Table 5.1

   For H = 1.56   B > 1.50 $C \cong 1.56$

4. Check the discharge for C = 1.56 and H = 1.56

   $q = 1.56 \bullet 1.56^{3/2} = 3.04 \cong 3.00 \text{ m}^3/\text{sec.m}$

5. Compute d_2

$$H = d_2 + (\frac{q}{d_2})^2 \frac{1}{2g}$$

$$1.56 = d_2 + (\frac{3}{d_2})^2 \frac{1}{19.62}$$

$$d_2 = 1.28 \text{ m}$$

6. Determine the energy head at the contracted section in respect to the datum line D-D (Fig. 5.5)

$$U_2 = \frac{3.00}{1.28} = 2.344 \text{ m/sec}$$

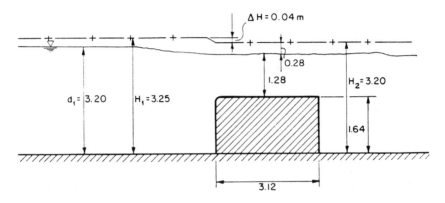

Figure 5.5. **Flow taking place on the broad-crested weir of Example 5.01.**

$$\frac{U_2^2}{2g} = 0.28$$

$$H_2 = h + d_2 + \frac{U^2}{2g} = 1.64 + 1.28 + 0.28 = 3.20 \text{ m}$$

7. Determine the energy head at the section 1-1 upstream of the structure

$$U_1 = \frac{3.00}{3.20} = 0.94 \text{ m/sec}$$

$$\frac{U_1^2}{2g} = 0.05$$

$$H_1 = 3.20 + 0.05 = 3.25 \text{ m}$$

8. Determine the head loss on the structure

$$H = 3.25 - 3.20 = 0.05 \text{ m}$$

9. Determine b

$$b = \frac{1.56}{0.50} = 3.12 \text{ m}$$

5.3 HYDRAULICS OF SIDE SPILLWAYS

Side spillways have been extensively used on large dams during the last three decades (Fig. 2.13). They have also been used as water level stabilizers in irrigation canals. Their construction is easy and relatively economical. However, their overall application is limited because the increase of discharge is not proportional to the crest length. Sometimes, levees are equipped with side spillways to be used as emergency fuse-plugs.

When the streamlines are perpendicular to the axis of the spillway the efficiency of the structure attains its maximum value. In case of side spillways, the streamlines are somewhat parallel to the spillway axis and the spillway discharge occurs after the flow deviates from its forward direction. A decrease in the efficiency of the structure is then observed. As a general rule it can be stated that a deviation of the streamlines creates additional head losses which directly affect the efficiency of the structure causing this deviation. When the discharge efficiency is a secondary consideration, the side spillway becomes the first choice. This consideration justifies its use on levees. Levees have more than adequate space for installation of a side spillway. Since the height of the levee is lower along the reach occupied by the side spillway, construction cost may be less than that of a full height section.

The hydraulic computation of side spillways can be performed graphically or analytically. These two approaches are explained in the following text.

5.3.1 Graphical Solution

An analysis is made of a side channel spillway located on one side of an irrigation canal with rectangular cross section. The crest of the spillway is assumed parallel to the canal bottom and flow in the canal is subcritical.

The problem reduces to:

1. The computation of the discharge Q_s of the spillway for a given crest length L, or

2. The computation of the crest length L of the spillway discharging Q_s.

A. Computation of the Crest Length, L, of a Side Spillway for Given Geometrical Characteristics and for the Discharge Q_s.

The crest length L of a side spillway can be computed as a function of the following data:

- the discharge, Q, the slope, S, and the geometric characteristics of the canal

- the discharge, Q_s, of the side spillway

- the dimension, C (Fig. 5.6).

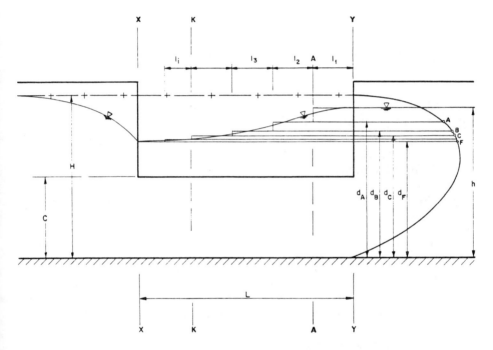

Figure 5.6. Graphical analysis of a side spillway.

Furthermore it is assumed that:

- the flow is uniform in the canal

- the flow in the canal is subcritical

- the discharge, Q_s, is computed by the formula

$$Q_s = mA (2gH)^{1/2}$$

211

where m is the discharge coefficient, assumed equal to 0.4.

- A, is the wetted area, and

- H, is the hydraulic head, assumed constant on the spillway which is a short structure.

Then the hydraulic head at section X is equal to the hydraulic head at section Y.

The hydraulic head at section X is known; it is the hydraulic head of the canal, the characteristics of which are known

$$H_X = H_y = H = h_y + \frac{Q_y^2}{2gA^2} \qquad (5.11)$$

where Q_y is the discharge of the canal at section Y

$$Q_y = Q - Q_s$$

Equation 5.11 yields the value of h_y.

The regime of flow in the canal being subcritical, the computation will begin from section Y at the downstream end of the side spillway, and proceed step by step toward the upstream end as follows:

1. Compute the depth of flow, d_y, at section Y situated at the downstream end of the side spillway.

2. Assume that the depth, d_y, does not substantially change on a crest of length ℓ_i. The choice of ℓ_i is a function of the characteristics of the flow. The designer is entitled to make the final decision regarding ℓ_i. In general, for small discharges, the value of ℓ_i is around 1.00 m, for larger discharges it can be chosen larger than 1.00 m.

3. Compute Q_{s1} as

$$Q_{s1} = m\ell_i \, d_y \, (2gH)^{1/2} \qquad (5.12)$$

4. At a distance ℓ_i from section Y the discharge of the canal is

$$Q = Q + Q_{s1}$$

5. Repeat the above operation up to

$$Q_s = \sum Q_{si}$$

and determine the length of the spillway as

$$L = \sum_{i=1}^{m} \ell_i$$

The operation is performed graphically in Fig. 5.6. The following steps are taken for the graphical solution:

1. Compute H_y as shown previously

2. Draw the

$$H = d + \frac{Q^2}{2g}$$

   curve downstream of the spillway: H being constant, d varies as a function of Q. With $Q - Q_s$ corresponding to a depth d_y at the section Y, d_y can be obtained directly from the graph.

3. Assume a crest length value $\ell = \ell_i$ and compute the corresponding discharge Q_{s1}, as

$$Q_{s1} = 0.40 \; \ell_i \; (d_y\text{-}C) \; (2gH)^{1/2}$$

4. At section A, located at a distance ℓ_i, from the section Y, the discharge of the canal is

$$Q_A = Q - Q_s + Q_{s1}$$

   where Q_s is the total discharge of the side spillway and Q_{s1} is the side spillway discharge between sections Y and A. Discharge Q_A corresponds to depth d_A, which can be obtained directly from the graph.

5. Assume that the depth, d_A, prevails along the crest length ℓ_2 and repeat the previous steps up to

$$Q_s = \sum_{i=1}^{m} Q_{si}$$

6. The crest length of the side spillway is

$$L = \sum_{i=1}^{m} \ell_i$$

213

Example 5.02

A rectangular channel carries $15.00 \text{ m}^3/\text{sec}$. Determine the crest length of a side spillway discharging $Q_s = 3.00 \text{ m}^3/\text{sec}$.

Data

Characteristics of the channel:

$S = 0.0005$

The width B of the channel $= 3.00 \text{ m}$

$n = 0.015$

Solution

1. Compute the flow depth in the channel, upstream of the spillway

$$Q = \frac{A}{n} R^{2/3} S^{1/2}$$

$$15 = \frac{3.00d}{0.015} (\frac{3d}{3.00+2d})^{2/3} 0.0005^{1/2}$$

$$d = 3.30 \text{ m}$$

2. Compute the flow depth in the channel downstream of the spillway

$$15.00 - 3.00 = \frac{3.00d}{0.015} (\frac{3d}{3.00+2d})^{2/3} 0.0005^{1/2}$$

$$d = 2.73 \text{ m}$$

Assume $C = 2.25 \text{ m}$

3. Verify the regime of flow

$$d_c = (\frac{q^2}{g})^{1/3} = [\frac{(15/3)^2}{9.81}]^{1/3} = 1.37 < 2.73 < 3.30$$

The regime of flow is subcritical upstream and downstream of the side spillway.

4. Compute the hydraulic head, H.

   The regime of flow being subcritical and the side spillway a short structure, the downstream hydraulic head will prevail for the entire length of the side spillway from its downstream end to its upstream beginning

$$H = dy + \frac{Q_y^2}{2A_y^2 \, g} = 2.73 + \frac{12^2}{(3 \cdot 2.73)^2} \frac{1}{19.62} = 2.84 \text{ m}$$

5. The solution of the problem is shown in Fig. 5.7.

   - Draw $H = 2.84 = d + (Q^2/(A^2 \cdot 2g))$ graph downstream of the side spillway

   - Locate $Q = 12.00 \text{ m}^3/\text{sec}$ on the diagram and determine the depth of flow

     $$d_y = 2.73 \text{ m}$$

   - Assume that $Q = 12.00 \text{ m}^3/\text{sec}$ prevails along a crest length of 0.20 m

   - The side spillway discharge over $\ell_1 = 0.20$ m is:

     $$Q_{s1} = 0.40 \,(2.73 - 2.25) \; \ell_1 \,(2.84 \cdot 2g)^{1/2}$$

     $$= 2.99 \,(2.73 - 2.25) \; \ell_1 = 0.29$$

     $$\ell_1 = 0.20 \text{ m}$$

   At the upstream end of the reach YA the discharge in the canal is:

   $$Q_A = 12 + 0.29 = 12.29 \text{ m}^3/\text{sec}$$

   Assume that $Q = 12.29 \text{ m}^3/\text{sec}$ prevails along a crest length of 0.50 m and $d_A = 2.72$ m

   - The side spillway discharge on $\ell_2 = 0.50$ m is:

   $$Q_{s2} = 2.99 \,(2.72 - 2.25) \; \ell_2 = 2.99 \,(2.72 - 2.25) \, 0.50 = 0.70 \text{ m}^3/\text{sec}$$

   At the upstream end of the reach AB the discharge in the canal is:

   $$Q_B = 12.29 + 0.70 = 12.99 \text{ m}^3/\text{sec}$$

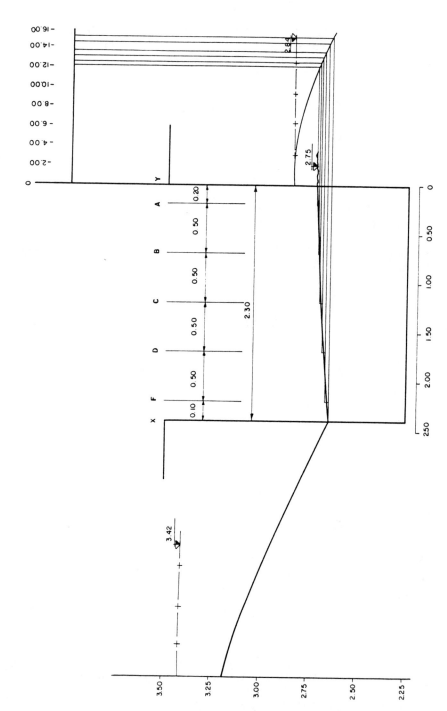

Figure 5.7. Flow on the spillway described in Example 5.02.

The results are compiled in Table 5.3.

TABLE 5.3

C = 2.25 m

Q m^3/sec	d m	ℓ_i m	Q_s m^3/sec	Q m^3/sec	Sections
12.00	2.73	0.20	0.29	12.29	Y - A
12.29	2.72	0.50	0.70	12.99	A - B
12.99	2.70	0.50	0.67	13.66	B - C
13.66	2.68	0.50	0.64	14.30	C - D
14.30	2.66	0.50	0.61	14.91	D - F
14.91	2.65	0.10	0.12	13.03	F - X

$$\sum_{i=1}^{6} \ell_i = 2.30 \text{ m}$$

$$\sum_{i=1}^{6} Q_{s_i} = 3.03 \cong 3.00 \text{ m}^3/\text{sec}$$

The crest length of the spillway is:

$$\sum_{i=1}^{6} \ell_i = 2.30 \text{ m}$$

B. Discharge Computation, Q_s, of a Side Spillway with Given Characteristics

This problem is solved by iteration. The discharge Q_s of the side spillway, is assumed constant and its crest length is computed by graphical method. If the computed crest length is found equal to the given one, the assumed Q_s is the answer to the problem. If not, a new assumption is selected and the computation repeated.

F. Sentürk

5.3.2 Analytical Solution

The analytical approach can also be used in calculating the hydraulics of side channel spillways. Assume the flow gradually varied over the spillway. The general relation defining gradually varied flow is:

$$\frac{dy}{dx} = \frac{S_b - S_E - \alpha\frac{qQ}{gA^2}}{1 - \alpha\frac{Q^2}{gA^2D}} \tag{5.13}$$

where

S_b is the slope of the bottom of the main channel

S_E is the slope of the energy line

α is the energy coefficient

Q is the discharge of the channel

q is the linear discharge in the channel

g is the gravitational acceleration, and

D is the hydraulic depth.

Assume that uniform flow prevails in the canal ($S_b = S_E$) and that $\alpha = 1$; then for a rectangular cross section, Eq. 5.13 reduces to Eq. 5.14 (Chow, V.T., 1959; Schmidt, 1954; Citrini, 1938)

$$\frac{dy}{dx} = \frac{Qy\,(-\frac{dQ}{dx})}{gb^2y^3 - Q^2} \tag{5.14}$$

where $$A = by$$

and $$q = \frac{dQ}{dx}$$

b, is the width of the rectangular cross section and y is the flow depth in this section. The discharge of the side spillway is

$$\frac{dQ_s}{dx} = q_s = m\,(2g)^{1/2}\,(y - C)^{3/2} = -\frac{dQ}{dx} \tag{5.15}$$

The specific energy of the flow is expressed by E; then the discharge at any section is

$$Q = by \, [2g \, (E - y)]^{1/2} \qquad (5.16)$$

Substituting this value in Eq. 5.14 yields

$$\frac{dy}{dx} = \frac{2m}{b} \frac{\sqrt{(E-Y)(Y-C)^3}}{3y - 2E} \qquad (5.17)$$

Integrating Eq. 5.17 and solving for x

$$x = \frac{b}{m} f \left(\frac{y}{E}\right) + ct \qquad (5.18)$$

where

$$f\left(\frac{y}{E}\right) = \left[\frac{2E - 3C}{E - C} \sqrt{\frac{E - Y}{Y - C}} - 3\sin^{-1} \sqrt{\frac{E - Y}{Y - C}} \right] \qquad (5.19)$$

Equation 5.19 shows the water surface profile of the side spillway.

5.3.3 Flow Profile Over Side Spillway

Possible flow profiles over side spillways are shown on Figs. 5.8, 5.9, 5.10, 5.11, and 5.12.

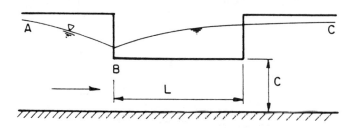

Figure 5.8. **Subcritical regime at the entrance and at the end of side spillway.**

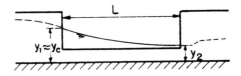

Figure 5.9. Critical depth at the entrance and supercritical at the end of side spillway.

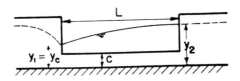

Figure 5.10. Critical depth at the entrance and subcritical at the end of side spillway.

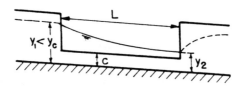

Figure 5.11. Supercritical flow over entire spillway.

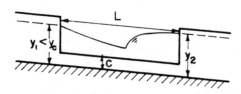

Figure 5.12. Supercritical flow at the entrance and subcritical at the end of side spillway.

Example 5.03

Design a side spillway where the flow is supercritical at the entrance and subcritical at the downstream end. The adjacent main channel is rectangular in section carrying a discharge of 40 m³/sec upstream of the side spillway. The width of the channel is 4.00 m; the discharge of the spillway is 10 m³/sec. It is required to determine the crest length, L, and the crest elevation of the spillway using elevation 0.00 for the bottom of the channel.

Data

$d = 2.10$ m at the entrance of the spillway

$d = 2.60$ m at the downstream exit of the spillway

Solution

1. Flow at the upstream end of the spillway

- Energy head $H_1 = d_1 + (U_1^2/2g) = 2.10 + \dfrac{100}{2.1^2\,2g} = 3.26$

- Critical water depth:

$$\frac{Q^2}{g} = \frac{A^3}{T} \qquad \frac{1600}{9.81} = \frac{4^3\,d^3}{4} \qquad d_{cr} = 2.17 \text{ m}$$

- Draw the curve

$$H_1 = 3.26 = d + \frac{Q^2}{(4d)^2\,2g}$$

H_1 is constant on the spillway and $d = 2.10 < 2.17$; then the flow is supercritical (Fig. 5.13). $C = 1.50$ m is determined by trial and error.

2. Determine the discharge of the spillway in the reach where supercritical flow prevails.

The computations are summarized in Table 5.4 and shown in Fig. 5.13. The problem is solved using the graphical method. The flow being supercritical, upstream hydraulic conditions govern the flow

$$Q_s = 0.40\,(d - C)\,(2gH_1)^{1/2} \qquad H_1 = 3.26 \text{ m} \qquad C = 1.50 \text{ m}$$

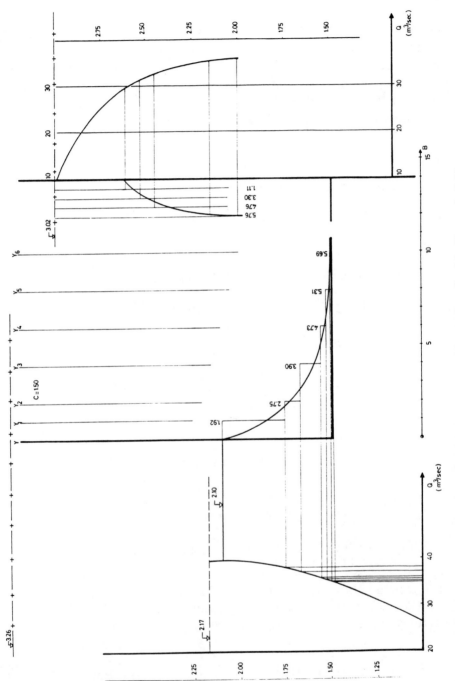

Figure 5.13. Flow on the spillway described in Example 5.03.

TABLE 5.4

Discharge of side spillway in supercritical reach

Q m³/sec	d m	ℓ m	Q_s m³/sec	Q m³/sec	Sections
40.00	2.10	1.00	1.92	38.08	$Y\text{-}Y_1$
38.08	1.76	1.00	0.83	37.25	$Y_1\text{-}Y_2$
37.25	1.68	2.00	1.15	36.10	$Y_2\text{-}Y_3$
36.10	1.63	2.00	0.83	35.27	$Y_3\text{-}Y_4$
35.27	1.59	2.00	0.58	34.69	$Y_4\text{-}Y_5$
34.69	1.56	2.00	0.38	34.31	$Y_5\text{-}Y_6$
34.31	1.54	2.00	0.26	34.05	$Y_6\text{-}Y_7$
34.05	1.52	2.00	0.13	33.92	$Y_7\text{-}Y_8$
33.92	1.50	2.00	0.00	33.92	$Y_8\text{-}Y_9$

3. Flow at the downstream end of the spillway

   • Energy head

$$H_2 = d_2 + (U_2^2/2g)$$

$$H_2 = 2.60 + \frac{7.5^2}{2.6^2\, 2g} = 3.02 \text{ m}$$

   • Critical water depth

$$\frac{Q^2}{g} = \frac{A^3}{T} \qquad \frac{4 \cdot 30^2}{9.81} = (4\, d_{cr})^3$$

$$d_{cr} = 1.79 \text{ m}$$

$$d = 2.60 > 1.79$$

then the flow is subcritical.

- Draw the curve

$$H_2 = 3.02 = d + \frac{Q^2}{(4d)^2} \frac{1}{2g}$$

H_2 is constant on the spillway where subcritical flow prevails.

- Determine the discharge of the spillway in the reach where subcritical flow prevails.

The computations are summarized in Table 5.5 and shown in Fig. 5.13. The flow being subcritical, downstream hydraulic conditions govern the flow.

TABLE 5.5

Discharge of side spillway in subcritical reach

$Q_s = 0.40 (d - C) (2gH_2)^{1/2}$ $\qquad$ $H_2 = 3.02$ m $\qquad$ C = 1.50 m

Q m³/sec	d m	ℓ m	Q_s m³/sec	Q m³/sec	Sections
30.00	2.61	0.50	1.71	31.71	X-X₁
31.71	2.53	0.50	1.59	33.30	X₁-X₂
33.30	2.45	0.50	1.46	34.76	X₂-X₃
34.76	2.15	0.50	1.00	35.76	X₃-X₄
35.76	2.00	0.50	0.77	-	-

Conclusion:*

Figure 5.14 shows the final water surface line on the spillway including a hydraulic jump. The positioning of the jump is determined assuming that supercritical flow predominates, and considering that the jump takes place where $Q_{sub} = Q_{sup}$. Backwater curves are ignored.

The crest length of the side spillway is found equal to:

$$\Sigma \ell = 6.82 \text{ m}$$

---

* See also Problem 5.03.

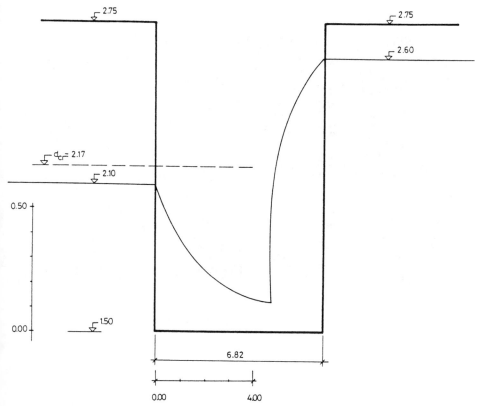

Figure 5.14. Hydraulic jump on the side spillway of Example 5.03.

5.3.4 Flow Downstream of the Side Spillway

When, in the approach channel, stream lines are perpendicular to the axis of the upstream structure they generally continue in the same direction in the downstream channel. In side spillways, even if the stream lines are perpendicular to the axis of the structure at the spillway entrance, they change direction so that the flow downstream is parallel to the centerline of the spillway (Fig 5.15). In hydraulic engineering it is known that each external factor affecting the flow causes an important amount of energy loss, of which change of the direction is a good example. Such flow creates high turbulence which is a basic factor in the energy loss. When the axis of the downstream channel is parallel to the crest of the spillway, its discharge is not constant and it increases as it moves downstream. Then the water surface line takes a certain profile along the discharge channel. A mathematical analysis yields a water surface profile as shown in Fig 5.16, but actual flow is as shown in Fig. 5.15.

Fig. 5.15. Flow in the downstream channel of a side spillway.

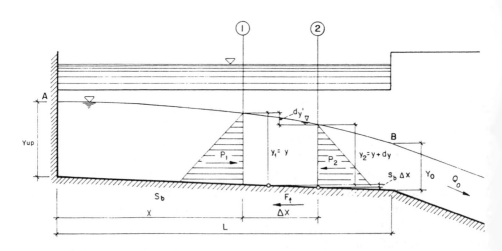

Figure 5.16. Flow taking place in the downstream channel in Example 5.04.

A. Flow with Increasing Discharge

A canal with gradually increasing discharge is shown schematically in Fig. 5.16 (Viparelli, 1952 and Chow, V.T., 1959). Hydraulic computation is conducted as follows:

Assumption No. 1: In the gradually increasing flow the linear discharge, q, is constant in its own unit.

Assumption No. 2: The bottom slope of the downstream channel is constant.

Under these conditions the characteristics of the mean flow can be defined. Let us consider two cross sections in the downstream channel, the distance separating them is assumed equal to dx. The upstream section will be labeled, 1-1, and the following section 2-2. The change in momentum between sections 1-1 and 2-2 is

$$\frac{\gamma}{g} (Q + dQ) (U + dU) - \frac{\gamma}{g} (QU) \qquad (5.20)$$

where Q and U denotes the discharge and the velocity in section 1-1. Equation 5.20 can be arranged and written as

$$QdU + (U + dU) dQ \qquad (5.21)$$

The change in momentum in unit time is equal to the total external force causing this change

$$F = \frac{dM}{dt}$$

which is the second law of Newton ($F = ma$).

Equation 5.21 is the momentum change; it is necessary to write the equivalent of the external force to complete this relationship. These external forces are:

• The gravitational force causing the flow,
• The friction force due to the flow, and
• The difference in the pressures acting on sections 1-1 and 2-2.

The gravitational and the pressure forces are accelerating forces and the friction forces are retarding forces.

a. The gravitational force

Let W represent the weight of water in the volume bound by sections 1-1 and 2-2; then its component in the direction of flow is

$$W \sin\theta = \gamma S_b (A + 0.5dA) dx = \gamma S_b A dx \qquad (5.22)$$

F. Sentürk

where S_b is the slope of the bottom of the canal

A is the wetted area between the two sections

γ is the specific weight of water, and

θ is equal to $\sin^{-1}S_b$

The terms containing differentials of second order are neglected in Eq. 5.22.

b. The frictional force

Assume h_f is the frictional loss between the two sections, then

$$h_f = S_E dx$$

where S_E is the slope of the energy line.
Let us assume that in the reach defined by dx a uniform flow takes place and can be defined by Manning's flow equation

$$S_E = \frac{U^2 n^2}{R^{4/3}} \tag{5.23}$$

The frictional force F_s is

$$F_s = \gamma (A+0.5 \, dA) S_E dx = \gamma \, AS_E dx \tag{5.24}$$

The term $\gamma 0.05 \, AS_E dx$ is neglected.

c. The pressure force

The pressure force is defined as:

$$P_1 - P_2 = -\gamma A dy \tag{5.25}$$

where

P_1 is the pressure acting on section 1-1

P_2 is the pressure acting on section 2-2, and

dy is the difference of depth between the two sections.*

---

* $\quad P_1 = \gamma \bar{z} A \quad P_2 = \gamma \, (\bar{z}+dy) \, A + \frac{\gamma}{2} \, dA dy = \gamma \, (\bar{z}+dy) \, A$

$P_1 - P_2 = -\gamma \, A dy$

where ($\bar{z}$) is the depth of centroid of A below the surface of flow. It is known that the hydrostatic presure can be defined as a function of ($\bar{z}$).

228

The momentum change between sections 1-1 and 2-2 is equal to the sum of the external forces:

$$\frac{dM}{dt} = P_1 - P_2 + W \sin\theta - F_s \qquad (5.26)$$

Substituting different terms by their values yields:

$$\frac{\gamma}{g} [QdU + (U+dU)\, dQ] = -\gamma\, A dy - \gamma\, A S_E\, dx + \gamma\, S_b\, A dx \qquad (5.27)$$

and rearranging*

$$dy = -\frac{1}{g}\left(UdU + \frac{U}{A} dQ\right) + (S_b - S_E)\, dx \qquad (5.28)$$

assume

$$q = \frac{dQ}{dx} \qquad\qquad dA = Tdy \qquad\qquad \frac{A}{T} = D$$

then

$$\frac{dy}{dx} = -\left(\frac{2Qq}{gA^2} - \frac{Q^2}{gA^2 A}\frac{dA}{dx}\right) + S_b - S_E$$

$$= -\left(\frac{2Qq}{gA^2} - \frac{Q^2}{2A^2 \frac{A}{T}}\frac{dy}{dx}\right) + S_b - S_E$$

and

$$\frac{dy}{dx}\left(1 - \frac{Q^2}{gA^2 D}\right) = -\frac{2Qq}{gA^2} + S_b - S_E$$

Introducing (α), the turbulence coefficient in the above relation yields

$$\frac{dy}{dx} = \frac{S_b - S_E - 2\alpha\dfrac{Qq}{gA^2}}{1 - \dfrac{\alpha Q^2}{gA^2 D}} \qquad (5.29)$$

This relation defines the water surface gradient of the flow with increasing discharge. As stated earlier a high turbulence prevails in the

---

* Differentials of second order are ignored.

downstream channel; the coefficient, α, is introduced in Eq. 5.29 to allow for the influence of turbulence. It is advisable to assume a value higher than 2.00 for the coefficient, α.

B. Flow with Decreasing Discharge

The relation defining a flow with decreasing discharge can be obtained directly from the energy equation:

$$H = z + y + \frac{\alpha Q^2}{2gA^2}$$

Differentiating with respect to x yields

$$\frac{dH}{dx} = \frac{dz}{dx} + \frac{dy}{dx} + \frac{\alpha}{2g}\left(\frac{2QdQ}{A^2dx} - \frac{2Q^2dA}{A^3dx}\right)$$

assuming

$$\frac{dH}{dx} = -S_E \qquad \frac{dz}{dx} = -S_b \qquad \text{and} \qquad \frac{dQ}{dx} = q$$

$$\frac{dA}{dx} = \frac{dA}{dy}\frac{dy}{dx} = \frac{Tdy}{dx}$$

and rearranging yields

$$\frac{dy}{dx} = \frac{S_b - S_E - \alpha\dfrac{Qq}{gA^2}}{1 - \alpha\dfrac{Q^2}{gA^2D}} \tag{5.30}$$

Equation 5.30 is the dynamic equation for spatially varied flow with decreasing discharge. The only difference between Eq. 5.29 and Eq. 5.30 is that the coefficient, 2, existing in Eq. 5.29 is not in Eq. 5.30.*

C. Method of Numerical Integration (**)

It is possible to use Eq. 5.29 to plot the flow profile for flow with increasing discharge and Eq. 5.30 for flow with decreasing discharge. However in engineering practice it is customary to draw the flow profile

---

*   If q = 0, Eq. 5.29 and Eq. 5.30 reduce to the dynamic equation for gradually varied flow of constant discharge.

**   This method is summarized from Creager, Justin, Hinds, 1945.

point by point using the method of numerical integration (Creager, Justin, Hinds, 1945). Equation 5.27 can be written in the following form:

$$\frac{\gamma}{g}\,[Q\Delta U + (U + \Delta U)\,\Delta Q] = -\int_{0}^{\Delta y} A\,dy + \gamma S_b \int_{0}^{\Delta x} A\,dx - \gamma S_E \int_{0}^{\Delta x} A\,dx \quad (5.31)$$

The differentials are substituted by finite increments. The sum of differentials of the form A dy is $\int_{0}^{\Delta y}$. Let

$$\int_{0}^{\Delta y} A\,dy = \bar{A}\Delta y$$

where $\bar{A}$ represents the average area; similarly

$$\int_{0}^{\Delta x} A\,dx = \bar{A}\Delta x$$

then Eq. 5.31 takes the form of Eq. 5.32:

$$\frac{\gamma}{g}\,[Q\Delta U + (U + \Delta U)\,\Delta Q] = -\gamma\Delta y\,\bar{A} + \gamma S_b\Delta x\,\bar{A} - \gamma S_E\Delta x\,\bar{A} \quad (5.32)$$

assuming $\bar{A} = (Q_1 + Q_2)/(U_1 + U_2)$, $Q = Q_1$ and $U + \Delta U = U_2$ and substituting in Eq. 5.32 and rearranging yields

$$\Delta y = -\frac{Q_1\,(U_1 + U_2)}{g\,(Q_1 + Q_2)}\,(\Delta U + \frac{U_2}{Q_1}\,\Delta Q) + S_b\Delta x - S_E\Delta x \quad (5.33)$$

Let dy' represent the drop of water level between sections 1-1 and 2-2; then

$$dy' = -dy + S_b dx$$

or

$$\Delta y' = -\Delta y + S_b\Delta x$$

Substituting in Eq. 5.33 yields

$$\Delta y' = \frac{\alpha Q_1\,(U_1 + U_2)}{g\,(Q_1 + Q_2)}\,(\Delta U + \frac{U_2}{Q_1}\,\Delta Q) + S_E\Delta x \quad (5.34)$$

Equation 5.34 can be used for the determination of the flow profile in a channel with increasing discharge. Substituting $q = 0$, $Q_1 = Q_2$, and $S_E = S_b$ yields the formula for uniform flow.

An application of Eq 5.34 is shown in Example 5.04.

Example 5.04

A flow takes place over a lateral spillway. The crest length of this spillway is L = 20.00 m and q = 1.00 m³/sec.m. The cross section of the downstream channel is rectangular with a bottom width equal to 10.00 m. The bottom slope of the channel is S_b = 0.10 and the bottom elevation at the upstream end is 50, furthermore, n = 0.014 and α = 1.00.

It is required to define the flow profile in the downstream channel and the characteristics of the flow taking place in it.

Solution

It is necessary to define a control section in the downstream channel. Upstream of the control section the regime of flow will be subcritical and downstream, the flow will be supercritical. It is also obvious that at this control section the regime of flow passes through critical and thus the critical depth can be computed, if the hydraulic conditions causing the flow are known. The computation will start at this section.

The control section can be virtual or natural. A weir or sill in the channel bottom can constitute a "natural" control section. If that does not occur, the control section must be determined according to hydraulic data.

Figure 5.16 illustrates the conditions outlined in Example 5.04.

The problem will be solved following the method suggested by Hinds, J., 1926. The geometry of the lateral spillway and its linear discharge, q, being given; it is possible to compute the corresponding critical depth in each section and draw its locus. When this locus which constitutes a fictitious water surface profile is obtained, its tangent parallel to the bottom slope defines this section. The control section is located at the point of tangency. Table 5.6 summarized the computations.

Column 1: Depth of flow

Column 2: Wetted area = Column 1 • T = Column 1 • 10.00

Column 3: Width of rectangular channel T = 10.00 m

Column 4: The critical depth is defined as

$U^2/g\,D = 1$, where D = A/T and T is the width of the free surface as shown in Fig. 5.17. Substituting this value of D in the above equation yields:

$$\frac{U^2}{2g} = \frac{A}{2T}$$

TABLE 5.6

Computation of the Critical Depth in the Downstream Channel

y	A	T	A/2T	U_{cr}	Q_{cr}	R_{cr}
m	m²	m	m	m/sec	m³/sec	m
1	2	3	4	5	6	7
0.05	0.50	10.00	0.025	0.70	0.35	0.050
0.10	1.00	"	0.050	0.99	0.99	0.098
0.15	1.50	"	0.075	1.21	1.82	0.145
0.20	2.00	"	0.100	1.40	2.80	0.192
0.25	2.50	"	0.125	1.57	3.92	0.234
0.30	3.00	"	0.150	1.72	5.16	0.283
0.35	3.50	"	0.175	1.85	6.48	0.327
0.40	4.00	"	0.200	1.98	7.92	0.370
0.50	5.00	"	0.250	2.21	11.05	0.454
0.60	6.00	"	0.300	2.43	14.58	0.536
0.70	7.00	"	0.350	2.62	18.34	0.614
0.80	8.00	"	0.400	2.80	22.40	0.690

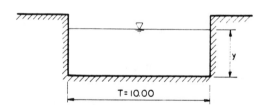

T= 10.00

Figure 5.17. Cross section of the canal.

Column 5:   Critical velocity $U_{cr} = (\frac{Ag}{T})^{1/2}$

Column 6:   Critical discharge: Column 2 • Column 5

$$Q_{cr} = AU_{cr}$$

Column 7:   Critical wetted perimeter $R = \dfrac{A}{T + 2y}$

233

Table 5.7 summarizes the computation for the determination of the control section.

TABLE 5.7

Computation for the Determination of the Control Section

x	Δx	Q	$Q_n + Q_{n+1}$	Y_{cr}	U_{cr}	$U_n + U_{n+1}$	ΔQ	ΔU	$\Delta y_m'$	R_{cr}	$\Delta x S_w$	$\Delta y'$	$\Sigma \Delta y'$
m	m	cms	cms	m	m/sec	m/sec	cms	m/sec	m	m	m10^{-3}	m	m
1	2	3	4	5	6	7	8	9	10	11	12	13	14
0													
2	2	2	2	0.16	1.25	1.25	2	1.25	-	0.155	-	-	-
4	2	4	6	0.24	1.67	2.92	2	0.42	0.207	0.229	7.80	0.215	0.215
6	2	6	10	0.33	1.82	3.49	2	0.15	0.151	0.310	6.19	0.160	0.275
8	2	8	14	0.40	1.99	3.81	2	0.17	0.139	0.370	5.844	0.145	0.520
10	2	10	18	0.47	2.13	4.12	2	0.14	0.125	0.430	5.484	0.130	0.650
12	2	12	22	0.53	2.28	4.41	2	0.15	0.124	0.479	5.442	0.129	0.779
14	2	14	26	0.58	2.41	4.69	2	0.13	0.117	0.520	5.449	0.122	0.901
16	2	16	30	0.64	2.50	4.91	2	0.09	0.104	0.567	5.220	0.109	1.010
18	2	18	34	0.69	2.61	5.11	2	0.11	0.107	0.606	5.207	0.112	1.022
20	2	20	38	0.74	2.70	5.31	2	0.09	0.100	0.645	5.128	0.105	1.277

cms = m^3/sec

Column 1:   Distance along the channel, m

Column 2:   Increment of distance, m

Column 3:   The inflow discharge = qx, m^3/sec

Column 4:   Sum of the discharges Q_1 of the previous sections and Q_2 of the section under consideration, m^3/sec

Column 5:   Critical depth in meters, interpolated from Table 5.6 corresponding to the discharge in Column 3

Column 6:   Critical velocity in m/sec, interpolated from Table 5.6 corresponding to the discharge in Column 3

Column 7:   Sum of the velocities prevailing in previous sections and in the actual section, m/sec

Column 8: $\Delta Q = Q_2 - Q_1$

Column 9: $\Delta U = U_2 - U_1$

Column 10: $\Delta y'_m = \dfrac{Q_1(U_1 + U_2)}{g(Q_1 + Q_2)} \left(\Delta U + \dfrac{U_2}{Q_1} \Delta Q\right)$

take Q_1 from Column 3

take $U_1 + U_2$ from Column 7

take $Q_1 + Q_2$ from Column 4

take ΔU from Column 9

take ΔQ from Column 8

Column 11:   Critical hydraulic radius in m, interpolated from Table 5.6 corresponding to the discharge in Column 3

Column 12: $S_w \Delta x = \dfrac{U^2 n^2 \Delta x}{R^{4/3}}$

Column 13: $\Delta y' = \Delta y'_m + S_w \Delta x$

Column 14: $\Sigma \Delta y'$

Using elevation 100 as the reference level for the flowline, the total losses can be taken from Table 5.7 and plotted on the section in Fig. 5.18a to show a fictitious flow line for the critical flow. Similarly, by plotting flow depths under their corresponding flowline locations, a fictitious bottom surface at critical flow can be developed. Both surface and bottom lines correspond to a fictitious channel. The point at which a line parallel to the bottom of existing channel is tangent to the fictitious line defines the control section. Upstream of this section the slope of the critical bottom is greater than the slope of the channel, thus the flow must be supercritical. Downstream the critical slope is less than the slope of the channel thus the flow must be subcritical. This is the basic condition for a section to be a control section (section 10 in Fig. 5.18a and Fig. 5.18b).

The hydraulic computation will start from this section and proceed upstream for supercritical flow and downstream for subcritical flow. They are summarized in Table 5.8.

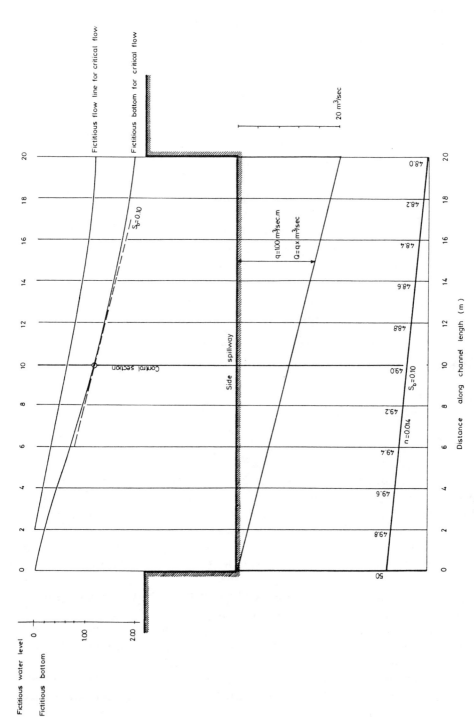

Figure 5.18a. Determination of the control section.

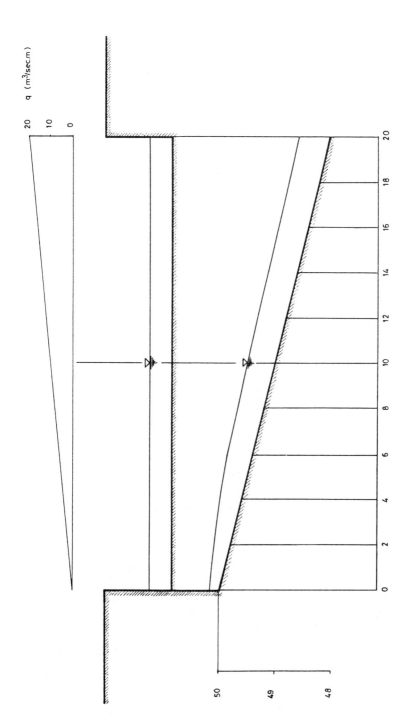

Figure 5.18b. Determination of the flow line.

TABLE 5.8

Computation of Flow Profile for Example 5.04

x	Δx	z_o	Δy'	y	A	Q	U	$Q_n + Q_{n+1}$	$U_n + U_{n+1}$	ΔQ	ΔU	$Δy'_m$	R	$S_w Δx$	Δy'	z
m	m	m	m	m	m²	m³/sec	m/sec	m³/sec	m/sec	m³/sec	m/sec	m	m	m	m	m
1	2	3	4	5	6	7	8	9	10	11	12	13	14	15	16	17

Upstream of section 10 - Subcritical regime

x	Δx	z_o	Δy'	y	A	Q	U	$Q_n + Q_{n+1}$	$U_n + U_{n+1}$	ΔQ	ΔU	$Δy'_m$	R	$S_w Δx$	Δy'	z
10	-	49.00	-	0.47	4.70	10	2.128	-	-	-	-	-	-	-	-	49.47
8	2.00	49.20	0.107	0.377	3.77	8	2.122	18	4.250	2	- 0.006	0.101	0.351	0.007	0.108	49.578
6	2.00	49.40	0.116	0.293	2.93	6	2.048	14	4.170	2	- 0.074	0.106	0.277	0.009	0.115	49.693
4	2.00	49.60	0.120	0.213	2.13	4	1.878	10	3.926	2	- 0.170	0.109	0.204	0.011	0.120	49.813
2	2.00	49.80	0.117	0.130	1.30	2	1.538	6	3.416	2	- 0.340	0.100	0.127	0.015	0.115	49.928
0	2.00	50.00														

Upstream of section 10 - Subcritical regime

x	Δx	z_o	Δy'	y	A	Q	U	$Q_n + Q_{n+1}$	$U_n + U_{n+1}$	ΔQ	ΔU	$Δy'_m$	R	$S_w Δx$	Δy'	z
10	-	49.00	-	0.47	4.70	10	-	10	-	-	-	-	-	-	-	49.47
12	2.00	48.80	0.175	0.495	4.95	12	2.424	22	4.552	2	0.296	0.165	0.450	0.007	0.172	49.295
14	2.00	48.60	0.170	0.525	5.25	14	2.667	26	5.091	2	0.243	0.165	0.475	0.008	0.173	49.125
16	2.00	48.40	0.160	0.565	5.65	16	2.832	30	5.499	2	0.165	0.149	0.508	0.008	0.157	48.965
18	2.00	48.20	0.170	0.595	5.95	18	3.025	34	5.857	2	0.193	0.160	0.532	0.008	0.168	48.795
20	2.00	48.00	0.180	0.615	6.15	20	3.252	38	6.277	2	0.227	0.178	0.548	0.009	0.187	48.615

The frictional losses are shown in Columns 14, 15 and 16. Since they are small in comparison to the remaining items, they can be considered negligible.

An arbitrary value is chosen for $\Delta y'$ which yields the water level z. In this example, the critical flow occurs at section 10 and so it naturally is chosen as the first point. The depth of flow is taken from Table 5.7. For upstream sections $\Delta y'$ will be added to the flow depth, then

$$z_n = z_{n-1} + \Delta y'$$

where z_{n-1} is shown in Column 17. The flow depth will be $(z_n - z_o)$; z_o is shown in Column 3 and is the bottom level of the channel. The new value is shown in Column 5.

The final value of $\Delta y'$ reached at Column 13 must be equal to the original value in Column 4. The method adopted to obtain the solution is shown in Table 5.8:

Column 1:   Distance along the channel

Column 2:   Increment of distance

Column 3: z_o, bottom level of the channel

Column 4: $(\Delta y')$, drop in water surface (an arbitrary value will be chosen)

Column 5: (y_n), water depth = y_{n-1} - (increment of bottom - $\Delta y'$)

Column 6: $A = 10y_n$

Column 7: $Q_n = Q_{n-1} - q\Delta x$

Column 8: $U_n = Q_n/A_n$

Column 9: $Q_n + Q_{n+1}$

Column 10: $U_n + U_{n+1}$

Column 11: $\Delta Q = Q_n - Q_{n-1}$

Column 12: $\Delta U = U_n - U_{n-1}$

Column 13: $\Delta y'_m = \dfrac{Q_{n-1}\,(U_n + U_{n-1})}{g(Q_n + Q_{n-1})}\left(\Delta U + \dfrac{U_n}{Q_{n-1}}\Delta Q\right)$

Column 14: $R_n = \dfrac{A_n}{P_n}$ $\qquad\qquad$ $P_n = 10 + 2y_n$

Column 15: $S_w\Delta x = \dfrac{U_n^2\, n^2 \Delta x}{R_n^{4/3}}$

Column 16: $\Delta y' = \Delta y_m' + S_w\Delta x$ (This value of $\Delta y'$ must be equal to the value chosen in Column 4, if not change $\Delta y'$ in Column 4 and repeat)

Column 17: $z_n = z_{n-1} \pm \Delta y_n'$ (Column 17 $\pm$ Column 16)

5.4   HYDRAULICS OF SHAFT SPILLWAYS

A shaft spillway consists of a vertical gallery with an entrance that is funnel-shaped, similar to a morning glory, hence the common nickname, *Morning Glory Spillway*. The vertical gallery connects with a slightly inclined outlet extending through or around the dam.

It is customary to consider the shaft spillway as being composed of three separate elements:

1. A spillway section

2. A transition section

3. Bends and downstream tunnel

The spillway and transition sections constitute the shaft. The shaft is followed by a bend and downstream tunnel which convey the water to the downstream main watercourse.

The spillway has the form of an inverted truncated cone. Its dimensions and special form have significant influence on discharge of the structure. The approach conditions are also of some importance. Where the approaching flow is not symmetrical, the efficiency of the spillway decreases and sometimes pulsating flow occurs which generates vibrations in the structure.

There are two types of shaft spillways: the standard crest spillway and the flat-crested spillway. Figure 5.19a shows a standard crest spillway and Fig. 5.19b a flat-crested spillway.

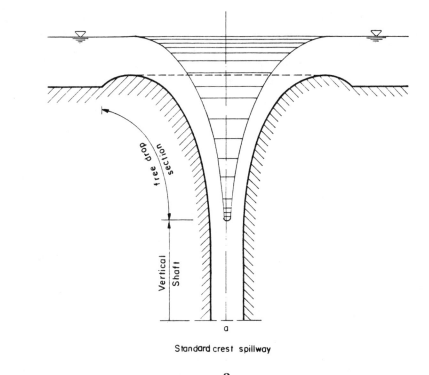

Standard crest spillway

a

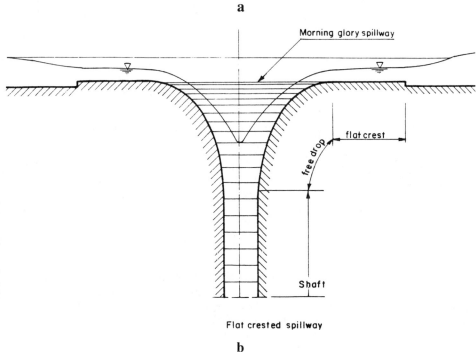

Flat crested spillway

b

Figure 5.19. Standard and flat-crested shaft spillways.

If the entrance of the shaft spillway consists of a free flow spillway as shown in Fig. 5.19a, it is called a standard crest spillway and if the entrance of the shaft spillway has the form of a broad-crested weir, as shown in Fig. 5.19b, it is called a flat-crested spillway.

The hydraulic head over the weir section influences the flow on the weir and in the shaft. Figure 5.20 shows different flow patterns taking place in a shaft spillway.

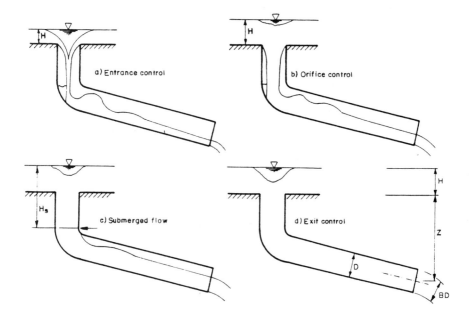

Figure 5.20. Flow over the entrance of a shaft spillway.

Let H be the hydraulic head over the weir. The discharge increases with the increasing values of H. The (a-g) part of the curve shown in Fig 5.21 illustrates this phenomenon. The water overflows from the weir, flows freely in the shaft, enters the slightly inclined tunnel and flows freely in it. The dimensions of the transition region and the tunnel are large enough to allow a free surface flow to take place. The only difference between a shaft spillway and a free flow spillway in this particular case is that the coefficient of discharge, C, in the shaft spillway has a slightly smaller value due to the geometry of weir.

If the hydraulic head, H, continues to increase and attains a value of H_o (Fig. 5.21), then the weir at the entrance acts as an orifice. The water profile at the entrance loses its original form and changes to a slightly concave curve (the g-h part, case 2, of the curve in Fig. 5.21). This alters the control condition. The shaft that controls the flow and the hydraulic head corresponding to this case is shown by H_a.

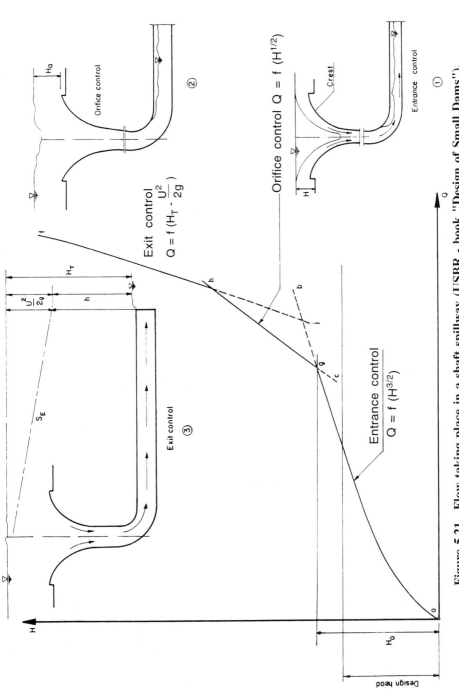

Figure 5.21. Flow taking place in a shaft spillway (USBR - book "Design of Small Dams").

If the hydraulic head increases further, the control section moves closer to the tunnel outlet to establish a new control section as shown in case 3, Fig. 5.21. In this case the hydraulic head is shown by H_T, (the h-f part of the curve in Fig. 5.21).

5.4.1 Flow at the Weir Entrance

As mentioned previously, there are two types of shaft spillways, the standard and the flat-crested weir. Case 1, Fig. 5.21, shows a standard crested spillway. In section, it resembles the usual free flow spillway, but in plan, its shape is circular. The broad-crested weir is also circular in plan but the spillway is no longer a free flow spillway. Both types are shown in Fig 5.19. The discharge coefficient for the flat-crested spillway is less than the discharge coefficient for the standard crest spillway. In engineering practice, flat crested is mostly used if shaft excavated in sound rock and standard crest in shaft, is conceived as a tower over the natural ground.

The discharge of shaft spillway is given by:

$$Q = mA(2gH)^{1/2} \qquad (5.35)$$

The coefficient, m, is smaller than the discharge coefficient of a straight free flow spillway. This is due to the geometry of the shaft which forces contraction of the flow and creates turbulence. Let $m(2g)^{1/2} = C$, then Eq. 5.35 takes the following form:

$$Q = CLH^{3/2} \qquad (5.36)$$

The definition of L and H is given by USBR and by Favre-Pugnet. The circumference of the circular weir at the entrance is (L)

$$L = 2\pi R_s$$

R_s is the radius of a circular sharp-crested weir as shown in Fig. 5.22, or

$$L = 2\pi R$$

where $R = R_s - X_s$ and H can be defined as H_o or H_s. C, must be defined according to adopted definitions. In this section, H_s and H_o are used and the corresponding coefficient of discharge is indicated as C_o. Equation 5.36 takes the form

$$Q = C_o (2\pi R_s) H_o^{3/2} \qquad (5.37)$$

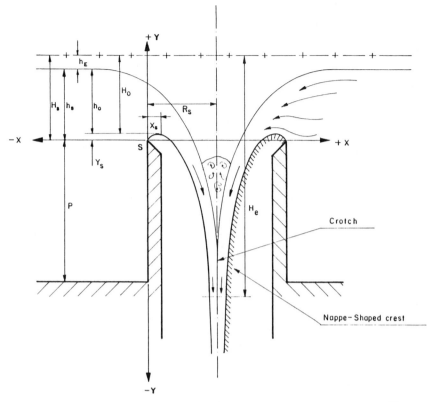

Figure 5.22. Nomenclature for shaft spillway (from USBR) .[*]

The coefficient of discharge C_0 is given in Fig. 5.23 as a function of P, R_S and H_0. The variation of these three variables determines the variation of the control section. Uncertainty exists in the definition of submergence. Different attempts have been made to determine its limit. They are reproduced as follows:

Wagner, 1954 $\dfrac{H_0}{R_S} = 0.45$

White-Pherson $\dfrac{H_0}{R_S} = 0.60$ and assumed variable between 0.45 and 0.60.

Blaisdell, 1962 $\dfrac{H_0}{R_S} = 0.47 \sim 0.49$

Lazarry $\dfrac{H_0}{R_S} = 0.50$

---

[*]   Design of Small Dams

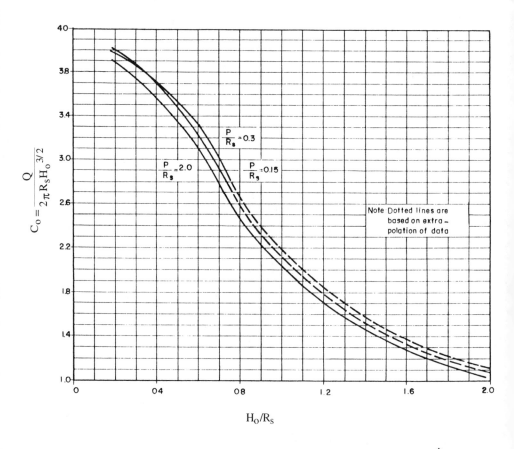

Figure 5.23. Coefficient of discharge C_0 of shaft spillway (USBR).[*]

For $\dfrac{H_o}{R_s} < 0.45$ free-flow prevails and weir control governs

For $0.45 < \dfrac{H_o}{R_s} < 1.00$ the weir is partly submerged and the controlling condition is that of a submerged weir

For $\dfrac{H_o}{R_s} \cong 1.0$ the weir is completely submerged

For $\dfrac{H_o}{R_s} > 1.0$ orifice flow is reached.

---

[*]   Design of Small Dams

These statements are true only when the approach conditions are uniform. In nature complete uniformity cannot be obtained and field data will show some variance from the result obtained from formulae.

The approach conditions for the shaft spillway of Alakir Dam, Turkey are outlined in Fig. 5.24. The velocity distribution is far from uniform. The best solution can only be obtained from a scale model investigation.

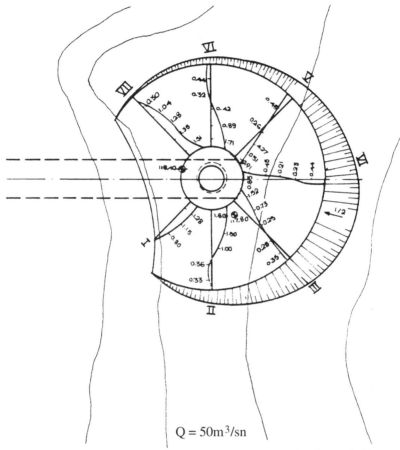

$$Q = 50 m^3/sn$$

Figure 5.24. Approach conditions in the Alakir Dam Spillway, Turkey.

The variation of C_0 is illustrated in Fig. 5.23. It is given for three different values of the ratio (P/R_s) where P is the height of the spillway tower. It is obvious that for constant values of (H_0/R_s), an increase of P corresponds to a decrease of C_0. In a free-flow straight-weir spillway, an increase of P corresponds to an increase of C_0. This occurs because of the difference in flow contraction characteristics.

Figure 5.25 shows the variation of C for hydraulic heads different from H_d.

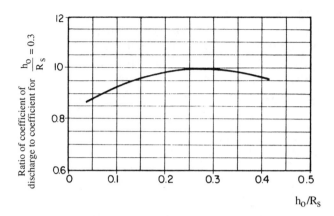

**Figure 5.25. Variation of C for hydraulic head different from H_d
(USBR).***

For $H > H_o$, subpressure occurs. The aeration in shaft spillways is an important problem to be solved carefully. If the lower nappe is not sufficiently aerated subpressures occur and pulsating flow can reach high fluctuations capable of completely destroying the spillway.

5.4.2 Flow in the Shaft

If the flow in the shaft is sufficiently aerated the control section will be at the entrance. The submerged flow boundary is reached when a discharge, Q, equals Q_o; then pressure flow takes place in the shaft . The dimensions of the shaft can be computed according to the following assumptions:

1. The equation of continuity prevails, $Q = AU$

2. Assuming negligible losses, the discharge formula for orifice flow can be used for computing the discharge in the shaft:

$$Q = A (2gH)^{1/2} = \pi r^2 (2gH)^{1/2}$$

where r is the radius of the shaft, and

$$r = k \frac{Q^{1/2}}{H^{1/4}} \tag{5.38}$$

where

$$k = \frac{1}{[\pi(2g)^{1/2}]^{1/2}}$$

---

*   Design of Small Dams

248

5.4.3 Flow at the Limit of Submergence

Figure 5.26 shows the hydraulic characteristics of a flow which takes place in a shaft spillway (Favre-Pugnet, 1959). The energy equation applied to this flow yields

$$H + z = 1.2 \left(\frac{Q^2}{2gA_1^2} \right) + \Delta H \qquad (5.39)$$

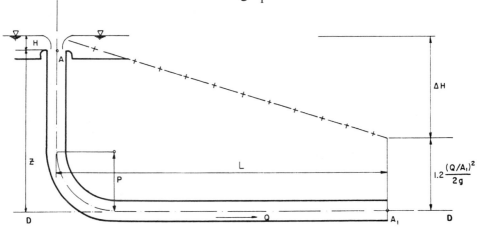

Figure 5.26. Hydraulic characteristics of a flow taking place in a shaft spillway

Due to high turbulence in the tunnel the energy coefficient (α) is taken equal to 1.2; ΔH is the total energy loss. The components of ΔH consist of the following:

a. Friction losses in the shaft and in the tunnel:

$$SL = L \left(n^2 / R_H^{4/3} \right) (Q^2 / A^2) \qquad (5.40)$$

where n is Manning's coefficient of friction (it is recommended that $n \geq 0.015$ be used because of the chemical and abrasive effect of water corrosion on the concrete over a period of time).

b. Entrance losses:

$$0.05 \frac{Q^2}{2gA^2} \qquad (5.41)$$

c. Losses at bends:

$$\left[0.13 + 1.8 \frac{r^{7/2}}{P} \right] \left(\frac{Q^2}{2gA^2} \right) \qquad (5.42)$$

249

where   P is the radius of the bend, and

r is the radius of the shaft.

d. Exit losses:

$$0.2 \left[1 - (\frac{A}{A_1})^2\right] (\frac{Q^2}{2gA^2}) \tag{5.43}$$

Where A is the cross sectional area of the tunnel and A_1 is the wetted area at the tunnel exit.

Substituting these values in Eq. 5.39 yields

$$H + z = \frac{Q^2}{K} \tag{5.44}$$

The relation expressing K can be written easily. It is known that for $H_o / R_s \geq 1.00$, orifice flow is reached. Let $H_o / R_s = 1.1$, then the control section is at the tunnel exit and the energy equation takes the following form, for

$$\frac{H_o}{R_s} = 1.1, \qquad C_o = 1.837$$

and

$$H = \frac{3}{2} (\frac{Q}{2\pi R_s})^{2/3} = BQ^{2/3} \tag{5.45}$$

Substituting this value in Eq. 5.44 yields

$$BQ^{2/3} + z = \frac{Q^2}{K} \tag{5.46}$$

This equation defines $Q = Q_{sub}$, the discharge of submerged flow.

5.4.4 Profile of the Lower Nappe in Shaft Spillways

The profile of the lower nappe of the flow taking place in a sharp-crested weir is given in Tables 5.9, 5.10, and 5.11. Figure 5.27 also shows the variation of (H_o/R_s) as a function of (P/R_s) and (H_s / H_o).

R_s is chosen by the designer. Using the selected R_s it is then possible to study variations of H_o and H_s. Figure 5.28 shows the upper nappe and the lower nappe profiles of the circular weir flow as a function of (X/H_s) and (Y/H_s).

The chart of Figure 5.30 has been prepared to calculate the increase of the radius R_s. The corrections shown in this figure have been prepared for a profile for $(H_s/R_s) = 0.30$ which defines the nappe nearest to the axis.

TABLE 5.9

Coordinates of lower nappe surface for different values of (H_S/R_S) when ($P/R_S = 2$)

$U^2/2g = 0$ and aerated nappe

$\dfrac{X}{H_S}$	$\dfrac{Y}{H_S}$ For portion of the profile above the weir crest								
0.000	0.0000	0.0000	0.0000	0.0000	0.0000	0.0000	0.0000	0.0000	0.0000
0.010	0.0150	0.0133	0.0128	0.0122	0.0116	0.0112	0.0104	0.0095	0.0070
0.020	0.0280	0.0250	0.0236	0.0225	0.0213	0.0202	0.0180	0.0159	0.0090
0.030	0.0395	0.0350	0.0327	0.0308	0.0289	0.0270	0.0231	0.0198	0.0085
0.040	0.0490	0.0435	0.0403	0.0377	0.0351	0.0324	0.0268	0.0220	0.0050
0.050	0.0575	0.0506	0.0471	0.0436	0.0402	0.0368	0.0292	0.0226	
0.100	0.0860	0.0762	0.0705	0.0642	0.0570	0.0482	0.0264	0.0089	
0.200	0.1105	0.0938	0.0819	0.0688	0.0521	0.0292			
0.300	0.1105	0.0850	0.0668	0.0446	0.0174				
0.400	0.0970	0.0620	0.0365	0.0060					
0.500	0.0700	0.0250							
0.600	0.0320								
$\dfrac{H_S}{R_S}$	0.00	0.20	0.30	0.40	0.50	0.60	0.80	1.00	2.00
$\dfrac{Y}{H_S}$	$\dfrac{X}{H_S}$ For portion of the profile below the weir crest								
0.000	0.668	0.554	0.487	0.413	0.334	0.262	0.158	0.116	0.048
-.020	0.705	0.592	0.526	0.452	0.369	0.293	0.185	0.145	0.074
-.040	0.742	0.627	0.563	0.487	0.400	0.320	0.212	0.165	0.088
-.060	0.777	0.660	0.596	0.519	0.428	0.342	0.232	0.182	0.100
-.080	0.808	0.692	0.628	0.549	0.454	0.363	0.250	0.197	0.110
-.100	0.838	0.722	0.657	0.577	0.478	0.381	0.266	0.210	0.118
-.200	-.978	0.860	0.790	0.698	0.575	0.459	0.326	0.260	0.144
-.300	1.100	0.976	0.900	0.797	0.648	0.518	0.368	0.296	0.160
-.400	1.207	1.079	1.000	0.880	0.706	0.562	0.400	0.322	0.168
-.500	1.308	1.172	1.087	0.951	0.753	0.598	0.427	0.342	0.173
-1.000	1.713	1.564	1.440	1.189	0.899	0.710	0.508	0.402	0.188
-2.000	2.302	2.126	1.891	1.381	1.025	0.810	0.572		
-3.000	2.778	2.559	2.119	1.468	1.086	0.853			
-4.000		2.914	2.201	1.500					
-5.000		3.178	2.227						
-6.000		3.405	2.232						

After Wagner (USBR)

TABLE 5.10

Coordinates of lower nappe surface for different values of (H$_S$/R$_S$) when (P/R$_S$ = 0.30)

(U^2/2g = 0)

$\dfrac{X}{H_s}$	$\dfrac{Y}{H_s}$	For portion of the profile above the weir crest.				
0.000	0.0000	0.0000	0.0000	0.0000	0.0000	0.0000
0.010	0.0130	0.0130	0.0120	0.0115	0.0110	0.0100
0.020	0.0245	0.0240	0.0225	0.0195	0.0180	0.0170
0.030	0.0340	0.0330	0.0300	0.0270	0.0240	0.0210
0.040	0.0415	0.0390	0.0365	0.0320	0.0285	0.0240
0.050	0.0495	0.0455	0.0420	0.0370	0.0325	0.0245
0.100	0.0740	0.0660	0.0575	0.0500	0.0395	0.0190
0.200	0.0885	0.0745	0.0575	0.0435	0.0180	
0.300	0.0780	0.0580	0.0340	0.0050		
0.400	0.0495	0.0240				
0.500	0.0090					
$\dfrac{H_S}{R_S}$	0.20	0.30	0.40	0.50	0.60	0.80
$\dfrac{Y}{H_s}$	$\dfrac{X}{H_s}$	For portion of the profile below the weir crest.				
0.000	0.519	0.455	0.384	0.310	0.238	0.144
-.020	0.560	0.495	0.423	0.345	0.272	0.174
-.040	0.598	0.532	0.458	0.376	0.300	0.198
-.060	0.632	0.567	0.491	0.406	0.324	0.220
-.080	0.664	0.600	0.522	0.432	0.348	0.238
-.100	0.693	0.631	0.552	0.456	0.368	0.254
-.200	0.831	0.763	0.677	0.558	0.451	0.317
-.300	0.953	0.880	0.779	0.634	0.510	0.362
-.400	1.060	0.981	0.867	0.692	0.556	0.396
-.500	1.156	1.072	0.938	0.745	0.595	0.424
-1.000	1.549	1.430	1.180	0.892	0.707	0.504
-2.000	2.120	1.892	1.380	1.022	0.810	0.569
-3.000	2.557	2.113	1.464	1.081	0.852	
-4.000	2.911	2.200	1.499			
-5.000	3.173	2.223				
-6.000	3.400					

After Wagner (USBR)

TABLE 5.11

Coordinates of lower nappe surface for different values of (H_S / R_S) when $(P/R_S = 0.15)$

$$(U^2/2g = 0)$$

$\dfrac{X}{H_S}$	$\dfrac{Y}{H_S}$	For portion of the profile above the weir crest				
0.000	0.0000	0.0000	0.0000	0.0000	0.0000	0.0000
0.010	0.0120	0.0115	0.0110	0.0105	0.0100	0.0090
0.020	0.0210	0.0195	0.0185	0.0170	0.0160	0.0140
0.030	0.0285	0.0265	0.0250	0.0225	0.0200	0.0165
0.040	0.0345	0.0325	0.0300	0.0265	0.0230	0.0170
0.050	0.0450	0.0375	0.0345	0.0300	0.0250	0.0170
0.100	0.0590	0.0535	0.0465	0.0375	0.0255	0.0065
0.200	0.0670	0.0560	0.0295	0.0200		
0.300	0.0520	0.0330	0.0100			
0.400	0.0210					
$\dfrac{H_S}{R_S}$	0.20	0.30	0.40	0.50	0.60	0.80

$\dfrac{Y}{H_S}$	$\dfrac{X}{H_S}$	For portion of the profile below the weir crest				
-0.000	0.454	0.392	0.325	0.253	0.189	0.116
-0.20	0.499	0.437	0.369	0.292	0.189	0.149
-0.40	0.540	0.478	0.407	0.328	0.259	0.174
-0.060	0.579	0.516	0.443	0.358	0.286	0.195
-0.080	0.615	0.550	0.476	0.386	0.310	0.213
-0.100	0.650	0.584	0.506	0.412	0.331	0.228
-0.200	0.795	0.729	0.639	0.516	0.413	0.293
-0.300	0.922	0.843	0.741	0.594	0.474	0.342
-0.400	1.029	0.947	0.828	0.656	0.523	0.381
-0.500	1.128	1.040	0.902	0.710	0.567	0.413
-1.000	1.525	1.420	1.164	0.878	0.696	0.498
-2.000	2.104	1.879	1.372	1.013	0.801	0.560
-3.000	2.550	2.105	1.457	1.077	0.840	
-4.000	2.904	2.180	1.487			
-5.000	3.169	2.207				
-6.000	3.396					

After Wagner (USBR)

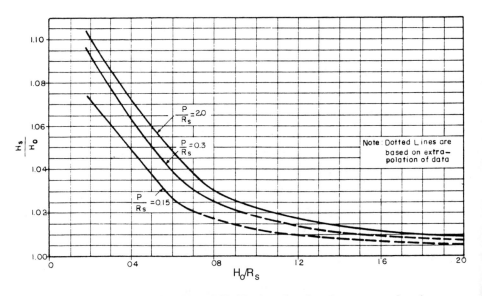

Figure 5.27. Relationship of H_O/R_S for circular sharp crested weirs (USBR).*

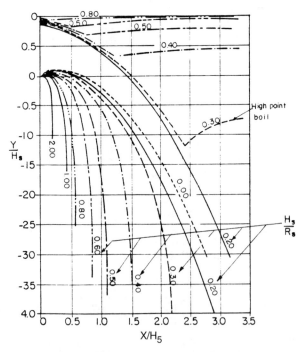

Figure 5.28. Upper and lower nappe profiles for circular weir (aerated nappe and negligible approach velocity) (USBR).*

---

* Design of Small Dams

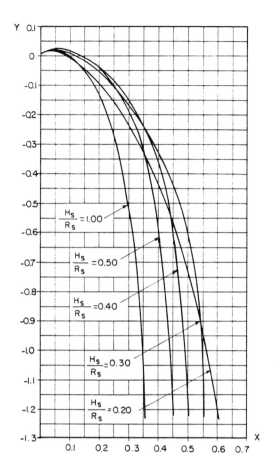

Figure 5.29. Comparison of lower nappe shapes for a circular weir for different heads. (USBR). *

In Fig. 5.29 the lower nappe profiles are shown as a function of (H_s / R_s) at different X distances from the axis of the shaft. For $(H_s / R_s) > 0.20 \sim 0.30$, a sudden increase of the lower nappe can be observed. For example, if the spillway is designed for $(H_s / R_s) = 0.20$ when $H_s / R_s > 0.20$, the nappe profiles may intersect each other and submergence occurs. For example, if $(H_s / R_s) = 0.30$, $Y = -0.60$ corresponds to a value of X = 0.46 m, but the abscissa of the lower nappe for $(H_s / R_s) = 0.20$ will be X = 0.50. Then the lower nappe is projected towards the spillway axis by a value equal to 0.50 − 0.46, causing an extra subpressure to be eliminated. To avoid this effect, R_s must be increased.

---

*   Design of Small Dams

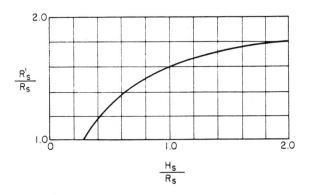

**Figure 5.30. Increased circular crest radius needed to minimize
subatmospheric pressure along the crest (USBR).***

Example 5.05

It is required to determine the maximum discharge capacity of the shaft
spillway shown in Fig. 5.31.

Data

- Crest elevation of the morning glory   100
- R_s 2.50 m
- P   10.00 m
- Diameter of the circular conduit   3.50 m
- Length of shaft   30.00 m
- Length of tunnel   150.00 m
- Diameter of the shaft   3.50 m
- n   0.017
- α 1.00
- Exit depth of flow (3.50/2=)   1.75 m
- Wetted area just downstream of the exit:

$$A_{exit} = \frac{A_{cond}}{2}$$

or

$$A_1 = \frac{A}{2}$$

---

* Design of Small Dams

256

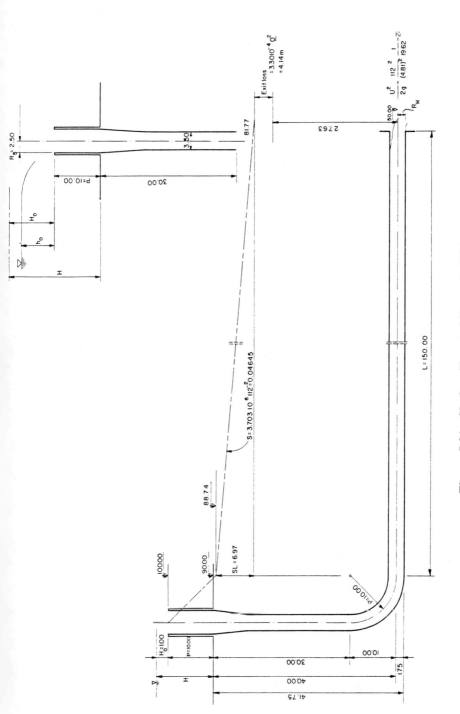

Figure 5.31. Shaft spillway of Example 5.05.

F. Sentürk

It is assumed that the circular conduit flows full but the flow is not supported and the exit depth is equal to the half of the diameter. Only the entrance, the end and the exit losses will be considered as local losses.

Solution

$$\frac{P}{R_s} = \frac{10}{2.5} = 4.00$$

The flow is controlled by the entrance if the spillway discharge is similar to a free flow spillway. In this case

$$\frac{H_o}{R_s} < 0.45$$

after Wagner. For

$$\frac{H_o}{R_s} > 0.45$$

orifice flow occurs. It is also possible for the control section to be located at the exit. These three different cases will examined separately. The smallest of the discharge corresponding to different flow pattern will be the capacity of the shaft spillway.

1. Flow equation

   Equation 5.39 governs the flow

$$H + z = \alpha \left(\frac{U_1^2}{2g}\right) + \Delta H + R_H \qquad 5.39$$

where

H   is the hydraulic head of the shaft spillway

U_1 is the velocity at the exit

ΔH is the sum of the local head losses

$R_H = \frac{D}{2} = \frac{3.50}{2} = 1.75$ m

D   = 3.50 m, is the tunnel diameter, and

α = 1.00

The exit velocity is

$$U_1 = \frac{Q}{A_1} \qquad A_1 = \frac{A}{2} = \frac{\pi D^2}{4} \qquad A = \frac{\pi\, 3.5^2}{4} \qquad A\frac{1}{2} = \frac{9.62}{2}\, m^2$$

where A is the area of the full conduit, and A_1 is the area of the conduit at the exit.

Assuming the system is flowing full, the friction loss can be computed using Manning's formula

$$SL = \frac{Ln^2}{R^{4/3}} \frac{Q^2}{A^2}$$

$$L = P + 30 + \frac{\pi\, D_{bend}}{4} + 150$$

where (Fig. 5.31)

$$P = 10.00, \text{ and } D_{bend} = 20.00\ m$$

Substituting in the above equation yields

$$L = 10.00 + 30.00 + \frac{\pi \cdot 20.00}{4} + 150 = 205.71\ m$$

and

$$S = \frac{0.017^2 Q^2}{R^{4/3} 9.62^2}$$

where

$$R = \frac{\pi D^2}{4\pi D} = \frac{D}{4} = \frac{3.50}{4} = 0.88\ m$$

then

$$S = \frac{0.017^2 Q^2}{0.88^{4/3} \cdot 9.62^2} = 3.703\ 10^{-6} Q^2$$

• Local Losses

    a. Friction loss

$$\Delta H_1 = SL = 3.703\ 10^{-6} Q^2 \cdot 205.71 = 7.62\ 10^{-4} Q^2$$

b. Head loss at the entrance

$$\Delta H_2 = 0.05 \frac{Q^2}{A^2 2g} \qquad 5.41$$

where

$$A = \pi R_s^2 = \pi 2.5^2 = 19.63 \text{ m}^2$$

then

$$\Delta H_2 = 0.05 \frac{Q^2}{19.63^2 \cdot 19.62} = 6.61 \; 10^{-6}Q^2$$

c. Head loss at the bend

$$\Delta h_3 = (0.13 + 1.8 \frac{r^{7/2}}{P}) \frac{Q^2}{2gA^2} \qquad 5.42$$

where $r = 1.75$ m $P = 10.00$ m $A = 9.62$ m^2

$$\Delta h_3 = (0.13 + 1.8 \frac{1.75^{7/2}}{10}) \frac{Q^2}{19.62 \cdot 9.62^2}$$

$$\Delta h_3 = 7.74 \; 10^{-4}Q^2$$

d. Exit loss

$$\Delta h_4 = 0.2 \; [1 - (\frac{A_1}{A})^2] \frac{Q^2}{2gA_1^2} \qquad 5.43$$

$$\Delta h_4 = 0.2 \; [1 - (\frac{A}{2A})^2] \frac{Q^2}{2gA_1^2}$$

$$A_1 = \frac{A}{2} = \frac{9.62}{2} = 4.81 \text{m}^2$$

$$\Delta h_4 = 0.2 \; (1 - 0.5^2) \frac{Q^2}{19.62 \cdot 4.81^2} = 3.30 \; 10^{-4}Q^2$$

e. Total loss

$$\Sigma \Delta H = (7.62 \; 10^{-4} + 6.61 \; 10^{-6} + 7.74 \; 10^{-4} + 3.3 \; 10^{-4}) \; Q^2$$
$$= 1.87 \; 10^{-3}Q^2$$

Substituting these values in Eq. 5.39 yields

$$H + z = \alpha \frac{U_1^2}{2g} + 1.87 \; 10^{-3}Q^2 + R_H \qquad 5.39$$

where

$$U_1 = \frac{Q}{A_1} = \frac{Q}{4.81}$$

Assuming that the datum line passes through the center of the exit section, Eq. 5.39 takes the following form

$$H + z = \alpha \frac{Q^2}{4.81^2} \frac{1}{2g} + 1.87 \; 10^{-3}Q^2 = (2.2 \; 10^{-3}\alpha + 1.87 \; 10^{-3}) \, Q^2$$

with $R_H = 0$

2. Study of different flow conditions

   a. Control at the exit

      In this case the flow equation is:

$$H = (2.20 \; 10^{-3}\alpha + 1.87 \; 10^{-3})Q^2 - 40.00$$

The discharge Q is computed accordingly

$$Q = \sqrt{\frac{H + 40.00}{2.2 \; 10^{-3}\alpha + 1.87 \; 10^{-3}}}$$

$$H = 10 + H_o \; (*)$$

According to Wagner $H_o/R_s \le 0.45$ is the condition for which free flow takes place in the entrance. In case that the crest profile is designed for

$$\frac{H_d}{R_s} \cong 1.25 \sim 1.30 \frac{H_s}{R_s}$$

or considering that y_s is negligible for the first approximation

$$\frac{H_d}{R_s} \cong 1.25 \sim 1.30 \frac{H_o}{R_s}$$

The subpressure can occur when $H > H_d$. This property (Fig. 5.29) must be taken into consideration while designing the spillway.

---

* y_s is assumed negligible for a first approximation. See Fig. 5.22 for the definition of symbols.

Assume for a first approximation

$$\frac{H_o}{R_s} = 0.40$$

then $\qquad\qquad H_o = 0.4 \cdot 2.50 = 1.00$ m

and $\qquad\qquad H = 10.00 + 1.00 = 11.00$ m

$$Q = \sqrt{\frac{11 + 40.00}{2.2 \ 10^{-3}\alpha + 1.87 \ 10^{-3}}}$$

The values of Q corresponding to different values of (α) are listed below

α	$Q(m^3/sec)$
1.00	112
1.10	109
1.20	106
1.40	102

$$\alpha = 1, \text{ then } Q \cong 112.00 \text{ m}^3/\text{sec}$$

Figure 5.31 shows the characteristics of the flow taking place in the shaft when crest control prevails. The flow equation is

$$H = (2.20 \ 10^{-3} + 1.87 \ 10^{-3})Q^2 - 40.00 = 4.07 \ 10^{-3}Q^2 - 40.00$$

b. Control at the entrance

$$\frac{P}{R_s} = \frac{10}{2.5} = 4.00 \qquad\qquad \frac{H_o}{R_s} = \frac{1.00}{2.5} = 0.40$$

then Fig. 5.23 yields

$$C_o \cong 3.50$$

The discharge of the glory hole is given by

$$Q = C_o H_o^{3/2}$$

$$L = \pi D = \pi 5.00 = 15.71 \text{ m}$$

$$H_o = 1.00m$$

$$Q = 3.50 \bullet 15.71 \bullet 1^{3/2} = 54.99 < 112 \ m^3/sec$$

c. Orifice control

$$Q = CAH^{1/2}$$

$$C = 0.68 \bullet 4.43 = 3.01$$

$$A = \frac{\pi D^2}{4} = \frac{\pi 5^2}{4} = 19.63$$

$$Q = 3.01 \bullet 19.63 \bullet 1^{1/2} = 59.09 \ m^3/sec$$

$$\frac{H_o}{R_s} = \frac{1}{2.5} = 0.4 < 0.45$$

then the orifice flow is avoided.

Conclusions:

The maximum discharge of the system is $Q = 112 \ m^3/sec$ and the control is at the exit. If free flow could have been developed at the entrance with a discharge Q, of less than $112 \ m^3/sec$, then the entrance would have been the control section; the discharge of the glory hole corresponding to $H_o = 1.00$, being $Q = 55 \ m^3/sec$, the Q_{max} to be expected is then $55 \ m^3/sec$.

The variation in the flow pattern is shown in Fig. 5.32 (see Problem 5.08 for design procedure).

The curve OAB represents the control at the entrance.

The curve AC represents the orifice flow, and

The curve BD represents the control at the exit.

For $H_o/R_s < 0.45$ the orifice flow is avoided, then for $H_o < 0.45 \bullet 2.50 = 1.13$ orifice flow cannot take place. Point A is situated at the intersection of the curve OAB and the curve AC corresponds to this case; then for $H_o < 1.13$ the portion AA' of the curve AC does not represent the flow.

For $H_o < 1.13$, free flow takes place on the spillway, the control is at the entrance $Q < 61.60 \ m^3/sec$

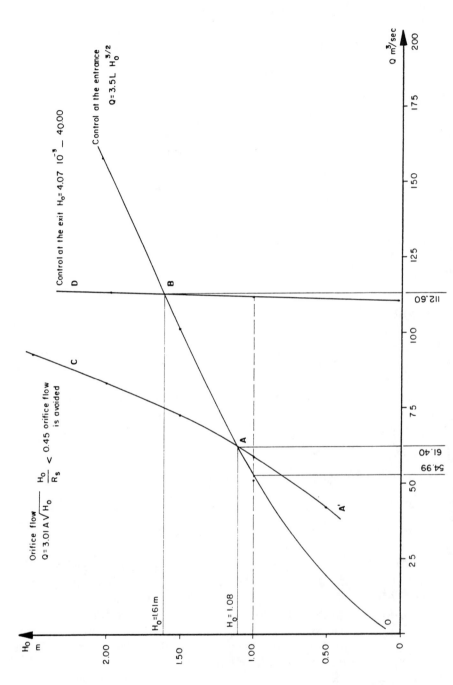

Figure 5.32. Flow pattern in the shaft of the spillway described in Example 5.05.

For $1.13 < H_o < 1.61$ orifice flow prevails $61.40 < Q < 112.60$ m³/sec

For $H_o > 1.61$, the control is at the exit $Q > 112.60$ m³/sec.

5.5 HYDRAULICS OF SIPHON SPILLWAYS

A siphon spillway is a tube with a relatively high discharge capacity and a high velocity of flow (see Fig. 5.33). It can be divided into 7 different parts:

1. Seal at inlet: lip of upper leg, submerged an arbitrary distance but should beset deeply enough to stop at the desired drawdown,

2. Upper leg: a transition providing a gradually contracting inlet area,

3. Siphon breaker: an air vent set at the summit with inlet end set at/or slightly below normal water surface,

4. Throat: the top section of the siphon,

5. Upper bend: follows immediately after the throat and prepares the priming,

6. Lower bend: follows the upper bend and direct the priming,

7. Outlet end: set at the desired elevation (converging or diverging tubes).

The hydraulics of siphon spillways consists of computing the discharge for given spillway dimensions or determining the dimensions of the structure for a given discharge.

Siphon spillways were intensively constructed around the 1950's. The linear discharge being relatively high, their cost is relatively low. Another advantage is the ability to operate without moving parts or mechanical devices. Certain disadvantages prevented their general application:

1. When priming does not occur the discharge is restricted and there is the danger of overtopping.

2. After priming, heavy vibration occurs which may cause loosening of the joints. This could allow air entry and the spillway stops functioning or operates erratically.

265

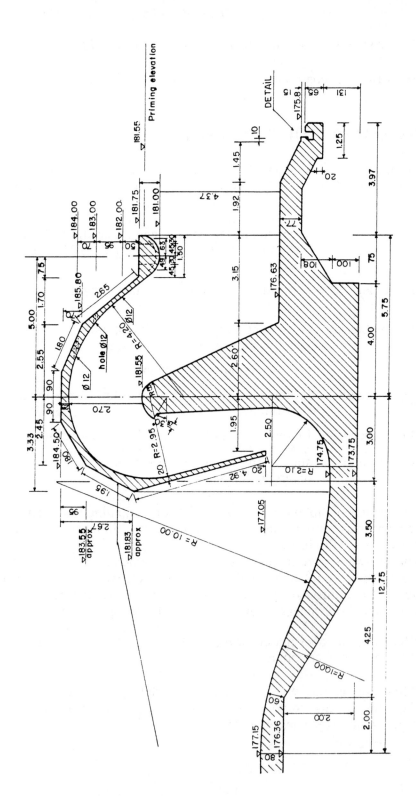

Figure 5.33. Cross section of an industrial siphon.

3. There is the possibility of blockage of the siphon by debris or ice. This can sometimes be prevented by using special devices.

Different computational approaches exist for solving the hydraulic problems relative to siphon spillways. A graphical solution developed by Şentürk, F., 1972 is given in the following section.

5.5.1 Hydraulic Computation for the Siphon Spillway

A. Graphical Solution

Figure 5.34 shows a graphical solution for determining the subatmospheric pressure head taking place in the conduit constituting the siphon when the geometric and hydraulic characteristics are given. The same

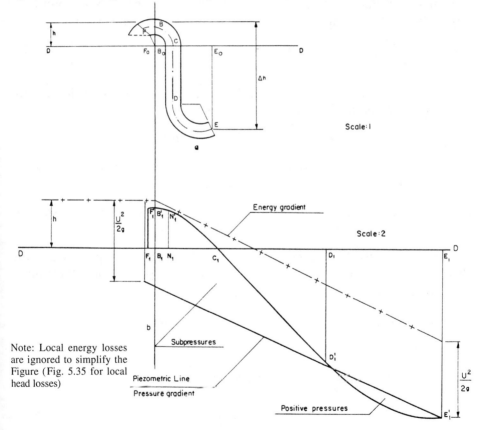

Figure 5.34. Graphical method for determining the subatmospheric pressure head in a siphon spillway.

267

method can be applied to determine the dimensions of the hydraulic structure for a given discharge and desired subatmospheric pressure (Şentürk, F., 1972).

The axis of the siphon is shown on the figure by the line FBCDE. Structural details are avoided. The entrance is short and water is directly introduced into the throat in such a way that frictional loss is avoided.

The flow around the throat is assumed irrotational, the siphon can be considered as a short structure in the same region, then the hydraulic head is assumed constant up to the centerline.

Let the horizontal straight line $F_1B_1C_1D_1E_1$ represent this line in such a way that

$$F_1B_1 = FB \quad B_1C_1 = BC \quad C_1D_1 = CD \quad \text{and} \quad D_1E_1 = DE$$

Furthermore assume that the horizontal straight line D-D becomes the base line beginning at point F_0 in Fig. 5.34a and respectively at point F_1 in Fig. 5.34b. The siphon axis can be reported on the base line as follows:

- Point F is represented by point F_1' on the perpendicular to the straight line from point F_1 such as $F_1F_1' = FF_0$.

- Point B is represented by point B_1' on the perpendicular to the straight line from point B_1 such as $B_1B_1' = BB_0$.

- Point C is represented by point C_1 on the straight line.

- Point D is represented by point D_1' on the perpendicular to the straight line from point D_1 such as $D_1D_1' = DC$.

- Point E is represented by point E_1' on the perpendicular to the straight line from point E_1 such as $E_1E_1' = EE_0$.

Then the curve $F_1'B_1'C_1D_1'E_1'$ represents the siphon in a plan where the distances to these points referred to the base line represent the elevation of the corresponding points on the siphon.

The problem can be solved step by step as follows:

1. Determine the local head losses

- Entrance and converging transition loss

$$\Delta h_1 = k_1 \frac{U^2}{2g} \tag{5.47}$$

$$k_1 = 0.2 \qquad k_1 = 0.1 \text{ for very well designed siphon and rounded corners}$$

- Bend losses

  For $r_c = 2.50D$

  $$\Delta h_2 = k_2 \frac{U^2}{2g} \qquad (5.48)$$

where

  $k_2 = 0.40 \cong 0.42$

  r_c = centerline radius, and

  D: the throat diameter

- Transition losses - diverging outlets

  $$\Delta h_3 = k_3 \left(\frac{U_1^2}{2g} - \frac{U_0^2}{2g} \right) \qquad (5.49)$$

where

  U_1 is the velocity at the entrance of the transition, and

  U_0 is the velocity at the exit from the transition

  $k_3 = 0.2$

- Transition losses - converging outlets

  $$\Delta h_4 = k_4 \left(\frac{U_0^2}{2g} - \frac{U_1^2}{2g} \right) \qquad (5.50)$$

  $$k_4 = 0.1$$

- Exit losses

  $$\Delta h_5 = \left(\frac{A_1}{A_0} \right)^2 \frac{U_1^2}{2g} \qquad (5.51)$$

where

  A_1 is the wetted area at the throat

  U_1 is the mean velocity at the throat

  A_0 is the wetted area at the outlet, and

  U_0 is the mean velocity at the outlet, (Fig. 5.35)

2. Determine the friction loss

In hypercritical flow the Manning coefficient of resistance n, is increased. For example, it is calculated equal to 0.03 in the downstream

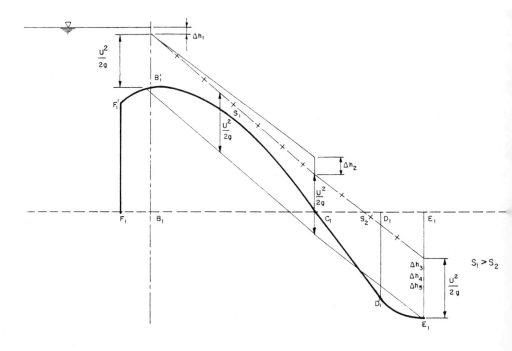

Figure 5.35. Graphical solution for a siphon.

channel of the Aigle Dam Spillway, France. For this reason and for a first approximation a conservative value of the friction loss is sometimes recommended:

$$\Delta h = 0.25 \frac{U_1^2}{2g} \qquad (5.52)$$

A better approximation can be made by computing the friction loss using Manning's equation adopting an n value greater than 0.017, comparing it with the value from Eq. 5.52 and adopting the greater of the two.

3. Compute the energy gradient and the mean velocity.

    Use Manning's equation. If n is not given, take $n = 0.017$

$$U = \frac{1}{n} (\frac{D}{4})^{2/3} S_{com}^{1/2}$$

where   n is the given or selected resistance coefficient
             D is the diameter of the circular cross section of the siphon,

and

S_{comp} is the computed energy gradient (Example 5.06)

4. Compute the discharge

$$Q = UA$$

5. Prepare the graph in Fig. 5.35.

   Plot the head losses on the diagram and draw the energy line to obtain the hydraulic heads. Draw a tangent to the curve $F_1'B_1'C_1$ etc., parallel to the energy line; the point of tangency is N_1' (Fig. 5.34). The maximum subatmospheric pressure head is obtained as equal to N_1N_1'.

6. Subatmospheric pressures occur where the representative line of the siphon axis is above the hydraulic pressure line.

7. The maximum subatmospheric pressure head occurs at or around point B_1'; it is generally given in engineering practice by the equation

$$NN_1' = 0.7\,h_{at} \tag{5.53}$$

This method is illustrated in Example 5.06 and in Problem 5.09, and in Problem 13.03.

B. Analytical Solution

Let r be the radius of the upper bend of the siphon spillway and U, the mean velocity at the throat. The flow in this reach of the siphon can be assumed irrotational then

$$Ur = \text{Constant} \tag{5.54}$$

Assume r_x, the radius of the stream line in the upper bend, at a point X and r_c, the radius of the crest of the throat. Equation 5.54 can be written as

$$U_x r_x = U_c r_c$$

$$U_x = \frac{U_c r_c}{r_x} \tag{5.55}$$

F. Sentürk

The discharge of a layer of thickness dr_x and of unit length is

$$dq = (\frac{U_c r_c}{r_x}) \, dr_x \qquad (5.56)$$

and

$$q = U_c r_c \int_{r_c}^{r_o} \frac{1}{r_x} \, dr_x = U_c r_c \, Ln\,(\frac{r_o}{r_c}) \qquad (5.57)$$

where

$$U_c = (2g\,h_{uc})^{1/2}$$

Substituting this value in Eq. 5.57 yields

$$Q = 4.43 \, b \, (h_{uc})^{1/2} \, r_c \, Ln(\frac{r_o}{r_c}) \qquad (5.58)$$

where   b is the width of the siphon

h_{uc} is the pressure at the throat. In general, it is described by Eq. 5.53, then

$$Q = 4.43 \, b \, (0.7h_{at})^{1/2} r_c \, Ln(\frac{r_o}{r_c}) \qquad (5.59)$$

h_{uc}, is represented by $NN_1^!$ in Fig. 5.34 and Fig. 5.35. Figure 5.33 illustrates a siphon spillway as designed for practical construction.

USBR recommends the following values of H_{at} as a function of the elevation above the sea level (Table 5.12).

TABLE 5.12

Allowable subatmospheric pressures for conduit flowing full

Elevation above the sea level		Allowable subatmospheric pressure	
ft	m	m	ft
0	0	6.71	22
2000	610	6.10	20
4000	1220	5.49	18
6000	1829	4.88	16
8000	2438	4.27	14

Example 5.06

1. It is required to determine the subatmospheric pressure head in the throat of the prefabricated concrete siphon shown in Fig. 5.36 and its discharge capacity.

2. What will be the discharge of this siphon if the subatmospheric pressure head at point C is -6.00 m.

Data

- The cross section of the siphon is square with B = 1.00 m
- n = 0.014
- Assume: $k_1 = 0.1$ $k_2 = 0.40$ $k_3 = k_4 = 0$
- Assume $A_1 = A_o$
- See Fig. 5.36 for geometrical characteristics.

Solution

The graphical method is used for solving this problem.

1. Determine the local head losses.

- Head loss at the entrance $\Delta h_1 = 0.1 \dfrac{U^2}{2g}$

- Bend loss $\Delta h_2 = 0.4 \dfrac{U^2}{2g}$

- Transition loss $\Delta h_3 = \Delta h_4 = 0$

- Exit loss $\Delta h_5 = (\dfrac{A_1}{A_o})^2 \dfrac{U^2}{2g} = \dfrac{U^2}{2g}$

Total head loss

$$\Sigma \Delta h = 1.50 \frac{U^2}{2g} = 0.0765 \, U^2$$

where, U is the mean velocity.

2. Compute the energy line slope

$$SL + 0.0765U^2 = \text{Total drop}$$
$$L = 3.93 + 4.00 + 3.14 = 11.07 \text{ m}$$

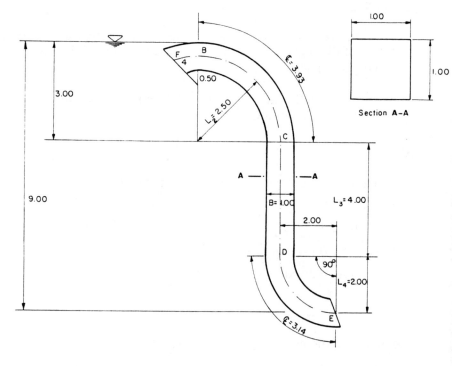

$$L = 3.93 + 4.00 + 3.14 = 11.07 \text{ m.}$$

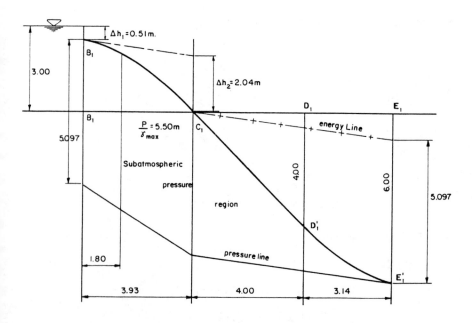

Figure 5.36. Geometric characteristics of the siphon in Example 5.06.

Total drop $= 3.00 + 4.00 + 2.00 = 9.00$ m

then $SL + 0.0675U^2 = 9.00$ $S = \dfrac{9.00 - 0.067U^2}{L}$

3. Compute the mean velocity using Manning formula

$$U = \frac{1}{n} R^{2/3} S^{1/2} \qquad R = \frac{A}{P} = \frac{1}{4}$$

$$U = \frac{1}{0.014} (0.25)^{2/3} \left(\frac{9.00 - 0.075U^2}{11.07}\right)^{1/2}$$

$$U = 9.985 \text{ m/sec}$$

4. Compute the discharge

$$Q = UA = 9.985 \bullet 1.00 = 9.985 \cong 10.00 \text{ m}^3/\text{sec}$$

5. Compute the energy line slope

$$S = \frac{9.00 - 0.0765 \bullet 9.985^2}{11.07} = 0.124$$

6. Compute the velocity head

$$\frac{U^2}{2g} = \frac{9.985^2}{19.62} \cong 5.10 \text{ m}$$

7. Compute the local head losses

$$\Delta h_1 = 0.1 \bullet 5.10 = 0.51 \text{ m}$$

$$\Delta h_2 = 0.4 \bullet 5.10 = 2.04 \text{ m}$$

$$\Delta h_3 = 5.10 \text{ m}$$

$$\Sigma \Delta h = 7.65 \text{ m}$$

$\Sigma \Delta h + SL$, must be equal to the total drop

$$7.65 + 0.124 \bullet 11.07 = 9.02 \cong 9.00 \text{ m}$$

8. Checking Q with Eq. 5.59 yields

$$Q = 4.43b(0.7 \, h_{at})^{1/2} \, b_c \, Ln\frac{r_o}{r_c}$$

$$r_o = 1.50 \text{ m} \qquad b = 1.00 \text{ m} \qquad r_c = 0.50 \qquad h_{at} = 6.00 \text{ m}$$

$$b_c = 1.00 \text{ m}$$

and

$$Q = 9.974 \cong 10 \text{ m}^3/\text{sec}$$

h_{at} will be checked with the graphical solution.

9. Compute the discharge corresponding to a subatmospheric pressure at point C equal to (-6.00 m).

Assume (x) the total drop, then the pressure at point C is (Fig. 5.36)

$$(x-3) - (\frac{x-0.0765U^2}{11.07})^{1/2} \, 7.14 = 6.00 \text{ m} \qquad\qquad (5.60)$$

and

$$U = \frac{1}{0.014} (0.25)^{2/3} (\frac{x-0.0765U^2}{11.07})^{1/2} = 28.3464 (\frac{x-0.076U^2}{11.07})^{1/2}$$

then

$$x = 0.0765U^2 + \frac{11.07U^2}{28.3464^2}$$

Substituting this value of x in Eq. 5.60 yields

$$0.0765U^2 + \frac{11.07 \, U^2}{28.3464^2} - 3.0 - (\frac{(0.0765U^2 + \frac{11.07 \, U^2}{28.3464^2}) - 0.0765 \, U^2}{11.07})^{1/2} \, 7.14 = 6.00$$

$$U^2 = 131.7 \qquad U = 11.476 \text{ m/sec} \qquad x = 11.889 \text{ m}$$

The discharge is

$$Q = 11.476 \bullet 1 \bullet 1 = 11.476 \text{ m}^3/\text{sec}$$

5.6   PROBLEMS

Problem 5.01

Design a broad crested weir with characteristics given.

Data

- $q = 10.00 \text{ m}^3/\text{sec.m}$
- $d_1 \geq 4.50 \text{ m}$
- $g = 9.81 \text{ m/sec}^2$
- Hint: See Example 5.01.

Problem 5.02

Design a side spillway with characteristics given. Use graphical solution.

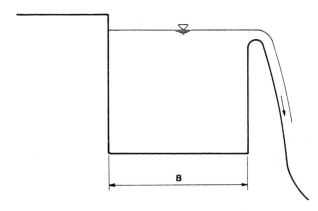

Fig. 1. Problem 5.02. Cross section of the canal.

Data

- The discharge in the canal is $Q = 20.00 \text{ m}^3/\text{sec}$
- The cross section of the canal: rectangular
- $B = 1.00 \text{ m}$
- The flow in the canal is uniform and subcritical
- $n = 0.024$
- $Q_s = 5.00 \text{ m3/sec}$
- $m = 0.40$
- $S = 0.0009$

Problem 5.03

Determine the water surface line over the side spillway, the side spillway discharge and the discharge of the canal upstream of the side spillway described in Problem 5.02 in case that the discharge of the canal downstream of the side spillway is 25 m^3/sec.

Data

The discharge in the canal is subcritical

- The cross section of the canal is shown in Fig. 1, Problem 5.02

- n = 0.024

- S = 0.0009

Problem 5.04

Verify Eq. 5.17.

Problem 5.05

Design a side spillway where the flow is subcritical at the entrance and supercritical at the downstream exit. The canal on which it is constructed is rectangular carrying a discharge of 49.52 m^3/sec upstream of the side spillway. The width of the canal is 4.00 m. The discharge of the spillway is 14.52 m^3/sec. It is required to determine the crest length L, and the crest elevation of the spillway, assuming 0, the bottom elevation of the canal and that the flow is at critical stage at the entrance section.

Data

C = 1.00 m

d = 1.25 m at the downstream end of the spillway

n = 0.014

m = 0.35

The backwater curves are ignored

Problem 5.06

A horizontal rectangular side-spillway channel has a free overflow outlet. The inflow is uniformly distributed along the channel with a rate of q per unit length of the channel. Derive the equation of the flow profile, by ignoring friction loss.

Problem 5.07

A flow takes place over a lateral spillway. The crest length of this spillway is L = 22.00 m and q = 1.50 m³/sec.m. The cross section of the downstream channel is rectangular with a bottom width equal to 5.00 m. The bottom slope of the channel is $S_b = 0.10$ and the bottom elevation at the upstream end is 50. Furthermore, n = 0.015 and $\alpha = 2.00$.

It is required to define the flow profile in the downstream channel and its flow characteristics.

Problem 5.08

Design a shaft spillway, so that for $H_o = 0.60$ m, the control section is at the entrance.

Data

$Q = 28.00$ m³/sec

Crest elevation of the morning glory   100

P   5.00 m

Only entrance, bent and exit losses should be considered as local losses.

Problem 5.09

- Determine the exit elevation and Q_{max} of the siphon described in Example 5.06 under the following conditions:

  1 - The priming occurs when h = 3.50 m

  2 - n = 0.017

  3 - $k_1 = 0.1$ $k_2 = 0.40$ $k_3 = k_4 = 0$

- Determine the maximum subpressure when the siphon discharges Q_{max}.

(Fig. 1, Problem 5.09 for geometric characteristics).

Problem 5.10

Determine the discharge of the siphon spillway described in Problem 5.09 by applying graphical and analytical methods and discuss the results.

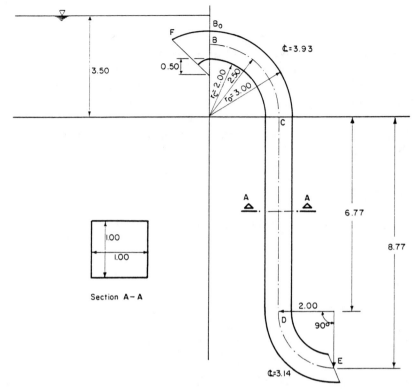

Fig. 1. Problem 5.09.

Data

- cross section of the siphon: square = 1 • 1 m^2
- The siphon is located at 10 meters above the sea level
- n = 0.017.

Problem 5.11

Derive Eq. 5.14 starting from the general equation for gradually varied flow:

$$\frac{dy}{dx} = \frac{S_b - S_E - \alpha\dfrac{qQ}{gA^2}}{1 - \alpha\dfrac{Q^2}{gA^2 D}}$$

Assume:

$$A = by \qquad \alpha = 2 \qquad S_b = S_E \qquad D = \frac{A}{T}\, y$$

280

Chapter 6

HYDRAULICS OF THE DOWNSTREAM CHANNEL

6.1 INTRODUCTION

Reservoir discharge flows over the spillway, downstream to the tailrace channel or basin and then passes through a terminal structure into the main water course. Various problems must be solved to prevent damaging effects of the high velocity flow in its path to the main river such as:

- Erosion of the downstream face of the spillway due to cavitation

- Increase of the depth of flow due to air entrainment

- Superelevation of the water surface due to cross waves

- Erosion of the downstream river bed caused by the residual energy of the flow downstream of the stilling basin or flip buckets.

These problems have been investigated since the middle of the 20th century and some are still under investigation. Solutions have been suggested in Europe and in the United States. Even today, however, it is difficult to state that satisfactory final solutions are attained.

Flood, sometimes increased by the accidental opening of the spillway gates, can cause heavy damage to the natural river bed. Therefore, a complete hydraulics study of the downstream channel must also include the research of the river bed degradation up to a selected section, which is sometimes located well downstream of the dam.

A typical example of erosion of downstream river bed occurred downstream of the Aswan Dam, Egypt. Two more examples are cited here to emphasize this phenomenon. The retaining walls along the Seyhan River downstream of the Seyhan Dam in Adana, a city in southern Turkey, collapsed two months after the release of water from the Seyhan Dam; and the local erosion immediately downstream of the Elmali Dam, Istanbul, Turkey, was so heavy that erosion began to occur under the foundations of the dam causing immediate hazard.

It is, therefore, understandable that river stability must be included in the study of hydraulics of the channel downstream of the spillway. A complete chapter of this book is dedicated to the study of these flow problems in the natural river bed.

6.2 HYDRAULIC PROBLEMS DOWNSTREAM OF SPILLWAYS

Sometimes, particularly in fill-type dams the discharge channel is designed in such a way that it conveys water far downstream from the dam. The channel slope is usually small, and the transition is gradual from the discharge channel to the natural watercourse.

In other situations, where the spillway is a low height weir, the downstream water level may be influenced by the upstream water level. Submerged flow conditions may take place and creates a downstream hydraulic jump. These two particular hydraulic phenomena are studied hereafter.

6.2.1 Submerged Flow

Figure 6.1 shows submerged flow taking place at a weir. The downstream water level is higher than the crest of the weir; accordingly, the upstream water level is super elevated. This is the simplest case of the downstream hydraulics of spillways. The discharge decreases as a function of the degree of submergence. Figure 6.2 shows a diagram for computing the discharge of submerged spillways (U.S.B.R.)

The diagram defines the flow as a function of

$$\frac{h_d + d}{H_E} \qquad \text{(abscissa), and}$$

$$\frac{h_d}{H_E} \qquad \text{(ordinate)}$$

d is the downstream flow depth

h_d is the water surface drop between upstream and downstream water levels, and

H_e is the energy head.

Four distinct regions are shown in Fig. 6.2:

1. Supercritical flow region

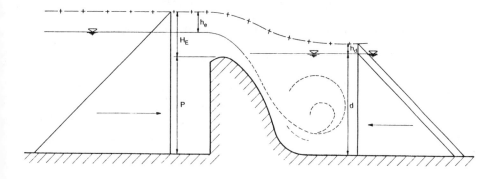

Figure 6.1. Submerged flow taking place at a weir.

h_d+d/H_E

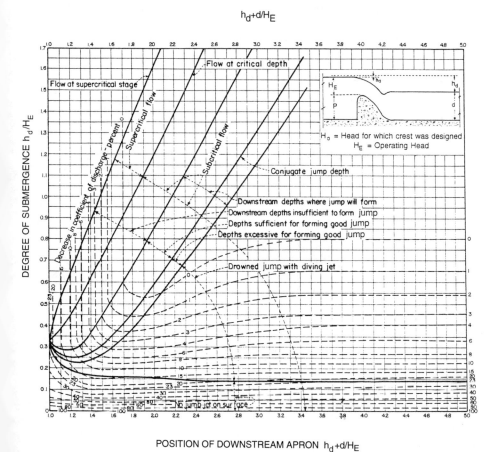

POSITION OF DOWNSTREAM APRON h_d+d/H_E

Figure 6.2.   Effect of downstream influences on flow over the weir crest (USBR; Design of Small Dams).

2. Subcritical flow, involving the hydraulic jump

3. Flow accompanied by a drowned jump with diving jet

4. Flow approaching complete submergence.

The decreases in the coefficient of discharge in percent are shown by dotted lines on the figure. The percentage is small in the first and second regions. It ranges from 0 to 20%. In the region where no jump exists, it ranges from 20% to 100%. An example is given to illustrate this change in the coefficient m.

Example 6.01

Considering the flow taking place at the weir in Fig. 6.1 assume that

$$d = 6.00 \text{ m} \qquad P = 3.00 \text{ m} \qquad h_d = 1.00 \text{ m}$$

Determine the decrease in the coefficient of discharge.

Solution

It can be seen from Fig. 6.2 that

$$H_E = d + h_d - P = 6.00 + 1.00 - 3.00 = 4.00 \text{ m}$$

furthermore

$$\frac{h_d + d}{H_E} = \frac{1.00 + 6.00}{4.00} = 1.75$$

$$\frac{h_d}{H_E} = \frac{1}{4} = 0.25$$

The representative point corresponding to these values shows that

• the flow is in the region of the drowned jump with diving jet, and

• the decrease in the coefficient of discharge is 6%.

If   m = 0.49, as shown in Chapter 3,

then   m = 0.49-0.06 • 0.49 = 0.46.

6.2.2 Hydraulics of Downstream Channel

The flow taking place in the downstream channel is hypercritical for the spillways of large dams and supercritical for weirs. In Fig. 6.3, the flow for partial opening of the gates which takes place downstream of the gated spillway of the Seyhan Dam, Turkey, is shown. The standing cross-waves and the air entrainment are apparent. It is difficult to formulate an expression for this type of flow. Different simplifications will be used for writing the flow equation.

Figure 6.3. Flow taking place in the downstream channel of the Seyhan Dam Spillway, Turkey.

First, assume that the flow is supercritical and gradually varied. This assumption gives a solution which may not represent the flow exactly. Second, the increase in the flow depth due to air entrainment will be added to the previously obtained result. The end solution requires serious consideration. It will give the lead designer the first approximation for the design of the downstream channel below the spillway.

Figure 6.4 shows the flow taking place in a steep sloped channel. Let z_1 at section 1-1 be measured from the datum line which passes through the lowest point of section 2-2, d_1 the flow depth, and $(U_1^2/2g)$ the velocity head at section 1-1. The friction head loss between the two sections is shown by Δh_1, the local head losses are not taken into consideration.

Assume the flow is gradually varied. Applying the energy equation to the flow taking place between the two sections yields

$$z_1 + h_1 + \frac{\alpha_1 U_1^2}{2g} = h_2 + \frac{\alpha_2 U_2^2}{2g} + \Delta h_1 \tag{6.1}$$

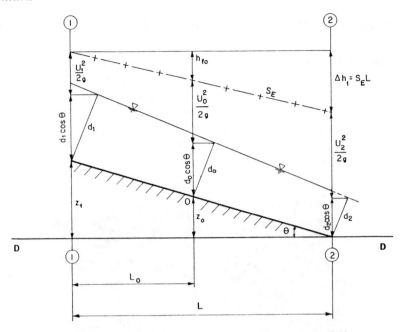

Figure 6.4. Flow in the steep-sloped channel.

where

$$z_1 = S_b L \qquad h_1 = d_1 \cos\theta \qquad h_2 = d_2 \cos\theta$$

$$U_1 = \frac{Q}{A_1} \qquad \Delta h_1 = S_E L \qquad U_2 = \frac{Q}{A_2}$$

and

S_b is the slope of the bottom of the channel

L is the distance separating the two cross sections

d_1 and d_2 are the depths of flow in sections 1-1 and 2-2

θ is the angle between the channel bottom and the horizontal

A_1 and A_2 are the wetted areas of sections 1-1 and 2-2

Q is the discharge of the channel, and

α_i is the energy coefficient (in this analysis α_i is assumed equal to unity).

Applying Manning's equation yields

$$S_E = \frac{U_m^2 n^2}{R^{4/3}} \tag{6.2}$$

where

S_E is the energy slope, and

$$U_m = \frac{U_1 + U_2}{2}$$

$$R_m = \frac{R_1 + R_2}{2}$$

R_1 and R_2 are hydraulic radii at sections 1-1 and 2-2, and

n is Manning's coefficient of resistance.

Substituting these values into Eq. 6.1 yields

$$S_b L + d_1 \cos\theta + \frac{Q^2}{A_1^2 \, 2g} = d_2 \cos\theta + \frac{Q^2}{A_2^2 \, 2g} + \frac{U_m^2 n^2}{R_m^{4/3}} L \qquad (6.3)$$

The left side of Eq. 6.3 is the available hydraulic head H, then

$$H = d_2 \cos\theta + \frac{Q^2}{A_2^2 \, 2g} + \frac{U_m^2 n^2}{R_m^{4/3}} L \qquad (6.4)$$

or

$$H = d_2 \cos\theta + \frac{Q^2}{A_2^2 \, 2g} + \frac{(\frac{U_1 + U_2}{2})^2}{(\frac{R_1 + R_2}{2})^{4/3}} n^2 L \qquad (6.5)$$

When the geometry of the channel is known, R = A/P can be determined (P is the wetted perimeter) and θ is from the problem data; n is a coefficient which must be chosen by the designer. Substituting these values into Eq. 6.5, d_2 can be computed easily. Example 6.02 gives an application of this procedure.

Maître, R. and Obolensky, S. (1961) have investigated head losses on the downstream channel of the Saint Etienne Cantalès Dam Spillway and the values based on their observations are given in Table 6.1.

The value of the Manning's coefficient of resistance, n, computed from the measured values on the Saint Etienne Cantalès Dam Spillway is n = 0.033.

TABLE 6.1

Flow characteristics observed in the downstream channel of the Saint Etienne Cantalès Dam Spillway, France.

Q = 230 m³/sec; Reservoir water elevation: 514.50

Stations	F x=10.10 m	E x=18.36 m	D x=32.76 m	C x=40.36 m	B x=50.36 m	A x=58.26 m
Elevation	503.60	492.30	482.65	480.90	480.20	480.20
$h_o \cos\theta$	1.10	0.70	1.00	1.10	1.20	1.60
$p/\gamma 2$	-0.40	0.40	0.70	0.75	0	0
Elevation of piezometric line	504.30	493.40	484.35	482.75	481.40	481.80
Measured values of $\alpha U_m^2 / 2g$, $\alpha=1.0$	7.65	18.30	22.90	22.90	21.20	19.40
Elevation of the energy line	511.90	511.70	507.25	505.65	502.60	501.20
Head losses	2.55	2.80	7.25	8.85	11.90	13.30

This value corresponds to channels covered with aquatic plants. The spillway's bottom surface is perfectly smooth concrete and the corresponding resistance coefficient might be around 0.013. The difference between these two values is due to the characteristics of the very rapid flow, which will be discussed in further detail. The increase is due to extra vortices generated by local roughness elements of the concrete surface and the existence of the undulating boundary layer. This problem cannot be solved on a scale model.

The only way to reach a solution is by taking in-situ measurements on prototypes. The velocity of flow in the downstream channel is very rapid. The flow entrains air which increases the velocity and decreases the specific weight of water and air mixture. The increase in the velocity also causes an increase of air friction which decreases the velocity. An equilibrium is reached at a velocity assumed to be around 34 m/sec. Figure 6.5 shows the energy head and the variation of the velocity at the toe of a spillway as a function of the difference of water level between upstream and downstream sides of the spillway.

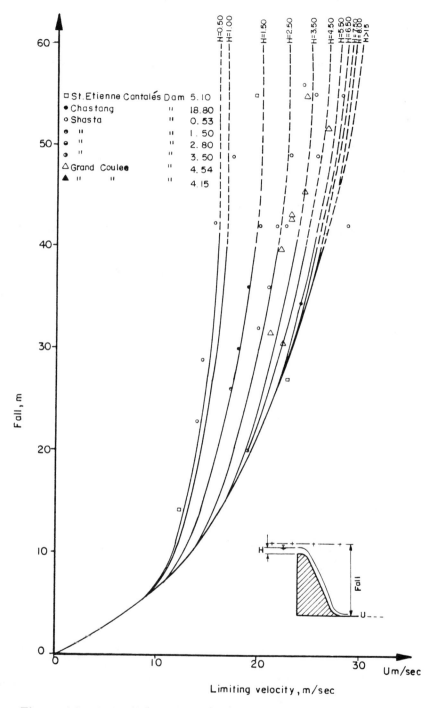

Figure 6.5. Curves for determination of velocity at the toe of spillway. USBR, "Hydraulic Design of Stilling Basins and Energy Dissipators"; 1958. European examples are added to the diagram.

Example 6.02

Hydraulic computation of downstream channel of the Kiralkizi Dam Spillway, Turkey.

Data

- Crest elevation of the spillway   802.25
- Water elevation in the reservoir   815.75
- Design flood maximum discharge   5,400 m³/sec
- The geometry of the downstream channel is given in Fig. 6.6 and in Table 6.2
- The cross section of the downstream channel is rectangular
- The flow depth in section 1-1 is 4.44 m
- $n = 0.014$

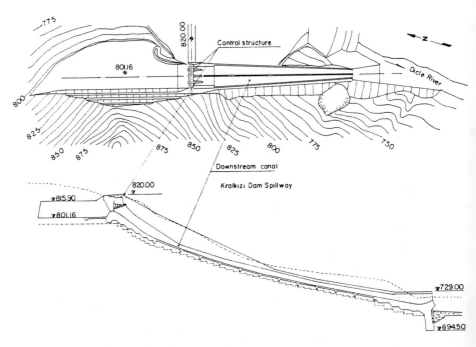

Figure 6.6. **Plan and longitudinal section of the Kiralkizi Dam Spillway, Turkey.**

TABLE 6.2

The Geometry of the Downstream Channel of the Kiralkizi Dam Spillway

Distances from spillway axis m	Elevation	Half width m	slope (*)
19.64	790.31	28.25	0.4800
40.00	780.94	28.25	0.4404
60.00	722.51	27.42	0.4345
75.00	766.67	26.63	0.3758
90.00	761.24	25.81	0.3494
100.00	757.83	25.32	0.3321
110.00	754.59	24.80	0.3151
125.00	750.05	24.01	0.2901
150.00	743.31	22.71	0.2494
168.36	738.26	21.75	0.2200
200.57	731.17	20.06	"
232.79	724.09	18.38	"
265.00	717.00	16.69	"

Solution

Figure 6.6 shows the plan and the longitudinal section of the Kiralkizi Dam Spillway. In this particular case the width and the slope of the channel vary continuously which causes a rather complicated flow pattern. The problem will be solved separately for each section where the new values can be substituted to suit the varying flow parameters, (Fig. 6.7).

Equation 6.5 can be applied to each section, separately:

$$H_n = d_{n+1} \cos\theta_{n+1} + \frac{Q^2}{A_{n+1}^2 \, 2g} + \frac{(\frac{U_n + U_{n+1}}{2})^2}{(\frac{R_n + R_{n+1}}{2})^{4/3}} n^2 L \qquad 6.5$$

Applying this equation to the Kiralkizi Dam Spillway yields

$$H_n = S_b L + d_n \cos\theta_n + \frac{U_n^2}{2g} \qquad \frac{U_n^2}{2g} = \frac{Q^2}{A_n^2 \, 2g} = \frac{Q^2}{(d_n \ell_n)^2 \, 2g}$$

---

* The values of the slopes are taken as the slope of the tangents to the circle along the invert at the given (x) distances.

291

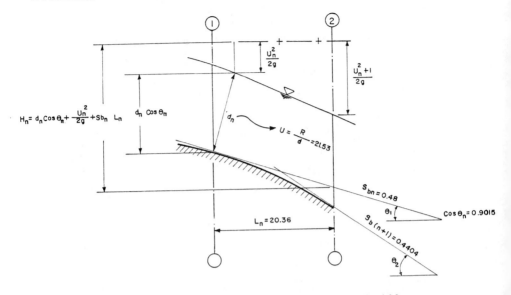

Figure 6.7. Definition of symbols for Example 6.02.

$$R_n = \frac{d_n \ell_n}{\ell_n + 2d_n}$$

$$R_{n+1} = \frac{d_{n+1}\,\ell_{n+1}}{\ell_{n+1} + 2d_{n+1}} \qquad \frac{U_{n+1}^2}{2g} = \frac{Q^2}{(d_{n+1}\,\ell_{n+1})^2\,2g}$$

where

ℓ_n is the width of cross section n-n

d_n is the flow depth in cross section n-n

θ_n is the angle of the tangent with the horizontal

ℓ_{n+1} is the width of cross section (n+1; n+1)

d_{n+1} is the flow depth in cross section (n+1; n+1)

θ_{n+1} is the angle of tangent with the horizontal

Q is the half of flood discharge, because the spillway is formed by two identical channels and only the flow in one of them is considered.

Then

$$H = S_b L + d_n \cos\theta + \frac{Q^2}{(d_n \, \ell_n)^2 \, 2g} = d_{n+1} \cos\theta_{n+1} + \frac{Q^2}{(\ell_{n+1} \, d_{n+1})^2 \, 2g} +$$

$$+ \frac{\dfrac{Q^2}{4}\left(\dfrac{1}{\ell_n d_n} + \dfrac{1}{\ell_{n+1} d_{n+1}}\right)^2}{\left[\dfrac{1}{2}\left(\dfrac{\ell_n d_n}{\ell_n + 2d_n} + \dfrac{\ell_{n+1} d_{n+1}}{\ell_{n+1} + 2d_{n+1}}\right)\right]^{4/3}} \; n^2 \, L \qquad (6.6)$$

Q, L, d_n, ℓ_{n+1} and ℓ_n are given. Then the left side of the equation is known and can be computed. The unknown in the equation is d_{n+1}. The computations are summarized in Table 6.3 and explained as follows:

Column 1:   Stations

Column 2:   Bottom slope at station (n) (see Fig. 6.7)

Column 3:   Bottom slope at station (n+1)

Column 4:   cosine of the angle θ_n corresponding to bottom slope S_{bn}

Column 5:   cosine of the angle θ_{n+1} corresponding to bottom slope $S_{b(n+1)}$

Column 6:   (L_i) horizontal distance between stations

Column 7:   (ΣL_i), cumulative distances to the axis of spillway

Column 8:   (d_n), flow depth at station (n)

Column 9:   (d_{n+1}), flow depth at station (n+1). Assume a value to (d_{n+1}) and refer to Column 18 and 19

Column 10: $d_n \cos\theta$ (column 4 • Column 8)

Column 11: $d_{n+1} \cos\theta_{n+1}$ (column 5 • Column 9)

Column 12:   (ℓ_n), width of the rectangular cross section (n)

Column 13:   (ℓ_{n+1}), width of the rectangular cross section (n+1)

Column 14: $U_n = (Q/A_n)$, A_n: Column 8 • Column 12 ($d_n \ell_n$)

Column 15: $U_{n+1} = (Q/A_{n+1})$, A_{n+1}: Column 9 • Column 13 ($d_{n+1} \ell_{n+1}$)

TABLE 6.3

Flow in the Downstream Channel of the Kiralkizi Dam Spillway

m = 0.014   Q = 2700 m³/sec

Station	S_{bn}	S_{bn+1}	$\cos\theta_n$	$\cos\theta_{n+1}$	L	$\sum_1^{13} L_i$	d_n	d_{n+1}	$d_n\cos\theta_n$	$d_{n+1}\cos\theta_{n+1}$	ℓ_n	ℓ_{n+1}	U_n	U_{n+1}	R_n	R_{n+1}	H_n	H_{n+1}	Bottom Elevation n+1	Bottom Elevation n	Energy Line Elevation E_n	Energy Line Elevation E_{n+1}
					m	m	m	m	m	m	m	m	m/sec	m/sec	m	m	m	m				
1	2	3	4	5	6	7	8	9	10	11	12	13	14	15	16	17	18	19	20	21	22	23
0-1	-	0.4800	-	0.9015	19.64	19.64	-	4.440	-	4.003	-	28.25	-	21.526	-	3.378	-	36.929	-	790.31	-	817.93
1-2	0.4800	0.4404	0.9015	0.9152	20.36	40.00	4.440	3.755	4.003	3.436	28.25	28.25	21.526	25.453	3.378	2.966	36.990	45.380	790.31	780.94	817.93	817.40
2-3	0.4404	0.4345	0.9152	0.9172	20.00	60.00	3.755	3.450	3.436	3.164	28.25	27.42	25.453	28.542	2.966	2.756	45.205	50.767	780.94	772.51	817.40	817.20
3-4	0.4345	0.3758	0.9172	0.9361	15.00	75.00	3.450	3.340	3.164	3.127	27.42	26.63	28.542	30.356	2.756	2.670	50.761	55.258	772.51	766.67	817.20	816.26
4-5	0.3758	0.3494	0.9361	0.9440	15.00	90.00	3.340	3.295	3.127	3.111	26.63	25.81	30.356	31.748	2.670	2.623	55.533	57.754	766.67	761.24	816.26	815.72
5-6	0.3494	0.3321	0.9440	0.9490	10.00	100.00	3.295	2.273	3.111	3.090	25.81	25.32	31.748	32.580	2.623	2.601	57.892	60.383	761.24	757.83	815.72	815.02
6-7	0.3321	0.3151	0.9490	0.9538	10.00	110.00	2.273	3.265	3.090	3.114	25.32	24.80	32.580	33.345	2.601	2.585	60.428	64.373	757.83	754.59	815.02	814.37
7-8	0.3151	0.2901	0.9538	0.9604	15.00	125.00	3.265	3.270	3.114	3.141	24.80	24.01	33.345	34.390	2.585	2.570	64.324	70.104	754.59	750.05	814.37	813.47
8-9	0.2901	0.2494	0.9604	0.9703	25.00	150.00	3.270	3.325	3.141	3.226	24.01	22.71	34.390	35.757	2.570	2.572	70.161	72.591	750.05	743.31	813.47	811.70
9-10	0.2494	0.2200	0.9703	0.9766	18.36	168.4	3.325	3.400	3.226	3.320	22.71	21.75	35.757	36.511	2.572	2.590	72.700	78.446	743.31	738.26	811.70	809.53
10-11	0.2200	0.2200	0.9766	0.9766	32.21	200.6	3.400	3.568	3.320	3.485	21.75	20.06	36.511	37.723	2.590	2.612	78.350	83.198	738.26	731.17	809.53	807.18
11-12	0.2200	0.2200	0.9766	0.9766	32.21	232.8	3.568	3.780	3.485	3.692	20.06	18.38	37.723	38.862	2.612	2.678	83.101	87.560	731.17	724.09	807.18	804.76
12-13	0.2200	0.2200	0.9766	0.9766	32.21	265.0	3.780	4.058	3.692	3.963	18.38	16.69	38.862	39.865	2.678	2.730	87.754		724.09	717.00	804.76	801.96

Column 16: $R_n = (A_n/P_n)$, $P_n = \ell_n + 2d_n$ (Column 8 + 2 • Column 12)

Column 17: $R_{n+1} = (A_{n+1}/P_{n+1})$, $P_{n+1} = \ell_{n+1} + 2 \cdot d_{n+1}$ (Column 9 + 2 • Column 13)

Column 18: $H_n = d_n \cos\theta_n + \dfrac{U_n^2}{2g} + S_{bm} \cdot L_n$ $\qquad\qquad S_{bm} = \dfrac{S_{bn} + S_{b(n+1)}}{2}$

$\qquad\qquad$ (Column 10 + $\dfrac{(\text{Column } 14)^2}{2g}$ + $\frac{1}{2}$(Column 2 + Column 3) • Column 6)

Column 19: $d_{n+1} \cos\theta_{n+1} + \dfrac{U_{n+1}^2}{2g} + \dfrac{(\frac{U_n + U_{n+1}}{2})^2}{(\frac{R_n + R_{n+1}}{2})^{4/3}} n^2 L_n$

$\qquad\qquad$ (Column 11 + $\dfrac{(\text{Column } 15)^2}{2g}$ + $\dfrac{\left[\frac{1}{2}(\text{Column } 14 + \text{Column } 15)\right]^2}{\left[\frac{1}{2}(\text{Column } 16 + \text{Column } 17)\right]^{4/3}}$

$\qquad$ • (n^2) • Column 6)

If $H_{n+1} \neq H_n$, change (d_{n+1}) in Column 9 in such a way that $H_{n+1} = H_n$ at the end of the trial and error.

Column 20: Bottom elevation, (station n)

Column 21: Bottom elevation, (station n+1)

Column 22: Energy line elevation E_n

$\qquad$ (Column 10) + $\dfrac{(\text{Column } 14)^2}{2g}$ + (Column 20)

Column 23: Energy line elevation E_{n+1}.

$\qquad$ (Column 11) + $\dfrac{(\text{Column } 15)^2}{2g}$ + (Column 21)

The energy line elevation can also be computed by adding the value at Column 21 to H_{n+1}. But in this case, the error involved by using S_{bm}, is directly reported to the energy line elevation.

Example:

$\qquad$ Station 7 $\qquad H_{n+1} = 60.383$

Bottom line elevation: 754.59

H_{n+1} + Bottom elevation = 60.383 + 754.59 = 814.973

This value is computed equal to 814.375 as shown in Column 23.
The result is summarized in Table 6.3 and shown in Fig. 6.8.

A. Buildup of the Boundary Layer; the Bauer's Procedure

In Example 6.02, the velocity of the flow at the exit of the downstream channel is computed as equal to 39.87 m/sec. It is known by field measurement that the maximum magnitude of the flow velocity at the exit is lower than this value due to air entrainment, the retardation due to air friction and the formation of the boundary layer.

The concept of boundary layer was developed at the turn of the 20th century. Various approaches have been suggested for solving the problem. One of the procedures, the one suggested by Bauer (1954) is simple and of easy application. Bauer's procedure has been developed for solving flow problems in steep channels. It can also be applied to spillway flow.

A flow taking place in a steep sloped channel is shown in Fig. 6.9. The boundary layer begins immediately downstream of the channel entrance and is developed in the direction of the flow. Bauer has suggested Eq. 6.7 for computing the thickness of the boundary layer:

$$\frac{\delta}{x} = \frac{0.024}{(\frac{x}{k})^{0.13}} \tag{6.7}$$

where

δ is the thickness of the boundary layer

x is the distance from the origin, and

k is the roughness height.

The flow takes place on the boundary layer, therefore, its thickness must be different from the thickness obtained as a result of normal hydraulic computation as outlined in Example 6.02. Example 6.03 shows an application of Bauer's procedure.

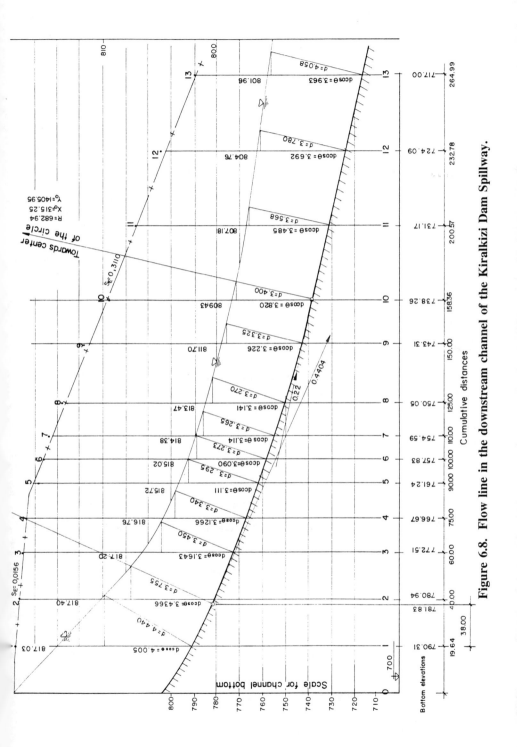

Figure 6.8. Flow line in the downstream channel of the Kiralkizi Dam Spillway.

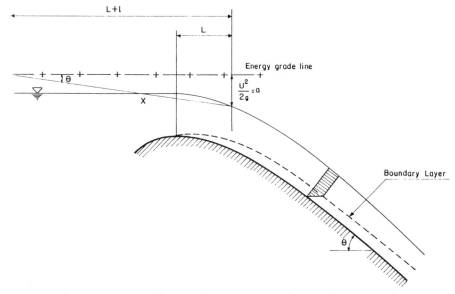

Figure 6.9. Formation of Boundary Layer (according to Bauer).

Example 6.03

Compute the thickness of the boundary layer and the flow depth for the flow defined in Example 6.02, using Bauer's procedure.

Solution

The solution is summarized in Table 6.4 and defined as follows:

Column 1:   number of stations

Column 2:   cosine of the angle of the tangent with horizontal

Column 3:   L, the distance to origin measured from point 0

Column 4:   The depth of flow and the velocity head at the station under consideration are measured in nature or computed, then ℓ is determined as follows:

$$\frac{U^2}{2g} = a = X \sin\theta$$

$$L + \ell = X \cos\theta$$

TABLE 6.4

Boundary Layer Computation (the Bauer's procedure)

$$k = 1.27 \cdot 10^{-4} \text{ m} \qquad d = 4.44 \text{ m}$$

	$\cos\theta$	$\sum_{i=1}^{13} L_i$	x	$\dfrac{x/k}{10^4}$	$\dfrac{\delta/x}{10^{-3}}$	δ	$\dfrac{U^2}{2g}$	U	ℓ	q	$d = \dfrac{q}{U}$	d	d_o	$\Delta = \dfrac{d}{d_o}$
1	2	3	4	5	6	7	8	9	10	11	12	13	14	15
1	0.9015	19.64	54.60	43.0	4.44	0.24	23.63	21.53	28.25	95.58	4.440	4.68	4.440	1.05
2	0.9152	40.00	76.02	59.90	4.26	0.32	30.64	24.52	28.25	95.58	3.898	4.22	3.755	1.12
3	0.9172	60.00	97.66	76.90	4.12	0.40	38.91	27.63	27.42	98.47	3.560	3.97	3.450	1.15
4	0.9361	75.00	111.71	88.00	4.05	0.45	39.29	27.77	26.63	101.39	3.651	4.10	3.340	1.23
5	0.9440	90.00	126.67	99.70	3.98	0.50	41.79	28.64	25.81	104.61	3.653	4.16	3.295	1.26
6	0.9490	100.00	136.57	108.0	3.95	0.54	43.05	29.06	25.32	106.64	3.670	4.20	3.273	1.29
7	0.9538	110.00	146.33	115.0	3.91	0.57	43.96	29.37	24.80	108.87	3.710	4.28	3.265	1.31
8	0.9604	125.00	160.95	127.0	3.86	0.62	44.84	29.66	24.01	112.45	3.790	4.41	3.270	1.35
9	0.9703	150.00	185.07	146.0	3.79	0.70	44.77	29.63	22.71	118.89	4.010	4.71	3.325	1.42
10	0.9766	168.36	202.68	160.0	3.75	0.76	43.79	29.24	21.75	124.14	4.240	5.00	3.400	1.47
11	0.9766	200.57	235.66	186.0	3.68	0.87	50.68	31.53	20.06	134.60	4.270	5.13	3.568	1.44
12	0.9766	232.78	268.64	212.0	3.61	0.97	57.77	33.69	18.38	146.90	4.360	5.33	3.780	1.41
13	0.9766	264.99	301.62	237.0	3.56	1.07	64.87	35.67	16.69	161.77	4.530	5.61	4.058	1.38

Then $\quad \ell = \dfrac{a}{\tan\theta} - L$

when ℓ is known, X, for station n+1 is:

$$X = \dfrac{L+\ell}{\cos\theta}$$

Column 5:   X/k, k, is the roughness height.

In concrete surfaces $k = 1 \sim 2 \; 10^{-4}$ m. In the particular case, $k = 1.27 \; 10^{-4}$ m

Column 6: $\dfrac{\delta}{X} = \dfrac{0.024}{(X/k)^{0.13}} = b$

Column 7: $\delta = bX$ m

Column 8: $\dfrac{U^2}{2g} = X \; \sin\theta = a$ m

Column 9: $U = (2ga)^{1/2}$ m/sec

Column 10: ℓ, the width of cross section, m

Column 11: $q = \dfrac{Q}{\ell}$ m³/sec. m

Column 12: $\dfrac{q}{U} = d_o$ m

Column 13: $d = 1.1 \; d_o$ m

It is assumed that boundary layer increases the depth of flow 10%.

Column 14:   See Column 9, Table 6.3

Column 15:   A comparison of flow depths obtained by the application of Bauer's procedure with the flow depth computed applying the gradually varied flow formulas is also given.

     Bauer's procedure yields greater flow depths and smaller flow velocities, which are more close to the reality. In fact it is known that a boundary value exists for the velocities taking place in steep sloped channels. This limiting velocity is about 32-35 m/sec and is varying with the temperature of water (Fig. 6.5).

U_{max} = 39.87 m/sec in gradually varied flow procedure

U_{max} = 35.67 m/sec in Bauer's procedure.

B. Buildup of Boundary Layer: the Corps of Engineers Procedure

The U.S. Army Corps of Engineer's procedure (1965) for solving the aeration problem of the rapid flow is summarized below.

The variation of the boundary layer thickness is given by

$$\frac{\delta}{L} = 0.08 \left(\frac{L}{k}\right)^{-0.233} \tag{6.8}$$

where

δ is the thickness of the boundary layer

L is the length of the wetted surface where the boundary layer takes place, and

k is the height of the roughness element.

Equation 6.8 can be employed in English and metric units equally.

The Corps of Engineers suggests:

$$k = 0.002 \text{ ft} = 6.096 \ 10^{-4} \text{ m} \tag{6.9}$$

for the roughness height of concrete surfaces. On the other hand, Simons, D.B. and Şentürk, F., (1992) suggested

$$\delta^* = 0.18\delta \tag{6.10}$$

where

δ^* is the displacement thickness.

The depth of flow under these conditions is given in Eq. 6.11

$$d_r = d + \delta^* \left(1 + \frac{2d}{b}\right) \tag{6.11}$$

where

d is the flow depth obtained applying uniform flow formulas, and

b is the width of the cross section.

Equation 6.11 applies to rapid flows. For U < 25 m/sec it is suggested to consider Eq. 6.12

$$d_r = d + \delta^*$$ (6.12)

For similar flows USBR (1957) suggests

$$Freeboard = 2 + 0.025U \, d^{1/3}$$ (6.13)

This relation is only valid in English units. Example 6.04 illustrates the Corps of Engineers' procedure.

Example 6.04

Apply the Corps of Engineers procedure to the flow described in Example 6.02.

Solution

The solution is shown in Table 6.5.

Column 1:   Station number

Column 2:   Bottom elevation of the stations (Column 20, Table 6.3)

Column 3:   Width of the cross section at the station (Column 13, Table 6.3) in meters

Column 4:   Width of the cross section at the station (Column 13, Table 6.3) in feet

Column 5:   Linear discharge (Column 11, Table 6.4) in m³/sec.m

Column 6:   Linear discharge (Column 11, Table 6.4) in cfs/ft

Column 7:   Depth of flow in m (Column 9, Table 6.3)

TABLE 6.5

Flow in the Downstream Channel of the Kiralkizi Dam Spillway

	z	b		q		d		L		δ	δ*	d_r	U		f	h		d_r
Station		m	ft	$\frac{m^3/sec}{m}$	cfs/ft	m	ft	m	ft	m	m	ft	fps	m/sec	ft	ft	m	m
1	2	3	4	5	6	7	8	9	10	11	12	13	14	15	16	17	18	19
1	790.31	28.25	92.68	95.58	1028.84	4.440	14.57	19.64	64.44	0.459	0.083	14.679	70.091	21.364	6.312	20.990	6.398	4.474
2	780.94	28.25	92.68	95.58	1028.84	3.755	12.32	40.00	131.23	0.792	0.143	12.500	82.304	25.086	6.822	19.322	5.889	3.810
3	772.51	27.42	89.96	98.47	1059.95	3.450	11.32	60.00	196.86	1.081	0.195	11.564	91.663	27.940	7.256	18.820	5.736	3.525
4	766.67	26.63	87.37	101.39	1091.38	3.340	10.96	75.00	246.06	1.283	0.231	11.249	97.066	29.586	7.532	18.781	5.725	3.429
5	761.24	25.81	84.68	104.61	1126.04	3.295	10.81	90.00	295.26	1.475	0.266	11.143	101.05	30.800	7.758	18.901	5.761	3.396
6	757.88	25.32	83.07	106.64	1147.89	3.273	10.74	100.00	328.08	1.600	0.288	11.102	103.39	31.514	7.895	19.000	5.790	3.384
7	754.59	24.80	81.36	108.87	1171.89	3.265	10.71	110.00	360.89	1.721	0.310	11.101	105.56	32.176	8.030	19.131	5.831	3.384
8	750.05	24.01	78.77	112.45	1210.43	3.270	10.73	125.00	410.10	1.898	0.342	11.165	108.42	33.045	8.220	19.385	5.909	3.403
9	743.31	22.71	74.51	118.89	1279.75	3.325	10.91	150.00	492.13	2.183	0.393	11.418	112.08	34.162	8.504	19.922	6.072	3.482
10	738.26	21.75	71.36	124.14	1336.26	3.400	11.15	168.36	552.36	2.385	0.429	11.376	114.08	34.771	8.693	20.407	6.220	3.570
11	731.17	20.06	65.81	134.60	1448.86	3.568	11.71	200.57	658.04	2.728	0.491	12.376	117.07	35.684	9.024	21.400	6.523	3.772
12	724.09	18.38	60.30	146.90	1581.21	3.780	12.40	232.78	763.71	3.058	0.550	13.177	120.00	36.576	9.379	22.556	6.875	4.016
13	717.00	16.69	54.76	161.77	1741.32	4.058	13.31	264.99	264.39	3.378	0.608	14.214	122.51	37.342	9.751	23.965	7.304	4.332

F. Sentürk

Column 8:   Depth of flow (Column 9, Table 6.3) in feet

Column 9:   Distance of the station to the origin (Column 6, Table 6.3) in meters

Column 10:   Distance of the station to the origin (Column 6, Table 6.3) in feet

Column 11:   (δ)   from Eq. 6.8 in feet

Column 12:   (δ^*) from Eq. 6.10 in feet

Column 13:   (d_r) from Eq. 6.11 in feet

Column 14: $U = \frac{q}{d}$, in fps

Column 15:   (U), in meters

Column 16:   Freeboard from Eq. 6.13, in feet

Column 17: $h = f + d_r$, (Column 16 + Column 13) in feet

Column 18:   (h), in meters

Column 19:   (d_r), in meters

C. Flow at the toe of Spillway, the U.S. Army Corps of Engineers Procedure *

Equation 6.8 is given for defining the thickness of the turbulent boundary layer (Corps of Engineers, 1955). The flux of the dissipated energy in the (x) direction in the boundary layer can be written as (Schlichting, 1960)

$$dF = 0.50 \, \rho \int_0^\delta u \, (U^2 - u^2) \, dy \qquad (6.14)$$

where

δ is the thickness of the boundary layer

u   is the velocity at a distance y from the boundary

---

* Summarized from Publication 111-18, US Corps of Engineers.

ρ is the specific mass of the fluid in motion, and

U is the mean velocity of the flow.

By definition, the energy thickness (δ_3) is the thickness of a layer of fluid with mean velocity, U, then

$$E_L = \frac{1}{2} \rho U^3 \delta_3 = \frac{1}{2} \rho \int_0^\delta u \, (U^2 - u^2) \, dy \qquad (6.15)$$

$$\delta_3 = \int_0^\delta \frac{u}{U} [1 - (\frac{u}{U})^2] \, dy \qquad (6.16)$$

Consider an energy flux taking place in the boundary layer with thickness δ_3 and with unit width. U, being the mean velocity of the flow the energy flux loss, E_L, is

$$E_L = \rho \bullet 1 \bullet \delta_3 \, U \frac{U^2}{2} = \rho \delta_3 \frac{U^3}{2} = (\gamma) \delta_3 U \frac{U^2}{2g} = (\gamma) \, q \frac{U^2}{2g}$$

and dividing it by $q\gamma$ yields

$$H_L = (\gamma) \, \delta_3 U \frac{U^2}{2g(\gamma)q} = \frac{\delta_3 U^3}{2gq} = \frac{U^2}{2g} \qquad (6.17)$$

where H_L is the energy loss in terms of feet of head. (δ_3) is given experimentally by Bauer as follows

$$\delta_3 = 0.22\delta \qquad (6.18)$$

and

$$\delta^* = 0.18\delta \qquad (6.19)$$

(Simons, D.B. and Şentürk, F., 1992). This method will be applied to a problem and explained in Example 6.05.

Example 6.05

Compute and define the flow characteristics taking place on the spillway shown in Fig. 6.10.

Data

H_d = 9.14 m	k = 6.096 10^{-4} m	m = 0.49
H = 106.68 m		

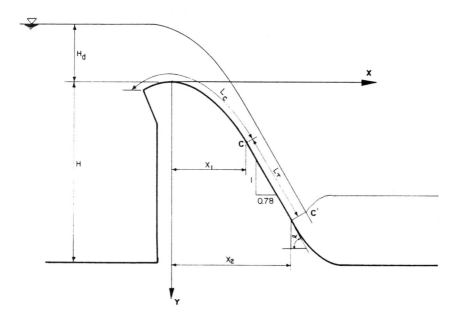

Figure 6.10. Profile of the Spillway in Example 6.05.

Slope of the downstream face $= \dfrac{1}{0.78}$

Profile of the ogee $\qquad X = 2\ H_d^{0.85}\ Y$

It is assumed that $H \le H_d$

Solution

1. Boundary geometry

    a. Length of curved crest, L_c

       The ogee equation is

$$Y = \frac{1}{2}\ \frac{X_1^{1.85}}{H_d^{0.85}}$$

and

$$\frac{dY}{dX} = \frac{1}{2}\ 1.85 \left[\frac{X_1^{0.85}}{H_d^{0.85}} \right] = \frac{1}{0.78} = 1.28$$

$$\frac{X_1}{H_d} = 1.47$$

X_1 is the abscissa of point C on the profile, the curve ends at this point and its tangent value is given as $(1/0.78)$. (L_c/H_d) values are given as a function of (X/H_d) in Fig. 6.11, then

$$\frac{L_c}{H_d} = 2.15 \qquad\qquad L_c = 19.65 \text{ m}$$

b. Length L_T of the straight line following the ogee

$$X_1 = 1.47 \text{ H}_d$$

$$Y_1 = \frac{1}{2}(1.47 \text{ H}_d)^{1.85}\frac{1}{H_d^{0.85}} = \frac{2.04}{2}\text{H}_d$$

$$\frac{Y_1}{H_d} = 1.02 \qquad\qquad Y_1 = 1.02 \text{ H}_d = 9.32 \text{ m}$$

$$Y_2 - Y_1 = 106.68 - 9.32 = 97.36 \text{ m}$$

$$\tan\alpha = \frac{1}{0.78} = 1.2821$$

$$\sin\alpha = 0.7885$$

$$L_T = \frac{Y_2 - Y_1}{\sin\alpha} = \frac{97.86}{0.7885} = 123.47 \text{ m}$$

c. Total length

$$L = L_c + L_T = 19.65 + 123.47 = 143.12 \text{ m}$$

2. Hydraulic Computation

a. Boundary layer thickness

$$\frac{\delta}{L} = 0.08 \left(\frac{L}{k}\right)^{-0.233} = 0.08 \left(\frac{143.12}{6.096\ 10^{-4}}\right)^{-0.233} \qquad 6.8$$

$$\delta = 0.64 \text{ m}$$

b. Energy thickness

$$\delta_3 = 0.22\ \delta \qquad\qquad 6.18$$

$$\delta_3 = 0.22 \bullet 0.64 = 0.141 \text{ m}$$

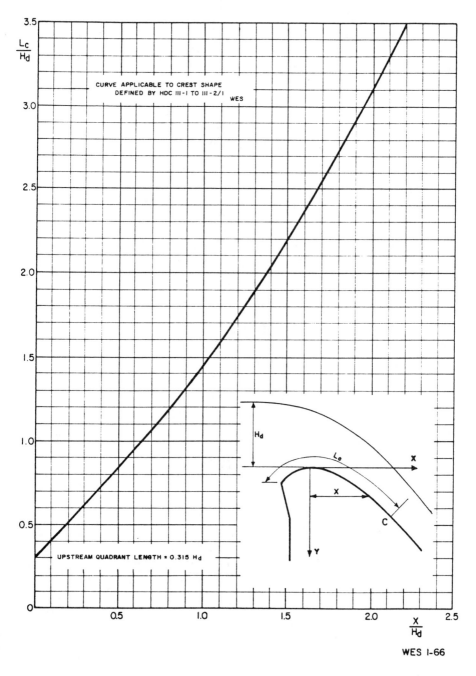

Figure 6.11. Values, L_c/H_d, as a function of X/H_d. U.S. Army
Engineers "Hydraulic Design Criteria" 111-18.

c. Linear discharge

$$q = 0.49\, H_d\, (2g\, H_d)^{1/2} = 0.49 \bullet 9.14\, (19.62 \bullet 9.14)^{1/2}$$
$$q = 59.97\ m^3/sec.m$$

d. Depth of flow and mean velocity of the flow at point C' (toe of spillway)

$$H_T = d\cos\alpha + (U^2/2g)$$

$$\cos\alpha = 0.6150$$

$$H_T = H + H_d = 106.68 + 9.14 = 115.82\ m$$

The solution of this equation is obtained with trial and error as shown in Table 6.6.

TABLE 6.6

U	$\dfrac{U^2}{2g}$	$H_T - \dfrac{U^2}{2g}$	$d = \dfrac{H_T - \dfrac{U^2}{2g}}{\cos\alpha}$	$q = Ud$
m/sec	m	m	m	m³/sec.m
47.55	115.24	0.5800	0.94	44.70
47.51	115.05	0.7741	1.26	59.80 ≈ 59.97

e. Energy loss on the spillway

$$H_L = \frac{\delta_3\, U^3}{2gq} = \frac{0.141 \bullet 47.51^3}{19.62 \bullet 59.97} = 12.85\ m$$

f. Energy head entering stilling basin

$$H_{c'} = H + H_d - H_L = 106.68 + 9.14 - 12.85 = 102.97\ m$$

g. Flow depth entering stilling basin

$$d_{c'} = d + \delta^* = 1.25 + 0.18 \bullet 0.64 = 1.37\ m$$

Then

$$U_{c'} = \frac{59.97}{1.37} = 43.77\ m/sec$$

Even this value of $U_{c'}$ is greater than the limiting value of 32 m/sec.

D. Air Injection in the Flow Taking Place on the Downstream Face of a Spillway to Prevent Erosion of the Concrete Surface by Cavitation

Cavitation phenomenon is frequent on the downstream surface of spillway where water moves with great speed. It was particularly observed in recent years on spillways such as Keban, Aslantas, etc. in Turkey. Figure 4.26 shows the damages observed on the Keban Dam Spillway in 1977.

To prevent such damages, it is necessary to introduce air between the lower nappe and the spillway surface. It is desired to determine the quantity of air and the location of air exits (Fig. 6.12).

Demiröz, E. (1985) work will be summarized to shed some light on the problem.

a. Draw the backwater curve on the spillway corresponding to design flood. Then the flow depth and the mean velocity are determined. Draw the line showing the variation of the Froude's number along the spillway.

b. The angle (α) between the horizontal and the downstream profile is known when the geometry of the spillway is known.

Some of the (α) angle used in practice are given in Table 6.7.

Curves determining the variation of $\beta = q_a/q$, where q_a is the unit air discharge and q the water discharge have been prepared for use in the computation (Demiröz, 1985). Figure 6.13 shows one of these curves as an example.

A function of A/A_o is defined as (β), F_r and t_e/h, where

A   is the exit area of the air chimney of the aerator

A_o is the wetted area of the downstream channel

F_r is the Froude number

t_e is shown in Fig. 6.14, and

h   is the flow depth.

Choose the curve corresponding to the spillway characteristics. It is known that the introduction of $2 \cong 3\%$ of air is sufficient to prevent a good deal of cavitation, but $7 \cong 8\%$ air supply is necessary to prevent cavitation.

c. When $\beta = (q_a/q)$ is determined, according to the criteria previously outlined, Fig. 6.13 provides the Froude number. Air will be injected at a position on the profile corresponding to this value of F_r. The

Figure 6.12. Aerated flow from the Foz do Areia Dam Spillway, discharging 2500 m^3/sec, Brazil. Courtesy of Pinto Nelson L. de S.

TABLE 6.7

Tangent of the downstream face of spillways (Turkish practice)

Name of the dam	Tan
Adigüzel	0.47
Altinkaya	0.38 - 0.666
Aslantas	0.294 - 0.35
Atatürk	0.14
Catalan	0.17
Cakmak	0.20
Dicle	0.166
Hasan Ugurlu	0.10 - 0.74
Ilisu	0.121
Karakaya	1.70
Kayraktepe	0.43 - 0.213
Keban	0.166
Kestel	0.35
Kiralkizi	0.26
Menzelet	0.28
Sir	0.195 - 0.46
Söylemez	0.67
Yedigöze	0.60
Arpacay	1.43

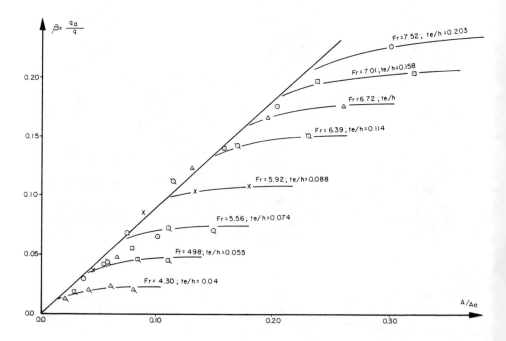

Figure 6.13. $\beta = q_a/q$ versus A/A_o.

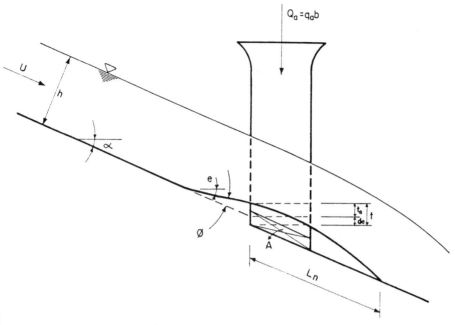

Figure 6.14. Definition of symbols.

characteristics of the aerator A, t_e, are also obtained from the diagram.

d. Check the accuracy of L_n, the length of the lower nappe of the flow along the air exit (Fig. 6.14). If the length is not sufficiently developed, then choose an adequate length for L_n and repeat the computation to obtain the desired (β).

Curves similar to the curves given in Figs. 6.13 and 6.15 must be prepared to perform this operation.

Pinto, N.L. de S., 1989, suggests the following procedure to compute Q_a:

1. Compute the effective air duct area per unit of chute, D

$$D = \frac{CA}{B}, \text{ m}^2/\text{m}$$

where

- C is the discharge coefficient of the formula for the air discharge through ducts

$$q_a B = CA \, (2 \frac{\Delta p}{\rho_a})^{1/2}$$

F. Sentürk

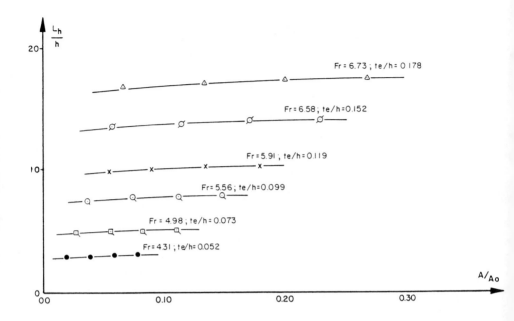

Figure 6.15. Variation of L_n/h in function of F_r and (t_e/h).

- B is the width of the chute, corresponding to the aerator (m).

- A is the control area of the air duct; most commonly the orifice area beneath the nappe (m^2) (Fig. 6.14).

- (Δp) is the difference between atmospheric pressure and average pressure below the jet, as measured along the vertical face of the steep ramp (N/m^2).

- ρ_a is the air density, kg/m^3.

2. Compute the area performance

$$\beta = \frac{Q_a}{Q} = \frac{q_a}{q}$$

where q is the water discharge per unit width of chute (m^3/sec.m); Pinto, N. L. de S., 1987, has shown that

$$\beta = f(F_r, \frac{t}{h}, \frac{D}{h})$$

314

where

- F_r is the Froude number $= \dfrac{U}{(gh)^{1/2}}$

- h is the unaerated water depth of the flow (m) at the aerator section (Example 6.2 for computational procedure)

- t is the ramp and/or step height (Fig. 6.14).

In case the total water head is maintained constant, to each value of F_r corresponds a unique pair of (D/h) and (t/h) values, then

$$\beta = \emptyset \, (F_r)$$

Experiments conducted on model showed that as a first approximation it was possible to write:

$$\beta = 0.29 \, (F_r - 1)^{0.62} \, (\frac{D}{h})^{0.59}$$

This relation shows a correlation factor of 97.62%. Pinto, N.L. de S. suggests that this equation "could be used for rough estimates of air entrainment at the initial stage of a project." To finalize the design, model study is necessary. Empirical relations have been suggested by different researchers for obtaining the q_a values such as, Shui-bo et al., 1980

$$q_a = 0.22 U_L L_n \tag{6.20}$$

where

q_a is the unit air discharge,

U_L is the mean velocity of the flow, and

L_n is the length of the lower nappe of the flow at the air exit.

6.3   RETURN OF FLOW TO THE MAIN WATER COURSE WITH ENERGY DISSIPATION

The flood water downstream of the discharge channel is, of course, returned to the main water course. Its velocity is large. In many cases it attains its limiting value, around $32 \cong 35$ m/sec, which can develop destructive energy. This energy must be damped to avoid damage at the downstream toe of the dam and in the main bed of the river. Heavy degradation is often observed in the river bed damaging retaining walls along the embankment and causing local erosion around bridge piers, etc. Therefore, this destructive energy must be dissipated. Two solutions are suggested for solving this problem

1. Dissipation of energy at downstream water level by stilling basins; and

2. Dissipation of energy at a level above the downstream water level by energy dissipators.

Dissipation of energy at the downstream water level is done by stilling basins which constitute the classic solution for solving the problem of the return of the discharge to the main water course downstream of dams. Up to the middle of the 20th century numerous stilling basins were constructed throughout the world. Today this solution is mainly used for weirs and small dams up to 6 0 - 7 0 m high. Exceptions, such as the stilling basin of the Atatürk Dam on Firat River, exist naturally (Height of Dam, 172 m). A tremendous amount of experimental work has been performed relative to stilling basins in the laboratories and in nature. General rules for the design of stilling basins is given in this chapter (see subsection 6.4.2).

Since the middle of the 20th century, it has been general practice for large dams to dissipate the discharge energy by structures built higher than the downstream water surface. These structures are called ski jumps and deflectors. General rules for the design of ski jumps is given in subsection 6.4.7.

6.4   STILLING BASINS

6.4.1 Direct Water Return Hydraulics

This solution is used only for very small dams and weirs where the river bed is sound rock. The energy is dissipated by an hydraulic jump forming on the apron downstream of the spillway. The following conditions must be fulfilled for such a solution:

- The Froude number at the return cross section must be smaller than $1.5 \cong 2.0$

- The angle between the center lines of the discharge channel and the river must be very small

$$\theta < 15°$$

- The bottom of the river must be resistant to erosion.

6.4.2 Stilling Basins Hydraulics

Stilling basins are the main devices for damping the active energy. They have been in use since the middle of the 20th century. The stilling basin is designed to dissipate the energy of the discharge by utilizing the development of the hydraulic jump within the basin. A large amount of experimental work has been performed by researchers such as Pavlowsky, Certussov, Ludin, Pietrowski, Woycicky, Bachmeteff, etc. (Table 6.8). Thus the length, the depth, the energy absorption and the forms of hydraulic jump can be mathematically approximated.

TABLE 6.8

Length of Hydraulic Jump

Name of authors	Formulas	
Pavlowsky	$\dfrac{L}{d_2 - d_1} = \dfrac{2.375(-1+\sqrt{1+F_{r1}^2})-2.50}{0.50(-3+\sqrt{1+8F_{r1}^2}}$	(6.22)
Certussov	$\dfrac{L}{d_2 - d_1} = \dfrac{20.6(F_{r1}-1)^{0.81}}{-3+\left(-1+8F_{r1}^2\right)^{1/2}}$	(6.23)
Ludin	$\dfrac{d_2 - d_1}{L} = \dfrac{1}{4.50} - \dfrac{1}{6F_{r1}}$	(6.24)
Pietrowski	$L = 5.9d_1 \qquad\qquad F_{r1} = 4.33d_2$	(6.25)
Woycicky	$L = (8 - 0.05\dfrac{d_2}{d_1})(d_2 - d_1)$	(6.26)
Mazmann	$L = \dfrac{(d_2 + d_1)(d_2 - d_1)}{2d_1}$	(6.27)
Bachmeteff	$L = [\dfrac{5.4d_1}{d_2} - 0.06] [(\dfrac{d_2}{d_1})^3 - 1]$	(6.28)
Safranetz	$\dfrac{L}{d_2 - d_1} = \dfrac{12}{-3 + (1 + 8F_{r1}^2)^{1/2}}$	(6.29)
Smetana	$\dfrac{L}{d_2 - d_1} = 6 \qquad$ or $\qquad L = 4.5d_2$	(6.30)

A. Depth of Hydraulic Jump

Figure 6.16 shows a jump taking place in an infinitely large horizontal channel.

Assume that the flow in the steep upstream channel is supercritical. The bottom slope of this channel is S_b and θ is the angle of the bottom with the horizontal floor of the downstream channel.

Furthermore, the flow depth in the upstream channel is d_1 and the velocity U_1. An hydraulic phenomenon, known for many years, takes place in this channel system. The depth of the rapid flow entering the horizontal

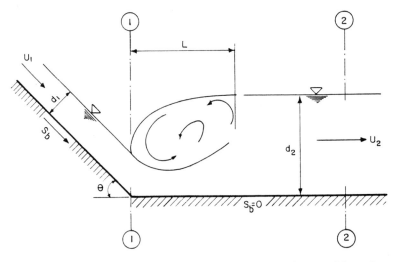

Figure 6.16. Definition of symbols in a hydraulic jump taking place in a horizontal channel.

reach increases suddenly up to a depth $d_2 > d_1$ and its velocity decreases to $U_2 < U_1$. This sudden increase is called the hydraulic jump. The position of the jump changes as a function of upstream flow conditions. For higher values of the upstream Froude number it is located further downstream; for smaller values of it, it is located closer to section 1-1 where the change of bottom slope occurs. For the simple jump described above, the relation between the flow depth upstream and downstream of the jump based on the pressure-momentum principle may be written as:

$$\frac{d_2}{d_1} = \frac{1}{2}\left[\left(1 + 8F_{r1}^2\right)^{1/2} - 1\right] \qquad (6.21)$$

where F_{r1} is the upstream Froude number. The ratio (d_2/d_1) is plotted with respect to the Froude number in Fig 6.17. The representative line, which is

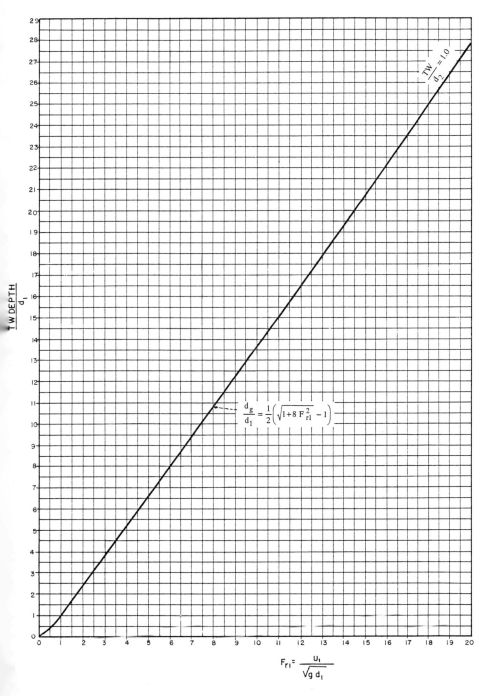

Figure 6.17. Ratio of tailwater depth to d_1 as a function of F_{r1}. (USBR, "Hydraulic Design of Stilling Basins and Energy Dissipators"-1958).

virtually straight except for the lower end, represents Eq. 6.21. For values of $F_{r1} < 2$ the straight line shows a slight curvature. The remaining experimental points show perfect agreement with the theoretical straight line.

B. Length of Hydraulic Jump

The theoretical computation for the length of the hydraulic jump is difficult. Its measurement also presents difficulties due to pulsating character of the phenomenon. Special measures must be taken when performing the measurements. Numerous formulas are suggested for computing the length of the hydraulic jump, but because of the great difficulties in its measurement and theoretical approach these formulas are rather approximate. Table 6.8 shows some of these formulas.

Figure 6.18 can be used to determine the length of the hydraulic jump. It is suggested by USBR (Peterka, A.J., 1964).

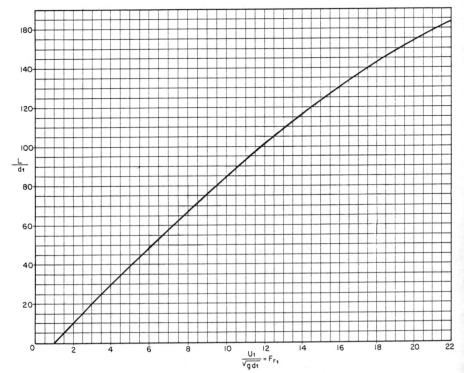

Figure 6.18. Length of jump in terms of d_1 (USBR).[*]

---

[*]   USBR, "Hydraulic Design of Stilling Basins and Energy Dissipators" - Engineering, Mono: 25,1958)

The results obtained by Bachmeteff and Matzke of the Zürich E.T.H. Laboratories and the Technical University of Berlin, respectively, are shown in Fig. 6.19. The suggested mean representative line is also plotted on the same diagram.

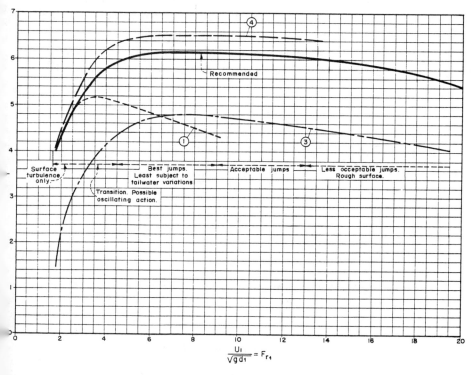

Figure 6.19. Length of Jump in Terms of d_2 (USBR).*

The curve labeled (3) was obtained by the Technical University of Berlin on a flume 0.5 m wide by 10 m long. The curve labeled (4) was determined from experiments performed at Zürich E.T.H. Laboratories, on a flume 0.6 m wide and 7.00 m long. It is obvious that the generalization of such results seems problematical. They constitute only suggestions for the orientation of the designer.

C. Location of the Hydraulic Jump

The location of the hydraulic jump is important for determining the length of the downstream lining of the channel where the jump may take

---

USBR, "Hydraulic Design of Stilling Basins and Energy Dissipators" - Engineering, Mono: 25, 1958)

place. If special measures are not taken to confine the jump to a given space the determination of the location of the jump becomes more important.

Figure 6.20 shows a jump taking place downstream of a sluice gate. The curve AG is the water surface line of the supercritical flow with a hydraulic head, H_u. The flow issuing from point A of the opening increases its depth rapidly according to downstream conditions and reaches a point, G, where the jump occurs. The horizontal distance, AG = L denotes the beginning of the jump. The length of the jump L_2 is the horizontal projection of GF.

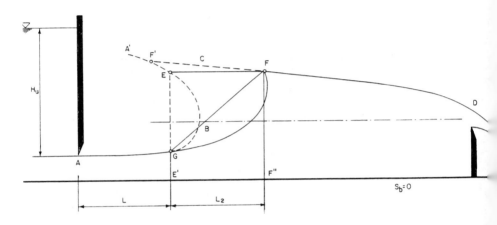

Figure 6.20. Location of the Jump.

Assume that the downstream channel is horizontal; then the subcritical flow taking place in it must be gradually varied. If the discharge and a point D on the water profile are known, it is possible to draw the flow line upstream of this point D by applying the gradually varied flow formulas.

H_u being known, the supercritical flow water surface profile can be computed and plotted. This profile is shown on Fig. 6.20 by the line AGBEA'. Subsequently the flow line of the supercritical flow is also shown on the same figure by the line DFCF'. Point F' is at the intersection of the two curves. Further, assume that the downstream depth of the jump is FF'' = d_2. Entering Fig. 6.19 with this value yields the length L_2 of the jump. This length is intercalated between the two curves (curve AGBA' and DFF' as shown by EF. The position of point E defines point G and point F, then the jump.

In reality, the downstream depth of the jump is different from FF'' and is virtually equal to EE'. A trial and error process yields the correct answer. I general the result is retained due to the pulsating, then variable character of the phenomenon.

Example 6.06

Locate the hydraulic jump, the characteristics of which are given in Fig. 6.21.

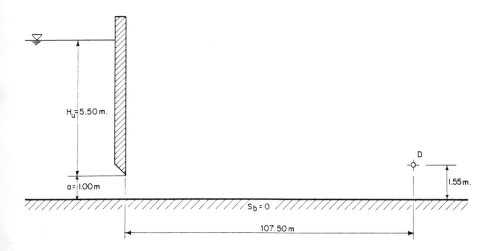

Figure 6.21. Characteristics of the jump in Example 6.06.

Data

- The width of the opening   = 5.00 m
- $Q = Ca(2gH_u)^{1/2}$ $C = 0.54$
- Location of point D is shown in Fig. 6.21
- $n = 0.014$.

Solution

Step 1: Compute q and U

$$Q = 0.54 \, (1.00 \cdot 5.00) \, (2g \cdot 5.50)^{1/2} = 28.05 \text{ m}^3/\text{sec}$$

$$q = \frac{28.05}{5.00} = 5.61 \text{ m}^3/\text{sec.m, and}$$

$$U = \frac{5.61}{1.00} = 5.61 \text{ m/sec}$$

Step 2: Draw the supercritical flow line by applying the gradually varied flow formula

$$d_1 + \frac{U_1^2}{2g} = d_2 + \frac{U_2^2}{2g} + \frac{n^2 U^2}{R^{4/3}} \ x$$

where

d_1 is the flow depth upstream of the jump

d_2 is the flow depth downstream of the jump

U_1 is the mean flow velocity at station 1-1

U_2 is the mean flow velocity at station 2-2

R is the mean hydraulic radius defined as

$$R = \frac{R_1 + R_2}{2}$$

Because d_1 is small compared to the width of the opening, (R) can be substituted by (d) and

$$R = \frac{d_1 + d_2}{2}$$

U is the mean velocity defined as

$$U = \frac{U_1 + U_2}{2}$$

and

x is the distance separating the stations 1-1 and 2-2.

Substituting these values into the energy equation yields

$$d_n + \frac{U_n^2}{2g} = d_{n+1} + \frac{U_{n+1}^2}{2g} + \frac{(\frac{U_n + U_{n+1}}{2})^2 n^2}{(\frac{d_n + d_{n+1}}{2})^{4/3}} \ x$$

or

$$d_n + \frac{1.60}{d_n^2} = d_{n+1} + \frac{1.60}{d_{n+1}^2} + 3.89 \ 10^{-3} \ \frac{(\frac{d_n + d_{n+1}}{d_n d_{n+1}})^2}{(d_n + d_{n+1})^{4/3}} \ x$$

The solution of this equation is shown in Table 6.9.

The equation of the water surface line of downstream subcritical flow is

$$d_n + \frac{1.60}{d_n^2} + 3.89 \ 10^{-3} \ \frac{(\frac{d_n + d_{n+1}}{d_n d_{n+1}})^2}{(d_n + d_{n+1})^{4/3}} \ x = d_{n+1} + \frac{U_{n+1}^2}{2g}$$

The solution of this equation is shown in Table 6.10.

TABLE 6.9

Stations	d_1	X	d_2
1	1.000	25	1.073
2	1.073	25	1.155
3	1.155	25	1.254
4	1.254	25	1.421
5	1.421	1.95	1.465
6	1.465	0.615	1.470
1	1.470	15	1.633
2	1.633	25	1.721
3	1.721	25	1.782
4	1.782	25	1.830
5	1.830	25	1.871

TABLE 6.10

Stations	d_1	X	d_2
1	1.5500	25	1.6840
2	1.6840	25	1.7548
3	1.7548	25	1.8085
4	1.8085	25	1.8533

These curves are shown in Fig. 6.22. The jump length is given in Fig. 6.18

$$F_{r1} = \frac{U_1}{(gd_1)^{1/2}}$$

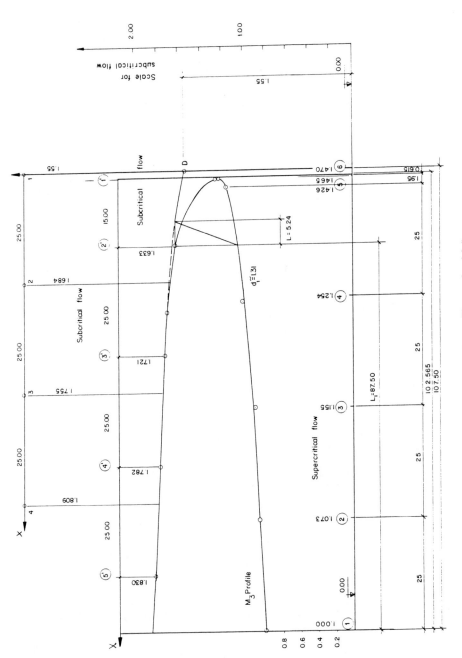

Figure 6.22. Flow lines (Example 6.06).

where

$$d_1 = 1.31 \text{ m} \qquad U_1 = \frac{5.61}{1.31} = 4.28 \text{ m/sec}$$

$$F_{r1} = \frac{4.28}{(9.81 \bullet 1.31)^{1/2}} = 1.19$$

and

$$\frac{L}{d_1} = 4$$

$$L = 1.31 \bullet 4.00 = 5.24 \text{ m}$$

Result

The jump begins 87.50 m downstream of the sluice gate:

$$L_1 = 87.50 \text{ m}$$

D. Energy Loss in Hydraulic Jump

Figure 6.23 shows a jump and the loss of energy due to it

$$\Delta E = E_1 - E_2 = \frac{(d_2 - d_1)^2}{4d_1 d_2} \qquad (6.31)$$

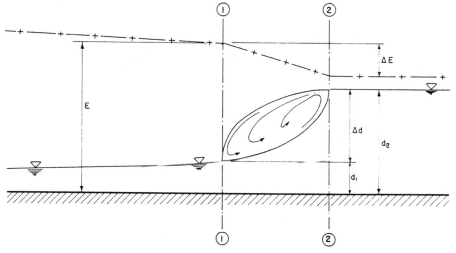

Figure 6.23. Energy loss in hydraulic jump.

The ratio $(\Delta E/E_1)$ is known as relative loss. The efficiency of a jump is defined as

$$\frac{E_2}{E_1} = \frac{(8F_{r1}^2 + 1)^{3/2} - 4F_{r1}^2 + 1}{8F_{r1}^2 (2 + F_{r1})^2} \tag{6.32}$$

The ratio $(\Delta d/E_1)$ is the relative height, it is equal to

$$\frac{\Delta d}{E_1} = \frac{d_2}{E_1} - \frac{d_1}{E_1} \tag{6.33}$$

where

(d_2/E_1) is the relative sequent depth, and

(d_1/E_1) is the relative initial depth.

The relative height can be expressed as a function of the Froude number as

$$\frac{\Delta d}{E_1} = \frac{(1 + 8F_{r1}^2)^{1/2} - 3}{F_{r1}^2 + 2} \tag{6.34}$$

These characteristics are shown in Fig. 6.24. Figure 6.25 visualizes the variation of $(E_1 - E_2)/E_1$ as a function of the Froude number.

E. The Flow Profile

In engineering practice, it is desirable to know the flow profile of a jump for establishing the freeboard and for designing the retaining walls. The computation of the pressures for use in structural design can only be finalized if the flow profile is known. The flow profile of a jump varies as a function of the type of jump. Thus a thorough definition of the type of jump is necessary.

F. Types of Jump

A classification of jumps on a horizontal floor as a function of the upstream Froude number is as follows:

1. $F_{r1} = 1$ critical flow, no jump occurs

2. $1 < F_{r1} < 1.7$ Undular jump: Undulations are observed on the water surface (Figs. 6.26 and 6.27). Simple stilling basins can be used for damping the excess energy.

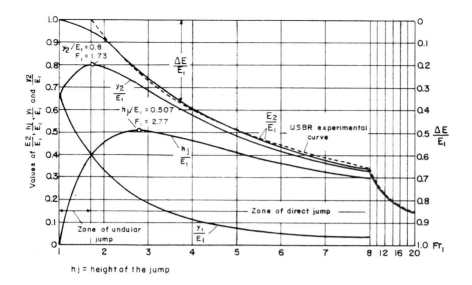

Figure 6.24. Characteristic curves of hydraulic jumps in horizontal
rectangular channels (Ven Te Chow, "Open Channel
Hydraulics", 1959).

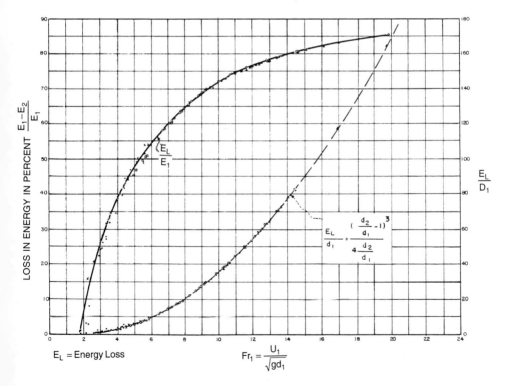

Figure 6.25. Energy loss in hydraulic jump (USBR).[*]

[*]   USBR, "Hydraulic Design of Stilling Basins and Energy Dissipators" - Engineering, Mono: 25,
      1958)

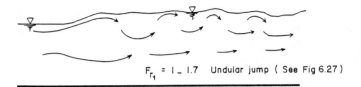

$F_{r_1} = 1 - 1.7$ Undular jump (See Fig 6.27)

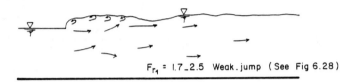

$F_{r_1} = 1.7 - 2.5$ Weak jump (See Fig 6.28)

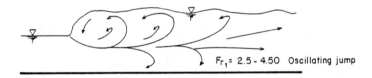

$F_{r_1} = 2.5 - 4.50$ Oscillating jump

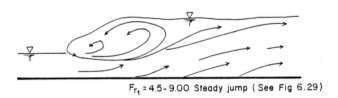

$F_{r_1} = 4.5 - 9.00$ Steady jump (See Fig 6.29)

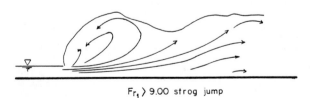

$F_{r_1} > 9.00$ strog jump

Figure 6.26. Types of hydraulic jump. (V.T. Chow).

Figure 6.27. Undular jump in nature. (the Selevir Dam, Turkey).

3. $1.7 < F_{r1} < 2.5$ Weak jump: Downstream water surface remains relatively smooth. Pulsations occur on the jump; the energy loss is low (Figs. 6.26 and 6.28).

4. $2.5 < F_{r1} < 4.5$ Oscillating jump: This kind of jump is, in general, observed downstream of medium height dam spillways.

                       If not controlled they can cause great damage along the downstream channel, (Fig. 6.26). A relatively efficient stilling basin must be used for damping the excess energy.

5. $4.5 < F_{r1} < 9$ Steady jump: The energy dissipation in this jump attains 70%. The downstream extremity of the surface roller and the point at which the high velocity jet tends to leave the flow occur practically

Figure 6.28. Weak jump in nature. (the Selevir Dam, Turkey).

at the same vertical section (Fig. 6.26 and Fig. 6.29). The excess energy can be damped by using engineering structures such as Basin II, USBR.

6. $F_{r1} \geq 9$ Strong jump: The energy dissipation in this kind of jump attains 85%. The surface of the downstream flow is rough. Ski jumps can be used to damp the excess energy.

The boundaries as defined are not clear cut. A certain amount of overlapping always occurs.

6.4.3 Types of Stilling Basins

A. Lined Downstream Channel - Type I

When the jump can be confined to the lined downstream channel no particular device is needed to overcome the damaging energy of the flow. The downstream channel acts as a stilling basin and it will be classified as a Type I Basin.

Figure 6.29. Steady jump in nature. (the Selevir Dam, Turkey).

Example 6.07

Water flowing at the toe of a weir has the following characteristics:

$$U_1 = 25.00 \text{ m/sec}$$

$$d_1 = 1.75 \text{ m}$$

the downstream channel is horizontal.

Determine the conjugate tailwater depth, the length of basin required to confine the jump, the effectiveness of the basin to dissipate energy and the type of jump to be expected if the jump occurs at the toe of the weir.

Solution

$$F_{r1} = \frac{25}{(9.81 \cdot 1.75)^{1/2}} = 6.03$$

entering Fig. 6.17 with $F_{r1} = 6.03 \cong 6.00$

$$\frac{d_2}{d_1} = 8$$

the conjugate tailwater depth:

$$d_2 = 8d_1 = 8 \bullet 1.75 = 14.00 \text{ m}$$

entering the recommended curve on Fig. 6.19 with $F_{r1} = 6$

$$\frac{L}{d_2} = 6.10$$

length of the basin for confining the jump

$$L = 6.10 \bullet 14.00 = 85.4 \cong 86.00 \text{ m}$$

Entering Fig. 6.25 with $F_{r1} = 6.00$ it is found that the absorbed energy is 55%. $F_{r1} = 6.00$ then the jump is a steady jump.

B. USBR Stilling Basin Type II, $F_{r1} \geq 4.50$

The USBR stilling basin Type II is an engineering structure designed to confine the entire length of the jump to the paved apron. The cost of the structure will be reduced because existing appurtenances can be used to modify the jump characteristics and so allow the designer to adjust its characteristics to suit the engineering needs. The following data can be adopted to develop an acceptable design:

1. The variation of the downstream depth

2. The definition of hydraulic conditions under which the basin may function.

Then

1. The length of the basin,

2. The flow profile in the basin, and

3. The engineering precautions to be adopted at the basin entrance and exit can be determined and used for design.

Figure 6.30 shows the longitudinal section of a Type II basin. The symbols used in the text are defined in the figure.

In Basin II, chute blocks are placed at the entrance with a dentated sill at the exit. Between the chute blocks and the sill the apron is free of any kind of obstruction. In general this type is used in the spillways of large dams where the upstream velocity has reached its boundary or close to it. An obstacle introduced to such a fast moving flow may create a vacuum behind it. Measures are taken in Basin II to avoid this difficulty.

As stated previously, the stilling basin is a structure intended to confine the jump to a given space. The jump characteristics are a function of the upstream Froude number and the downstream hydraulic conditions, which means that downstream conditions must be clearly defined. If upstream

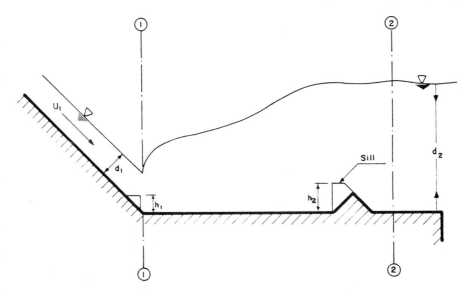

Figure 6.30. Schematic longitudinal section of Basin II.

conditions are known, then (d_2/d_1) can be obtained from Eq. 6.21. If d_1, under the defined conditions in Fig. 6.16, is known, then d_2 can be computed. For large values of d_2 the jump can take place immediately downstream of the toe, but for smaller values of d_2 it is projected far downstream and out of the basin.

In the United States, stilling basins were tested on scale models; in-situ measurements were also made to check the results and to complete the assembly of information. Figure 6.31 reflects the results obtained.

Figure 6.31 has a heavy solid line which shows the variation of (d_2/d_1) as a function of F_{r1}. Other dashed lines on the chart reflect different ratios of d_2/d_1. When $(d_2/d_1) \geq 1.2$, construction becomes costly and difficult, the

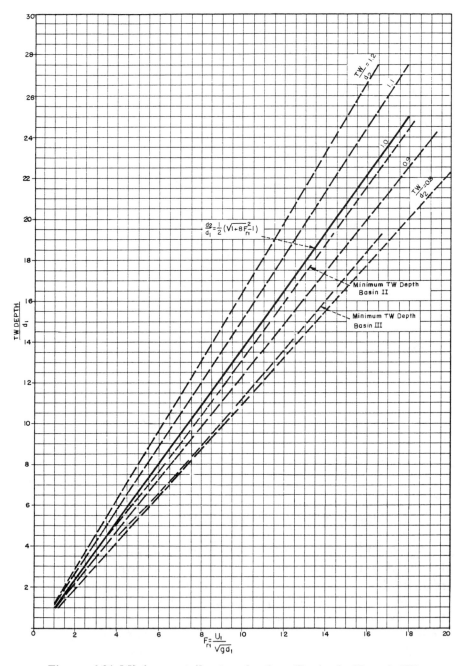

Figure 6.31. Minimum tailwater depth.   (Basin I, II and III). (USBR).*

---

*   USBR, "Hydraulic Design of Stilling Basins and Energy Dissipators" - Engineering, Monograph: 25-1958)

heavy dashed line, labelled "Minimum T.W. Depth, Basin II," has been used as the lower design limit for many structures. It is recommended that the limits for stilling basin design stay between the minimum T.W. Depth, Basin II and the line based on Eq. 6.21.

C. SAF Stilling Basin *

The efficiency of SAF stilling basin is high for $1.7 < F_{r1} < 17$. Example of basins exist for $F_{r1} > 17$, but they are rather rare. Blocks are positioned at the entrance and in the middle of the basin so that the tailwater depth is reduced. Laboratory tests show a reduction of 15%; but it is suggested that some margin be provided to allow for the boundary conditions and to reduce the incidence of cavitation. Figure 6.32 illustrates the design conditions for the basin and Fig. 6.33 shows a diagram for hydraulic computation.

6.4.4 Guidelines for Design

The design of a stilling basin begins by the determination of U_1 and d_1 and subsequently F_{r1}. The value of the Froude number so determined, yields the value of d_2 using Fig. 6.31. If the tailwater depth in nature is equal to or greater than this value, the problem is solved. If not, the designer may deepen the floor of the basin to fulfill this basic condition. Tailwater depth is a very important factor in the design of stilling basin and a number of guidelines may be considered.

- For the tailwater depth smaller than d_2 the jump does not occur in the stilling basin; the supercritical flow sweeps out of the basin and the jump is formed downstream of it. Damage to the downstream stream bed cannot be prevented in such cases. Scale model tests have shown that when the tailwater depth is equal to d_2 the jump is always confined within the boundaries of the stilling basin. Furthermore, a 3% reduction in the tailwater depth can be tolerated. This tolerance is reflected in Fig. 6.31.

- Stilling basin designs are generally tested on a scale model prior to making a final decision on shape and size. When analyzing the results of the model tests, it is important to consider the effect of the retardation phenomenon on the model and in nature. The retardation

---

*   SAF: St. Antonny Falls

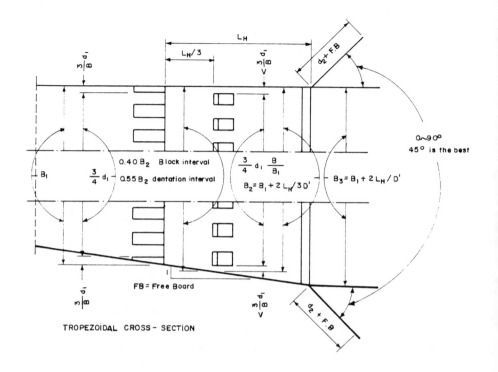

TROPEZOIDAL CROSS- SECTION

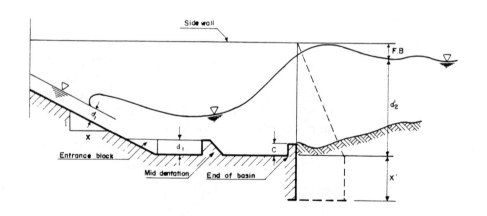

Figure 6.32. The SAF basin. Fred W. Blaisdel, "The SAF Stilling Basin," U.S. Soil Conservation Service, Report SCS-TP-79, 1949.

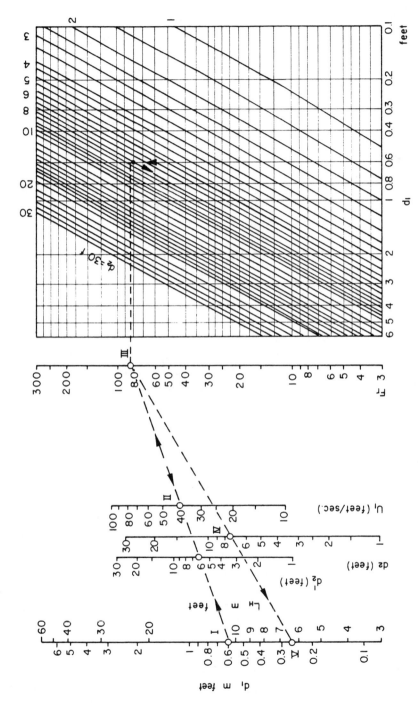

Figure 6.33. Diagram for hydraulic computation (U.S. Department of Agriculture, The SAF Stilling Basin - 1959).

phenomenon may be described as the lag in variation of the water
depth relative to the variation of the discharge. For example, if the
tailwater depth in a particular case corresponding to a discharge, Q_o,
is d_o and if Q_o decreases to $Q_1 < Q_o$ in "t" time, the decrease of the
tailwater depth from d_o to d_1 corresponding to Q_1 takes "kt" time, k,
being greater than unity. If measurements on the model and in nature
are performed in a time t' < kt, the obtained results may be erroneous.
This phenomenon is due to the retardation in U following Q and in d
following U (Şentürk, F., 1969). It is recommended that a retardation
margin of 5 - 7% be used to allow for this gap.

In model investigations, a rule of thumb is to begin with a depth equal to
"0.93 tailwater depth" and to increase this value to d_2 during a time interval
10% greater than the duration of Q_{max}.

- Generally, the tailwater depth is computed using the chart given in
Fig. 6.31 or by Eq. 6.21. It is assumed that the return period of Q_{max}
is very large or the probability of its occurrence is very small.
Extrapolation is used to reach this very high value of Q_{max}. There are
different procedures in use to obtain an acceptable Q_{max} value.
However, the difference between the actual value of Q_{max} and that
obtained by extrapolation may be significant. In order to allow for
error, it is recommended that the tailwater depth be equal to $0.95d_2$.
Because of the uncertainty of the computed tailwater depth, d_2, USA
literature suggests using the more conservative factor of $1.05d_2$.

- It is also recommended that the time factor be considered in the
computation of the tailwater depth. Experience has shown that in the
years following the completion of the dam heavy degradation occurs
in the downstream river bed. This degradation must be determined as
a function of time and be considered in the computation of the
tailwater depth (see Chapter 12) and in building the scale model.
Coordination between the designer and the model engineer is highly
recommended. The importance of the determination of the tailwater
depth cannot be overemphasized. A primary error cannot be
corrected after the completion of the stilling basin.

- Upstream water depth of the stilling basin.

The dimensions of the basin are defined as a function of the velocity of
the flow entering the basin. Therefore, this velocity must be computed with
high accuracy. Theoretical approaches and scale model tests are not highly
satisfactory due to the inability of the scale model to reproduce effectively

the air entrainment phenomenon. The characteristics of the flow taking place in a steep-sloped channel are enumerated as follows:

1. The air entrainment begins for $U \geq 10$ m/sec. Actually air is also entrained for smaller values of U but in engineering practice its effect is considered negligible. The air entrainment's first action is to increase the velocity of the flow (DSI - HI 500; Straub, G.L. and Lamb, O.P., 1953).

2. Entrained air decreases the density of the liquid-air mixture; thus the flow velocity is increased. Eventually, an equilibrium is reached.

3. The increase of the flow velocity increases the friction of the fluid in motion with air, its effect is to decrease the flow velocity.

4. In hypercritical flow in a channel with rigid boundaries friction increases, essentially, because each element of roughness creates a small cavitation effect beyond it. Measurements on prototypes have shown that Manning's roughness coefficient exceeds measurements on the models.

Investigations and prototype measurements have been performed in many countries. Figure 6.5 illustrates the USBR suggested U_1 values. Results of tests performed on prototypes in Europe and Turkey are also shown on this diagram. This figure shows that a boundary indeed exists for the velocities of flow taking place on a channel with steep slopes. This limiting value is suggested to be $32 \cong 35$ m/sec

$$U_{max} \leq 32 \cong 35 \text{ m/sec}$$

Naturally this boundary value varies with the viscosity of the fluid and with the temperature of water. U_1, being determined, d_1 is

$$d_1 = \frac{Q}{U_1}$$

• Length of the basin

The length of the basin is given in Fig. 6.34. This value of the length of jump is smaller than the natural length of the jump. This result is due to the existence of chute blocks and sill.

• Water profile in the basin

The water profile is further illustrated in Fig. 6.35. When the water profile is known the pressure profile can be determined as shown on the same figure.

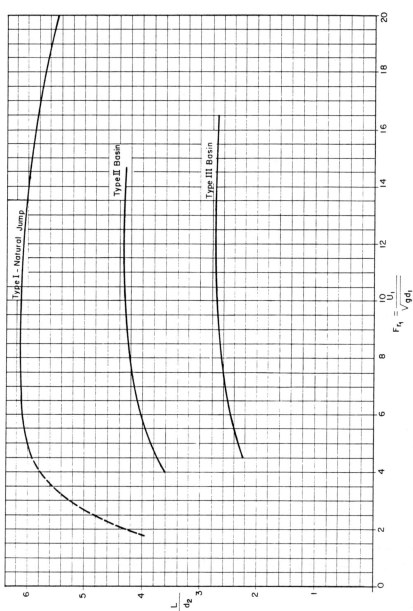

Figure 6.34. Length of jump on horizontal floor. (Basins I, II and III.) (USBR) *

* USBR, "Hydraulic Design of Stilling Basins and Energy Dissipators - 1958"

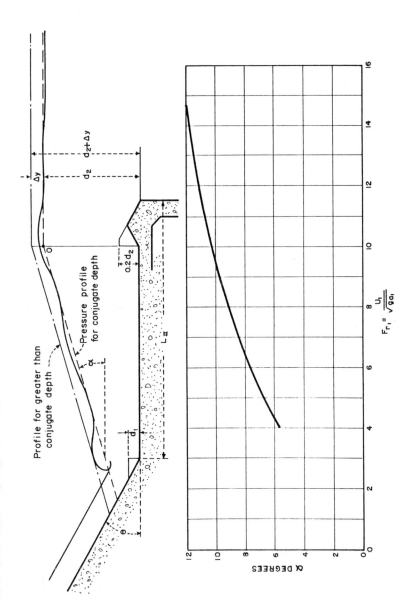

Figure 6.35. Approximate water surface and pressure profiles, Basin II. (USBR) *

---

* USBR, "Hydraulic Design of Stilling Basins and Energy Dissipators - 1958"

6.4.5 Design of Basin II

The following steps must be taken in designing the Basin II.

1. Locate the bottom of the basin at such level that the tailwater depth is equal to d_2 or as a maximum, to $1.05d_2$, American practice, or $0.93d_2$, European practice.

$$\text{Tailwater depth} = (0.93 - 1.05)d_2 \qquad (6.35)$$

2. Check the dimensions of the design on a scale model, if $F_{r1} \leq 4$. The Type II basin is efficient only for $F_{r1} \geq 4.5$.

3. Use Fig. 6.5 or similar computational procedure for determining U_1.

4. Compute the length of basin using Fig 6.34.

5. Figure 6.36 is suggested for determining the dimensions of chute blocks and dentated sill. If the method shown in the figure does not result in the requisite number of blocks, then the block intervals can be modified.

6. Consider d_2 in the design of the dentated sill and use Fig. 6.36 as the design guide.

7. The chute blocks and the sill blocks are not required to line up with each other.

8. A circular transition can be used for very small values of θ at the intersection of the channel with the horizontal apron.

The Basin II works efficiently for height differentials smaller than 70.00 m and $q \leq 150$ m^3/sec.

Example 6.08

The crest of an overflow dam is 45 m above the horizontal floor of the stilling basin. The angle, θ, between the horizontal and the downstream slope of the dam is 38°; thus its tangent is equal to 0.78. The hydraulic head of the spillway is 15.00 m and the linear discharge is

$$q = 0.495h \, (2gh)^{1/2} = 127 \text{ m}^3/\text{sec.m}$$

Design a Type II basin satisfying these conditions.

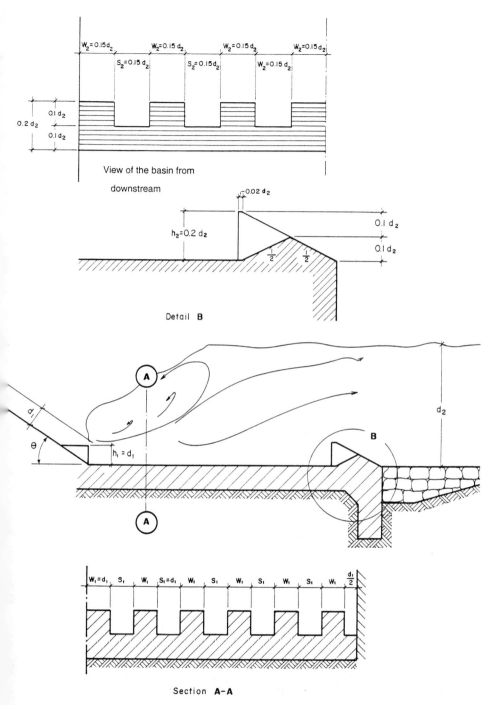

View of the basin from downstream

Detail **B**

Section **A–A**

Figure 6.36. **Recommended proportions (Basin II). (USBR).**[*]

[*] USBR, ("Hydraulic Design of Stilling Basins and Energy Dissipators" - Engineering, Mono: 25, 1958).

Solution

1. Figure 6.5 yields U_1 for

    total drop $= 45 + 15 = 60.00$ m

    $\quad$ H $\quad = 15.00$ m

    $\quad$ U_1 $\quad = 30.00$ m/sec < 32.00 m/sec

2. Compute d_1

$$d_1 = \frac{127}{30} = 4.23 \text{ m}$$

3. Compute the Froude number

$$F_{r1} = \frac{U_1}{(gd_1)^{1/2}} = \frac{30}{(9.81 \cdot 4.23)^{1/2}} = 4.66 > 4.5$$

A Type II Basin is the Solution for the Problem

4. Determine the tailwater depth using the solid line in Fig. 6.31

$$\frac{T.W.}{d_1} = 6.20 \text{ m} \qquad T.W. = d_2 = 6.20 \cdot 4.23 = 26.23 \text{ m}$$

Using Fig. 6.31, factor of safety is 2%. This margin of safety is too small. Assume a factor of safety equal to 7%. Choose a point on solid dotted line to suit this value, then

$$\frac{T.W.}{d_1} = 6.00 \qquad T.W. = 6.00 \cdot 4.23 = 25.38 \text{ m}$$

Adding 10% to this figure, the stilling basin apron should be dimensioned to develop a tailwater depth of

$$1.1 \cdot 25.38 = 27.92 \cong 28.00 \text{ m}$$

or $\qquad\qquad\qquad\qquad\qquad\qquad = 1.0674 \, d_2$

5. Determine the length, width and spacing of the chute blocks according to Fig. 6.36. The height, width and clear spacing between blocks are each 4.23 m. The end space is 2.115 m.

6. Using Fig. 6.36 the height of the dentated sill is determined as equal to $0.2d_2$ or 5.60, and the width and spacing as each $0.15d_2 = 3.93$ m.

7. Using Fig. 6.34, determine the length of the basin as

$$\frac{L}{d_2} = 3.80 \qquad L = 3.80 \bullet d_2 = 3.80 \bullet 28 = 106.40 = 106 \text{ m}$$

Figure 6.37 shows the final dimensions of the Type II Basin described in Example 6.08.

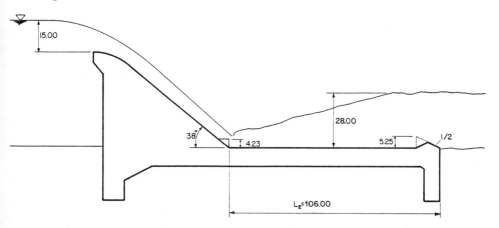

Figure 6.37. Type II Basin in Example 6.08.

6.4.6 Design of SAF Basin

The following rules are to be taken into consideration for designing the SAF basin.

1. Length of basin

For $F_{r1} > 3$ the length of basin is given by Eq. 6.36

$$L_H = 4.50 \; \frac{d_2}{F_{r1}^{0.76}} \tag{6.36}$$

2. The height of blocks at the entrance and at the middle of the apron is d_1 and the distance separating them is $(3/4 \; d_1)$. The blocks can bc in alignment in plan.

3. The design of the blocks in middle of the apron may follow the following rules:

• the distance of the blocks to the entrance: $\dfrac{L_H}{3}$

F. Sentürk

- the distance of the blocks from the side walls:

$$\min \frac{3}{8} d_1$$

- the blocks may cover $40 \cong 50\%$ of the basin width

- when the basin is trapezoidal in plan the above given proportion must be increased accordingly

4. The height of blocks in dentated sill must be

$$c = 0.070\ d_2 \tag{6.37}$$

5. The minimum value of the tailwater depth must be

$$d_2 = 1.40\ F_{r1}^{0.90}\ d_1 \tag{6.38}$$

6. The recommended value of the freeboard is

$$\text{Freeboard} = \frac{1}{3} d_2 \tag{6.39}$$

7. The side walls downstream of the basin will begin immediately downstream of the dentated sill, have a slope of 1/1 and form an angle of 45° with the main direction of the flow.

8. The remaining details must be solved by the designer according, to local conditions.

Example 6.09

Design a SAF stilling basin the hydraulic characteristics of which are given below:

$$d_1 = 0.183\ \text{m} \qquad U_1 = 12.2\ \text{m} \qquad \theta = 38°$$

Solution

1. Compute the Froude number as

$$F_{r1} = (U_1^2/gd_1)^{1/2} = 82.90^{1/2} = 9.10$$

2. Compute d_2 as

$$d_2 = \frac{d_1}{2}[-1 + (8\ F_{r1}^2 - 1)^{1/2}] \tag{6.21}$$

$$d_2 = \frac{0.183}{2}[-1 + (8 \bullet 83 + 1)^{1/2}] = 2.27\ \text{m} = 7.45\ \text{ft}$$

3. Compute the tailwater depth as

$$\text{T.W.} = 1.40 \ F_{r1}^{0.90} d_1 \qquad\qquad 6.38$$

$$\text{T.W.} = 1.40 \cdot 9.10^{0.90} \cdot 0.183 = 1.87 \text{ m} = 6.1 \text{ ft} < d_2$$

4. Compute the freeboard as

$$\text{Freeboard} = \frac{1}{3} d_2 \qquad\qquad 6.39$$

$$= \frac{2.27}{3} = 0.76 \text{ m} = 2.48 \text{ ft}$$

5. Compute L_H as

$$L_H = 4.5 \ (d_2 F_{r1}^{-0.76}) = 4.5 \ \frac{7.45}{9.10^{0.76}} = 6.26 \text{ ft} = 1.91 \text{ m}$$

6. Compute the height of blocks as

$$c = 0.07 d_2 \qquad\qquad 6.37$$

$$c = 0.07 \cdot 2.27 = 0.16 \text{ m} = 0.52 \text{ ft}$$

These dimensions can be checked using the chart in Fig. 6.32. Figure 6.38 shows the designed SAF basin (see also Problem 6.09).

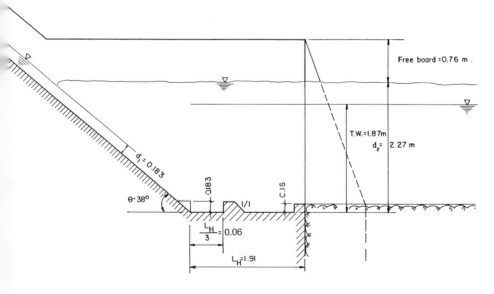

Figure 6.38. SAF Basin in Example 6.09.

349

6.4.7 Hydraulics of Ski Jumps

Numerous ski jumps, deflectors, etc. have been built since 1950 and the efficiency of the system for dissipating the active energy of water is found to be remarkable.

A ski jump is an engineering structure intended to project water coming with great velocity from the spillway far downstream from the dam. The rapid flow is mixed with air as it flow down the steep-sloped face of the spillway. The air is entrained with the flow, a boil occurs in the trajectory of the flow somewhere above the downstream water surface destroying the active energy and water falls into the river like rainfall. If the boil is not enough to dissipate the energy of flow completely its impact can cause erosion in the river bed, but usually far enough of the toe of the dam so that the erosion can not endanger the dam.

Different types of ski jumps are in use in engineering practice such as

- simple jump (flip buckets)

- deflectors

- double action deflectors

- mixed buckets.

A. Simple Jump

The simple jump or flip bucket has been frequently used in spillways since the middle of the 20th century. It consists of a concave section at the toe of the spillway section as shown in Fig. 6.39.

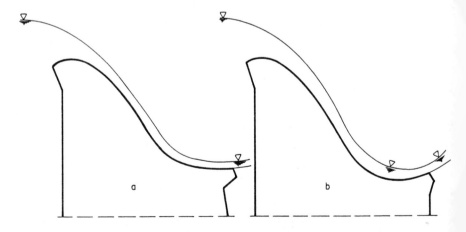

Figure 6.39. Types of flip buckets.

The nappe is directed upward and projected freely into the air. Advantages and disadvantages of the ski jumps are discussed below.

a. The advantages of the ski jumps

When foundation rock is sound enough this solution is very economical. The length of the lined or protected downstream channel is reduced and the foundation excavation is performed without any important groundwater problem. The construction of large stilling basins, large retaining walls, preventive measures to be taken against heavy erosion and foundation excavation problems can be avoided. Figure 6.40 shows the flip bucket for a very large dam, the flip bucket of the Keban Dam Spillway. This solution has eliminated the need for a very large stilling basin that would have been required to damp the energy of 17,000 m³/sec of water at the toe of a dam 205 m high.

Figure 6.41 shows the flip bucket in operation.

-   The dissipation of the destructive energy of spillway discharge in the air is shown in Fig 6.42.

-   In a narrow valley, the powerhouse of the hydroelectric plant can be located at the toe of the dam underneath the flip bucket. A good example of this type of arrangement is shown in Fig 6.43.

Figure 6.40. Flip bucket of the Keban Dam Spillway, Turkey.

Figure 6.41. The ski jump of the Keban Dam Spillway in operation.

Figure 6.42. The damping of the destructive energy of water in the air. The Aigle Dam Spillway (France) in action (Courtesy of Coyne and Bellier).

The water is projected far enough downstream so that erosion in the stream bed does not constitute any danger to the dam.

b. Disadvantages of ski jump

-   When the foundation material is not sufficiently sound, the vibration of the ski jump discharge can cause consolidation of the soil after construction is completed, which can lead to significant settlement and possible failure. Special precautions must be taken in designing the foundation which may cause this type of structure to be economical.

-   For small discharges the flip bucket acts like a stilling basin and dissipate the dynamic energy of the discharge under submerged conditions. Figure 6.44 shows such a case.

-   When the weather is cold the flip bucket projects water into the air in such a way that small spherical ice particles are formed in space. These spheres are projected downstream with great energy. If they strike structures such as powerhouses, walls or aprons they cause heavy damage; when deposited on highways they block traffic; and also make it difficult to walk.

**Figure 6.43. The hydroelectric power plant beneath the ski jump.
(The Monteynard Dam Spillway, France).**

353

**Figure 6.44. The flip bucket of the Seyhan Dam Spillway acting like
a stilling basin and dissipating the active energy.**

- In hot weather small particles of water projected with great energy
can strike the side slopes of the valley downstream and penetrate deeply
enough to cause landslides.

These advantages and disadvantages of the ski jump must be compared
with the advantages and disadvantages of stilling basins. The final decision
can be made at the end of a complete study of each particular case.

c. Design of simple ski jumps

Figure 6.45 shows a ski jump spillway. Its hydraulic design requires the
determination of:

- The elevation of point 0

- The determination of angle θ

- The determination of the radius of the circle which forms the jump

- The solution to be adopted for small discharges.

The first three unknowns are established as a function of the distance, x,
which is chosen so that any dangerous erosion occurs, at the toe of the
spillway. When x is determined, Eq. 3.11 yields the value of y, with the
angle-θ, given, or vice-versa. The designer generally chooses y as a
function of the topography and geology and computes the value of θ
accordingly. When θ is known, the center can be determined and the bucket

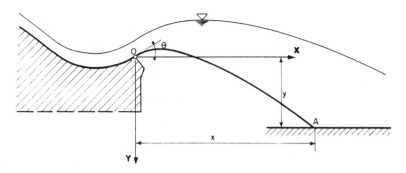

Figure 6.45. Definition of symbols in a ski jump.

radius, R, of the flip bucket be computed. USBR suggests the chart shown in Fig. 6.46a and 6.46b, for determining R, for submerged bucket dissipators. This diagram gives a preliminary value for R.

Scale model tests are necessary for the final adjustments. The following rules can be adopted for designing a ski jump:

1. Determine the exit elevation of the ski jump

Figure 6.47 shows two possible solutions for the choice of this elevation.

Solution No: 1

> The bottom elevation of the spillway downstream channel decreases up to a minimum elevation, 0, then stays constant. This design solves two problems:
>
> •   Small discharges are discharged freely downstream without additional head loss
>
> •   When the flow carries ice blocks or solid floating objects such as branches, etc., these objects are propelled downstream without causing significant damage.

The only disadvantage of this solution is that the required x distance shown on Fig. 6.45 cannot be attained even if the existing elevation is increased.

Solution No: 2

> The bottom elevation of the bucket is raised to (a).
>
> The design resolves the disadvantage of Solution 1; the distance, x, can be attained by the choice of an adequate exit elevation (position 3). The disadvantage of this design is that the flip bucket acts as a

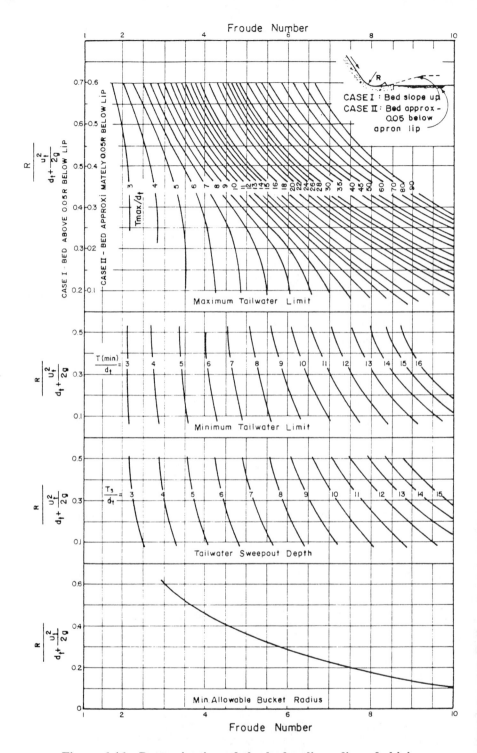

Figure 6.46a. Determination of the hydraulic radius of ski jump. (USBR).*

---

*   Design of Small Dams

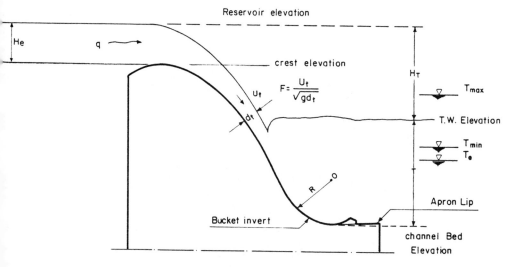

Figure 6.46b. Determination of the hydraulic radius of ski jump. (USBR).[*]

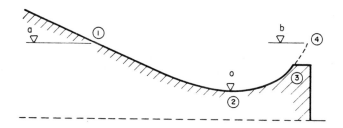

Figure 6.47. Different solutions for the exit level.

stilling basin and the jump is eliminated; water falls freely downstream from the exit causing great damage, if the elevation at 3 is raised higher to elevation at 4, the distance x increases and the stilling action of the flip bucket also increases, continuing this way with larger discharges (Fig. 6.48).

In both solution No. 1 and No. 2, the velocity decreases downstream of elevation 2. Two examples are given to illustrate this case (Maître, R., Obolensky, S., 1958).

• The Saint Etienne Cantalès Dam Spillway

In the 23.90 m horizontal exit channel downstream of elevation 2 the velocity decreases from 21.20 m/sec to 19.50 m/sec.

---

[*]   Design of Small Dams

Figure 6.48. Energy dissipating effect of a ski jump.

• The Chastang Dam Spillway

In the 12.80 m horizontal exit channel downstream of elevation 2 the velocity decreases from 24.20 m/sec to 22.80 m/sec (Example 6.11).

It is recommended that a velocity, U, equal to 0.9U, (U computed), be chosen at the bucket exit. $U_{computed}$ is the velocity defined by Fig. 6.5 or by direct computation using gradually varied flow formulas.

2. Design the exit as shown in Fig. 6.49 in form of a cantilever so that, even when the jump does not occur, water be projected far downstream of the foundations of the ski jump.

3. Do not forget that the form of the nappe in nature can be different from the computed form.

Figure 6.50 shows schematically the natural and the computed nappe. The air friction has compressed the water jet in the space and the distance x is reduced from x_1 to x_2. It is suggested that the following be chosen

$$x_2 = 0.90x_1 \qquad (6.40)$$

for a first approximation. It is very difficult and expensive to solve this problem on a scale model due to the difficulty of simulating the air friction. Figure 6.51 shows the flow nappe of the Chastang Dam Spillway. The influence of air friction is visible in this photograph.

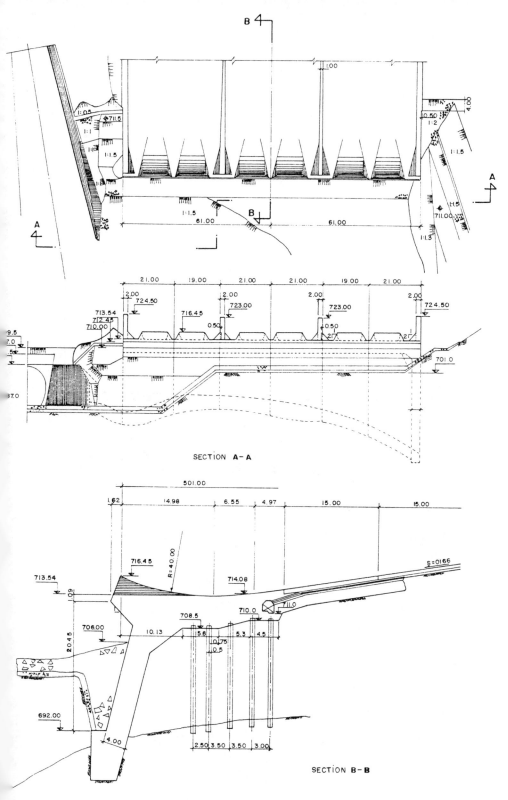

Figure 6.49. Terminal structure of a spillway downstream channel. (Courtesy of DSI).

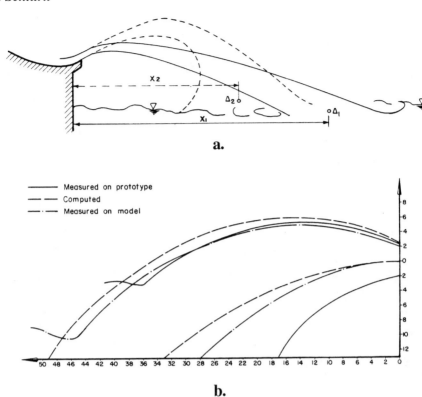

a.

Measured on prototype
Computed
Measured on model

b.

Figure 6.50. Theoretical and natural form of the flow nappe
downstream of the terminal structure.

Example 6.10

Design a ski jump the characteristics of which are given below, (Fig. 6.52)

$$Q = 2500 \text{ m}^3/\text{sec}$$

Width of the channel at the exit = 20.00 m

x = 100.00 m (Fig. 6.45)

U = 32.00 m/sec

S_b = 1.66

Solution

1. Use Eq. 3.11 for determining y. Assume $\theta = 45°$

$$U_o = 0.9U = 0.9 \cdot 32.00 = 28.80 \text{ m/sec}$$

Figure 6.51. Spilling flood water from the terminal structure of the Chastang Dam Spillway. (France).

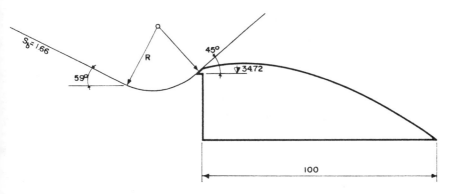

Figure 6.52. Flip Bucket in Example 6.10.

$$y = \frac{g}{2\,U_0^2 \cos^2 \theta}\, x^2 - x \tan\theta$$

Choose x as

$$x = \frac{100}{0.90} \cong 111 \text{ m}$$

$$y = \frac{g}{2 \cdot 28.80^2 \cos^2\theta} \ 111^2 - 111 \tan 45° = 34.72 \text{ m}$$

2. Determine R using Fig. 6.46

   U at the entrance to bucket = 32 m/sec

   U at the exit from the bucket = 28.80 m/sec

$$U_m = \frac{32.00 + 28.80}{2} = 30.40 \text{ m/sec}$$

$$d = \frac{2500}{30.40 \cdot 20} = 4.112 \text{ m}$$

$$F_r = \frac{30.40}{(4.112g)^{1/2}} = 4.787$$

Enter Fig. 6.46 with this value, then

$$(\frac{R}{d + U_m^2/2g}) = 0.38$$

and

$$R = 19.462 \text{ m}$$

Choose R = 20.00 m as a first approximation.

3. Extend the exit towards downstream as shown in Fig. 6.49.

Figure 6.53 shows the preliminary design of the ski jump.

Example 6.11

Compare the velocities measured on the spillway channel of the Saint Etienne Cantalès and the Chastang Dam Spillways.

Solution

The channel of the Saint Etienne Cantalès Dam Spillway is shown in Table 6.11, Fig. 6.54, and the channel of the Chastang Dam Spillway is shown in Table 6.12, Fig. 6.55.

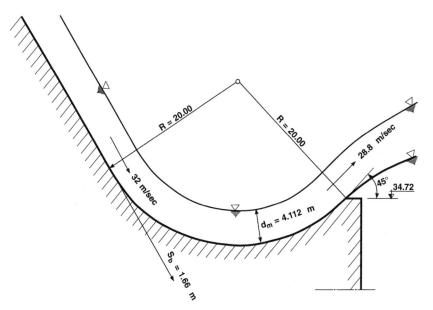

Figure 6.53. Solution of Example 6.10.

TABLE 6.11

Velocity measured on the downstream channel of the Saint Etienne Cantalès Dam Spillway

Station	Wetted cross section m²	Mean flow depth m	Measured mean velocity m/sec
A	11.90	1.62	20.40
B	11.35	1.32	19.50
C	10.95	1.17	21.20
D	10.95	1.08	21.20
E	12.25	1.10	18.95
F	18.95	1.65	14.50

Characteristics of this spillway:

Total drop = 37.00 m

$q = 71.00 \text{ m}^3/\text{sec.m}$

Theoretical velocity at the exit = 26.8 m/sec

$$\frac{U_{\text{meas.}}}{U_{\text{com.}}} = \frac{14.5}{26.8} = 0.54$$

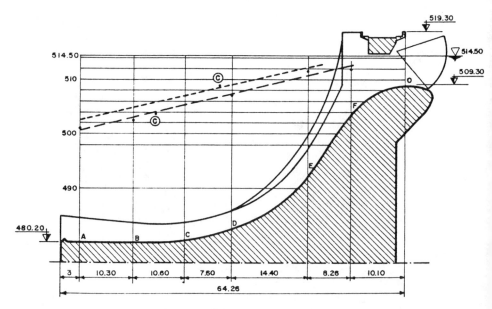

Figure 6.54. The Saint Etienne Cantalès Dam Spillway (France).

TABLE 6.12

**Velocities measured on the downstream channel of
the Chastang Dam Spillway**

Station	Wetted cross section m²	Mean flow depth m	Measured mean velocity m/sec
B	25.45	1.87	23.57
C	25.35	1.86	23.67
D	25.32	1.86	23.70
E	24.79	1.82	24.20
F	26.30	1.94	22.80

Characteristics of this spillway:

Total drop: 42.00 m

$$q = 147 \text{ m}^3/\text{sec.m}$$

Theoretical velocity = 28.7 m/sec

$$\frac{U_{meas.}}{U_{com.}} = \frac{22.80}{28.70} = 0.79$$

The two examples taken from in-situ measurements show the decrease in the velocities due to air friction (see also Table 6.1).

The designer may take this phenomenon into consideration when preparing the design.

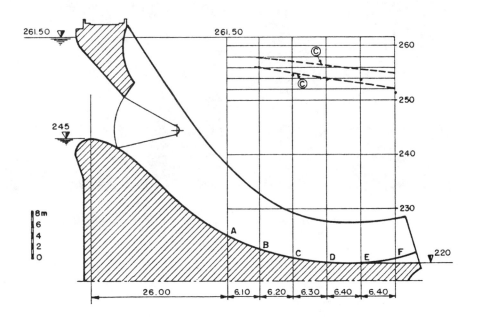

Figure 6.55. The Chastang Dam Spillway (France).

B. Deflectors

Deflector is a special type of flip bucket designed to change the direction of the water jet in plan. This phenomenon is accompanied by high vibrations and the design must carefully consider the side effects produced by such vibrations. A mathematical analysis of deflectors is tiresome, very costly and its accuracy is not high compared to physical model studies. Consequently deflectors are designed effectively using scale model studies. Figure 6.56 shows the deflector of the Castello do Bode Dam Spillway, Portugal.

**Figure 6.56. Deflector working, the Castello do Bode Dam Spillway,
Portugal. (Courtesy of Coyne and Bellier).**

The design of deflectors is performed by considering the following:

Step 1: Choose an adequate type of deflector.

Different type of deflectors exist. These are tested in nature and on model. When the axis of the downstream channel forms an angle greater than 20° with the axis of the downstream river there is a strong probability that the deflected jet will strike the valley side slope. This could cause landslides and even blockage of the main water course. The deflector should be so designed that the jet's impact is placed in the middle of the river.

If the valley is wide enough, the nappe spreads. On the other hand in a narrow valley the nappe is elongated. Figure 6.57 shows the elongated nappe of the Karakaya Dam Spillway, Turkey (see also Fig. 6.51).

In both cases the engineer may study the return currents due to the jet entering the downstream water surface and try to eliminate their damaging effect.

Figure 6.57. The Karakaya Dam Spillway.

Step 2: Choose a deflection angle smaller than 15°.

When the deflection is greater than 20°, the impact of the nappe will be close enough to the valley side slopes as to cause significant erosion and damage. The chosen value for the deflection angle is arbitrary.

Step 3: It is important that the transition be gradual from the direction of the spillway axis to the change in direction created by the deflector.

The radius of the transition curves may begin from infinity and decrease to the required value gradually. This rule is important in case that the transition presents a surface of rotation; then the loss of energy is minimized and the jet will be well formed. If not, abrupt changes in water surface can occur and disturb the uniformity of the flow. In preparing the design of such a hydraulic structure the energy must be dissipated at the section chosen by the designer and not upstream or downstream of it.

Step 4: For small discharges consider a breach at which the flow takes place without appreciable energy loss.

The dimensions at this particular section are a function of the discharge. Figure 6.57 shows an example of a deflector operation where the discharge is small. The flow passes through this section without interference by the deflectors; for greater discharges the flow will be deviated by them.

Step 5: Investigate carefully the foundation subgrade for the deflectors.

The subgrade is affected by vibrations generated by the flow. If the foundation subgrade is weathered rock or loose material, deflectors are to be avoided.

These five steps may be useful in preparing the preliminary design for deflectors. Model tests are necessary for final design.

C. Special Types of Deflectors

The flow nappe is guided by the deflectors. There are special types of spillways that are designed to spread the nappe across the valley width. They are called spreading type of spillways or spoon spillways. Flow passing over the spoon is spread and distributed into the width of the valley. This distribution of the discharge varies according to the shape of the spoon and minimizes its active energy.

Another type of spillway using deflectors generates a helicoidal nappe turning it along its axis and thus dissipating energy effectively. Figure 6.58 shows such a special type of spillway.

Figure 6.58. Special type of deflector.

6.5 PROBLEMS

Problem 6.01

Assuming $d \cos\theta = d$ in Example 6.02; solve the problem given in the example and compare the results.

Problem 6.02

Determine the flow characteristics in the channel described in Problem 6.01 using Bauer's procedure. Compare the results.

Data

$$k = 1.27 \ 10^{-4} \ m$$

Problem 6.03

Determine the flow characteristics in the channel described in Problem 6.01 using the Corps of Engineers procedure. Compare the results.

Problem 6.04

Compare the results obtained in Problem 6.01, 6.02 and 6.03.

Problem 6.05

Compute the air demand required to prevent cavitation damage in a spillway.

Data

The mean velocity along the spillway face

$$U_L = 35 \text{ m/sec}$$

The depth of flow at the air exit is equal to 3.00 m

Problem 6.06

Compute the length of the jump, the characteristics of which are given in Example 6.06, using Table 6.8 and compare the results.

Problem 6.07

Determine the loss of energy in a jump, the characteristics of which are given, on a horizontal floor.

Data

$$U_1 = 6.00 \text{ m/sec} \qquad d_1 = 0.50 \text{ m}$$

Determine the location of the jump.

Problem 6.08

The crest of an overflow dam is 50 m above the horizontal floor of the stilling basin. The angle (θ) between the horizontal and the downstream slope of the dam is 60°. The hydraulic head of the spillway is 10.00 m and the coefficient of discharge m = 0.495.

Design a Type II stilling basin satisfying these conditions (Example 6.08).

Problem 6.09

Design a SAF stilling basin corresponding to the conditions outlined in Example 6.09.

Discuss the result and compare with Type II basin.

Problem 6.10

Design the flip bucket of a spillway the characteristics of which are given as:

Data

$$Q = 3,000 \text{ m}^3/\text{sec}$$

$$x = 125.00 \text{ m}$$

Width of the channel at the exit = 22.00 m

$$U = 31.00 \text{ m/sec}$$

$$S_b = 0.732$$

Water surface elevation in the river = 0
(Example 6.10)

Problem 6.11

Determine the position of point C in Fig. 1, Problem 6.11, in such a way that the jump is confined in the space shown on the same figure ($\ell = 180.00$ m).

F. Sentürk

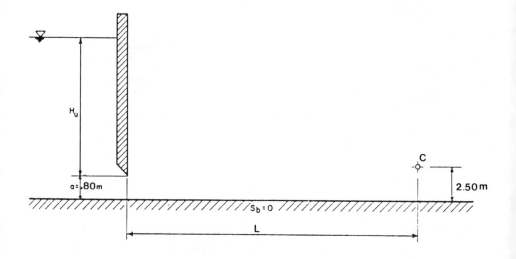

Fig. 1. Problem 6.11.

Data

Orifice opening: 0.80 m

$q = 10 \text{ m}^3/\text{sec.m}$

$n = 0.014.$

Chapter 7

STATIC AND DYNAMIC
LOADS ON SPILLWAY

7.1   INTRODUCTION

The spillway is a dam over which the flood waters spill. The forces that act on dams are the same as those acting on spillways. In addition to these forces, the dynamic loads due to the motion of water are to be taken in consideration in studying the stability of spillways.

In general the forces involved in a stability analysis can be divided into two distinct groups:

- Static forces

- Dynamic forces

The following is a list of principal forces acting on a spillway.

STATIC FORCES:

1.   Vertical loads due to the construction material of the spillway.

2.   Vertical loads due to water pressure.

3.   Horizontal pressures due to reservoir water load.

4.   Horizontal pressures due to tailwater.

5.   Pressures due to silt loads.

6.   Ice pressure.

7.   Temperature stresses.

DYNAMIC FORCES:

1.   Vertical loads due to earthquake acceleration acting on the structure.

2. Vertical loads due to earthquake forces acting on the water in the reservoir and on the tailwater.

3 Horizontal inertia forces due to earthquake acceleration acting on the structure.

4. Horizontal increase in pressure due to earthquake acceleration acting on the reservoir water and on the tailwater.

5. Wave action.

6. Dynamic forces exerted by flowing water on the overflow crest.

7. Dynamic forces exerted by flowing water on the bottom of the downstream channel.

8. Uplift pressures.

9. Pressures due to percolating water.

10. Pressures due to wind action.

11. Dynamic forces exerted by flowing water on flip buckets and deflectors.

12. Dynamic forces due to vibrations of the ski jump exit section.

7.2 STATIC LOADS

7.2.1 Forces Due to the Weight of the Construction Material

The construction material used in spillways is usually concrete. Masonry was also used in early spillways. Vertical loads due to the weight of concrete or masonry can be computed easily. They are based on the unit weight of the material and its volume. The weight of the construction material varies with the material chosen. If concrete is used for constructing the spillway, its weight is usually taken as 2.4 T/m^3 (150 pounds per cubic foot) for design purposes. The concrete being assumed impermeable, the density 2.4 T/m^3 is used without subtracting 1 T/m^3 due to submergence.

In earth and rockfill dams the computation of the vertical load due to the weight of the fill material is more complicated. The density of the dry material may be considered separately from the density of the wet material. The phreatic water surface line is determined and shown on the cross section of the impervious core (see Chapter 9). The volume of the fill material above the phreatic flow line is multiplied by the dry density and the volume of the fill material below it by the dry density minus one and added for obtaining the vertical load due to the weight of the construction material.

Different kinds of fill material are used such as sand, cobbles, rock, silt, clay, etc., for forming the dam embankment. The density varies from one material to the other. The appropriate density must be applied for each respective material for determining the vertical loads. The density of the chosen construction material is determined carefully under different loading conditions. For preliminary estimate of their mean values, standard soil mechanics books may be used (Hough, 1957; USBR, Design of Small Dams; Post, G., Londe, P., 1953). Table 7.1 reproduces one of these standards as an example. In Turkey, it is recommended that 0.95 of the computed load be used as the design load.

7.2.2   External Water Pressure, Pressure due to Silt Load

A.   External Water Pressure

Certain assumptions regarding the external water pressures include,

1.   The sudden increase in water level in the reservoir is not taken into consideration for determining the hydrostatic pressures.

2.   The phreatic water line stays constant during sudden drops in the water level in the reservoir.

3.   For sudden drawdown conditions, it is recommended that one-third of the original water level be considered as a load against the retaining walls.

Hydrostatic pressure is computed according to procedures outlined in hydraulic text books. The intensity of liquid pressure (p) at depth (h) below water surface is:

$$p = \gamma_\omega h \qquad (7.1)$$

where (γ_ω) is the density of water. On an inclined surface of unit width the hydrostatic force can be computed as follows (Fig. 7.1).

The inclined surface AB supports the volume of water defined by ABC; the width of the prism being equal to unity. The weight of water contained in the prism ABC is

$$F_V = \gamma_\omega \frac{hL}{2} \qquad (7.2)$$

TABLE 7.1

Average Properties of Soil Voids * (Hough, B. K.)

Typical name of the material	Particle size and gradation — Approx. size range — mm D_{max}	mm D_{min}	Approx. mm D_{10}	Void ratio e_{max} loose	e_{cr}	e_{min} dense	Porosity (%) n_{max} loose	n_{min} dense	Dry weight min loose	max dense	Unit weight (T/m³) wet wt. min loose	max dense	sub.wt min loose	max dense
1. Granular material														
a. Equal spheres (theoretical value)	-	-	-	0.92	-	0.35	47.6	26.0	-	-	-	-	-	-
b. Standard Ottawa sand	0.84	0.59	0.67	0.80	0.75	0.50	44.0	33.0	1.47	1.76	1.49	2.10	0.91	1.10
c. Clean, uniform sand (fine or medium)	-	-	-	1.00	0.80	0.40	50.0	29.0	1.33	1.89	1.35	2.18	0.83	1.18
d. Uniform, inorganic silt	0.05	0.005	0.012	1.10	-	0.40	52.0	29.0	1.28	1.89	1.30	2.18	0.82	1.18
2. Well graded material														
a. Silty sand	2.00	0.005	0.020	0.90	-	0.30	47.0	23.0	1.39	2.03	1.41	2.27	0.86	1.27
b. Clean, fine to coarse sand	2.00	0.05	0.090	0.95	0.70	0.20	49.0	17.0	1.36	2.21	1.38	2.37	0.85	1.37
c. Micaceous sand	-	-	-	1.20	-	0.40	55.0	29.0	1.22	1.92	1.23	2.21	0.77	1.20
d. Silty sand and gravel	100.0	0.005	0.020	0.85	-	0.14	46.0	12.0	1.43	2.34	1.44	2.48	0.90	1.48
3. Mixed soils														
a. Sandy or silty clay	2.00	0.001	0.003	1.80	-	0.25	64.0	20.0	0.96	2.16	1.60	2.35	0.81	1.38
b. Skip-graded silty clay with stone or rock frag.	2.50	0.001	-	1.00	-	0.20	50.0	17.0	1.35	2.24	1.84	2.42	0.85	1.48
c. Well-graded gravel, sand, silt and clay mixture	2.50	0.001	0.002	0.70	-	0.13	41.0	11.0	1.60	2.37	2.00	2.50	0.99	1.58
4. Clay soils														
a. Clay (30-50% clay sizes)	0.05	0.5 μ	0.001	2.40	-	0.50	71.0	33.0	0.80	1.79	1.51	2.13	0.50	1.13
b. Colloidal clay	0.01	10 A°	-	12.0	-	0.60	92.0	37.0	0.21	1.70	1.14	2.05	0.13	1.05
5. Organic soils														
a. Organic silts	-	-	-	3.00	-	0.55	75.0	35.0	0.64	1.76	1.39	2.10	0.40	1.10
b. Organic clay (30-50% clay size)	-	-	-	4.40	-	0.70	81.0	41.0	0.48	1.60	1.30	2.00	0.29	1.00

* Granular materials may reach e_{max} when dry or only slightly moist, clays can reach e_{max} only when fully saturated.
Tabulation is based on C = 2.65 for granular soil, C = 2.7 for clays, and C = 2.6 for organic soils.

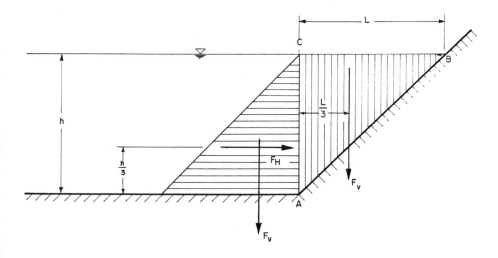

Figure 7.1. Hydrostatic force acting on the inclined surface AB.

where   L   is the horizontal dimension of the prism

       h   is the depth of water, and

      γ_ω is the specific weight of water.

The horizontal component of the hydraulic pressure is

$$F_H = \gamma_\omega \frac{h^2}{2} \tag{7.3}$$

This force acts at a distance (h/3) from the bottom of the reservoir. The hydrostatic pressures acting on a spillway are shown in Fig. 7.2.

The hydraulic structure (Fig. 7.2) is in equilibrium under the influence of the forces F_{H1}, F_{H2}, F_{V2}, F_{V3}, G, F_U and the foundation reaction. G is the weight of the structure plus the weight of water passing over it; F_{H1} is the horizontal component of hydrostatic pressure of water in the reservoir. The vertical component does not exist due to vertical upstream face of the spillway. F_{H2} is the horizontal component of hydrostatic pressure of tailwater; F_{V2} is the vertical component. F_U is the uplift pressure.

The forces F_{H3} and F_{V3} are due to the nappe supported by the spillway body. These forces may be positive or negative due to the existence of subpressure over the profile (see Chapters 3 and 4).

The calculation for static forces acting on a spillway is given in Example 7.01.

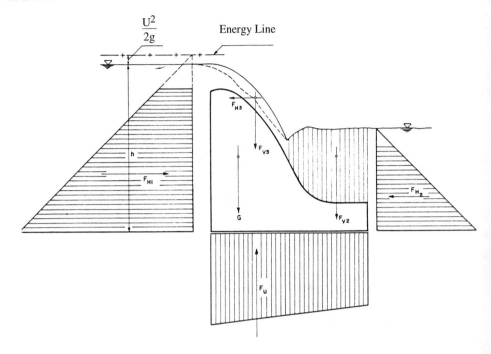

Figure 7.2. Horizontal Hydrostatic Pressures, F_{H1}, F_{H2}, due to Water in the Reservoir and to Tailwater.

B. Silt Pressure

The silt is carried into the reservoir in the form of a density current and as the suspended load (see Chapter 8). It is deposited against the upstream face of the dam. When it is in suspension it acts as a liquid and the liquid solid mixture has a density varying between 1.3 kg/m³ and 1.60 kg/m³. The pressure of consolidated silt is evaluated using the Rankine's formula:

$$H_s = \frac{\gamma_s' h^2}{2} \left(\frac{1 - \sin\varnothing}{1 + \sin\varnothing} \right) \qquad (7.4)$$

where

H_s is the horizontal component of silt loads

γ_s' is the submerged specific weight of material

h is the thickness of the material, and

$\varnothing$ is the angle of internal friction

The angle of internal fraction for submerged material is given in classical books (Simons, D.B. and Şentürk, F., 1992). The unit specific weight of submerged silt can be estimated using Eq. 7.5

$$\gamma'_s = \gamma(\frac{\lambda-1}{\lambda}) = \gamma - \gamma_\omega(1-k) \tag{7.5}$$

where

    γ'_s is the submerged weight of material in T/m^3

    γ is the unit dry weight of the material in T/m^3

    γ_ω is the unit weight of water in T/m^3

    k   is the percentage of voids, expressed as a decimal

    λ is the specific gravity of the solid particle (mean value 2.67)

γ, can be estimated from samples of silt deposits in neighboring reservoirs. The unit weight of saturated silt is

$$\gamma_{ss} = \gamma + k\gamma_\omega = \gamma_\omega + \gamma \frac{\lambda - 1}{\lambda} \tag{7.6}$$

where

    γ_{ss} is the saturated weight of solid material.

Example 7.01

Determine the foundation reactions in the gravity dam shown in Fig. 7.3.

Solution

The following forces act on the gravity dam in Fig. 7.3, where a slice of 1.00 m thick is shown

    ΣA is the resultant of gravitational forces

    ΣP is the resultant of hydrostatic pressures

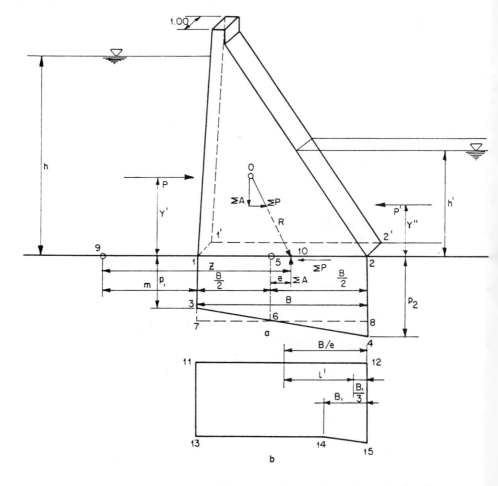

Figure 7.3. Gravity dam subject to hydrostatic and gravitational forces.

1, 2, 3, 4   is the diagram of foundation pressures or reactions (this diagram does not include the uplift pressure)

11, 12, 13, 14, 15   is the diagram of foundation pressures or reactions (this diagram includes uplift pressures)

1, 2, 8, 7   is the foundation reaction diagram if the reaction is uniform.

The shape of the dam in Fig. 7.3 can vary, but the forces acting upon it and the foundation reactions do not change.

Assume that the foundation area is 1-2-1'-2' and the center of gravity is located at point 5. If 1-2-1'-2' is rectangular, then point 5 divides the section in two equal parts, and

$$1 - 5 = 2 - 5 = \frac{B}{2} \qquad (7.7)$$

The dam generates two kind of stresses on the foundation:

• stresses due to gravitational pressures, and

• stresses due to bending moments.

Stresses due to homogeneous gravitational pressure:

$$\frac{\Sigma A}{A_o} \qquad (7.8)$$

Stresses due to eccentric loading:

$$\frac{\Sigma(py) \pm \Sigma(Ax)}{I} \frac{B}{2} \qquad (7.9)$$

where A_o is the area of the base.
        Uplift is not included at this time.
        Equation 7.9 can also be written as

$$\pm \frac{\Sigma(A)e}{I} \frac{B}{2} \qquad (7.10)$$

where A_o is the area 1-1'-2-2' (Fig. 7.3).

             A   is the weight of the dam body.

     A_a and A_y are the weight of water (Fig 7.4a).

             Ax   is the moment about the center of gravity of the base, point 5. It is positive in clockwise, negative in counter clockwise direction.

             P_y is the moment about the center of gravity of the base, point 5. It is positive in clockwise, negative in counter clockwise direction.

             x,x'   are the horizontal moment lever arms measured from the center of gravity of the base.

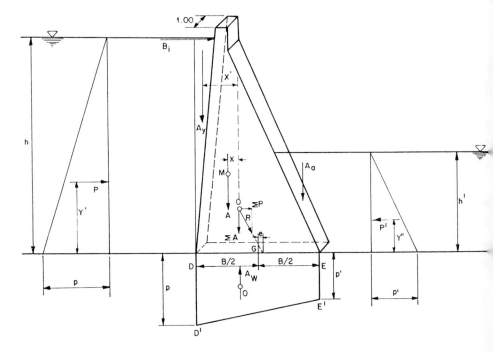

Figure 7.4a. Definition of forces without uplift.

I   is the moment of inertia of the base about its center of gravity.

y', y"   are the vertical distances to the points where the moments are computed, from the center of gravity, and

e   is the eccentricity of the sum of all the loads, or the distance from the center of gravity, 5, to the point of intersection, 10, of the resultant R, with the base (Fig. 7.3).

The moment about an arbitrary point (9) is

$$Z \sum A = \sum Px_9 + \sum Ax_9$$

$$Z = \frac{\sum Px_9 + \sum Ax_9}{\sum A} \qquad (7.11)$$

and

$$e = Z - (\frac{B}{2} + m) \qquad (7.12)$$

(p) is equal to the sum of the stresses due to the gravitational pressure and the stresses due to eccentric loading, then

$$p = \frac{\Sigma A}{A_o} \pm \frac{\Sigma Py + \Sigma Ax}{I} \frac{B}{2} \qquad (7.13)$$

or

$$p = \frac{\Sigma A}{A_o} \pm \frac{\Sigma Ae}{I} \frac{B}{2} \qquad (7.14)$$

Applying Eq. 7.13 to the toe of the dam yields

$$p_1 = \frac{\Sigma A}{A_o} - (\frac{\Sigma Py + \Sigma Ax}{I}) \frac{B}{2} \qquad (7.15)$$

and

$$p_2 = \frac{\Sigma A}{A_o} + (\frac{\Sigma Py + \Sigma Ax}{I}) \frac{B}{2} \qquad (7.16)$$

or

$$p_1 = \frac{\Sigma A}{A_o} - \frac{\Sigma Ae}{I} \frac{B}{2} = \Sigma A (\frac{1}{A_o} - \frac{e}{I} \frac{B}{2}) \qquad (7.17)$$

and

$$p_2 = \frac{\Sigma A}{A_o} + \frac{\Sigma Ae}{I} \frac{B}{2} = \Sigma A (\frac{1}{A_o} + \frac{e}{I} \frac{B}{2}) \qquad (7.18)$$

The total pressure diagram is then as shown in Fig. 7.3 (trapezoid 1-2-3-4). Applying Eqs. 7.15 and 7.16 to a gravity dam as an example, gives,

The base is the rectangle 1-1' -2-2'

The dimension 1-1' is equal to unity

The dimension 1-2 is assumed equal to B

The projection 5 of the center of gravity of the base, on 1-2, divides B into two equal parts

$$A_o = 1 \cdot B \qquad\qquad I = B^3/12$$

Substituting these values in Eqs. 7.17 and 7.18 yields

$$p_1 = \Sigma A \left(\frac{1}{B} - \frac{e}{B^3} 12 \frac{B}{2}\right) = \frac{\Sigma A}{B} \left(1 - \frac{6e}{B}\right) \tag{7.19}$$

and

$$p_2 = \frac{\Sigma A}{B} \left(1 + \frac{6e}{B}\right) \tag{7.20}$$

When the reservoir is empty, then (e) is to the left of the center of gravity and

$$p_1 = \frac{\Sigma A}{B} \left(1 + \frac{6e}{B}\right) \tag{7.21}$$

$$p_2 = \frac{\Sigma A}{B} \left(1 - \frac{6e}{B}\right) \tag{7.22}$$

Uplift acts on the foundation reaction as follows,

1. The reactions p and p' shown in Fig. 7.4a are smaller, respectively, than the reactions p_1 and p_2 shown in Fig. 7.3. Apply Eqs. 7.19 and 7.20.

2. Figure 7.3 shows, $p > p_1$, on the foundation reaction diagram 11-12-13-14-15 with 11-13 > 1-7 and 11-13 < 12-15. The line 14-15 is parallel to (3-4).

Figure 7.3 represents a case in which the uplift pressure is greater than the pressure at the heel of the dam, p_1. The foundation pressure will have to be revised due to the excessive uplift. This is done by assuming a horizontal crack extending from the upstream face toward the downstream face to a point where the vertical stress of the adjusted diagram equals the uplift pressure, p_3, as shown below.

Then

$$p_1 = p_3 = p_{(11-13)} \quad \text{(Fig. 7.3 and 7.4b)}.$$

and

$$p_2 = p_{(12-15)} \tag{Fig. 7.3}$$

$$p_2 = \frac{2(\Sigma A - p_3 B)}{B_1} + p_3$$

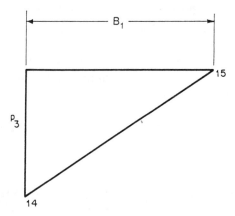

Figure 7.4b. Uplift pressure diagram.

7.2.3 Ice Pressure

When the temperature stays below freezing, the surface of a reservoir is covered with ice. Thickness of the ice cover increases with the decrease of the temperature of atmosphere. Melting ice exerts a thrust on the dam or spillway or gates. This thrust attains high values; the designer, then, has to take this thrust into account to complete the equilibrium analysis of the structure.

Ice adheres to the surface of the wall or the inclined slope of the dam. This adherence is defined by an adherence factor varying with the nature of the surface. During melting period the ice pressure increases. The ice being incapable of supporting bending moment breaks and large pieces of ice are directed towards the dam body or the piers of the spillway. This phenomenon is described by Korzhavin, K.N. (1972) as follows:

> "Coming in contact with the ice cover the vertical cutting edge of the pier apron penetrates into the ice sheet about 0.1 - 0.5 m causing local crumpling and subsequent failure, with two or three concussion cracks forming the so-called ice cantilevers that break off at a distance equal to 3 to 6 ice thicknesses.
>
> The decay pattern of an ice floc is mainly governed by its kinetic energy as well as by the shape, dimensions and material of the pier. With a sufficiently strong ice cover the width of the pier section subjected to ice pressure is not more than 0.5 - 0.8 of the maximum width of the pier. An inclined apron is more effective

in cutting the ice sheet from below and inducing failure by shear, thus facilitating ice passage.

The solution to the problem of ice impact on the structure requires an insight into the mechanical and physical properties of ice. But it is possible to state that:

- the ultimate compressive and bending strength of river ice is reduced from 1.5 -3 times by the beginning of the break-up period.

- owing to local crumpling the ice pressure on the structure may be increased 1.8 to 2 times.

- the ultimate strength both in flexure and compression is reduced inversely to blocks' dimensions."

A. Computation of Thickness of Ice

Empirical formulas exist for computing the thickness of ice. The Stefan's formula is reproduced as an example:

$$\varepsilon = (2KT.t/Le)^{1/2} \qquad (7.23)$$

where

ε: thickness of ice

K: heat conductivity of ice at 0 C° (457.92 cal/cm per day)

L: a parameter equal to 80 cal/gr

e: a parameter, equal to 0.99 gr/sm³ at 0 C°

t: temperature, in C° Celsius, and

T: duration of freezing period

B. Computation of Ice Pressure

An approximation is given by USBR for computing ice pressure when ice thickness is known. This chart prepared by Rose shows the thrust in kips for thicknesses up to 1.20 m for air temperature rise of 5°, 10° or 15 F° per hour, respectively (curves A, B and C), (Fig. 7.5).

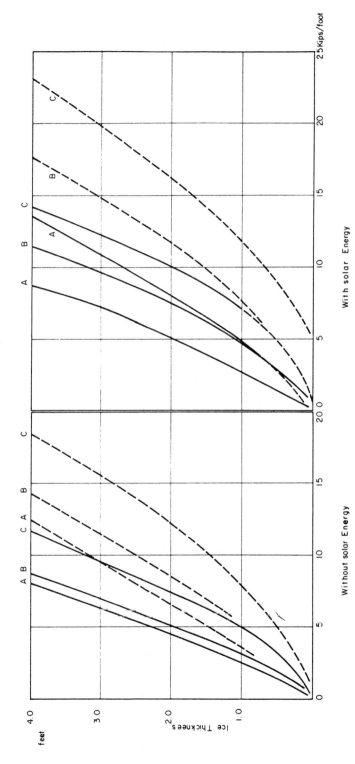

Figure 7.5. Ice thrust in relation to ice thickness, air temperature rise and restraint (USBR).

Example 7.02

Compute the ice thickness in Karakuz Dam Reservoir.

Data

- Duration of the freezing period, T: 127 days
- Minimum temperature at dam site: variable between 0° and -35 C°.

Solution

$$C° = \frac{tT}{2} \text{ (day)}$$

$$C°day = \frac{127 \cdot 35}{2} = 2,222.5 \text{ (mean value)}$$

$$\varepsilon = (\frac{2 \cdot 457.92 \cdot 2,222.5}{80 \cdot 0.99})^{1/2} = 160 \text{ cm}$$

Example 7.03

Compute the ice thrust in Karakuz Dam (Example 7.02).

Solution

Table 7.2 summarizes the result obtained by using Fig. 7.5.

TABLE 7.2

Ice thrust in the Karakuz Dam (Increase of temperature: 15 F°)
No lateral restraint, solar energy considered.

Ice thickness ft	Ice thrust kips per linear foot
1	7.50
2	10.50
3	12.50
4	14.00

7.2.4   Temperature Stresses

Temperature changes produce stresses due to restraints that the abutments impose upon any attempt of the dam to change its dimensions. It is obvious that the temperature stresses are important in concrete dams and have a less significant effect in fill-type dams. The existence of such stresses is pointed out in this chapter but their quantitative computations are beyond the scope of this book.

7.3   DYNAMIC LOADS ON SPILLWAYS

Dynamic loads acting on spillways are enumerated in section 7.1. Their numerical computation is given in this section.

7.3.1   Uplift Forces

Water flowing in a porous medium is subject to a head loss which can be computed using Darcy's law. When the hydraulic head is known at a given point, it can be calculated at a second point as a function of the distance separating these two points. A privileged path that water follows is the interface between the base of the spillway and the subgrade. Lane suggested a method for determining the head loss in this case.

These two methods are given as follows:

A.   Lane's Procedure

Lane suggests a simple method for computing the uplift

$$L = CH \tag{7.24}$$

where

L   is the path followed by the water

C   is a coefficient, and

H   is the effective hydraulic head at a given point.

Lane assumed that if the path followed by the water forms an angle greater than 45°, L will be considered as though the angle is smaller than or equal to 45°, L is reduced to L/3.

Lane's coefficient C is given in Table 7.3. The values considered in this table are selected from those commonly used in Europe. Different authors suggested different values. Differences are as much as 25%.

The application of this method is given in Example 7.04.

TABLE 7.3

Values of Lane's coefficient C

Nature of the foundation	C
Sound rock exempt of fissures, the base of spillway is unified with the foundation by means of contact injections	1.00
Sound rock without fissures	2.00
Rock with fissures and cracks, with cracks blocked with clay	2.50
Clay	2.00-2.50
Rock with fissures and cracks	3.00 - 5.00
Natural soil	6.00
Sand	5.00 - 7.00
Gravel	11.00

Example 7.04

Draw the uplift diagram of the Kiralkizi Dam Spillway.

Data

Maximum water elevation in the reservoir   815.75

Apron elevation downstream of the spillway   781.83

- The cross section of the spillway is shown in Fig. 7.6
- Nature of the foundation soil: sound rock
- Contact injection is applied
- Depth of water at the toe of the spillway is 2.25 m for Q_{max}

Solution

- The foundation rock is compact and contact injection blocks the circulation of percolation water, then the coefficient C is chosen equal to 1.00,

$$L = CH = H$$

390

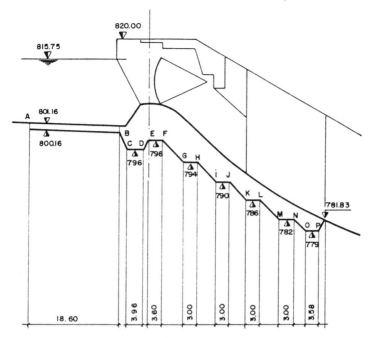

Figure 7.6. Cross section of the Kiralkizi Dam Spillway.

- Determination of most dangerous loading condition

  In gated spillways the most dangerous loading condition corresponds to the case where the gates are closed and reservoir full.

  In case that the gates are open, water flowing in the discharge channel will reduce the load. The flow depth corresponding to Q_{max} is equal to 2.25 m, then the hydraulic load is

  $$H = 815.75 - (781.83 + 2.25) = 31.67 \text{ m}$$

  For closed gates and a full reservoir, the hydraulic load becomes

  $$H = 815.75 - 781.83 = 33.92 > 31.67$$

- The existing percolation length in the structure is

  $$L = 40.71 \text{ m} \quad \text{(Table 7.4)}$$

- The foundation of spillway is equipped with drains.

TABLE 7.4

Uplift Forces in the Kiralkizi Dam Spillway

Stations	Distances	Drop of pressures	Elevations geomet.	Elevations comp.	Final pres.
	m	m	m	m	m
1	2	3	4	5	6
A			800.16	815.75	15.59
	18.60 H	5.1659			
B			800.16	810.58	10.42
	4.16 V	3.4662			
C			796.00	807.12	11.12
	3.96 H	1.0998			
D			796.00	806.02	10.02
	2.00 V	1.6664			
E			798.00	804.35	6.35
	3.50 H	0.9721			
F			798.00	803.38	5.38
	4.00 V	3.3328			
G			794.00	800.05	6.05
	3.00 H	0.8332			
H			794.00	799.21	5.21
	4.00 V	3.3328			
I			790.00	795.88	5.88
	3.00 H	0.8332			
J			790.00	795.05	5.05
	4.00 V	3.3328			
K			786.00	791.71	5.71
	3.00 H	0.8332			
L			786.00	790.88	4.88
	4.00 V	3.3328			
M			782.00	787.55	5.55
	3.00 H	0.8332			
N			782.00	786.72	4.72
	2.42 V	2.0164			
O			779.58	784.70	5.12
	3.58 H	0.9943			
P			779.58	783.70	4.12
	2.25 V	1.8747			
R			789.73	781.83	0
		33.9199			

Vertical percolation length = 26.83   Total length = 40.71

The influence of these drains is not included in the computation of the percolation length and is considered to be a margin of safety.

Then the energy gradient is:

$$S - \frac{33.92}{40.71} = 0.8332$$

The uplift computation is summarized in Table 7.4. The uplift diagram is shown in Fig 7.7.

The definition of columns of Table 7.4 is as follows:

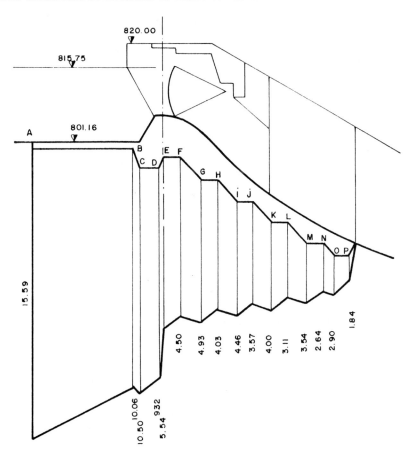

Figure 7.7. Uplift diagram of the Kiralkizi Dam Spillway.

Column 1:   Stations

Column 2:   Distance between stations

Column 3:   S • distance between stations (Column 2)

Column 4:   Elevation of stations

Column 5:   Column 4 - Column 3

Column 6:   Column 5 - Column 4

B.   Flow Net Procedure

The uplift can also be determined by flow net analysis. The method is explained in books on hydrodynamics. Only a summary is given in this chapter.

Darcy's law for steady flow at low rates through a granular non-cohesive material is

$$Q = -kA \frac{dh}{ds} \tag{7.25}$$

or

$$U = -k \frac{dh}{ds} \tag{7.26}$$

where

Q is the flow rate, m^3/sec

U is the mean velocity, m/sec

A is the wetted area of flow path, m^2

dh is the difference in head between two stream lines

ds is the distance over which the head loss prevails

$\frac{dh}{ds}$ is the piezometric gradient along the flow path, and

k is the coefficient of permeability, m/sec.

This equation is valid for

$$R_e = \frac{UD}{\mu} < 1 \tag{7.27}$$

where

R_e is the Reynolds number related to the grain size

D is the representative grain diameter of the medium where the flow takes place, and

ν is the kinematic viscosity.

Equation 7.25 meets the requirements of potential flow for

$$\emptyset = -kh \tag{7.28}$$

then

$$U = -k \frac{dh}{ds} = \frac{d\emptyset}{ds} \tag{7.29}$$

Equation 7.29 defines Darcy's law of flow in a porous medium. Flow net analysis is therefore applicable. For flow in contact with a solid boundary, such as the base of a concrete dam and its foundation, and through a material which is isotropic as regards permeability, such as a uniform sand material for example, the flow net is determined by the form of the confined boundaries. Then the base of a dam resting on the foundation constitutes one stream line. Figure 7.8 shows a dam and its foundation. The hydraulic head is H, which exists between the upstream and downstream water levels. This head drives the flow taking place through the foundation soil. Assume that the $d\emptyset$ is constant between two stream lines, then it can be obtained by dividing H by the number of stream lines. If this number is shown as (n), the head (ΔH) corresponding to each channel is

$$\Delta H = \frac{H}{n} \tag{7.30}$$

n, is an arbitrary number chosen by the engineer.

The method suggested by Casagrande for obtaining the flow net is outlined as follows:

Step 1: Define the boundaries of the flow. These boundaries constitute the limiting stream lines. The base of the dam is, for example, a boundary. A second example can be the existence of an impermeable clay layer underground.

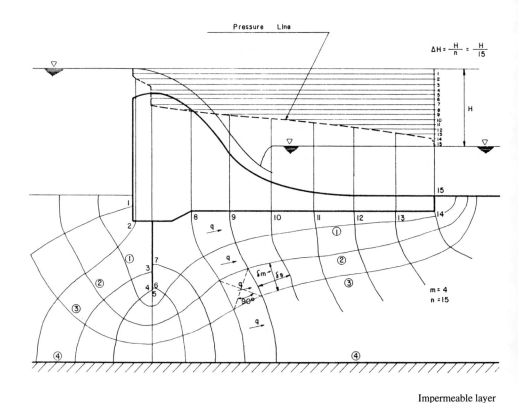

Figure 7.8. Percolation under a concrete dam.

Step 2: The flow takes place between the impermeable layer and the concrete base of the dam. Divide this space into (n) parts.

Step 3: Draw the equipotential lines, similar to the equipotential lines and stream lines as shown in Fig 7.8. These lines are orthogonal to each other.

Step 4: If the polygons formed by the intersection of the stream lines and equipotential lines are not squares, change (n) in such a way that the final geometric form is roughly orthogonal.

Equipotential lines are the lines perpendicular to the stream lines.

Casagrande suggests the use of an eraser and a pencil at the same time. The designer may erase and correct the stream and equipotential lines so that an acceptable flow net is obtained. Both stream lines and equipotential lines are deformed around corners such as at point 5. The final squares are approximate as shown in Fig 7.8.

When the flow net is ready, the discharge through the area where it takes place can be computed as follows.

The rate of flow through a unit square of one channel per unit width of dam is:

$$q = -kA \frac{dh}{ds} \qquad 7.25$$

Using finite increments instead of differential values, Eq. 7.25 can be written as

$$q = -kA \frac{\Delta H}{\Delta S} = -kA \frac{H/n}{\Delta S} \qquad (7.31)$$

The area of the unit square is

$$A = 1 \bullet \Delta n$$

where (Δn) is the distance between two stream lines. Substituting A in Eq. 7.31 yields

$$q = -k\Delta n \frac{H/n}{\Delta S}$$

but $\Delta n = \Delta S$ in a square, then

$$q = -k \frac{H}{n} \qquad (7.32)$$

The total discharge is

$$q = -k \frac{m}{n} H \qquad (7.33)$$

where m represents the total number of channels.

The procedure is approximate. For this reason it is customary to use four to five stream lines in solving flow net problems. In fill-type dams, if the coefficient of permeability is k_v in the vertical direction and k_h in the horizontal direction, the scale is distorted by a rate

$$(k_v/k_h)^{1/2}$$

and the dam is drawn using a new scale. The flow net is drawn on this figure and transposed to the true scale for obtaining the final flow net. This theoretical approach is given only for reference. Its application is laborious.

Example 7.05

If, in Fig. 7.8, H = 10.00 m and k = 60 m/day estimate the rate of seepage per unit width of dam.

Solution

$$Q = k\frac{m}{n}H = 0.60\,\frac{4}{15}\,10 = 1.6 \text{ m}^3/\text{day}$$

If the net is drawn with m = 5 channels, n = 19
If the net is drawn with m = 6 channels, n = 22

and

$$q = 0.6\,\frac{5}{19}\,10 = 1.579 \text{ m}^3/\text{day}$$

$$q = 0.6\,\frac{6}{22}\,10 = 1.636 \text{ m}^3/\text{day}$$

It is clear that the discrepancy is small.

7.3.2   Wave Pressure

Waves generated by winds or other causes act on the dam body and particularly on spillway gates. The pressure due to wave action must be taken into consideration in the design of the gates. U.S. Army Corps of Engineers (HDS 310-1) suggests a method of computation for determining this pressure. The wave direction is assumed perpendicular to the gate axis

$$h_o = \frac{\pi z_a^2}{\lambda}\,\text{coth}\,\frac{2\pi d}{\lambda} \tag{7.34}$$

$$a = \frac{z_a}{\cosh \frac{2\pi d}{\lambda}} \tag{7.35}$$

where

h_o is the height of the mean level of the standing wave above still water in meters or feet

λ is a characteristic wave length, in meters or feet

d is the depth of water in meters or feet

a is the bottom pressure parameter in meters and feet and, z_a is the wave height in meters or feet

Figure 7.9 shows the diagram of wave pressure.

Example 7.06

The wave height in the Kiralkizi Dam Reservoir is 1.17 m. If the wave length is 15.50 m and d = 110.00 m, compute the wave pressure on the gate of the Kiralkizi Dam Spillway.

Solution

Using Eq. 7.34, h_o can be computed as

$$h_o = \frac{\pi z_a^2}{\lambda} \coth \left(\frac{2\pi d}{\lambda}\right) \tag{7.34}$$

where

$z_a = 1.17$ m

$\lambda = 15.50$ m

$d = 110.00$ m

Then

$$h_o = \frac{3.14 \cdot 1.17^2}{15.50} \coth \left(\frac{2 \cdot 3.14 \cdot 110.00}{15.50}\right) = 0.28 \text{ m}$$

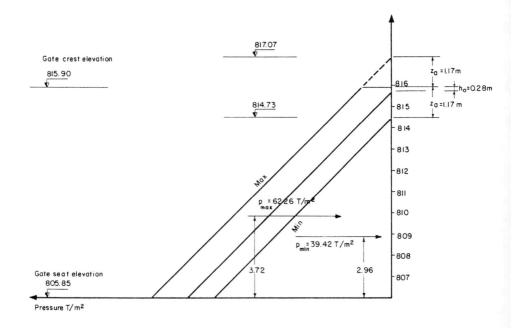

Figure 7.9. Pressure on gate including the wave action.

and

$$a = \frac{z_a}{\cosh \left(\dfrac{2\pi d}{\lambda}\right)} = \frac{1.17}{\cosh \dfrac{(2 \cdot 3.4 \cdot 110)}{15.5}} \cong 0$$

The variation of wave pressure is shown in Fig. 7.9.

7.3.3   Earthquake Forces

A.   Horizontal Thrust

The increase in water pressure acting on the upstream face of the dam, caused by horizontal earthquake acceleration, is given by the Westergaard's formula:

$$p_e = C\alpha\gamma h \qquad (7.36)$$

where

p_e is the water pressure increase due to earthquake, T/m^3

α is the earthquake acceleration/acceleration of gravity

γ is the specific weight of water, T/m^3

400

h   is the depth of reservoir at the section under consideration, in meters

C   is a coefficient depending upon the slope of the dam face and upon the elevation of the point under consideration.

Equation 7.36 is plotted in Fig 7.10. Variation of C is given in Fig. 7.11 as a function of the slope of the upstream face and in Fig. 7.12 as a function of y/h.

USBR suggests the use of Eq. 7.37 for computing the coefficient C (Design of Small Dams)

$$C = \frac{C_*}{2} [\frac{y}{h}(2 - \frac{y}{h}) + \sqrt{\frac{h}{h}(2 - \frac{y}{h})} \,]$$

$$\frac{C_*}{2} = 0.365 \quad \text{for} \quad \psi = 0^\circ \quad \text{(Fig. 7.11)} \tag{7.37}$$

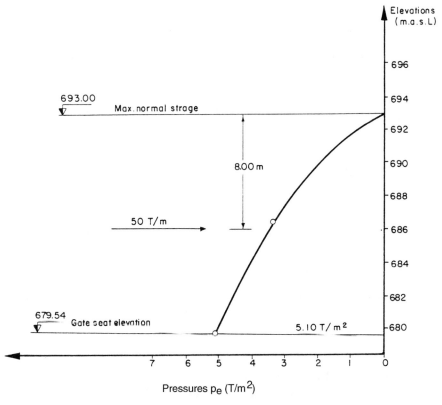

Figure 7.10. **Horizontal earthquake thrust on the gate of the Karakaya Dam**

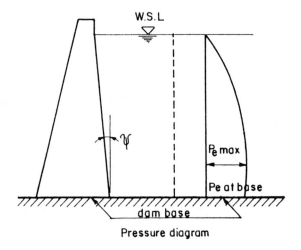

W.S.L

P_e max

Pe at base

dam base

Pressure diagram

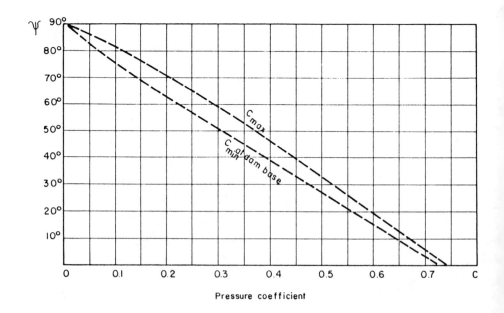

ψ

90°
80°
70°
60°
50°
40°
30°
20°
10°

0   0.1   0.2   0.3   0.4   0.5   0.6   0.7   C

C_{max}

C_{min} at dam base

Pressure coefficient

Figure 7.11. Variation of C as a function of the slope of the upstream face of the dam. (USBR).

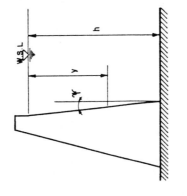

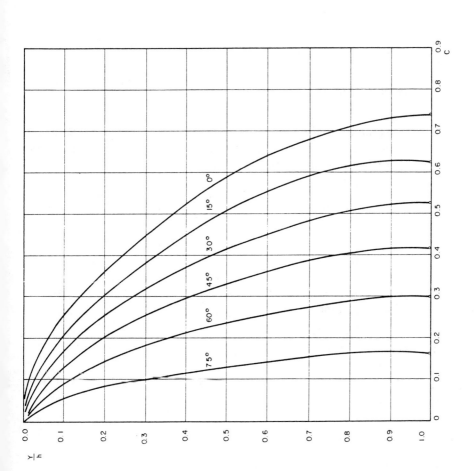

Figure 7.12. Variation of C as a function of (y/h) and the angle (ψ) (USBR).

where y is the water depth at the section considered.
   The value of the total horizontal force, F, acting on the gate is

$$F = 0.726 p_e y \qquad (7.38)$$

This force acts on the gate at a distance h_o below the free surface

$$h_o = 0.59y \qquad (7.39)$$

The total overturning moment, M, above that elevation is

$$M = 0.299 p_e y^2 \qquad (7.40)$$

   If the dam is formed by a combination of a vertical and a sloping upstream face, USBR suggests the following procedure:

- Consider the upstream face of the dam as vertical only when,

$$\text{The height of the vertical part} \geq \frac{\text{the height of the dam}}{3}$$

- Use the slope of the line joining the point of intersection of the water surface with the upstream face to the point of intersection of the foundation with the upstream face when,

$$\text{The height of the vertical part} < \frac{\text{The height of the dam}}{3}$$

B. Vertical Thrust

   Modify the weight of water and the weight of the construction material by the appropriate acceleration.

Example 7.07

   Compute the horizontal thrust of earthquake on the Karakaya arch dam spillway gate.

Data

   h = 173.00

   Height of gate = y = 13.50 $\alpha = 0.15$

Solution

$$y/h = 0.078$$

$$C = 0.365 \, (0.15 + 0.39) = 0.197 \qquad 7.37$$

$$p_e = 0.197 \bullet 0.15 \bullet 172 = 5.10 \text{ T/m}^2 \qquad 7.36$$

The resulting diagram of earthquake induced water pressure acting on the face of the gate is shown in Fig 7.10.

The resultant of the pressure diagram for unit width is

$$F = 0.726 p_e y \qquad 7.38$$

$$F = 0.726 \bullet 5.1 \bullet 13.5 \cong 50 \text{ T/m} \qquad 7.38$$

where

y   is the height of the gate.

This force acts on the gate at a distance from the water surface equal to:

$$h_o = 0.59 \, y = 0.59 \bullet 13.5 = 8.00 \text{ m} \qquad 7.39$$

7.3.4   Tractive Force due to Flow

Water flowing in a channel exerts a force on the bottom and walls, in the direction of the flow. A reaction to this force is developed according to the principle of action and reaction. The tractive force is given by

$$\tau = \gamma RS \text{ kg/m}^2 \text{ (lb/ft}^2) \qquad (7.41)$$

where

γ is the specific weight of flowing water

R   is the hydraulic radius, and

S   is the slope of the energy line.

In a rectangular channel, if the depth of water, d, is negligible in comparison with the bottom width then d can be substituted for R and Eq. 7.41 takes the form

$$\tau = \gamma dS \text{ kg/m}^2 \text{ (lb/ft}^2)$$

This force becomes important for the steep sloped spillway discharge channel.

Example 7.08

Compute the tractive force in a spillway downstream discharge channel, the characteristics of which are given.

Data

$$S = 0.81 \qquad\qquad d = 2.50 \text{ m}$$

Solution

The tractive force per linear width is

$$\tau = 1 \bullet 2.50 \bullet 0.81 = 2.03 \text{ T/m}^2$$

7.4   PROBLEMS

Problem 7.01

Compute the loads acting on a gravity dam the characteristics of which are given as

Data

Height of dam	100.00 m
Slope of the upstream face	Vertical
Slope of the downstream face	1/0.7
Upstream water depth	95.00 m
Crest width	5.00 m
Downstream water depth	22.00 m
Depth of upstream silt deposit	15.00 m
Ø	10°
Specific dry weight of silt deposit	2.16 T/m³
n (porosity)	0.20 %

Problem 7.02

Compute the ice thickness in the reservoir of a dam with the following characteristics:

- Duration of the freezing period   100 days
- Minimum temperature at the dam site 0°, and -40°

Problem 7.03

Compute the ice thrust on the dam described in Problem 7.02.

Data

Increase of temperature   15 F°

No lateral restraint

Solar energy must be taken in consideration

Problem 7.04

Draw the uplift diagram for the Kiralkizi Dam spillway for:

- Maximum reservoir elevation   820
- Lane coefficient, C   1.00
- The cross section of the Kiralkizi Dam spillway is given in Fig 7.6

Problem 7.05

Determine the leakage rate in the dam shown in Fig. 1, Problem 7.05. The flow net is shown in the figure.

Data

$k = 1.5 \ 10^{-4}$ m/day $H = 90.00$ m

Problem 7.06

Show the variation in the flow net of the dam described in Problem 7.05 when tailwater elevation 25.00 m is reached.

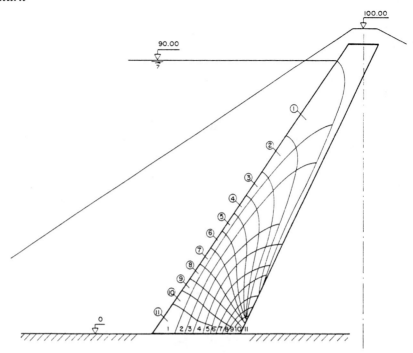

Fig. 1. Problem 7.05.

Problem 7.07

Determine the leakage rate in the dam, the characteristics of which are given below, using flow net procedure (Fig. 1, Problem 7.07).

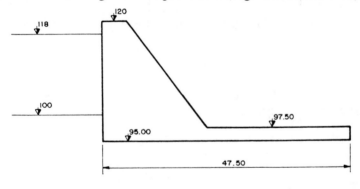

Fig. 1. Problem 7.07.

Data

$k = 1.5 \ 10^{-3}$ m/day

$C = 2.50$

Problem 7.08

Make some suggestions for reducing the leakage rate by 15% at the dam shown in Problem 7.07?

Problem 7.09

The wave height in a reservoir is 2.00 m. If the wave length is 10.00 m and the depth of water, $d = 20.00$ m, compute the wave pressure against the gate of this dam.

Data

Gate seat elevation	10.00
Gate crest elevation	20.00

Problem 7.10

Compute the horizontal thrust of earthquake on the gate of a dam, the characteristics of which are given as,

Data

- h   200.00 m
- height of the gate   17.00 m
- α (earthquake acceleration)   0.17

Problem 7.11

Compute the vertical thrust of the earthquake on the gate of the dam, the characteristics of which are given as,

Data

The acceleration factor 0.15

$\gamma_s = 2.4 \text{ T/m}^3$

See Fig. 1, Problem 7.11 for geometric characteristics of the dam body.

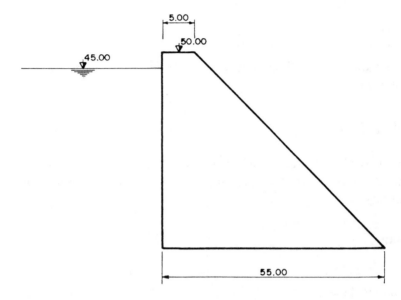

Fig. 1. Problem 7.11.

Chapter 8

HYDRAULICS OF RESERVOIRS

8.1   INTRODUCTION

The size and depth of reservoirs can vary from the very small ones on the steeply sloped valleys, to the very large ones, where the impoundment area covers a large flat plain. The reservoir impacts the dam, the spillway and appurtenances in many ways.

Wind blowing on the surface of the reservoir generates waves capable of destroying the structure by overtopping. Wave height, wave run-up and wave set-up of given feasibility of occurrence must be estimated for protection from destructive action of the waves.

Density currents transport sediments which are deposited immediately upstream of the dam body creating the extra loads and obstructing water intakes and bottom valves. Solid particles entering penstocks destroy the turbines particularly in case of high heads. Beach erosion provides extra sediment, silting of the reservoir and decrease of its economic life.

A full reservoir saturates an embankment dam and creates pore pressures which vary with the head of water. Pore pressures must be considered in designing the dam to avoid dangerous slides during the steady-state condition or the rapid drawdown.

The percolation of reservoir water can also cause leakage through seepage paths which, if uncontrolled, can destroy the dam. It is clear that the reservoir is a viable entity the engineers have to deal with. The first step to be taken is its engineering definition.

8.1.1 Area-Capacity Curves

The area-capacity curve is necessary for defining the storage capacity of a reservoir. An area-capacity curve is obtained by planimetering the area enclosed within each contour line in the reservoir area (Fig. 8.1). The graphical plotting of area and capacity curves relates the surface area and the storage capacity of the reservoir to the elevation of the water surface (Fig. 8.1).

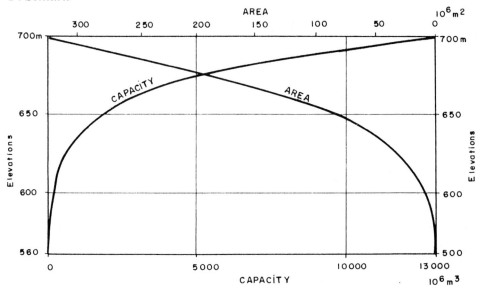

Figure 8.1. Area and capacity curves for the Karakaya Dam Reservoir on the Euphrates River, Turkey.

8.2 WIND, TIDES AND WAVES IN RESERVOIRS

Wind blowing on a free water surface generates waves. The wave height is a function of the wind velocity, the wind duration, the depth of water and the distance of water surface over which the wind blow. This length is called fetch and it will be identified by "F". Waves striking a rigid surface, the upstream face of a dam for example, are stopped by this surface and their crest height is above the still water level of the reservoir. Where the barrier is inclined the wave run-up is added to the wave height. If the wind is blowing directly toward the barrier, wave set-up must also be added to the total. The wave set-up or wind tide is the tilting of the reservoir surface above the horizontal water surface level on the leeward side and below it on the windward side. The wave run-up is observed when a wave strikes the upstream slope of a dam. The wave runs up to slope to a height above the normal water level. When the wind blows strongly enough, wave set-up, wave run-up and wave height occur simultaneously. The wave height is measured from trough to crest. At any given location and for constant water depth, wind speed, wind duration, fetch and wave height are not constant and vary from one wave to the other. The wave height attains a maximum value, decreases and then increases again to reach approximately the same height. This group of waves is called the wave train. Sverdrup and Munch

(1947) made an important contribution to the computation of wave height by introducing the concept of significant wave. A wave train can then be considered as formed by constant significant waves.

"The significant wave is defined statistically as the wave having the average height and period of the highest one-third of the total number of waves observed."

Many of the following analyses are based on the significant wave concept. Figure 8.2 shows a dimensionless graph showing relationships among fetch, wind velocity, significant wave height, wave period, wave steepness and wind duration in deep water (subsection 8.2.3).

The computation of wave height, wave set-up and wave run-up is given below.

8.2.1 Computation of Fetch

Fetch is the continuous area of water over which the wind blows in essentially the same direction. Fetch length is horizontal distance over which the wind blows across the reservoir surface. In this section Fetch will be used synonymously with Fetch Length. Since the wind can blow from different directions, they are all considered in determining the "effective fetch." The effective fetch is defined in Example 8.01.

Example 8.01

Compute the effective fetch on the reservoir shown in Fig. 8.3.

Solution

The computation of the effective fetch is shown in Fig. 8.3. It is self-explanatory.

8.2.2 Computation of Wind Velocity

The second factor in the computation of wave height is the wind velocity. The wind velocity obtained from meteorological reports, which give wind velocities over land, can be increased by 20% to reflect the velocities over the reservoir surface. Saville et al., 1962, suggested such an approach as given in Table 8.1.

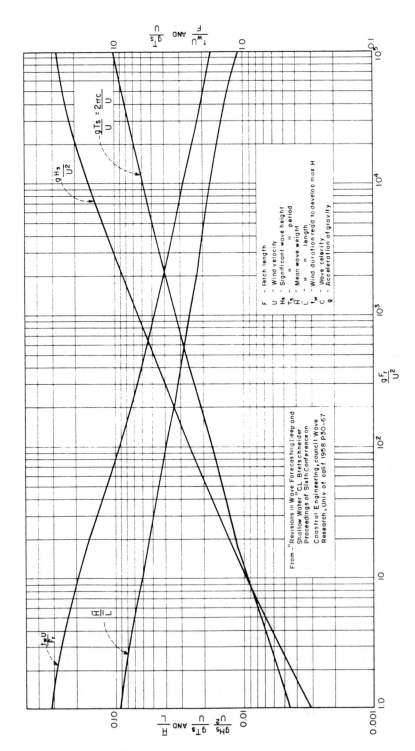

Figure 8.2. Dimensionless graph showing relationships among fetch, wind velocity, wave height, wave period, wave steepness and wind duraiton in deep water. (After Bretschneider.)

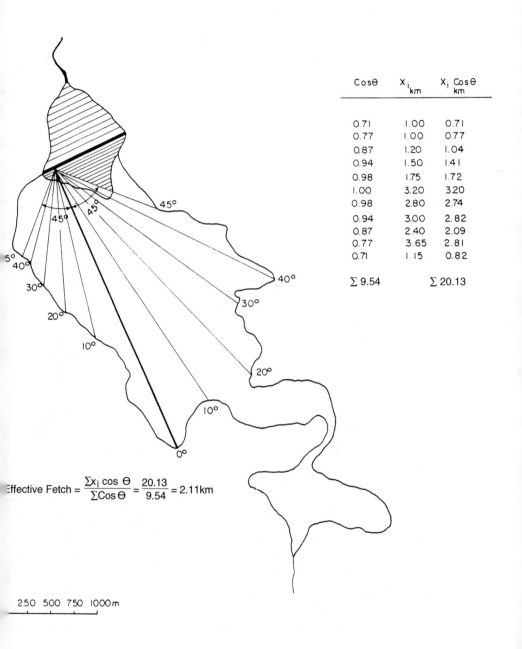

Cos θ	X_i km	X_i Cos θ km
0.71	1.00	0.71
0.77	1.00	0.77
0.87	1.20	1.04
0.94	1.50	1.41
0.98	1.75	1.72
1.00	3.20	3.20
0.98	2.80	2.74
0.94	3.00	2.82
0.87	2.40	2.09
0.77	3.65	2.81
0.71	1.15	0.82
Σ 9.54		Σ 20.13

$$\text{Effective Fetch} = \frac{\Sigma x_i \cos \theta}{\Sigma \cos \theta} = \frac{20.13}{9.54} = 2.11 \text{km}$$

250 500 750 1000m

Figure 8.3. Computation of effective fetch.

TABLE 8.1

Relationship Between Wind over Land to that over Water

Fetch	(km)	0.85	1.675	3.350	6.705	10.00
	(miles)	0.50	1.000	2.000	4.000	6.00
$\dfrac{U_{water}}{U_{land}}$		1.08	1.13	1.21	1.28	1.31

8.2.3 Computation of Wave Height

The wave height is a basic factor in the design of the dam freeboard, especially for embankment dams. The problem has been seriously investigated by researchers since the beginning of the 20th century. A brief historical background of wave height computation is given by Falvey,and Henry (1974).

"Since wind is one of the most prevalent generators of waves, it is natural that correlations between wind velocities and waves have been sought for a long time. Bretschneider in 1955, using the methods of dimensional analysis, was one of the first to provide a useful tool for correlation of wave heights with wind velocities and fetch lengths. His result, a FETCH GRAPH FOR DEEP WATER, forms the basis for most of the currently used prediction methods. However, the efforts to find a model for the air-water energy transfer process began as early as 1887 when Helmholz and Kelvin studied the oscillations set up at the interface between two fluids of different densities. In the period 1924 - 1927, the tangential and normal stress energy transfer mechanisms were studied. These theories were investigated chiefly by Jeffreys, Sverdrup, and Munch. In 1957, Phillips proposed a mechanism whereby turbulent variations in the wind initiate the formation of waves. These theories and some new concepts regarding nonlinear interactions were examined by Kinsman in 1965. Another approach was given by Barnett in 1968 which utilizes transfer mechanisms and is probably as accurate as any proposed to date."

The most commonly used wave height formulas are enumerated as follows.

A. U.S. Bureau of Reclamation Formula

The main sources of information regarding wave heights are visual observations, in-situ measurements and correlations with measurements performed in laboratories. The formulas obtained are empirical and difficult to generalize. Thomas Stefenson's (1874) work is the basis for these relationships. Later work was completed by Gaillard (1904) and Molitor, D.A., (1935). Although both of them directed their research at seawalls, their formulas have given satisfactory results when applied to man-made lakes or reservoirs.

Molitor's equations for computing the wave height follow.

For   F < 20 miles (E)   F < 32 kms (M)

$$z_d = 0.17(UF)^{1/2} + 2.5 - F^{1/4} \quad z_d = 0.032(UF)^{1/2} + 0.75 - 0.27F^{1/4} \quad (8.1)$$

and for   F > 20 miles (E)   F > 32 kms (M)

$$z_d = 0.17(UF)^{1/2} \qquad\qquad z_d = 0.032\,(UF)^{1/2} \quad (M) \qquad\qquad (8.2)$$

where

z_d is the wave height in meters or feet

U   is the wind velocity in miles per hour, or km/hr, and

F   is the fetch in miles or in kms

Figure 8.4 based on Eqs. 8.1 and 8.2 is prepared for determining the wave heights for wind velocities from 20 to 70 miles per hour for fetches of 2 to 40 miles. Figure 8.5 represents the wave height and wave pressure as a function of the fetch.

Example 8.02

Determine the wave height for

$$U = 34 \text{ miles/hr}$$

$$F = 10 \text{ miles}$$

using Fig. 8.4

Solution

- Determine the curve corresponding to F = 10 miles on the graph given in Fig. 8.4 (third black curve from the bottom of the diagram).

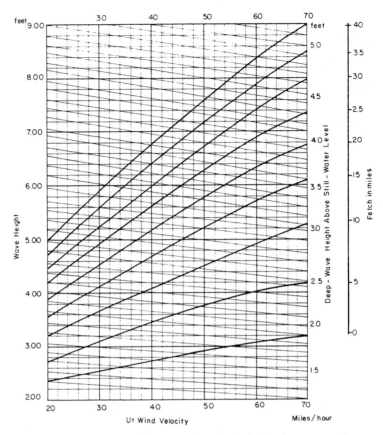

Figure 8.4. Curves for determining the height of waves. (Reeves, A.B. - Thomas, C.W., ICID 1954 Second Congress, "Considerations Affecting US Practice in Providing Free-Board for Irrigation Works").

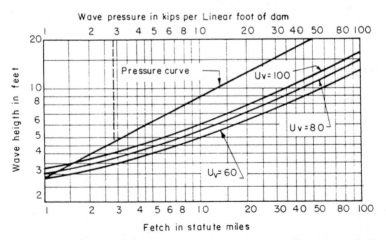

Figure 8.5. Wave height and wave pressure as a function of the fetch.[*]

---

[*] Molitor, D.A., "Wave pressures on Sea-walls and breakwaters." (T.A.S.C.E. 1935, V.100)

- Determine the intersection of this curve with the parallel line to the ordinate axis issued from U = 34 miles/hr

- Read the wave height on the ordinate axis as

$$z_d = 3.80 \text{ ft}$$

- Read the deep-wave height above still water level by following the oblique straight line issued from $z_d = 3.80$ ft as

$$z_{dd} = 1.98 \text{ feet}$$

- The shallow-wave height above still water is given by Gaillard and Molitor as

$$z_{ds} = z_{dd} \left(1 + \frac{U}{420 + U}\right) \tag{8.3}$$

$$z_{ds} = 1.98 \left(1 + \frac{34}{420 + 34}\right) = 2.13 \text{ feet}$$

B. Ocean Formula

$$z_d = 0.034 \, U_r^{1.06} \, F^{0.47} \tag{8.4}$$

where

z_d is the average height of the highest one-third of waves, the significant wave height in feet

U_r is the wind velocity in miles per hour. It is assumed that the wind is blowing 24.6 ft. (7.50 m) above the water surface, and

F is the fetch in miles.

This equation is shown graphically in Fig. 8.6.

C. Stefenson Formula

This formula is one of the most commonly used relationships

for   F < 18 km   or   F < 10 nautical miles

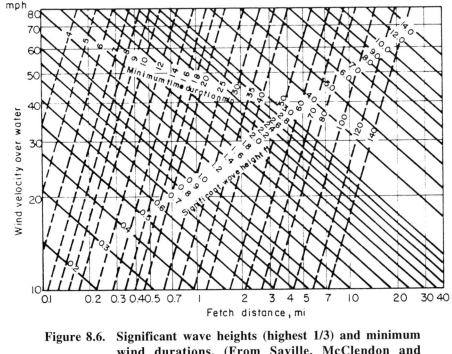

Figure 8.6. Significant wave heights (highest 1/3) and minimum wind durations. (From Saville, McClendon and Cochran.)

$$z_d = 0.75 + 0.34F^{1/2} - 0.26F^{1/4} \quad \text{(metric units)}$$

$$z_d = 1.5F^{1/2} - 2.5F^{1/4} \qquad \text{(English units)} \qquad (8.5)$$

For   F > 18 km   or   F > 10 nautical miles

$$z_d = 0.34F^{1/2} \qquad \text{(metric units)}$$

$$z_d = 1.50F^{1/4} \qquad \text{(English units)} \qquad (8.6)$$

where z_d is the wave height, and

F is the fetch.

D. U.S. Army Corps of Engineers Formula

The US Army Corps of Engineers suggestion for the computation of wave height is given in Fig. 8.7.

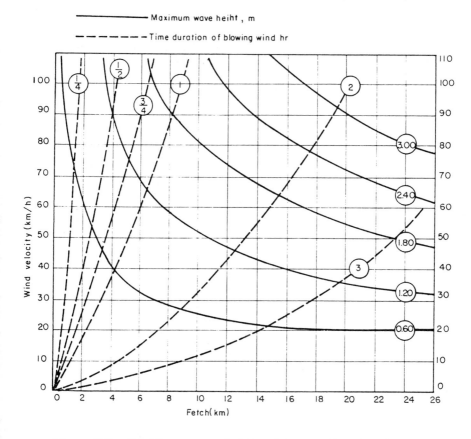

Figure 8.7. Significant wave heights (highest 1/3) and minimum wind duration (From the US Army Corps of Engineers).

E. Falvey's Formula

$$\overline{H}_{1/3} = (3.1 \; 10^{-4} \; U_{10}^2 + 1.6 \; 10^{-2} \; U_{10})F^{1/2} \tag{8.7}$$

where

$\overline{H}_{1/3}$ is the average height of highest one-third of the waves, the significant wave height

U_{10} is the horizontal wave velocity (m/sec) at 10 m height above still water surface, and

F is the fetch in km.

Equation 8.7 is valid for wind velocity, referenced to the 10 m height, between 10 m/sec and 50 m/sec and for fetch lengths between 2 kms and 200 kms.

8.2.4 Computation of Wave Run-up

When a wave strikes the upstream slope of a dam it runs up this slope to a height above its open-water height. This run-up height is called the wave run-up which can be estimated only by empirical formulas. One of the most used formulas is shown graphically in Fig. 8.8.

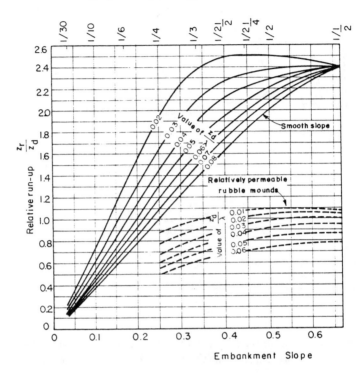

Figure 8.8. **Wave run-up ratios versus wave steepness and embankment slopes (From Saville, McClendon and Cochran).**

In Fig. 8.8 the wave period t_r and the wave length (λ) are given by the following formulas

$$t_r = 0.46 \ U_r^{0.44} \ F^{0.28} \tag{8.8}$$

$$\lambda = 5.12 \ t_r^2 \tag{8.9}$$

where

λ is the wave length in feet

t_r is the wave period

U_r is the wind velocity in miles per hour, and

F is the fetch in miles

8.2.5 Computation of Wave Set-up

The Zuider-Zee formula is generally used for computing the wave set-up

$$z_s = \frac{U_r^2 F}{63,000 \, d} \cos \varnothing \qquad (8.10a)$$

$$z_s = \frac{U^2 F}{1,400 \, d} \cos \varnothing \qquad (8.10b)$$

where

z_s is in feet, U is in miles/hr, F is in miles, and d is in feet in formula 8.10b, and

$\varnothing$ is the angle between the effective fetch and the actual blowing wind

U_r is the wind velocity in km/hr

F   is the fetch in km, and

d   is the mean reservoir depth in the direction of the effective fetch in meter in Eq. 8.10a.

The value of z_s can be particularly important for large fetches and shallow reservoirs.

Example 8.03

Compute the wave height, the wave set-up and the wave run-up in the reservoir shown in Fig. 8.9.

Data

F   = 30 km

U_r = 20 m/sec

$\varnothing$ = 30°

Embankment slope = 1/3 = 0.34

d   = 30 m

Solution

Step 1:   Compute the wave height as follows.

     1. Ocean formula

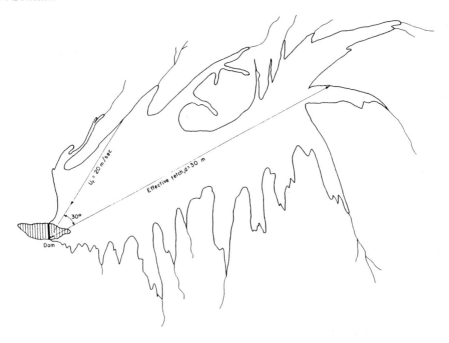

Figure 8.9. Characteristics of reservoir in Example 8.02.

$$z_d = 0.034 \, U_r^{1.06} \, F^{0.47} \qquad\qquad 8.4$$

$U_r = 20$ m/sec $= 44.74$ miles/hr

$F = 30$ km $= 18.64$ miles

$z_d = 0.034 \cdot 44.74^{1.06} \cdot 18.64^{0.47} = 7.55$ feet
$\quad\;\; = 2.30$ m

2. Stefenson formula

$F = 30$ km > 18 km

$$z_d = 0.34 \, F^{1/2} \qquad\qquad 8.6$$

$z_d = 0.34 \, (30)^{1/2} = 1.86$ m

3. Molitor formula

$F = 30$ km

$$z_d = 0.032 \, (\frac{20 \cdot 3{,}600}{10^3} \, 30)^{1/2} = 1.49 \text{ m}$$

4. Corps of Engineers formula

$$U_r = \frac{20 \cdot 3{,}600}{1{,}000} = 72 \text{ km/hr}$$

The value of F is outside the limits of Fig. 8.7.

5. Falvey's formula

$$z_d = (3.1 \ 10^{-4} \cdot 20^2 + 1.6 \ 10^{-2} \cdot 20)(30)^{1/2} = 2.43 \text{ m}$$

The mean value of z_d is then

$$z_d = \frac{2.27 + 1.86 + 1.49 + 2.43}{4} = 2.01 \approx 2.00 \text{ m}$$

Step 2: Compute the wave set-up as

$$z_s = \frac{U_r^2 F}{63{,}000d} \cos \varnothing \qquad \text{8.10a}$$

$$z_s = \frac{72^2 \cdot 30}{63{,}000 \cdot 30} \cos \varnothing = 0.07 \text{ m} \qquad \varnothing = 30°$$

Step 3: Compute the wave run-up as

$$t_r = 0.46 \ U_r^{0.44} \ F^{0.28} \qquad \text{8.8}$$

$$t_r = 0.46 \cdot 44.74^{0.44} \cdot 18.64^{0.28} = 5.55$$

$$\lambda = 5.12 \ t_r^2$$

and

$$\frac{z_d}{\lambda} = \frac{2.00 \cdot 3.28}{157.71} = 0.0416$$

Entering Fig. 8.8 with $z_s/\lambda = 0.0416$ and embankment slope = 0.34 yields

$$\frac{z_r}{z_d} = 1.82$$

and

$$z_r = 1.82 \cdot 2.00 = 3.64 \text{ m}$$

Step 4:   Compute the maximum wave action as

$$2.00 + 0.07 + 3.64 = 5.71 \text{ m}$$

8.2.6 Computation of Freeboard

The freeboard is the difference in elevation between the crest of the dam and the maximum level attained by the wave during the spilling of the design flood.

This definition implies

$$\text{Freeboard} = z_d + z_r + z_s$$

where

z_d is the wave height

z_r is the wave run-up, and

z_s is the wave set-up.

The total $(z_d + z_r + z_s)$ must be added to the highest water level in the reservoir during the spilling period. It is known that the magnitude of the design flood is chosen by the designer to satisfy the dam safety completely. It is also known that the probability of the occurrence of this maximum flood is very small. Design for the extreme maximum floods is very costly. To allow a more economical design, part of the freeboard can be used for the head developed by the peak design flood (see Chapter 4). The percentage of freeboard used is a function of the accuracy of the estimate of the anticipated maximum flood runoff and the time period of peak characteristics of the design hydrograph. In an uncontrolled spillway the water level will rise in proportion to the head required for the flood discharge. Overtopping by wave action with the water surface at flood peak must be prevented by allowing the adequate freeboard. In a controlled spillway the gates may be operated to prevent the extra rise of water, but consideration must be given to the possibility of the failure of automatic gates to operate or human failure. Therefore, the freeboard must be considered starting at the top of the gates.

Freeboards values for dams world-wide are listed in Table 8.2.

8.3   BEACH EROSION

Beach erosion is caused by the action of three factors:

1.   The drop of water surface level in the reservoir. Sudden drops are often the origin of landslides.

TABLE 8.2

Free-board Allowances on Dams

Name of dam		Type of dam		Height of dam ft	Free-board ft
Hoover (Boulder)	USA	Concrete,	gravi.	726	10.6
Shasta	"	"	"	602	12.5
Grand Coulee	"	"	"	550	23.0
Owyhee	"	"	"	417	5.0
Arrowrock	"	"	"	349.5	6.0
Friant	"	"	"	319	3.25
Elephant Butte	"	"	"	301.2	7.0
Thief Valley	"	"	buttress	73	10.0
Roosevelt	"	Masonry,	gravi.	280	8.0
Green Mountain	"	Earth		309	10.0
Alcova	"	"		265	10.0
Alamogordo	"	"		148	10.0
Belle Fourche	"	"		122	15.0
Fresno	"	"		111	22.0
Oymapinar	Turkey	Concrete,	arch	606.80	3.28
Karakaya	"	"		567.44	16.40
Sariyar	"	"	gravity	354	16.40
Gezende	"	"	arch	246	6.56
Keban	"	Earth		678.96	9.84
Kayraktepe	"	"		642.88	16.73
Altinkaya	"	"		639.60	16.40
Atatürk	"	"		603.52	22.96
Adigüzel	"	"		475.60	21.32
Kiralkizi	"	"		426.40	13.94
Naugarh	India	"		100	5.0
Komaki	Japan	Concrete		239.93	10.0

2. The mechanical action of waves generated by navigation.

3. The mechanical action of wind and wind generated waves.

8.3.1 Wave Action;

Waves generated by winds or by barges in motion on the lake strike the shore and their energy is dissipated by the beach shore. Beach material and beach topography are the major factors affecting beach erosion. Waves with similar characteristics have different erosive action on a cliff or on a bar where a portion of its energy is absorbed by friction. It has been observed that the damping of the pore pressure due to the oscillation of the water surface generated by wave action is very rapid. Where the beach material is fine silt and clay the variation of the inner pressure attains 10 times the grain diameter. Sometimes the cohesive forces prevent the damping effect when the hydraulic gradient is steep. It has been observed also that slides of up to one cubic meter have been caused by the sudden impact of waves generated by a moving ship near the shore.

Where the beach material is sand, coarse sand, gravel and cobbles the hydraulic gradient due to the action of waves influences the ground water flow to a greater depth and its damaging action is more important. Each beach front adjusts itself to absorb the energy of the wave attack.

A second mechanical action, the waves breaking obliquely along a shore line, is to induce a longshore or littoral current. This littoral current can be as effective as a current taking place in a river and causes active erosion. Adding the agitating action of the breaking waves, beach erosion can reach significant levels. Unfortunately, only qualitative information is available on the rate of beach drift that can be expected under a given condition of wave attack and beach material.

8.3.2 Action of Water Surface Drop in the Reservoir

The drop of the water surface level in reservoirs may be seasonal or accidental. The seasonal drop is periodic. It occurs once each year and its amplitude is a function of the rain and the catchment area each year. An accidental drop is due to damage or an emergency at the dam which causes the reservoir to empty rapidly. The seasonal variation of water surface level at the Keban Dam, Turkey averages 40.00 m. The velocity of the dropdown is around 25 cm per day. An illustrative example is reproduced as follows.

These observations have been made at various reservoirs in Switzerland. The distance from the beach measured inland and readjusted for different cohesion values are listed below (Bruschin, J. and Dysli, M., 1973).

Cohesion (cm^2/sec)	0.01	0.1	1.0	10.00	100	1000
Duration	Distances measured inland for drop of 15 cm					
1 day	0.7 m	2.2 m	7 m	2.2 m	70.0 m	220 m
1 month	4 m	12 m	40 m	125 m	400 m	1250 m

Figure 8.10 reproduces a diagram showing these distances as a function of the coefficient of cohesion.

The surface level drop in the beach soil is an indication of the probability of landslides. A diagram similar to the one given in Fig. 8.10 must be prepared for each particular case.

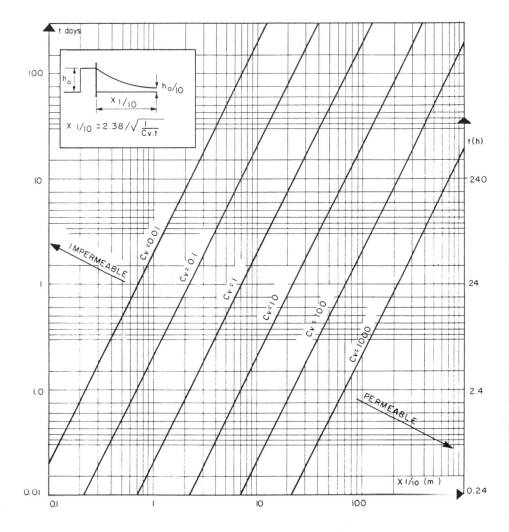

Figure 8.10. Sudden drop in a reservoir water surface level, distances (x) as a function of the coefficient of cohesion C_v and time. The drop of pressure is $-h_0/10$ (Bruschin, J. and Dysli, M., 1973).

429

8.4   RESERVOIR DEPOSITS - RESERVOIR SILTING

Considering an ephemeral stream flowing from a steep unstable watershed to a flat plain causing the formation of an alluvial fan, the sediment from the stream adds to the deposition each year (Fig. 8.11). A stable channel does not exist and floods can only be partially carried. The lands adjacent to the stream are flooded regularly and the sediment is deposited in the channel and on its adjacent banks. If flooding is prevented by levees, deposition continue to accumulate at the bed and on the flood plain, between the channel and its levees, wherever the slope and velocity of the system are inadequate to carry the sediment load. In general, the coarse materials are deposited upstream. Deposition and sorting take place along the channel as a function of the slope and the discharge.

Figure 8.11.   Backwater deposit in the Damsa Reservoir, Turkey.

Deposition of the sediment in the reservoir is as follows. The coarse material is deposited in the backwater and at the inflow to the reservoir (Figs. 8.11 and 8.12). The fine particles are transported a much greater distance into the reservoir often as a density current. These fine materials are deposited as illustrated in Fig. 8.12 and in the photo in Fig. 8.13.

The bed advances in waves as shown by 1-2 and 3 in Fig. 8.12 along the longitudinal profile. The density current is stopped at the dam and deposition of fine material may occur just upstream of the face of the dam. The path of the density current can be followed by echo sounding devices that can identify the deposits of the fine sediments on the bottom of the reservoir. Figure 8.14 shows the deposits which appear schematically in

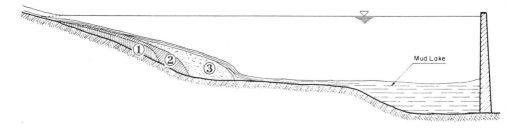

Figure 8.12.   Silting of a reservoir.

Figure 8.13. Silt deposition in a reservoir (completely emptied); the Dogancay Reservoir, Turkey.

Fig. 8.12. The complete drop of water surface shows the deposited sediment at the head water. A layer of fine sediment covers the first gravel deposits and the fine sediments are themselves covered by another gravel layer.

Note that siltation of reservoirs may occur simultaneously from both the upstream and downstream directions.

The upstream deposit forms in layers corresponding to the time of deposition and its downstream slope is the submerged angle of repose of the deposited sediment. The deposition procedure can be complicated further by rate and extent of reservoir drawdown. This deposition works its way downstream until it reaches the dam (Fig. 8.15). Sediment deposits in the

431

Figure 8.14. Sediment deposit at the backwater of a reservoir.

Figure 8.15. Reservoir silting proceeding towards the dam. Note the vegetation over the silt banks. The Aksu Dam Reservoir, Turkey.

live storage component of the reservoir reduce its effectiveness for flood control, power generation, etc. Table 8.3 reviews the silting characteristics of some reservoirs.

The silting process of a reservoir is further detailed in the following section.

TABLE 8.3

Silting of Reservoirs

Country	River	Reservoir	Catchment area square miles 10^3	Height of dam ft	Storage capacity foot-acres 10^4	Mean annual runoff foot-acres 10^6	Silt Deposited			% of water supply by volume
							Total foot-acres 10^{-3}	Annual foot-acres 10^2	% of original capacity per annum	
Egypt	Nile	Assuan	620	174	440	66	137			
USA	Colorado	Boulder	167	726	3050	15	112	75	2.00	0.18
USA	Columbia	G.Coulee	74	550	965	80	31	24	7.30	1.28
USA	Rio Grande	Elephant Butte	26	306	264	1	416	162	0.61	1.62
USA	Salt	Roosevelt	6	284	152	0.94	123	46	0.30	5.46
Australia	Murrum B.	Burrin Jack	5	200	77	1.15	3.9	2.9	0.04	0.28
Pakistan	Kushdill	Kushdill Khan	0.6	66	2.70	-	5.04	0.97	0.35	-
USA	Sweet Water	Sweet Water	0.18	90	3.6	0.02	0.68	6.8	68.1	1.20
Italy	-	Cismon	0.19	144	0.4	-	0.40	0.5	0.36	0.41
France	Drac	Pont du Loup	0.29	-	0.2	-	1.22	1.1	46.3	-
Burma	-	Meiktala Lake	0.24	-	3.03	-	7.09	1.8	0.59	-
Switzerland	Rhone	Genoa Lake	0.28	-	7210	-	13.9	17	0.62	-
Algier	Meckerre	Cheurfa	1.16	180	1.0	0.23	4.5	2.0	3.73	0.44
Germany	Begrenzer	Bodensee	0.32	-	3940	0.78	2.5	1.1	0.0002	35.0

8.4.1 Backwater Deposit

Backwater deposit is the material deposited in the backwater reach of the stream. This deposition begins at some distance upstream of the reservoir and continues into the reservoir. When the water level in the reservoir drops, the backwater deposit is eroded and transported to the reservoir. It is renewed during high water, forming an abrupt front progressing into the reservoir as shown in Fig. 8.11. This is how the filling process of reservoirs works. The continuous renewal and addition of sediment layers may be considered according to the following phases.

A. Local effects   1. Aggradation of bed

   (out of reservoir)   2. Loss of waterway capacity

                         3. Increased flood discharge

B. Upstream effects   1. Change in base level for tributaries (Fig. 8.16)

                         2. Deposition in tributaries near confluences

                         3. Aggradation causing a perched river channel which can change the alignment of the main channel

C. Downstream effect   1. Progressive filling of the reservoir

   (in the reservoir)   2. Intensive growth of phreatophytes, e.g. salt cedars (Fig. 8.15).

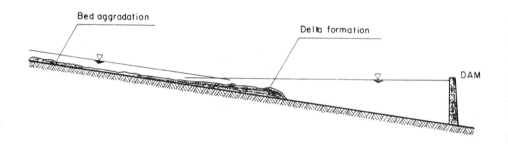

Figure 8.16. The change of base level of the tributary; delta formation.

The growth of plants on the deposited sediment in a reservoir, is very thick, impeding the flow of water and causing the sediment to deposit further. Newly deposited sediment acts as a fertilizer and permits further growth. This is a chain reaction. Examples of backwater deposits progressing downstream in the reservoir with growth of phreatophytes can be observed everywhere: Pecos River (New Mexico, USA), McMillan Reservoir, Aksu Reservoir (Antalya, Turkey), etc. (Fig. 8.15).

During the dry period when water surface level drops in reservoirs, particularly during the first years of reservoir operation, the river bed slope becomes very steep and severe degradation occurs. Then, different layers of deposited sediments can be seen in the eroded section (Fig. 8.14).

8.4.2 Delta Formation

Deltas are formed at the backwater or at the inflows of tributaries into the reservoir. Figure 8.13 shows an example of the first case, in a small reservoir, and Fig. 8.16 illustrates a typical delta formation of a tributary into a large reservoir. The normal water surface level can be seen in the photograph at the intersection of fine sediment deposit and coarse sediment deposit. During high water, fine sediments are transported rapidly through the delta in the form of bed load and density current and progress towards the dam.

A. Hydraulic Condition of the Delta Formation

The mathematical prediction of delta formation is difficult due to the multitude of involved parameters. Yücel, Ö, and Graf, W.H., 1973 suggested a mathematical approach which is summarized here.

A bi-dimensional flow is adopted for computing the sediment deposition in the reservoir.

1. A backwater profile is computed at the river entrance into the reservoir. As a consequence of the increase of flow depth, sediment will be deposited.

2. The deposited sediment will alter the backwater ensuring new sediment deposition. This process tends to attain an equilibrium which ends in the formation of a delta.

The assumptions adopted for finalizing the computations are:

- steady-state flow conditions prevail

- any circulation of secondary flow alters the flow pattern

- bed roughness remains constant

- sediment material consists of bed load, lateral injection of sediment is ignored

- deposited sediment material is not subject to consolidation.

The volume of deposited bed material is computed by using one sediment transport formulas given as

- Schoklitsch Formula

$$q_{bw} = \frac{7,000}{D_s^{1/2}} S^{3/2} (q - q_c) \tag{8.11}$$

$$q_c = 0.26 \left(\frac{\gamma_s}{\gamma}\right)^{5/3} \frac{D_s^{3/2}}{S^{7/6}} \tag{8.12}$$

- Meyer-Peter et al Formula

$$\gamma \left(\frac{k_s}{k_r}\right)^{3/2} R_s S = 0.047 \ \gamma_s' \ D_s + 0.25 \ q'^{2/3} \ \rho^{1/3} \tag{8.13}$$

- Einstein Formula

$$0.045 \ \frac{q_{bw}}{\rho_s g} \sqrt{\frac{\rho}{\rho_s - \rho} \frac{1}{gD_s^3}} = \exp\left(-0.391 \ \frac{\rho_s - \rho}{\rho} \frac{D_s}{SR}\right) \tag{8.14}$$

where

q_{bw} is the bed load discharge by weight per unit width and per unit time

q_{bw}' is the submerged weight of the transported sediment

D_s is the diameter of sediment

S is the slope

q is the actual water discharge per unit width

q_c is the discharge per unit width of channel with slope S_c where sediment transport begins

γ_s' is the specific weight of submerged sediment

γ is the specific weight of water

k_s is the roughness coefficient according to Strickler

k_r is the roughness coefficient due to skin friction

R_s is the hydraulic radius for the boundaries over which sediment is transported

ρ_s is the specific mass of solid particles

ρ is the specific mass of water, and

R is the hydraulic radius.

Graf, W.H., 1975, suggests the following approximations for obtaining a solution of the problem. "An effective slope, S, is to be chosen; this is the average of the energy slope, S_E and the channel bottom slope S_b, or

$$S = \frac{S_E + S_b}{2}$$

The numerical values of the constants would alter, though the mathematical form of the Eqs. 8.11 - 8.13, and 8.14 will remain constant. No information exists on this matter."

Graf suggests the use of Hjulström's diagram instead of Eq. 8.12 for computing q_c. Yücel and Graf., 1963, suggest the computation of the backwater curve first. For each section the bed load deposition is determined using one of the above given sediment transport formulas. The difference in transport capacities between two successive sections is deposited in area between the section under consideration. This difference is

$$\Delta q_{bwi} = q_{bwi} - q_{bw(i-1)}$$

The thickness of the deposition per unit time in a given reach is then

$$\delta_{si} = \frac{\Delta q_{bwi}}{L_i}$$

The quantity (δ_{si}) is responsible for a change in the bottom of the watercourse which in turn requires a new backwater profile calculation.

For empirical predictions of delta formation, Borland's (1971) approach is summarized.

- The topset slope is approximately equal to half of the original channel slope (Fig. 8.17).

The topset and forset beds on a delta formation are defined in Fig. 8.18.

- The forset slope is 6.5 times the top set slope.

- Borland states that "location of the pivot point between the topset and forset slopes depends on the reservoir operation and the existing channel slopes in the delta area." The upstream part of the delta is set at the point where the maximum water surface intersects the original stream bed. Data for the portion formed by the topset beds which extend from this limit to the pivot point are used to begin the reservoir delta profile computations.

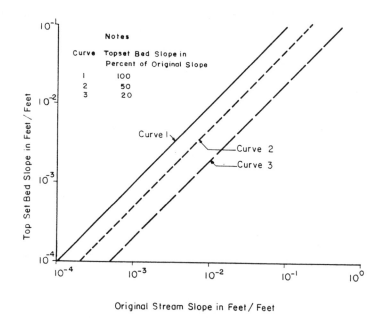

Figure 8.17. Relation of topset bed slope to original channel slope of reservoir delta. (Borland, M., 1971).

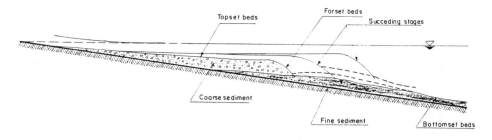

Figure 8.18. Profile of a typical reservoir delta.

Then a value for the forset slope is assumed, the topset slope being equal to (forset slope/6.50). When the location of pivot point and the topset and forset slopes are known the delta can be approximated. The volume of the delta as obtained must agree with the volume of coarse sediment carries by the inflowing delta stream.

8.4.3 Bottom Deposits

The bottom deposit is that material deposited along the bottom of the reservoir as bed load, suspended load and density current.

Consider a large reservoir with clear water. Assume that during a flood the river carries heavy sediment of all kinds, including coarse bed load, suspended load and clay in suspension, as density current. The upstream end of the reservoir becomes turbid, but the turbidity stops at a given boundary and processes very slowly downstream. Upstream of this boundary the reservoir surface is turbid and downstream of it, it is clear. The turbid waters of the river move into the reservoir and advance downstream in the form of a density current. The turbulence tries to disseminate sediment in the upstream volume of the reservoir. This creates turbidity on the outside of the density current. The density current can be detected from outside the reservoir by use of special floating device. A floating body is placed in the density current which acts as the transporting agent. A visual guide floating on the water surface and connected with the main body of the reservoir shows the path followed by the density current. At the Hirfanli Reservoir, Kizilirmak River, Turkey, it was possible to follow the contours of the existing meanders of the Kizilirmak River along the bottom of the reservoir with perfect fidelity by means of the floating device.

A second method of following the density current in its course is the use of echo sounding devices. These experiments have shown that the density current follows the main bed of the river on the bottom of the reservoir accumulating the fine sediment in suspension or clay flocules upstream of the dam body. These materials are settled there and consolidated in time adding an extra bed load on the dam body.

The reservoir is thus filled from upstream to downstream and vice-versa. The estimate of storage loss in the reservoirs can be obtained by converting weights to volumes. This procedure is not very exact due to the change of volume as a function of compaction. On the other hand, flocules do not settle rapidly. The earth dam at Demirköprü Reservoir, Turkey, is an interesting example. The upstream impervious blanket had developed fissures and leakage increased ten fold. Frogmen investigated the upstream slope of the dam, and observed a thick layer of clay flocules in suspension on the slope and some settlement on the upstream slope. The reservoir had been in operation for 15 years.

McHenry (1974), McHenry and Dendy (1964) and McHenry et al. (1969) have measured the density of continuously submerged sediments. These sediments were mostly fines with clay ($< 2\mu$) contends upwards of 30% - 40% and negligible sand contend. They found that as additional sediment accumulated the density increased with time until it reached a maximum value of 46 pcf - 76 pcf. The variation was due to the nature of the sediment involved. The depth of water above the sediment was 2.5 ft to 6 ft. For larger depths, the density of deposited sediments was relatively constant. The consolidation characteristics of sediment under water must be known to enable the designer to compute the loss of volume of the reservoir and the time involved.

8.4.4 Density of Deposits

The specific weight of sediment deposited in a reservoir is subject to change due to progressive consolidation. The upstream deposits are formed of coarse materials, the downstream deposits are composed of rather fine materials. The sediments settle as a function of size, shape and specific weight. The finer particles stay in suspension for a relatively long time.

Lane and Koelzer (1943 and 1953) suggested a method of estimating the specific weight of sediment deposited in a reservoir. They utilized the following formula

$$\gamma_m = \gamma_1 X_1 + (\gamma_2 + B_2 \log t)X_2 + (\gamma_3 + B_3 \log t) X_3 \qquad (8.15)$$

where

γ_m is the average specific weight of sediment in the reservoir at the end of any given period t.

γ_1 is the initial specific weight of the bed material

γ_2 is the initial specific weight of silt

γ_3 is the initial specific weight of clay

B_1 is a constant for the rate of compaction of sands ($B_1 = 0$)

B_2 is a constant for the rate of compaction of clays

and

X_1, X_2, X_3 are the percent of each class in the deposited material.

The values of B_i are given in Table. 8.4.

TABLE 8.4

Values of Constants B_i in Eq. 8.15

Reservoir operation	γ_1 Sand B_1 pcf		γ_2 Silt B_2 pcf		γ_3 Clay B_3 pcf	
Sediment submerged	93	0	65	5.7	30	16.0
Normal reservoir drawdown	93	0	74	2.7	46	10.7
Considerable res. drawdown	93	0	79	1.0	60	6.0
Reservoir emptied	93	0	82	0.0	78	0.0

8.5   DENSITY CURRENTS

The definition of density current has been given by researchers and different institutions as follows.

National Bureau of Standards

"A density current is the movement, without loss of identity by turbulent mixing at the boundary surface, of a stream of fluid, under, through or over a body of fluid, with which it is miscible and the density of which varies from that of the current, the density difference being a function of the differences in temperature, salt content or silt content of the two fluids."

Bell, 1942

"The densest water naturally sinks to the bottom where, if there exists a slope, it may continue to flow until its progress is checked by a dam or some other obstacle. A moving stratum of this kind is called a density current because its slightly greater density gives it the power of motion."

Brown, 1943

"A density current has been defined as a gravity flow of a fluid under, over or through a fluid of approximately equal density. Density currents have been referred to as stratified flows, density flows, under flows and tunneling of silty water."

Gould

"A turbidity current may be rather simply defined as a gravity flow of turbid water through, under or over water of different density."

U.S. Bureau of Reclamation

"The term density current, refers to a flow of water maintained by gravity forces through a main body of water such as Lake Mead and remaining separated therefrom because of the difference in density between the current and the lake."

Figure 8.19 shows a density current forming at the upstream end of a reservoir. The plunge point is apparent due to the existence of floating

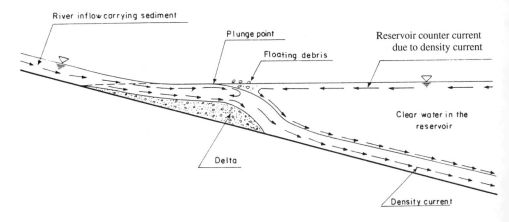

**Figure 8.19. Upstream end of reservoir showing formation of density
current.**

debris. The reservoir circulation induced by the density current is in reality
feeble. Nizéry (1973), reports that he has seen no counter current at the
upstream side of the plunge point in Sautet Reservoir, France, and the
author, also has observed complete immobility of water in the Hirfanli
Reservoir although feeble current exists at a certain depth near the upstream
end of the reservoir. Density current continues down to the dam, with
suspended material accumulating there. Above this layer flocules can be
observed. The mathematical approach for studying density current has been
reported by Wood (1967) and Fietz and Wood (1967). A brief summary of
hydraulic considerations is described here.

8.5.1 Hydraulics of Density Current

The average velocity of the head of density current determines the
opening time of bottom valves for the evacuation of the material transported
by the density current. Kuenen (1952) has developed a relationship which
can be used for the computation of this velocity

$$U = K \left(\frac{\rho_{susp} - \rho}{\rho} \delta \sin \varnothing \right)^{1/2} \qquad (8.16)$$

where

     δ is the thickness of the density current

     K is a constant, and

     $\varnothing$ is the angle of tilt.

Equation 8.16 is valid for $\emptyset > 20°$, for the density current in the reservoir, $K = 280$ cm$^{1/2}$/sec. Utilizing the turbulence theory, Hinze (1960) suggests K as a value varying between 280 and 560 cm$^{1/2}$/sec.

Scheidegger (1970) and Allen (1969) remark that U is not only the velocity of the head of the density current but the velocity of the density current itself.

8.5.2 Resistance Laws for Uniform Underflow

Figure 8.20 shows a laminar density current between parallel boundaries. The lower boundary is fixed and the upper boundary is in motion. The boundary separating the two fluids is called the **interface**. The hydrodynamic properties of the interface is not taken into consideration. The depth of the density current is assumed small in comparison with the depth

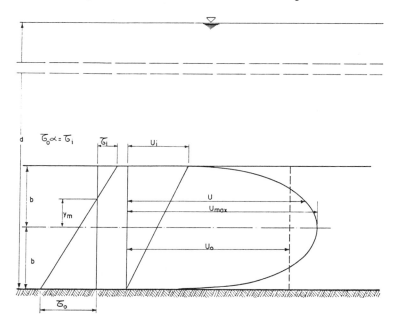

Figure 8.20. Laminar flow between parallel boundaries with the upper boundary in motion.

of the clear liquid above and therefore the mixing process of the two layers is avoided. The flow of the density current causes a shear on the bottom shown by (τ_o) and a shear on the lighter fluid shown by (τ_i). Then the density current may overcome or equalize a total shear of ($\tau_o + \tau_i$) in order to move between the two boundaries. This motion is caused by the extra weight of the turbid flow; assume $\Delta\rho g = \Delta g$ the difference in density between

F. Sentürk

the two media; the denser one is flowing more rapidly than the light one. Under these assumptions the equilibrium relation can be written as

$$\tau_o + \tau_i = (\Delta\rho g)\, dS \tag{8.17}$$

where

g is the gravitational acceleration

d is the depth of flow, and

S is the slope.

Considering Fig. 8.20 it is possible to write

$$\tau_i = \alpha\,\tau_o \tag{8.18}$$

with $\alpha < 1$.* Equation 8.18 assumes that the interfacial shear stress is a constant ratio of the bottom shear. It depends only on the vertical location of the maximum velocity between the bottom and the interface. This assumption can only be valid if 2b is very small in comparison to (d).

Substituting (τ_i) from Eq. 8.18 into Eq. 8.17 yields

$$\tau_o (1 + \alpha) = (\Delta\rho g)\, dS \tag{8.19}$$

and

$$\tau_o = \frac{(\Delta\rho g)d}{1 + \alpha} S \tag{8.20}$$

Considering Fig. 8.20 it is possible to write

$$\frac{\alpha\tau_o}{\tau_o} = \frac{b - y_m}{b + y_m}$$

and at the limit where d = 2b

$$\alpha = \frac{1 - \dfrac{y_m}{b}}{1 + \dfrac{y_m}{b}} = \frac{1 - 2\dfrac{y_m}{d}}{1 + 2\dfrac{y_m}{d}} \tag{8.21}$$

for flow between parallel stationary plates $y_m = 0$ and for open channel flow $\tau_i = 0$.

---

* or $\tau_o = \alpha\tau_i$ with $\alpha > 1$ as shown on Fig. 8.20.

444

Then the limiting values of (α) are

$\alpha = 0$ for open channel flow

$\alpha = 1$ for flow between parallel stationary plates

Darcy-Weissbach relation is

$$\tau_o = \frac{f}{4}\rho\frac{U_o^2}{2} = \frac{f}{4}\gamma\frac{U_o^2}{2g} \tag{8.22}$$

where f is Darcy-Weissbach friction factor

$$f = \frac{8gRS}{U_o^2} \tag{8.23}$$

Eliminating t_o between the two relations 8.20 and 8.22 yields

$$U_o = \sqrt{\frac{8\frac{\Delta\rho}{\rho}g}{f}}\sqrt{\frac{d}{1+\alpha}S} \tag{8.24}$$

Equation 8.24 is similar to Chézy's; considering that

$$C \sim \sqrt{\frac{8\frac{\Delta\rho}{\rho}}{f}g} \qquad R \sim \frac{d}{1+\alpha}$$

Then $d/(1 + \alpha)$ is the effective two dimensional hydraulic radius of the density current. The density current can be laminar or turbulent; the two cases are studied as follows.

A. Laminar flow

Ippen and Harleman (1952) have shown that the general form of the laminar velocity distribution can be expressed as a function of a parameter J

$$J = \frac{F^2}{RS} \tag{8.25}$$

where F is the densimetric Froude number defined as

$$F = \frac{U_o}{\sqrt{\frac{\Delta\rho}{f}gd}} \tag{8.26}$$

R, is the Reynolds number, defined as

$$R = \frac{U_o d}{n} \tag{8.27}$$

and U_o is the mean velocity as defined in Fig. 8.20. The Froude number can be written as a function of Darcy-Weissbach friction factor (f) (Simons, D. B. and Şentürk, F., 1992) and

$$F_r = (\frac{8}{f})^{1/2} S^{1/2} \tag{8.28}$$

Considering Eq. 8.24 it is possible to write

$$F^2 = \frac{8S}{\rho(1+\alpha)} \quad \text{or} \quad f(1+\alpha) F^2 = 8S \tag{8.29}$$

and eliminating S between Eq. 8.25 and 8.29 yields

$$f(1+\alpha) = \frac{8}{JR} \tag{8.30}$$

Equation 8.30 is interpreted as follows:

1. Flow between parallel stationary plates, $J = 1/12$

$$f(1+\alpha) = 2f = \frac{96}{R} \tag{8.31}$$

2. Laminar free surface flow, $J = 1/3$

$$f(1+\alpha) = f = \frac{24}{R} \tag{8.32}$$

3. Laminar density current, $J = 0.138$

$$f(1+\alpha) = 1.64f = \frac{58}{R} \tag{8.33}$$

Substituting this value in Eq. 8.24 yields

$$U_o = \sqrt{\frac{\frac{8\Delta}{\rho}g}{f(1+\alpha)}} \sqrt{dS} = \frac{\sqrt{R}}{2.7} \sqrt{\frac{\Delta\rho}{\rho} gds} \tag{8.34}$$

Equation 8.34 has been checked satisfactorily by Bata and Kneževich (1953), Reynaud (1951), and Bonnefille et Godet (1959).

Example 8.04

Determine the mean velocity of a laminar density current under the following conditions:

$S = 0.0004$	$d = 10.0$ ft	$\alpha = 0.75$	$f = 0.010$

$$R = 500 \qquad \frac{\Delta\rho}{\rho} = 0.0007$$

Solution

$$U_o = [\frac{8 \cdot 0.0007 \cdot 32.2}{0.01 \, (1 + 0.75)}]^{1/2} \, (10 \cdot 0.0004)^{1/2} = 0.203 \text{ fps} = 0.062 \text{ m/sec}$$

B. Turbulent Flow

In case that the density current is turbulent an analytical approach is not possible. But the above analysis gives a way of thinking for solving the problem. Bata and Kneževich (1953) have shown that for Reynolds numbers between 1,000 and 25,000, $y = 0.7d$, where y is the depth at which U_{max} occurs. On the other hand the experiments have indicated an average value of the friction factor of the interface to be $f_i = 0.01$ for hydraulically smooth fixed boundaries. Keulogan (1949) has found $f_i = 0.007$ based on approximate calculations of the resistance due to wave motion and mixing at the interface. The previous approach yields $f + f_i = f(1 + \alpha)$ and $f_i = \alpha f = 0.43f$. For $f = 0.020$, $f_i = 0.0086$. Then, using Eq. 8.24, U_o can be computed.

8.6 ADDITIONAL LOAD FROM MUD LAKES IMMEDI-ATELY UPSTREAM OF DAMS

Sediments are transported by density current and deposited immediately upstream of dams causing an additional load. It is then necessary to compute the amount of sediment transported by density current.

The relation between the sediment concentration and the difference in specific gravity is

$$\frac{\Delta\gamma}{\gamma} = \frac{C}{100} \, (1 - \frac{S_f}{S_s}) \qquad (8.35)$$

where

C is the percentage of concentration of sediment by weight

S_f is the specific gravity of water, and

S_s is the specific gravity of sediment.

Equation 8.35 can be used to determine C, then the sediment transport. This sediment deposited upstream of the dam becomes consolidate. It then has a specific weight given by Eq. 8.15; and the Rankine formula is then used for computing the additional load.

Example 8.05

Compute the weight of sediment transported by a density current the characteristics of which are given, as,

Data

$$S = 0.009 \qquad \frac{\Delta\rho}{\rho} = 0.0005 \qquad f = 0.010 \qquad \alpha = 0.43$$

$$\mu = 1.05 \; 10^{-5} \; \text{ft}^2/\text{sec}$$

$$\text{width of the current} = 1,000 \text{ ft} \qquad d = 15 \text{ ft}$$

Solution

• Equation 8.24 will be used for computing the mean velocity

$$U_o = \sqrt{\frac{8(0.0005)32.2}{0.010}} \sqrt{\frac{15 \cdot 0.009}{1.43}} = 0.35 \text{ fps}$$

• Regime of flow

$$\frac{U_o d}{v} = 5 \; 10^5$$

the flow is turbulent

• The rate of transported sediment

Using Eq. 8.35

$$\frac{\Delta\gamma}{\gamma} = 0.0005 = \frac{C}{100} \left(1 - \frac{62.4}{166}\right)$$

$$C = 0.08\%$$

$$Q_w = 15 \bullet 1{,}000 \bullet 0.35 = 5250 \text{ cfs}$$

$$Q_s = 5250 \bullet 0.0008 \bullet 62.4 = 262 \text{ 1b/sec}$$

$$Q_s = 262 \bullet 86400 \bullet 0.454 \; 10^{-3} = 10277 \text{ metric T/day}$$

8.6.1 Rate of Silting and Trap Efficiency of Reservoirs

Sediment is introduced in a reservoir by the watercourses entering it and by the rain water carrying eroded material from the watershed. This material is deposited in the reservoir. Fine materials carried up to the dam by density currents can be evacuated by bottom valves or through head structures conveying water to power plants, to irrigation areas or simply to the main watercourse downstream of the dam for preserving wild life. The natural retention capacity for sediments of the reservoir is called **trap efficiency** of the reservoir. The annual capacity loss of a reservoir can be estimated as follows

$$C_L V = T_E Q_s \tag{8.36}$$

where

C_L is the reservoir capacity loss (percent per year)

V is the original volume of the reservoir

T_E is the trap efficiency (per cent), and

Q_s is the annual sediment yield of the drainage area (m³/year or acre feet per year)

Equation 8.36 can be transformed to

$$C_L = \frac{T_E Q_s}{V} \tag{8.37}$$

It is clear that the trap efficiency is a function of the topography of the reservoir site; thus it must be determined for each reservoir separately. The amount of precision in the computation or determination of C_L will govern the accuracy of the forecast of reservoir life, which is a factor directly influencing the choice of the dam site.

Many approaches have been suggested for a correct determination of the trap efficiency. Brown (1944) proposed the use of the diagram given in Fig. 8.21 together with Eq. 8.38

$$T_E = 100 \left(1 - \frac{1}{1 + K\dfrac{C}{W}}\right) \tag{8.38}$$

449

where

K is the numerical coefficient equal to 0.046 for the lower curve in Fig. 8.21 and K = 1 for the upper curve. The value proposed for the design curve is K equal to 0.1, K = 0.10.

C is the original reservoir storage capacity (acre feet), and

W is the watershed area.

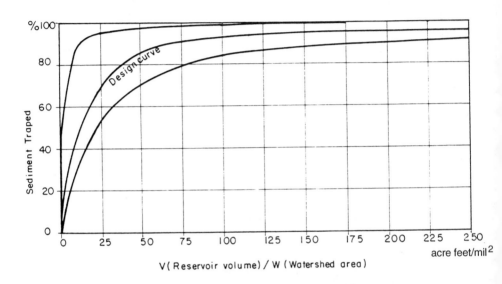

Figure 8.21. Trap efficiency curves from Brown.

The designer may choose among the curves suggested in Fig. 8.21, for an approximation. It is possible to begin with the value selected and proceed further by choosing a value for K varying between 1 and 0.046.

Brune (1953) suggests the curves of Fig. 8.22 for the evaluation of T_E and Churchill (1948) suggests the curve in Fig. 8.23.

Borland (1971) suggests the following procedure for the computation of the trap efficiency. He has taken as a basis the method proposed by Churchill:

Step 1: Compute the capacity of the reservoir at mean operating pool elevation for the period considered.

Step 2: Compute the daily average inflow during the period under consideration.

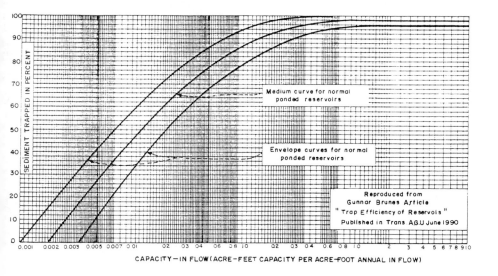

Figure 8.22. Trap efficiency curves from Brune.

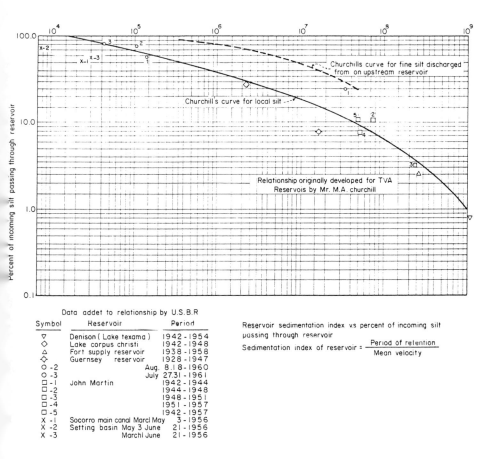

Data addet to relationship by U.S.B.R

Symbol	Reservoir	Period
▽	Denison (Lake texama)	1942-1954
◇	Lake corpus christi	1942-1948
△	Fort supply reservoir	1938-1958
◇	Guernsey reservoir	1928-1947
O -2		Aug. 8.18-1960
O -3		July 27.31-1961
□ -1	John Martin	1942-1944
□ -2		1944-1948
□ -3		1948-1951
□ -4		1951-1957
□ -5		1942-1957
X -1	Socorro main canal Marcl May	3-1956
X -2	Setting basin May 3 June	21-1956
X -3	Marchl June	21-1956

Reservoir sedimentation index vs percent of incoming silt passing through reservoir

$$\text{Sedimentation index of reservoir} = \frac{\text{Period of retention}}{\text{Mean velocity}}$$

Figure 8.23. Trap efficiency curve for reservoirs from Churchill (1948).

Step 3: Compute the period of retention or the capacity (in cf) divided by the inflow rate (in cfs).

Step 4: Compute the reservoir length in feet at mean operation pool level.

Step 5: Compute the mean velocity equal to the inflow divided by the average cross section area (in ft²). Compute the average cross sectional area by dividing the reservoir capacity by the reservoir length.

Step 6: Compute the sedimentation index which is equal to the period of retention divided by the mean velocity."

After the sedimentation index has been computed the percent of incoming silt passing through the reservoir can be obtained using Fig 8.23.

8.7 PROBLEMS

Problem 8.01

Compute the effective fetch of a reservoir shown in Fig. 1, Problem 8.01.

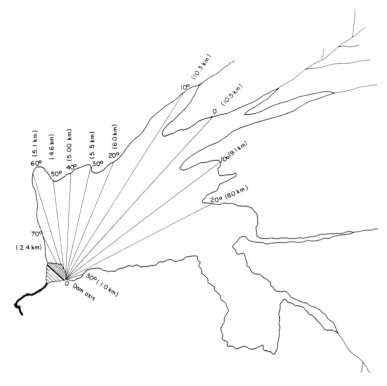

Fig. 1. Problem 8.01.

Problem 8.02

Compute the wave height, the wave set-up and the wave run-up in the reservoir the characteristics of which are given.

Data

$F = 45$ km $U = U_r = 45$ m/sec $\emptyset = 22°$

Embankment slope = 1/3 (vert/hor)

d = 25.00 m

Problem 8.03

What happens if the direction of the wind changes so that $\emptyset = 0$ in the reservoir of Problem 8.02.

Problem 8.04

Determine the variation of the mean velocity of a density current as a function of its thickness.

Data

- The cross section is assumed very large

- $R = 500$

- $\alpha = 0.50$

- $\dfrac{\Delta\rho}{\rho} = 0.0007$

- $S = 0.0004$

Problem 8.05

The length of a reservoir is 110 miles. Using conditions defined in Problem 8.04 determines the time necessary for the density current to reach the dam.

Data

- The mean depth of the first 10 miles in the reservoir is 30 ft.

- The mean depth of the following 50 miles in the reservoir is 90 ft.

- The mean depth of the following 50 miles in the reservoir is 210 ft.

Problem 8.06

Compute the thickness of sediment accumulated upstream of a dam in three years beginning from the time of closing of the bottom outlet valves.

Data

- The cross section of the reservoir upstream of the dam is assumed rectangular with base length equal to 1,000 ft.

- $S = 0.0001$ $\qquad$ $S_s = 166 \text{ lb/ft}^3$ $\qquad$ $S_f = 63.4 \text{ lb/ft}^3$

- $\dfrac{\Delta\rho}{\rho} = 0.0003$ $\qquad$ $f = 0.01$ $\qquad$ $\alpha = 0.45$ for turbulent flow

  $\qquad\qquad\qquad\qquad$ $d = 12.0 \text{ ft}$ $\qquad$ $v = 10^{-5}\text{ft}^2/\text{sec}$

- Width of the current $= 1,000$ ft

- The density current is assumed existent throughout the year.

Problem 8.07

Compute the trap efficiency of the reservoir the characteristics of which are given.

Data

- $K_o = 90$

where K_o is:

$$K_o = \frac{\text{Annual sediment yield of the drainage area}}{\text{surface of the drainage area}}$$

Hint: Use Brown's curve.

Chapter 9

HYDRAULICS OF BODIES OF DAMS

9.1   INTRODUCTION

The body of an embankment dam is subject to the following hydraulic actions:

1. Erosive action of waves on the upstream face.

2. Erosive action of the seepage through the impervious core.

3. Loss of water by leakage.

4. Erosive action of wind and waves on the downstream face of the reservoir .

First, the designer must evaluate as accurately as possible the effect of these actions and then try to choose the measures necessary to solve the problems. Dimensions of an embankment of the dam are established by application of the laws of static equilibrium and then checked for hydraulic conditions. When the static equilibrium and hydraulic stability are assured, the body of the dam can be designed. The hydraulic computation includes the following steps:

- Computation of the freeboard to prevent overtopping (for both fill and concrete type dams).

- Determination of the riprap size and thickness to prevent wave erosion of the upstream face of embankment dams.

- Determination of crest width for embankment dams.

- Determination of filter thickness and the material gradation of filters in embankment dams.

- Determination of the leakage loss through the impervious core in embankment dams.

• Determination of riprap stability downstream of the spillway.

The leakage through the impervious core is computed by applying the method of flow net analysis or by using empirical formulas. Both of these approaches are approximate. During operation of the dam, the loss from leakage is continuously checked and the variation in hydrostatic pressures inside the impervious core is measured. The osculation devices used for measurement purposes and interpretation of measurements are also presented in this chapter, together with hydraulic computations.

9.2   COMPUTATION OF FREEBOARD

The freeboard in an embankment or a concrete dam (gravity or arch type dam) must be greater than the total of,

1. The rise of water (in meters or feet) over the still water level, z_d

2. The wave run-up (in meters or feet), z_r

3. The wave set-up (in meters or feet), z_s.

So

$$\text{Freeboard} \geq z_d + z_r + z_s \qquad (9.1)$$

Equation 9.2 is also used for the computation of freeboard:

$$F = 0.75\, z_d + \frac{U_d^2}{2g} \qquad (9.2)$$

where

F   is the freeboard in meters

U_d is the wave celerity in meters per sec.

z_d is the wave height in meters.

The wave celerity U_d is given by Gaillard as

$$U_d = 1.5 + 2z_d \qquad (9.3a)$$

where

U_d is in meter per second.*

A minimum value of 3.00 m is, in general, used for freeboard. If Eq. 9.1, for example, give values less than 3.00 m, the freeboard may be chosen equal to 3.00 m (Table 8.2).

Example 9.01

Compute the freeboard of the dam defined in Example 8.03.

Solution

The maximum wave action is computed according to the general rule outlined in Example 8.03 and found to be equal to 5.71 m. In this example, this value will be checked against the empirical Eqs. 9.2 and 9.3b.

- Computation of the mean wave celerity

$$U_d = 1.50 + 2z_d$$

$$U_d = 1.50 + 2 \cdot 2 = 5.50 \text{ m/sec}$$

where

$$z_d = 2.00 \text{ m/sec (Example 8.03)}$$

- Computation of freeboard

$$F = 0.75z_d + \frac{U_d^2}{2g} \qquad\qquad 9.2$$

---

*   Different equations are used throughout the world for computing freeboards. As an example, the Arizona, USA code is given by

$$H = 1.5 \, (1.5d^{0.5} + 2.5 - d^{0.25}) \qquad\qquad (9.3b)$$

where

H   is the vertical dimension of freeboard in feet

d   is the distance in miles from the dam to the most remote point along the shore of the reservoir, measured across the open water. Equation 9.3b is written using the English systems of units.

$$F = 0.75 \cdot 2 + \frac{5.5^2}{19.62} = 3.04 \text{ m}$$

- Computation of freeboard by Eq. 9.3b

$$H = 1.5 \, (1.5 \cdot 18.64^{0.5} + 2.5 - 18.64^{0.25})$$

$$= 10.35 \text{ ft} = 3.15 \text{ m}$$

Since the two results are smaller than 5.71 (Example 8.03), the freeboard used will be 5.71 m.

9.3 DETERMINATION OF CREST WIDTH

The minimum crest width is based upon the required thickness of the top of the impervious core plus the additional width required to resist erosion. A standard method of computation does not exist. The following empirical considerations are suggested:

1. The minimum width of the crest must be 10.00 m

2. If the dam is located in a region of high seismicity, increase the width by 20%.

3. Knappen suggests Eq. 9.4 for the computation of the crest width

$$b = 1.65 \, h^{1/2} \tag{9.4}$$

where

   b   is the crest width in meters, and

   h   is the dam height in meters.

4. USBR's suggestion is

$$b = \frac{h}{3} + 10 \tag{9.5}$$

where b and h are in feet.

5. Preece(1938) suggestion is

$$b = 1.10h^{1/2} + 1 \tag{9.6}$$

Equation 9.4 is recommended for small dams only; the resulting b value being too large for large dams. Table 9.1 summarizes different crest widths of dams throughout the world.

TABLE 9.1

Crest Width of Dams Throughout the World

Name of Dam	Country	Height m	Crest Width m
Atatürk	Turkey	172	15.00
Keban	"	207	12.00
Hasan Uğurlu	"	175	12.00
Kiliçkaya	"	140	12.00
Kiralkizi	"	130	12.00
Salt Springs	USA	100	4.57
Trinity	"	164	12.00
Stevens	"	129	15.00
Oroville	"	224	24.50
Navajo	"	124	9.14
Fort Peck	"	82	30.50

9.4 SLOPE PROTECTION

The upstream slope of an earth fill dam is subject to mechanical action of waves and winds and the downstream slope is subject to action of winds, and heavy rains. Protection of both the upstream and downstream slopes of embankment dams is necessary to assure their stability.

Different types of protective coverings exist; some of them are:

- Dumped stone riprap

- Hand placed riprap

- Grouted riprap

- Concrete blocks

- Bituminous paving

- Planting on slopes

- Concrete slabs

Two that are frequently used in dam construction are dumped stone riprap for the upstream slope and planting or dumped stone riprap for the downstream slope. The other coverings are used only in special cases. During the first half of the 20th century, in Europe, concrete slabs and bituminous paving were used for medium size dams. In recent dam constructions, the use of earth moving equipment such as bulldozers and large tractors have reduced the cost of dumped stone riprap so that it is commonly used as the slope protection cover.

In this section, the computation of only the thickness of the protective dumped riprap layer and the stone sizes is given. Construction techniques are not the concern of this book.

9.4.1 Weight of Rock Used to Protect Slopes of Dams

General requirements for selecting the thickness and the stone size of riprap are reproduced here according to "Design Manual, Soil Mechanics, Foundation and Earth Structure, USBR".

1. For embankment slopes between 1/2 and 1/4(V/H), dumped riprap shall meet the following requirements (Table 9.2):

TABLE 9.2

Characteristics of Dumped Riprap

Wave Height ft (m)	Average Rock Size, D_{50} inches (m)	Max Rock Size lbs (kg)	Layer Thickness inches (m)
0 - 1 (0 - 0.30)	8 (0.20)	100 (45.35)	12 (0.30)
1 - 2 (0.3 - 0.6)	10 (0.25)	200 (90.70)	15 (0.38)
2 - 4 (0.6 - 1.2)	12 (0.31)	500 (226.75)	18 (0.46)
4 - 6 (1.2 - 1.8)	15 (0.38)	1500 (680.25)	24 (0.61)
6 - 8 (1.8 - 2.4)	18 (0.46)	2500 (1134)	30 (0.76)
8 - 10 (2.4 - 3.0)	24 (0.61)	4000 (1814)	36 (0.91)

2. Riprap shall be well graded from a maximum size, at least 1.5 times average rock size, to 1 inch spells suitable to fill voids between rocks.

3. Riprap blanket shall extend at least 8 ft (2.4 m) below lowest low water (this value changes with different authors, from 3 ft to 8 ft).

4. Under the most extreme icing and temperature changes, rock should meet soundness and density requirements for concrete aggregate.

5. Filter shall be provided between the riprap and the embankment soils to meet the following criteria (Table 9.3).

TABLE 9.3

Max Wave Height ft (m)	Filter D_{85} Size, at least inches (cm)
0 - 4 (0 - 1.200)	$1 - 1\frac{1}{2}$ (2.54 - 3.81)
4 - 10 (1.20 - 3.00)	$1\frac{1}{2} - 2$ (3.81 - 5.08)

6. Minimum thickness of single layer filters are as follows (Table 9.4):

TABLE 9.4

Minimum Thickness of Single Layer Filters

Maximum Wave Height ft (m)	Filter Thickness inches (cm)
0 - 4 (0.0 - 1.2)	6 (15.24)
4 - 8 (1.2 - 2.4)	9 (22.86)
8 - 12 (2.4 - 3.6)	12 (30.48)

Double filter layers should be at least 6 inches thick.

Different empirical formulas have been proposed for determining the weight of armor stone for breakwaters (Simons, D.B. and Şentürk, F., 1992) and (Hudson, 1961)

$$W_s = \frac{\gamma_s H^3}{K_D (S_s - 1)^3 \cot\alpha} \tag{9.7}$$

where

W_s is the weight of stone forming the breakwater

γ_s is the specific weight of stone

$$S_s = \frac{\gamma_s}{\gamma}$$

H is the maximum wave height

α is the angle of breakwater slope with the horizontal, and

K_D is an experimental damage coefficient.

Equation 9.7 is applicable to breakwaters. A different formula is used for riprap:

$$W_{50} = \frac{\gamma_s H^n}{K_{rr} (S_s - 1) \cot\alpha} \tag{9.8}$$

where

W_{50} is the average stone weight (lb)

n   is the exponent of H which ranges generally from about 2.0 to 3.0

K_{rr} is the riprap coefficient, which ranges generally from 1.8 to 3.5, and

α is the upstream slope angle.

The values of n and K_{rr} are given in Table 9.5.

TABLE 9.5

Average Stone Weight W_{50}, various values of n and K_{rr} in lb and (kg)

Wave height n	K_{rr}	Average stone weight, W_{50}		
		2 (0.60) ft (m)	4 (1.20) ft (m)	8 (2.40) ft (m)
2.0	1.82	30 (13.6)	116 (52.6)	473 (214.6)
2.5	3.50	22 (10.4)	121 (54.9)	700 (317.5)
2.6	3.20	26 (11.8)	154 (69.9)	940 (426.4)
2.7	3.20	27 (12.2)	174 (78.9)	1160 (526)
3.0	5.50	98 (44.5)	153 (69.4)	1255 (569)

| Notice | $S_s = 2.60$ | $\gamma_s = 162.24$ lb/cf | | $\cot\alpha = 3$ |

The suggestion of U.S. Corps of Engineers (1949) is given in Table 9.2.

It is also suggested that the rock used for dumped riprap should grade from 50% of the thickness of the layer to $D_{50} = 4$ inches.

9.4.2 Thickness of Dumped Stone Riprap Layer

A rigorous mathematical approach for computing the thickness of the dumped stone riprap layer does not exist. However, it is suggested that for estimating purposes, the depth of dumped riprap be equal to 60% of the height of the waves.

Figure 9.1 also can be used for the estimation of this depth (Table 9.2). The minimum thickness of the riprap layer suggested by TVA is

$$e = CU_d \tag{9.9}$$

where

C is a coefficient (Table 9.6)

U_d is the wave celerity defined by Eq. 9.3a

US Army Corps of Engineers suggestion is summarized in Table 9.7.

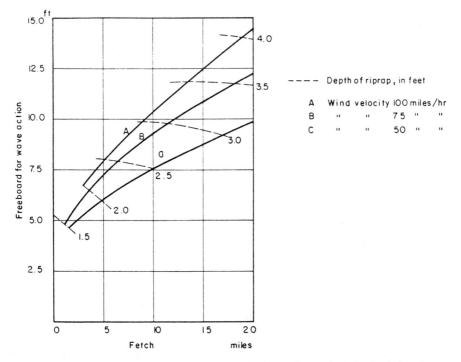

Figure 9.1. Thickness of the dumped riprap layer and the freeboard as a function of the fetch (Davis, C.V., 1952).

F. Sentürk

TABLE 9.6

Coefficient C in Equation 9.9.

Upstream slope of dam	Coefficient C as a function of the specific weight of the rock used		
V/H	2.5 T/m^3	2.65 T/m^3	2.80 T/m^3
1/12	0.024	0.022	0.020
1/4	0.027	0.024	0.022
1/3	0.028	0.025	0.023
1/2	0.031	0.028	0.026
1/1.5	0.036	0.032	0.030
1/1	0.047	0.041	0.038

TABLE 9.7

Thickness of the riprap layer the US Corps of Engineers (1949).

Wave height m	Min. thickness of the protective riprap layer	
	m	D$_{50}$ (m)
0.00< wave height < 0.30	0.30	0.20
0.30< wave height < 0.60	0.40	0.25
0.60< wave height < 1.20	0.45	0.30
1.20< wave height < 1.80	0.55	0.40
1.80< wave height < 2.40	0.70	0.45
2.40< wave height < 3.00	0.80	0.55

9.4.3 Riprap Stability Downstream of Stilling Basins

Two suggestions are reproduced here for computing the weight of rock required for protection of stilling basins from downstream erosion.

464

A. ASCE Task Committee on Preparation of a Sedimentation Manual Approach

ASCE Task Committee (1972) using the Isbach's relation suggested

$$W_{50} = \frac{4.1 \ 10^{-5} \ S_s U^6}{(S_s - 1)^3 \cos^3 \emptyset} \tag{9.10}$$

where

S_s is the specific weight of riprap (dimensionless)

$$S_s = \frac{\gamma_s}{\gamma}$$

γ_s is the specific weight of the rock (T/m^3, kg/m^3, lb/ft^3)

γ is the specific weight of water (T/m^3, kg/m^3, lb/ft^3)

$\emptyset$ is the angle of repose of the riprap.

Assuming spherical the form of riprap, W_{50} can be defined as

$$W_{50} = \gamma_s \frac{\pi}{6} \ D_{50}^3 \tag{9.11}$$

where

D_{50} is the median rock size in feet, and γ_s is in lb/ft^3. Substituting W_{50} by its value from Eq. 9.11 yields:

$$\frac{0.347U^2}{(S_s - 1)gD_{50}} = \cos\emptyset \tag{9.12}$$

where g is the gravitational acceleration in ft/sec^2. For $\emptyset = 0$, Eq. 9.12 reduces to

$$\frac{0.347U^2}{(S_s - 1)gD_{50}} = 1 \tag{9.13}$$

B. Bureau of Reclamation's Approach

This approach is summarized in Fig. 9.2. It is particularly recommended for stilling basin protection.

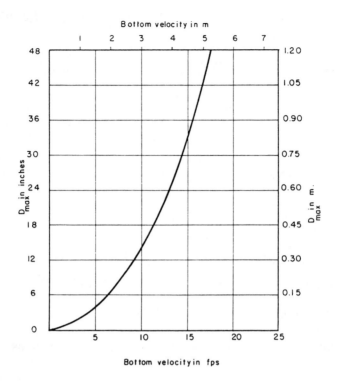

Figure 9.2. Rock size for USBR design.

If U_s shows the velocity against the stone, the curve in Fig. 9.2 can be approximated by

$$U_s^2 = 49.1 \, (S_s - 1) \, D_{100} \qquad (9.14)$$

Then, using Eq. 9.14, D_{100} can be computed for values of U_s. The reference velocity, U_r, is

$$U_r = 8.5 \, U_* \qquad (*) \qquad (9.15)$$

and

$$\frac{U_r}{U_s} = \frac{3.4 \, (0.9581 \log (\frac{y_o}{k}) + 1)}{1_n \, (12.21 \, \frac{y_o}{k})}$$

---

* See Simons, D.B. and Şentürk, F. (1992)

For values of (y_o / k) between 10° and 10^6, the value of (U_r / U_s) is nearly 1.4; then

$$U_s = \frac{U_r}{1.4} \qquad (9.16)$$

(y_o) is the depth of flow and (k) is the representative rock size.

Example 9.02

Determine the size of the riprap downstream of a stilling basin with the following characteristics. The cross section at the basin is assumed to be very large.

Data

- $S = 0.015$

  S is the slope of the energy line downstream of the stilling basin

- $y_o = 5.00$ m $= 16.4$ ft

  y_o is the depth of water downstream of the stilling basin.

- $S_s = 2.65$

Solution

USBR's relation is used for solving this problem

$$U_* = (gRS)^{1/2} = (32.2 \bullet 16.4 \bullet 0.015)^{1/2} = 2.814 \text{ fps}$$

and

$$U_r = 8.5 \bullet 2.814 = 23.92 \text{ fps}$$

The cross section being very large $R = y_o$.

- Use Eq. 9.16 for determining U_s

$$U_s = \frac{U_r}{1.4} = 17.09 \text{ ft}$$

- Determine D_{100} using Eq. 9.14

$$U_s^2 = 49.1 (S_s - 1) D_{100} \qquad 9.14$$

$$D_{100} = \frac{17.09^2}{49.1 \cdot 1.65} = 3.604 \text{ ft} = 1.1 \text{ m}$$

- Checking for (y_o/k) yields

$$\frac{y_o}{k} = \frac{16.4}{3.604} = 4.55$$

$1 < 4.55 < 10^6$ then Eq. 9.14 is valid.

9.5 LEAKAGE THROUGH DAM BODY

A number of methods, including an impervious core, a concrete face slab, are employed in rockfill dams to create a water barrier. In earth fill dams, an impervious core, an upstream impervious blanket are among the methods used in the semi-permeable fill to act as a water seal. Sometime the embankment material itself may be sufficiently impervious so that no additional water barrier is required. However a certain amount of water usually manages to seep through and is drained through the downstream zones of the embankment. The loss of water by leakage is generally small or negligible. In the Demirköprü Dam, for example, the discharge by leakage is around 300 lt/sec. In the Hirfanli Dam the leakage is 150 lt/sec and in the Cubuk Dam only some 20 lt/sec. In large dams like Keban Dam, Atatürk Dam, etc., this value is higher and attains as much as 1.50 m^3/sec.

An embankment dam requires an engineering review of the two specific problems:

a. Loss of water by leakage

b. A stability analysis based upon the determination of the seepage line.

These two aspects are studied further here.

9.5.1 Seepage Through Impervious Core

The seepage through the impervious core can be estimated by empirical methods or by flow net analysis (see subsection 7.3.1). The flow net analysis is used for determining the seepage line and for estimating the discharge as given in subsection 7.3.1.

A. Position of Seepage Line

Figure 9.3 illustrates the position of the seepage line in the impervious core of a dam.

The depth of water upstream of the impervious core is denoted by h. The upstream water line intersects the upstream slope at point B. Water flows through the impervious core following the seepage line BC. The fill material is dry above the seepage line and wet below it. The specific weight of the dry material being different from the specific weight of the wet material, the position of the seepage line is very important in the stability analysis of fill-type dams. It is then also very important to determinate its position accurately. Casagrande (1937) has shown that the seepage line follows very nearly a parabola called the fundamental parabola. Above the fundamental parabola only the dry weight of the fill material acts as a determining factor of stability, but below it the hydrostatic pressure must be added to the wet weight of the fill material. The definition of symbols shown in Fig. 9.3 follows,

B_2: the intersection of the fundamental parabola with the water surface

A: the downstream toe of the impervious core

C: the intersection of the seepage line with the downstream slope of the impervious core

d: the horizontal distance between point A and B_2

a: the distance AC; below point C the downstream slope is wet and dry above it

h: hydraulic head at point A (the depth of water upstream of the impervious core)

α: the internal angle formed by the downstream slope of the impervious core with the horizontal base

m: the horizontal projection of the wetted upstream slope

k: Darcy's coefficient of permeability.

The fundamental parabola is defined for the first time by Kozeny and it is called the Kozeny's parabola. In general it closely approximates the seepage line for $\alpha = 180°$ (Fig. 9.4).

Applying Eq. 7.25 for 1.00 m length of the impervious core yields:

$$Q = kh \frac{dh}{ds} \qquad 7.25$$

$$A = 1 \cdot h$$

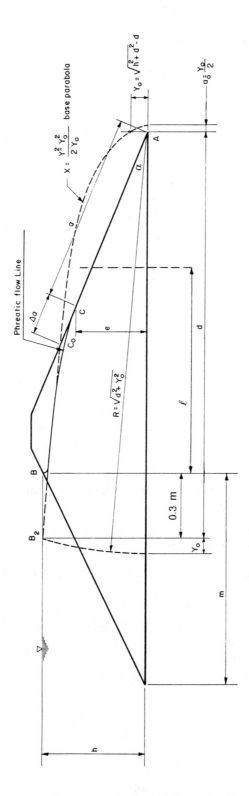

Figure 9.3. Seepage through the impervious core of a fill dam.

as the wetted area; then

$$q = ky \frac{dy}{dx} \tag{9.17}$$

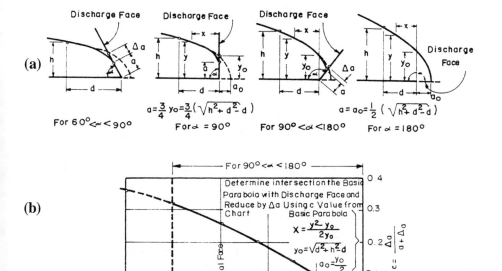

Figure 9.4. Diagram for determining Δa and a.

Kozeny has shown that the fundamental parabola can be expressed as

$$x = \frac{y^2 - y_0^2}{2y_0} \tag{9.18}$$

The focus of this parabola is at point A; the intersection of the perpendicular issued from this point to the base line with the parabola is called y_0 (Fig 9.3). Then the coordinates of the point B_2 are $y = h$ and $x = d$; d, is obtained by multiplying the base width of the impervious core by 0.7. Substituting these values in Eq. 9.18 yields

$$y_0 = (h^2 + d^2)^{1/2} - d \tag{9.19}$$

y_0 can also be obtained graphically as shown in Fig. 9.3. The polar equation defining the parabola can be written as

$$r = \frac{p}{1 - \cos\varnothing} \tag{9.20}$$

where

      r is the distance between any point on the parabola to the pole

      $p = y_o$

and

$$\emptyset = \alpha$$

For

$$r = a + \Delta a \qquad a + \Delta a = \frac{y_o}{1 - \cos\alpha} \qquad (9.21)$$

$\Delta a = CC_o$, C_o is the intersection of the fundamental parabola with the downstream slope of the impervious core; a, varies with (α). This variation is shown in Fig. 9.4.

    Kozeny's parabola is thus obtained by the application of the method outlined above; then the seepage line is obtained starting from the point C. This line is orthogonal to the upstream face of the impervious core at point B based upon the properties of the stream lines. It is possible to show that the flow in impervious core follows the Darcy's law, that the flow is irrotational and the stream lines are orthogonal to equipotential lines. The upstream slope of the dam being an equipotential line the upstream lines are perpendicular to it (Valentine, 1959); and the seepage line being a stream line, it is perpendicular to the upstream face of the dam, becoming a particular equipotential line.

Example 9.03

    Determine the seepage line in the impervious core of the Kiralkizi Dam. The data of the problem is given and defined in Fig. 9.5.

Solution

    The solution of the problem follows:

1. 1/e, the slope of the upstream face $e = 4.00$ m

2. K_1: crest elevation $K_1 = 819.00$

3. K_2: Upstream water elevation $K_2 = 815.75$

4. K_3: Downstream water elevation $K_3 = 715.00$

5. K_4: Bottom elevation $K_4 = 705.00$

6. t: Half width of the crest $t = 4.00$ m

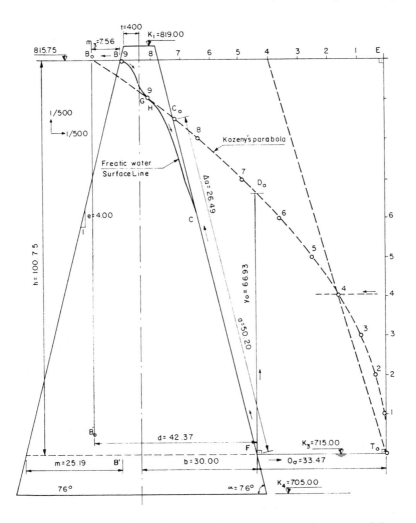

Figure 9.5. The position of seepage line in the impervious core of the Kiralkizi Dam.

7. $\tan\alpha = e$ (not the "e" of Fig. 9.3) $\qquad\qquad \tan\alpha = 4.00$

8. $b = \dfrac{K_1 - K_3}{e} + t \qquad\qquad b = \dfrac{819.00 - 715.00}{4.00} + 4.00 \qquad = 30.00\ \text{m}$

9. $h = K_2 - K_3 \qquad\qquad h = 815.75 - 715.00 = 100.75\ \text{m}$

10. $m = \dfrac{h}{e} \qquad\qquad m = \dfrac{100.75}{4} = 25.19\ \text{m}$

11. $m_3 = 0.3 \bullet m \qquad\qquad m_3 = 0.30 \bullet 25.19 = 7.56\ \text{m}$

12. $d = b + m_3 + t + \dfrac{3.25}{4}$ $\qquad$ $d = 30.00 + 7.56 + 4.00 + 0.81 = 42.37$ m

13. $y_o = (h^2 + d^2)^{1/2} - d$ $\qquad$ $y_o = (100.75^2 + 42.37^2)^{1/2} - 42.37 = 66.93$ m

14. $C = \dfrac{\Delta a}{a + \Delta a}$ $\qquad$ $C = 0$ (Fig. 9.4)

15. $\Delta a = \dfrac{y_o}{1 - \cos\alpha} - a$ $\qquad$ $\Delta a = \dfrac{66.93}{1 - \cos(76)} - a$ $\qquad\qquad$ 9.21

16. a is obtained from step 14 and 15 combined

  $a = 61.80$ m

  $\Delta a = 26.49$ m

B. The Flow Net

The theory of flow net is not given in full detail here; it can be found in all the classical hydrodynamic books (Valentine, 1959). Application of it is detailed subsequently.

1. Flow nets can be drawn only for the irrotational flow taking place in an homogeneous media.

2. For a given boundary condition, only one flow net can be obtained. The boundary conditions for the irrotational flow taking place in an impervious core of a dam which can be assumed to be an homogeneous medium, are:

   • The seepage line
   • The impervious foundation of the dam.

3. The flow net is composed of stream lines and lines orthogonal to them, the orthogonal lines are called the **equipotential lines**. The two systems of lines form a system of square shapes called **flow net**. The dimension of squares are chosen by designer. The smaller the dimensions, the more perfect the squares become. Equipotential lines intersect the boundaries orthogonally.

4. In the uniform and parallel flow the squares equal each other. In diverging flow the dimension of squares increase and in converging flow they decreases as shown in Fig. 9.6.

A practical approach for drawing the flow net is given in section 7.3.

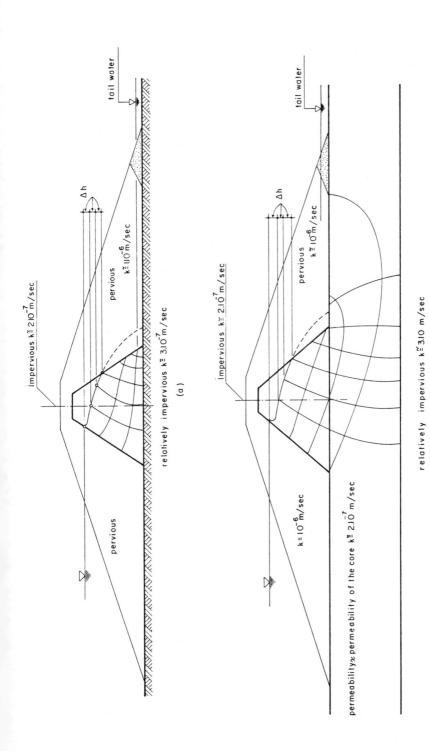

Figure 9.6. Flow net in an earthfill dam. (a - shallow foundation, b - deep foundation).

Example 9.04

Draw the flow net in the impervious core of the Kiralkizi Dam, Turkey, with the condition described in the Example 9.03.

Solution

The solution is obtained by direct application of the method outlined above and shown in Fig. 9.7. It is self-explanatory.

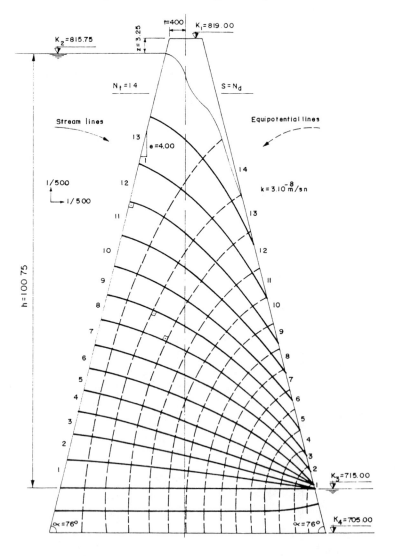

Figure 9.7. Flow net in the impervious core of the Kiralkizi Dam, Turkey.

C. Discharge by Leakage through Impervious Core

a. Flow net analysis

The flow net constitutes a tool for obtaining the discharge through the impervious core. It is explained in full detail in section 7.3.

b. Computation of seepage discharge using mathematical relations

Darcy's law is expressed by Eq 9.17

$$q = ky \frac{dy}{dx} \qquad 9.17$$

and the fundamental or base parabola by Eq. 9.18

$$x = \frac{y^2 - y_0^2}{2y_0} \qquad 9.18$$

where y_0 is

$$y_0 = (h^2 + d^2)^{1/2} - d \qquad 9.19$$

Equation 9.18 yields

$$y = (2xy_0 + y_0^2)^{1/2} \qquad (9.22)$$

Differentiating Eq. 9.18 and substituting y by its value taken from Eq. 9.22 yields

$$\frac{dy}{dx} = \frac{y_0}{\left(2xy_0 + y_0^2\right)^{1/2}}$$

and

$$q = k(2xy_0 + y_0^2)^{1/2} \frac{y_0}{\left(2xy_0 + y_0^2\right)^{1/2}} \qquad (9.23)$$

$$q = ky_0 \qquad (9.24)$$

This equation is valid for $\alpha = 180°$. For $30° < \alpha < 180°$

$$q = k(h^2 + d^2)^{1/2} - d \qquad (9.25)$$

and for $\alpha < 30°$, then

$$q = k \, a \, \sin^2\alpha \tag{9.26}$$

where

$$a = (h^2 + d^2)^{1/2} - (d^2 - h^2 \cot^2\alpha)^{1/2}$$

Using Fig. 9.3 it is possible to show that

$$e \cong \frac{h}{3} \tag{9.27}$$

then

$$q = k \frac{(h-e)}{\ell} \frac{(h+e)}{2} = \frac{k}{2} \frac{h^2 - e^2}{\ell} \tag{9.28}$$

Equation 9.26 is Darcy's law applied to the flow taking place in the impervious core

$$\frac{dy}{dx} = \frac{h-e}{\ell}$$

and

$$A = \frac{h+e}{2}$$

where ℓ is the mean length of the seepage (Fig. 9.3) which can be expressed as

$$\ell = 2(h+f)\cot\alpha + B - 0.7 \cot\alpha - \frac{e}{2}\cot\alpha = (1.3h + 2f - \frac{e}{2})\cot\alpha + B \tag{9.29}$$

in which f is the vertical distance from the head water to the top of impervious core and B is its crest width.

Substituting the value of (e) from Eq. 9.27 into Eq. 9.28 yields

$$q = \frac{k}{2} \frac{h^2 - \frac{h^2}{9}}{\ell} = 4k \frac{h^2}{9\ell} \tag{9.30}$$

Example 9.05

Compute the seepage discharge through the impervious core of the Kiralkizi Dam.

Data

• The flow net in the impervious core is shown in Fig. 9.7

• $k = 3 \ 10^{-8}$ m/sec

Solution

1. Upstream and downstream slopes of the impervious core: 1/0.25

2. Crest elevation $K_1 = 819.00$

3. Upstream water elevation $K_2 = 815.75$

4. Downstream water elevation $K_3 = 715.00$

5. Foundation elevation $K_4 = 705.00$

6. Crest length $B = 8.00$ m

7. $\tan(\alpha) = 4$ $\alpha = 76°$

8. $b = \dfrac{K_1 - K_3}{\tan \alpha} + \dfrac{B}{2}$ $b = \dfrac{819 - 715}{4} + 4 = 30.00$ m

9. $h = K_2 - K_3$ $h = 815.75 - 715.00 = 100.75$ m

10. $m = \dfrac{h}{\tan \alpha}$ $m = \dfrac{100.75}{4} = 25, 19$

11. $m_3 = 0.30 \bullet m$ $m_3 = 0.3 \bullet 25.19 = 7.56$ m

12. $d = b + b + m_3 - m$ $d = 30.00 + 30.00 + 7.56 - 25.19$
                                $= 42.37$ m

13. $y_o = (h^2 + d^2)^{1/2} - d$ $y_o = (100.75^2 + 43.72^2)^{1/2} - 42.37$
                                $= 66.93$ m

14. $C = \dfrac{\Delta a}{a + \Delta a}$ $C = 0.30$ (Fig. 9.4)

15. $\Delta a = \dfrac{y_0}{1-\cos\alpha} - a$ $\Delta a = \dfrac{66.93}{1-\cos 76} - a$ 9.21

16. (a) is obtained from $a = 61.80$ m $\Delta a = 26.49$ m
    step 14 and 15 combined

17. $f = K_1 - K_2$ $f = 819.00 - 815.75 = 3.25$ m

18. $h = K_2 - K_3$ $h = 815.75 - 715.00 = 100.75$ m

19. $e \approx \dfrac{h}{3}$ 9.27 $e = \dfrac{100.75}{3} = 33.58$ m

20. $\ell = (1.3h + 2f - \dfrac{e}{2})\cot\alpha + B$ $\ell = (1.3 \bullet 100.75 + 2 \bullet 3.25 - \dfrac{33.58}{2})\,0.25$

    $+ 8.00 = 38.17$ m

21. $q = 4k\,\dfrac{h^2}{9\,\ell}$ 9.30 $q = 4\dfrac{3\;10^{-8} \bullet 100.75^2}{9 \bullet 38.17}$

    $= 3.55\;10^{-6}$ m^3/sec.m

- Using flow net analysis the discharge is computed as follows:

    - Upstream water elevation $K_2 = 815.75$

    - Downstream water elevation $K_3 = 715.00$

    - Darcy's coefficient of permeability $k = 3\;10^{-8}$ m/sec

    - Total number of channels $m = 14$
      (Fig. 9.7)

    - Total number of equipotentials
      channels (Fig. 9.7) $n = 15$

    - $h = K_2 - K_3$ $h = 815.75 - 715.00$
      $= 100.75$

    - $q = k\,\dfrac{m}{n}\,h$ $q = 3\;10^{-8} \bullet \dfrac{14}{15} \bullet 100.75$
      $= 2.82\;10^{-6}$ m^3/sec

The linear discharge is then

    - $q_1 = 3.55\;10^{-6}$ m^3/sec. m   from Eq. 9.28

• $q_2 = 2.82 \ 10^{-6} \ m^3/sec. \ m$ from flow net analysis

• Mean discharge $q = \frac{q_1 + q_2}{2} = 3.19 \ 10^{-6} \ m^3/sec.m$

9.5.2 Filters in Fill-Type Dams

When the slope protection for fill type dams is effected by dumped riprap, wave action has a double impact; and their action is double:

1. Pressure is exerted on the sloping surface

2. The receding waves exert a negative pressure on the slope

This receding phase is dangerous because fine material of the earth fill is transported into the reservoir by suction.

This phenomenon can occur on the upstream and downstream sides of the impervious core. When the reservoir is full, seepage transports the fine material through the core. During rapid drawdown, the fine material is transported to the upstream side. In order to prevent the destruction of the embankment, filters are intercalated between the riprap and the embankment on each face of the impervious core.

A. Composition of Filters

Filters must meet two fundamental requirements:

• permeability and

• stability

The filter material must be fine enough to prevent the base material from escaping through it and must be more permeable than the base material. To reach this goal, filters are formed by layers of different permeability. In general two layers are adopted. The one which is next to the impervious core is called "fine filter" and the other, adjacent to the earth fill, "coarse filter." Terzaghi and Bertram (1940), USBR (1947) and US Corps of Engineers (1948), tried to classify and define filters. Two types of filters are suggested by them:

1. Uniform filter, and

2. Zoned filter

- **Uniform Filter**

Equation 9.31 defines the uniform filter

$$5 < \frac{D_{50}}{B_{50}} < 10 \qquad (9.31)$$

where B_{50} denotes the particle size of the base material and D_{50} is the median particle size of the filter. In the uniform filter, B_{50} is the median particle size of the impervious core.

- **Zoned filter**

The zoned filter is formed by two layers, fine filter near the core and coarse filter next to it. USBR (1947) defined zoned filter as follows:

$$12 < \frac{D_{50}}{B_{50}} < 58 \qquad (9.32)$$

$$12 < \frac{D_{50}}{B_{50}} < 40 \qquad (9.33)$$

Very fine material (finer than as defined by Tyler no. 200 sieve) must be less than 5% and the coarser material in the coarse filter must be smaller than 65.70 mm. Then a large segregation can be avoided during the construction of the filter. If a drain pipe penetrates the filter, the diameter of the fill material which must be placed around it, must be smaller than

$$\frac{D_{85}}{2} \qquad (9.34)$$

U.S. Corps of Engineers suggestion is somewhat different; it is reproduced here

$$\frac{D_{15}}{B_{85}} < 5 \qquad (9.35a)$$

$$\frac{D_{15}}{B_{15}} < 20 \qquad (9.35b)$$

and

$$\frac{D_{50}}{B_{50}} < 25 \qquad (9.35c)$$

and

$$\frac{D_{85}}{\text{distance between two drain pipes}} > 1 \qquad (9.36)$$

TVA suggestion is:

$$\frac{D_{15}}{B_{85}} < 4 \cong 5 \qquad (9.37a)$$

$$\frac{D_{15}}{B_{15}} > 4 \cong 5 \qquad (9.37b)$$

and Terzaghi suggested

$$\frac{D_{60}}{B_{10}} < 2 \qquad (9.38)$$

For construction purposes, the thickness of the filter zone must be greater than 1.00 m; but this thickness must be hydraulically checked.

B. Thickness of Filters

It is necessary to know the discharge capacity of the filter zone in order to calculate their thickness. Darcy's law will be used for computing the thickness of filters

$$q = k \frac{H}{\ell} A \qquad (9.39)$$

where

q is the linear discharge of the filter, m^3/sec.m

k is the coefficient of permeability of the filter, m/sec

H is the hydraulic head upstream of the filter, m

ℓ is the length of the filter, m

A is the wetted area of the filter (for unit length of filter), m, and

d is the depth of downstream flow, m.

These parameters are shown in Fig. 9.8.

The coefficient k, in Eq 9.39, must be computed due to the fact that the filter is a man-made construction element; then any measured value of k in

483

nature can be used for it. The most used formulas for computing k are reproduced below.

Allen Formula

$$k = D_{10}^2 \qquad\qquad (9.40)$$

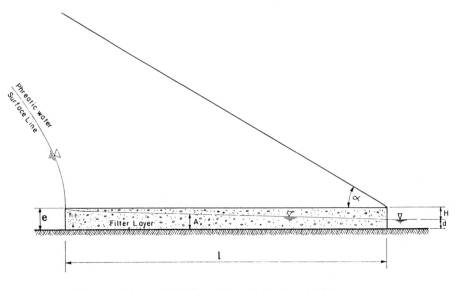

Figure 9.8. Flow through horizontal filter.

where

D_{10} is in cm and

k is in m/sec.

Hazen formula is derived for $D_{10} = 0.10$ to 3 mm and is very approximate.

A. Casagrande Formula (1948)

$$k_i = 1.4 \, k_{0.85} \, i^2 \qquad\qquad (9.41)$$

where

k_i is the coefficient of permeability of a filter with void ratio i

$k_{0.85}$ is the coefficient of permeability of a filter with void ratio 0.85,

and

    i   is the void ratio of the filter.

Terzaghi Formula (1943)

$$k_i = C \frac{0.87i - 0.13}{(1+i)^{23}} D_{10}^2 \qquad (9.42)$$

where

    C is a coefficient equal to C = 4.6 for angular particles and C = 8.0 for rounded and smooth particles.

Fair and Hatch Formula (1933)

$$k = \frac{1}{m[\frac{(1-p)^2}{p^3}(\frac{\theta}{100}\Sigma \frac{P}{D_m})^2]} \qquad (9.43)$$

This formula can be used for rounded and for angular particles.

p   is the porosity

m   is a factor which has a value of 5 in practice

θ is the shape factor. It is equal to 6 for spherical particles and 7.7 for angular particles

P   is the percent in weight for each sieve opening and

D_m is the geometric mean of the particle forming the mixture of the filter.

In this formula

$$D_m = (D_1 D_2)^{1/2} \qquad (9.44)$$

D_1 and D_2 are two consecutive sieve openings.
    When the coefficient of permeability is computed this way, assume that

$$H = e - d$$

On the other hand the filter is assumed evacuating the following discharge safely

$$q' = 2q \qquad (9.45)$$

q, being the discharge by seepage through the impervious core.

Darcy's law yields

$$k \frac{H}{\ell} (\frac{e + d}{2}) = 2q = q'$$

and

$$k \frac{e - d}{\ell} \frac{e + d}{2} = 2q$$

or

$$e^2 - d^2 = 4 \frac{q\ell}{k}$$

and assuming d negligible compared with e

$$e = 2 (\frac{q\ell}{k})^{1/2} \qquad (9.46)$$

where e is the thickness of the filter. It is advisable to choose this value of e for the approximate thickness of the filter. Table 9.8 summarizes a rough approximation of average conditions in the field for measured values of k. These values can be used as the first approximation for filter thickness.

Example 9.06

The seepage line in the impervious core of the Kiralkizi Dam is shown in Fig 9.9.
The seepage discharge is found equal to

$$q = 3.19 \; 10^{-6} \; m^3/sec$$

Compute the thickness of the filter.

Data

The particle size of the first filter layer is

$$D_{20} = 0.05 \qquad\qquad B_{85} = 0.03$$

TABLE 9.8

Approximate Permeability Coefficients of Various Soils
Based on 20% Size

20% size mm	k 10^{-4} cm/sec	k 10^{-4} ft/min	US Bureau of Soil Classification
0.005	0.030	0.059	
0.01	0.105	0.206	Coarse clay
0.02	0.400	0.787	Fine silt
0.03	0.850	1.675	Coarse silt
0.04	1.750	3.450	
0.05	2.800	5.510	
0.06	4.60	9.06	
0.07	6.50	12.80	
0.08	9.00	17.75	Very fine sand
0.09	14.00	27.60	
0.10	17.50	34.50	
0.12	26.00	51.30	
0.14	38.00	75.00	
0.16	51.00	100.00	Fine sand
0.18	68.50	135.00	
0.20	89.00	175.00	
0.25	140.00	276.00	
0.30	220.00	434.00	
0.35	320.00	630.00	
0.40	450.00	886.00	Medium sand
0.45	580.00	1142.00	
0.50	750.00	1480.00	
0.60	1100.00	2160.00	
0.70	1600.00	3160.00	
0.80	2150.00	4240.00	Coarse sand
0.90	2800.00	5520.00	Fine gravel
1.00	3600.00	7100.00	
2.00	18000.00	35400.00	

The particle size of the second filter size is

$$D_{20} = 0.20 \qquad\qquad B_{15} = 0.01$$

Solution

1. Assume that the first layer has a 20% size equal to 0.05 mm with a coefficient of permeability equal to

$$k = 2.8\ 10^{-4}\ \text{m/sec}\ \ (\text{Table 9.8})$$

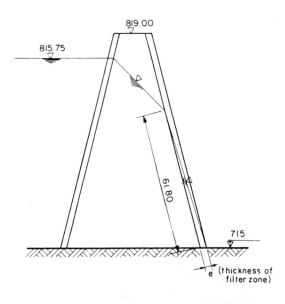

Figure 9.9. Impervious core of the Kiralkizi Dam.

2. Equation 9.46 yields

$$e = 2\ (\frac{3.19\ 10^{-6} \bullet 61.80}{2.8\ 10^{-4}})^{1/2} = 1.68\ \text{m} \qquad\qquad 9.46$$

3. Assume that the second layer has a 20% size equal to 0.20 mm with a coefficient of permeability equal to (Table 9.8)

$$k = 89\ 10^{-4}\ \text{m/sec}$$

Then

$$e = 2 \left(\frac{3.19 \ 10^{-6} \cdot 61.80}{89 \ 10^{-4}}\right)^{1/2} = 0.30 \ m$$

For construction purposes a minimum thickness of e = 1.00 m is necessary for each layer, then the total thickness is 2.68; use 2.75 m.

4. This result is checked by direct application of Darcy formula.

- First layer:

$$q = kAS$$

$$S = \frac{1.68}{61.8} = 0.027$$

$$q' = 2.8 \ 10^{-4} \cdot 0.027A = 3.19 \ 10^{-6} \cdot 2$$
$$= 6.38 \ 10^{-6}$$

$$A = 0.844 \ m^2$$

A is the median value (Fig. 9.8), then

$$2A = 1 \cdot e$$

$$e = 0.844 \cdot 2 = 1.688 \ m$$

- Second layer:

$$S = \frac{0.30}{61.8} = 0.004854$$

$$q' = 89 \ 10^{-4} \cdot 0.004854A = 6.38 \ 10^{-6}$$

$$A = 0.148 \ m^2$$

$$e = 0.148 \cdot 2 \approx 0.30$$

5. Applying TVA's suggestion yields,

- First layer:

$$\frac{D_{15}}{B_{85}} < 4 \cong 5 \qquad\qquad 9.37a$$

$$0.20 \leq (4 \approx 5) \ 0.03 = 0.12 \approx 0.20$$

$$\frac{D_{15}}{B_{15}} \geq 4 \sim 5 \qquad\qquad 9.37b$$

$$0.20 \geq (4 \approx 5) \, 0.01 = 0.04 \approx 0.05 \quad (\text{see Fig. 9.10})$$

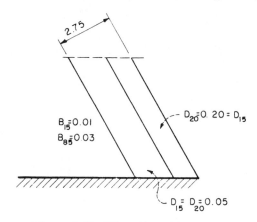

Figure 9.10. Filter disposition.

D_{15} is assumed approximately equal to D_{20}

- Second layer:

$$\frac{D_{15}}{B_{85}} < 4 \sim 5 \qquad 0.20 < (4 \sim 5) \, 0.08 = 0.32$$

$$\frac{D_{15}}{B_{15}} > 4 \sim 5 \qquad 0.20 \leq (4 \sim 5) \, 0.05 = 0.20$$

9.6 MEASURING DEVICES FOR OBSERVATION OF DAM PERFORMANCE

A dam is an engineering structure designed for supporting huge loads. It is deformable, and this deformation takes place according to certain assumptions which are used in preparing the design. These assumptions cover two different fields:

1. Hydraulic field

2. Static field

The designer must follow the different phases in the behavior of the dam, first during the construction period, second during the impounding period and third, during many years of reservoir operation.

Deformation of the dam is twofold: inner deformation and surface deformation. The foundation settlement which causes inner deformation creates highly dangerous leakage paths. These paths must be checked carefully and their existence must be observed and located so that necessary precautions can be taken. Three kinds of measuring devices for observation of dam performance are in use:

1. Piezometers for measuring pore water pressure and measuring wells for locating the seepage line in the impervious core.

2. Instruments for measuring horizontal movements, foundation settlements and embankment compression.

3. Surface marks for measuring horizontal movement and settlement.

Periodic measurements are necessary using each of these devices to monitor the deformation of the embankment and particularly the evolution of pore pressures. Figure 9.11 shows the evolution of pore pressures in the Seyhan Dam, Turkey. It also shows the efficiency of the impervious blanket. Figure 9.12 illustrates the variation of pore pressures in the same embankment; it is evident that the pore pressures follow the variation of the water level in the reservoir.

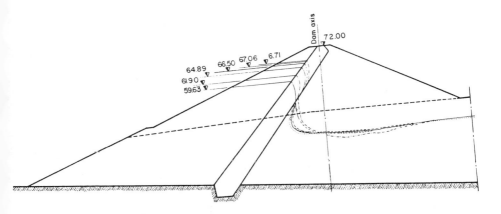

Figure 9.11. Variation of the pore pressure in the cross section of the Seyhan Dam showing the efficiency of the impervious blanket.

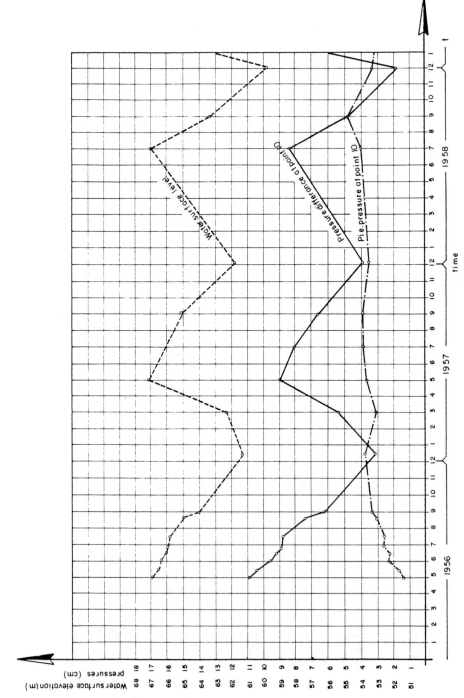

Figure 9.12. Evolution of pore pressure in function of time in the Seyhan Dam showing the influence of the reservoir water level.

Piezometers, showing the variation of pore pressure in the embankment, must be placed in such a way that they can indicate the behavior of the dam, the variation of permeability of the impervious core and foundation, especially when the foundation is semi-permeable. Figure 9.13 illustrates the positioning of piezometer tips in the Seyhan Dam.

If the head loss between the reservoir water level and some of the piezometers downstream of the impervious core, or in the foundation decreases, accompanied by an increase in the leakage, the efficiency of the impervious core must be checked carefully. An investigation must be conducted to determine whether a crack has developed in the impervious core. Figure 9.14 shows the evolution of the settlement in the Demirköprü Dam, during construction.

In the diagram, the settlements are homogeneous and any discontinuity becomes apparent. An abrupt increase in the settlement can be an indication of a break in the embankment or in the foundation. The particular settlement device showing this abrupt change will also show the location of the discontinuity in the structure.

It is clear that the positioning and the use of measuring devices are of primary importance. More details are given, particularly about the piezometers which relate directly to the hydraulic behavior of the dam.

9.6.1 Measuring Devices

A. Piezometers

Piezometers are devices for measuring the pore pressure. They are used in foundations and embankment for measuring the variation of these pressures which are indicators of the hydraulic behavior of dams.

Two kinds of piezometers have been used and are in use in fill-type dams:

1. Hydraulic piezometers

2. Electric piezometers

The hydraulic piezometers give the water pressure directly; in electric piezometers the pressure exerted deflects a calibrated membrane and the deflection is measured electrically. Two types of electrical piezometers are in use in the world: Maihac type piezometers and Telemac type piezometers.

Figure 9.15 shows the Maihac type electric piezometer; and Fig. 9.16 shows a hydraulic type piezometer which is used world wide.

Piezometers are connected by plastic pipes to a terminal well where manometers are located for easy reading. Figure 9.17 shows the installation of the plastic pipes and Fig. 9.18 shows the manometers.

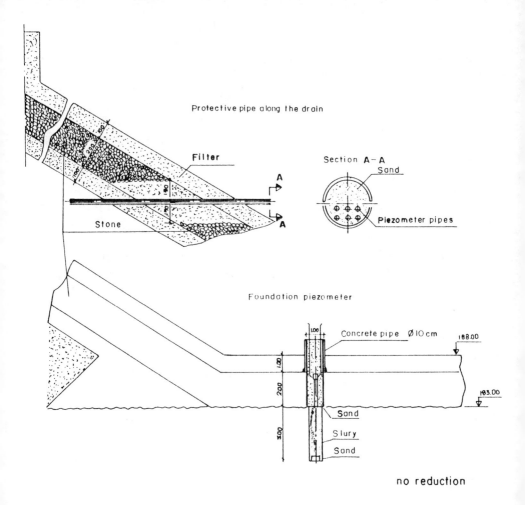

Figure 9.13 Measuring device in the Seyhan Dam (hydraulic piezometer).

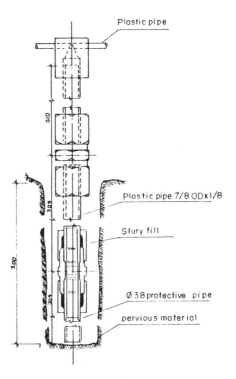

Foundation treatment

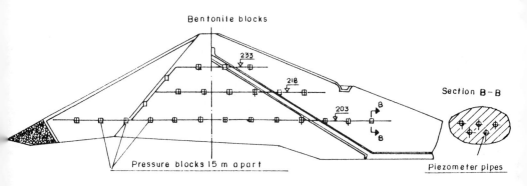

Section B–B

Figure 9.13. Continued.

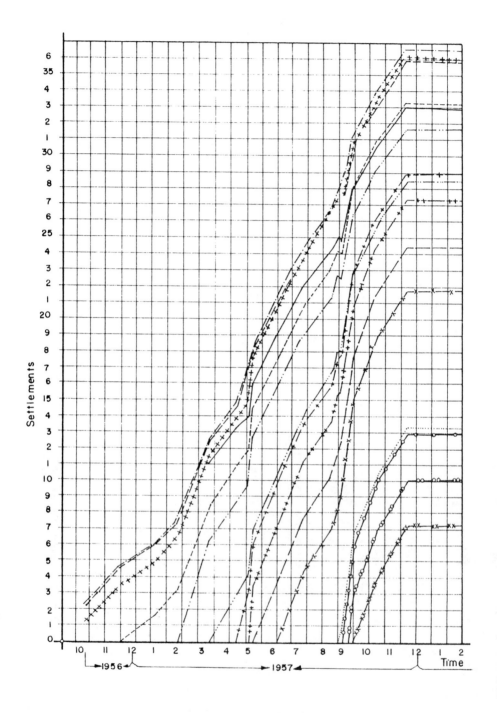

Figure 9.14. Settlement readings during the construction of the Demirköprü Dam, Turkey.

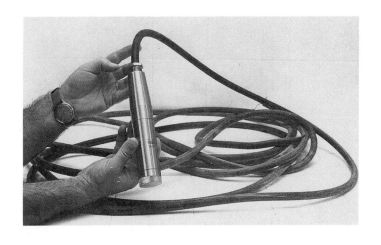

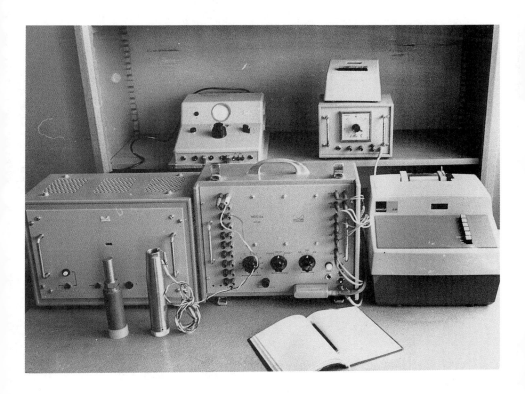

Figure 9.15. MAIHAC type electric piezometer.

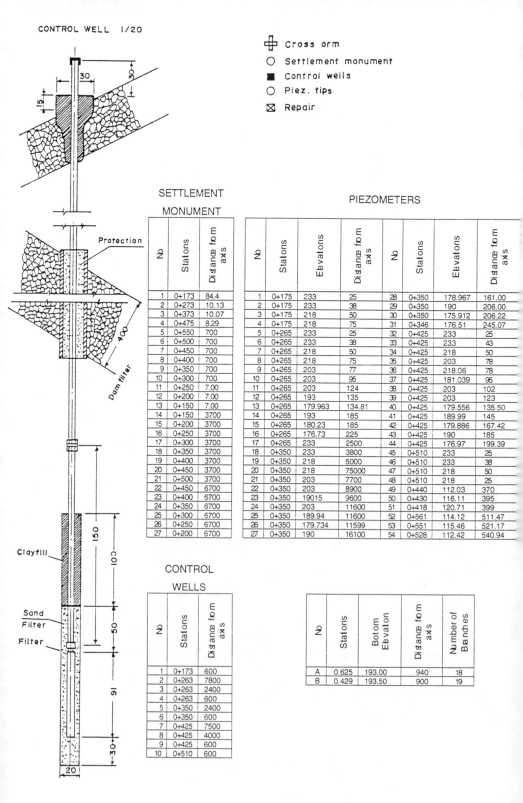

Figure 9.16. Hydraulic piezometer.

Figure 9.17. Installation of plastic tubes.

Figure 9.18. Manometers.

The technique for installation of the plastic tubes can be found in special instructions delivered by the manufacturer. Figure 9.19 gives instructions for the use of the piezometer system.

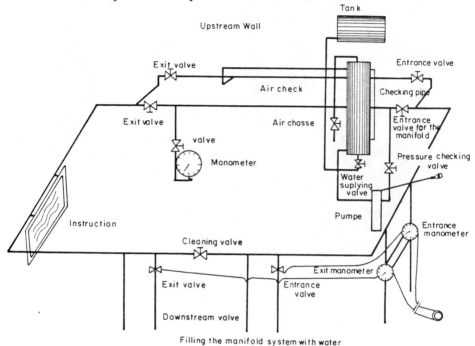

a) Close all valves first, then
1 - Open the cleaning valve
2 - Open the exit valve
3 - Open the control valve
4 - Open the entrance valve
5 - Pump water in
Continue until the air in the system is expelled
b) Close manifold system entrance valve
c) Cleaning of the system

Figure 9.19. Instruction for use of manifold piezometer system.

B. Settlement Measuring Devices

The settlement measuring device is shown in Fig. 9.20. The system is placed in the embankment during the embankment construction. Cross-arms are placed at a pre-set elevation from which sediment measurements are made. The measurement is performed with a special instrument called a "torpedo." Different kinds of torpedoes exist in the market and special instructions are given by the manufacturer. (Figure 9.20 shows a cross-arm currently used in practice).

C. Measurement of Leakage Discharge

Water coming from leakage through the impervious core is collected by drains and discharged to a main drainage channel. The discharge by leakage

Figure 9.20. Settlement measuring device.

is measured at the downstream end of this channel by an appropriate sharp-crested weir.

D. Measurement Wells

These ordinary wells located at specific locations in the dam embankment and its foundation are for checking piezometers readings. Figure 9.21 shows such a well.

E. Surface Marks for Measuring Horizontal Movement and Settlement of Embankment

Figures 9.22 and 9.23 show such a device which is generally located on the crest of the dam. These devices are used to measure displacement by means of triangulation.

Figure 9.21. Simple manomter well.

Figure 9.22. Surface mark for measuring the horizontal movement of
the dam.

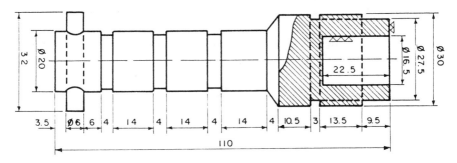

Figure 9.23. Detail of the mark.

9.7 PROBLEMS

Problem 9.01

Compute the discharge from leakage through the impervious core of the Kiralkizi Dam assuming that the downstream water elevation is at 705.00.

Data

See Example 9.05

Problem 9.02

Show the decrease of discharge versus the increase of downstream water level in the Kiralkizi Dam.

Data

See Fig. 1, Problem 9.02

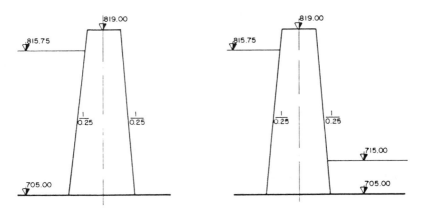

Figure 1. Problem 9.02.

F. Sentürk

Problem 9.03

Determine the filter thickness and the filter gradation for filter shown in Fig. 1, Problem 9.03.

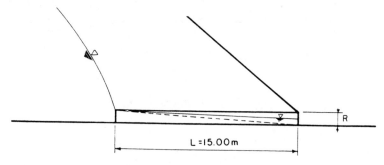

Figure 1. Problem 9.03.

Data

$q = 5 \ 10^{-5} \ m^3/sec.m.$

$L = 15.00 \ m$

Chapter 10

HYDRAULICS OF
DIVERSION STRUCTURES

10.1   INTRODUCTION

Large dams are generally located on a water course which is not ephemeral and which carries large quantities of water. Under these conditions the foundation excavation presents a difficult problem; it will be done either under water or "in the dry." The usual procedure for major structures is to divert the river and perform the excavation when the river bed is exposed, so-called "in the dry." Actually, river bed excavation, even with the main watercourse diverted, is never completely without water problems. Constant pumping is required to keep the working area free of water.

The diversion is a complex engineering work; it is composed of:

1. An upstream cofferdam,

2. A downstream cofferdam,

3. One or more diversion tunnels.

The upstream cofferdam blocks the river channel and directs the water into the tunnel(s). The tunnel entrance is located upstream of the cofferdam, bypasses the excavation site and exits below the downstream cofferdam. The downstream cofferdam prevents the river water from entering the excavation from below.

Before starting the detail design, a number of questions must be answered by the designer.

1. What will the maximum discharge of the river be during the diversion period?

2. The discharge being determined, the diversion system can be designed. An upstream cofferdam height is selected along with the size and number of diversion tunnels. It is possible to alter the height of the upstream cofferdam by varying the number and diameter of the diversion tunnels. What can be the optimum solution of the problem?

3. How to choose between one tunnel with a large diameter and two or more tunnels with smaller diameters.

4. How to prevent pulsating flow taking place in the diversion tunnel while passing from free surface flow to pressure flow?

5. What happens when, during flood time, one of the diversion tunnels is blocked?

These questions will be answered in the following paragraphs.

10.2 DETERMINATION OF THE DESIGN DISCHARGE

The choice of the design discharge of a diversion system is somewhat arbitrary. The following recommendations may be taken into consideration.

- For a construction site located far away from an urban area

$$Q_d = Q_{15}$$

Q_d: design discharge

Q_{15}: Discharge with 15 year return period

- For a construction site located near an urban area

$$Q_d = Q_{25}$$

- For a construction site located near an urban area and in a wide valley

$$Q_d = Q_{25}$$

The dimensions obtained for the tunnel(s) may be checked for

$$Q_{50}$$

- For a construction site located near an urban area and on a very large river

$$Q_d = Q_{50}$$

The dimensions obtained for the tunnel(s) may be checked for

$$Q_{60}$$

These figures are rather empirical and vary from country to country.
For a very large river, Q_{50} is of major proportions. For example, on the Keban project, the diversion system used

$$Q_{50} = 7000 \text{ m}^3/\text{sec}$$

10.3   DETERMINATION OF THE NUMBER OF DIVERSION TUNNELS

The choice of the number of diversion tunnels depends primarily on practical considerations. The following recommendations may be taken into account.

- D_{min} and D_{max} of diversion tunnels

  The diameter of the tunnel is a function not only of the discharge requirements, but also the type of the diversion system and construction requirements.

- D_{min} is a function of the discharge and the energy slope, but it has a minimum practical limit which is around 3.00 m, or $D \geq 3.00$ m. If D is smaller than this value, many construction difficulties arise. For instance, the injection pump cannot be, in general, brought into the tunnel because of its size and the necessary contact injection cannot be performed economically.

- D_{max} is, in general, around 7.00 m. In exceptional cases D_{max} can be chosen greater than this value. The Keban diversion tunnel is an example. In this particular case $D = 14.46 \sim 15.46$ m (horse shoe type).

  The diversion tunnels for Atatürk Dam are each $D = 8.00$ m, with a triple tunnel solution chosen instead of increasing the diameter of the tunnels. Geologic conditions must be included with the above considerations. Then, under suitable geologic conditions, the diameter of the diversion tunnel can be chosen between 3.00 m and 7.00 m and the number of tunnels determined as a function of the discharge.

  If the diameter of the diversion tunnel is greater than 5.00 m, and for lined tunnels it is recommended that two tunnels be selected rather than one for ease of construction and for better operation and maintenance access. Lining of one tunnel can be applied while the second tunnel is being excavated. During operation, the maintenance of one tunnel can be done while the second tunnel continues diverting water.

  If a diversion system carries a small quantity of water and a single tunnel is used, the length of the tunnel may pose a problem for transportation of material. Two tracks cannot pass each other due to the lack of space. This problem can be solved by enlarging the tunnel at intervals; this is a rather questionable solution, although in many instances it has been adopted in Turkey to save time. An example is the diversion tunnel for Gökçekaya Dam, Turkey. This diversion system was designed with a single tunnel for a different reason; the original tunnel had collapsed and an entire year was lost. A similar example will be explained in more detail in the following paragraphs.

10.4 OPTIMIZATION ANALYSIS OF DIVERSION STRUC-TURE'S DIMENSIONS

The optimization analysis determines the height of the upstream cofferdam and the characteristics of the diversion tunnels such as their diameter and the number of tunnels to be chosen. The following steps may be taken for solving the problem.

Step 1: Choose the number of tunnels, taking into consideration the data given in section 10.3; then assume a value for the diameter. In general, at the beginning of the step-by-step method, only one tunnel is considered and an arbitrary D_1 value is assigned for its diameter. Then the function

$$Q = f(D_1)$$

is determined.

Step 2: Choose the design discharge as shown in section 10.2; and route the flood as a function of D_1. This procedure determines the water depth above the diversion tunnel entrance. Add a freeboard allowance of 2.00 m to obtain the crest elevation of the upstream cofferdam.

Step 3: Determine the height of the cofferdam corresponding to various values of the diversion tunnel's diameter

$$\text{Cofferdam height} = \phi\,(D)$$

Step 4: Estimate the cost of the diversion system corresponding to each D value and prepare a diagram as shown in Fig. 10.1.

This diagram shows a minimum value curve which is the solution for the problem. With the optimization curve as a basis, alternate solutions can be considered including a possible change in the number of diversion tunnels according to section 10.3.

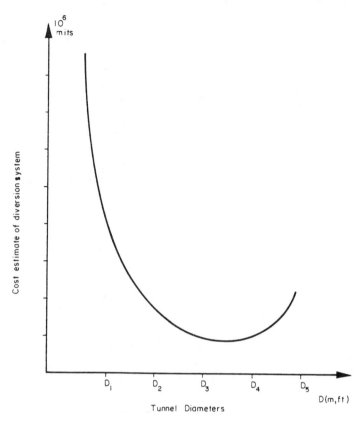

Figure 10.1. Optimization analysis of diversion structure dimensions.

10.5   HYDRAULICS OF DIVERSION STRUCTURES

10.5.1   Flow Taking Place in the Diversion Tunnel

The hydraulics of closed conduits presents some peculiarities which will be discussed in this section. Figure 10.2 summarizes different flow characteristics taking place in a rather short, closed conduit which is described as follows:

a. $S_h < S_c$ and $\dfrac{H}{D} < 1.2$

S_h, being the slope of the tunnel bottom includes all values of S_b smaller than S_c, and S_c is the critical value of the slope corresponding to the design flood.

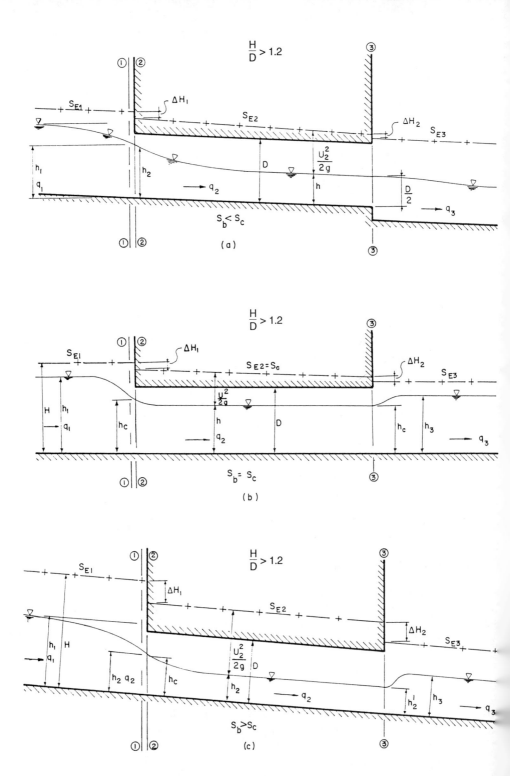

Figure 10.2. Flow characteristics taking place in a short closed conduit.

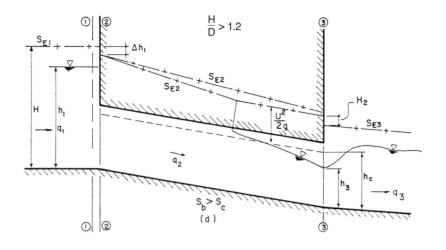

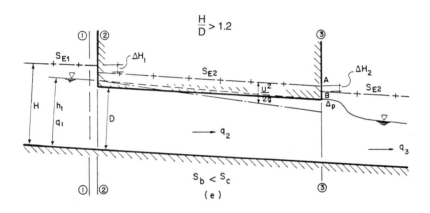

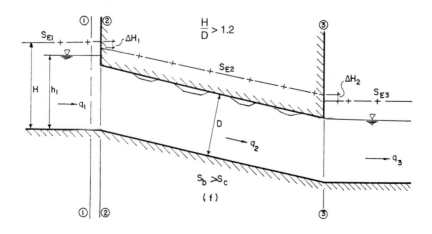

Figure 10.2. Continued.

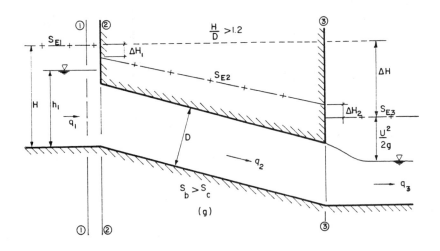

Figure 10.2. Continued.

H, being the hydraulic head upstream of the diversion tunnel, assumes the following basic condition

$$\frac{H}{D} < 1.2$$

where D is the diameter of the diversion tunnel.

Assume that a free surface flow is taking place in the tunnel. At the tunnel entrance, the water surface is lowered and the length of the closed conduit determines whether the flow is uniform or gradually varied. At the exit, the downstream flow condition prevails. The condition is illustrated in Fig. 10.2a. The function, $q = f(h)$, can be easily determined for free surface flow. The cross section of the conduit is assumed to be rectangular. The variation of (q) as a function of (h) is a parabola which can be drawn easily (Ex. 10.01). Where H is the hydraulic head at section 1-1 and S_{E1}, the slope of the energy line is at that section. It is easy to determine the depth, h_1, of the flow directly using the diagram. The same diagram also yields the critical depth corresponding to q_c. By assuming ΔH_1, the head loss is at the entrance. Different values of ΔH_1 are listed in Table 10.1.

Assume S_{E2} is the slope of the energy line in the tunnel, and

$$q_2 > q_1 \qquad\qquad h_c < h_2 < h_1$$

$$S_{E2} > S_{E1} \qquad \text{and} \qquad S_c > S_{E2}$$

512

TABLE 10.1

Head Losses Taking Place at Hydraulic Structures such as Tunnels, Culverts, Siphons etc.

1.	Entrance losses (includes the contraction loss at the entrance).	$0.2 \dfrac{U^2}{2g}$	(10.01)
2.	Friction loss	Manning's formula can be used.	
3.	Transition losses (diverging transition)	$0.2 \left(\dfrac{U_1^2}{2g} - \dfrac{U_2^2}{2g} \right)$	(10.02)
4.	Transition losses (converging transition) *	$0.1 \left(\dfrac{U_2^2}{2g} - \dfrac{U_1^2}{2g} \right)$	(10.03)
5.	Exit losses	$\dfrac{U^2}{2g}$	(10.04)
6.	Bend losses (Fig. 10.3 for K values)	$K \dfrac{U^2}{2g}$	(10.05)

TABLE 10.2

Head Loss Coefficient as a Function of the Convergence and Divergence Angles

Angle	2°	5°	10°	12°	15°	20°	25°	30°	60°
Coefficient	0.03	0.04	0.08	0.10	0.16	0.31	0.40	0.49	0.72

If uniform flow takes place in the closed conduit

$$\frac{U_2^2}{2g} = Ct \quad \text{and}$$

$$S_w = S_{E2}$$

where S_w is the slope of the water surface line.

The tunnel can terminate in a drop, as shown in Fig. 10.2a or the invert can be continuous as shown in Fig 10.2b. If the downstream flow depth is greater than the critical depth, the critical depth can not be reached and water surface level rises smoothly to the downstream water surface level.

---

* The transition region is defined as a function of the divergence or convergence angle. The head losses corresponding to each case are listed in Table 10.2. These values are rather arbitrary. The values given in Table 10.1 can be used as a first approximation.

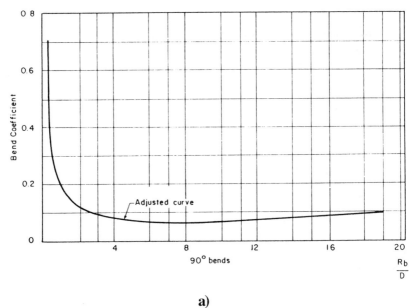

a)

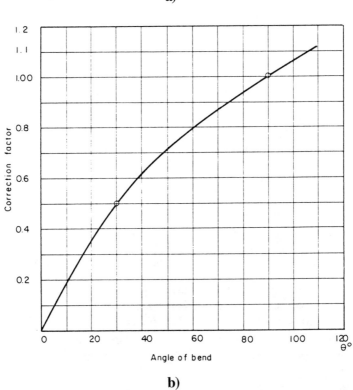

b)

Figure 10.3. Head losses at bends. Bend coefficient with relative radius for **90°** bends of circular cross section (USBR, Design of Small Dams).

In many cases, for economic reasons, the tunnel is designed in such a way that the water level reaches the critical position at the downstream exit. In this condition, the water depth immediately downstream of the exit section can be less than the depth of the downstream flow; then the water depth increases and reaches the water depth of the downstream flow.

In culverts, the closed conduit is too short and uniform flow cannot take place in it. In this condition the flow is shown in Fig. 10.2a. The flow is gradually varied and the velocity head increases along the tunnel.

The hydraulic computation of closed conduits begins by determining the downstream flow conditions. The downstream flow depth is computed as a function of the discharge Q. Then ΔH_2 is determined using the values given in Table 10.1, and energy gradient computed accordingly. S_{E2} is, in general, determined using Manning's equation, then $S_{E2}L$ is obtained. ΔH_1 is added to this value to obtain the upstream hydraulic head and the upstream water surface level.

b. $S_b = S_c$ $\dfrac{H}{D} < 1.2$

Flow in Tunnel

This flow is illustrated in Fig 10.2b. The flow depth reaches critical depth immediately downstream of the entrance and uniform flow prevails. The energy line, the water surface line and the bottom of the tunnel are parallel to each other. The flow depth at the exit increases or decreases according to downstream flow conditions. A gradual increase in the water surface line is shown in the figure.

Flow in Culvert

In culverts, the water surface line decreases and reaches the critical depth upstream of the entrance. This phenomenon is called the suction effect of the culvert. If the water depth continues to decrease in the culvert, a gradually varied flow takes place in it.

c. $S_b > S_c$ $\dfrac{H}{D} < 1.2$ and free flow in the culvert

Flow in Tunnel

This flow is illustrated in Fig. 10.2c. The flow is supercritical in the tunnel. The water surface line follows an S2 curve outside of the entrance and attains uniform flow level at section 1-1. It is rather difficult to determine the location of the section at which the flow depth attains the critical depth. For this reason it is customary to assume that critical depth occurs at section 1-1 at the entrance. Downstream flow conditions affect the

515

position of the water surface line. In general, the depth of flow increases gradually between sections 2-2 and 3-3 and leaves the tunnel at a depth h_2', greater than h_2. If the tunnel is long enough, $h_2' = h_c$.

It is possible to observe the hydraulic jump downstream of the exit as a function of downstream hydraulic conditions.

Flow in Culvert

Similar flow characteristics can be observed in the culvert but the flow in the culvert is mostly gradually varied.

d. $S_b > S_c$ $\qquad\qquad\qquad\qquad \dfrac{H}{D} > 1.2$ pressure flow

Flow in the Tunnel and Culvert

With H/D being greater than 1.2, the tunnel entrance is submerged. When $S_b > S_c$, then the flow becomes free surfaced at a given section which varies with the characteristics of the flow and the geometric properties of the conduit. The determination of this point will be explained in the following paragraphs. If the tunnel is long enough, the water surface line reaches the supercritical uniform flow water line.

If culverts are not long enough, then the critical depth will be reached at the exit section (Figs. 10.2 d and 10.2g).

e. $S_b < S_c$ $\qquad\qquad\qquad\qquad \dfrac{H}{D} > 1.2$ pressure flow

Flow in Tunnel and Culvert

In this case the pressure flow is continuous in the tunnel. Subpressures may occur at the exit section. This condition will be explained in detail in the following paragraphs. The subpressure is indicated by (Δp) in Fig. 10.2e.

f. $S_b > S_c$ $\qquad\qquad\qquad\qquad 1.2 < \dfrac{H}{D} < 1.5$ pressure flow

Flow in Tunnel and Culvert

This flow condition is the one encountered most often in the field. It must be discussed carefully due to the fact that air can be introduced in the tunnel causing disturbances. These disturbances reveal themselves in the form of dangerous pulsating flow. Such a flow exists in Keban Dam Tunnel and it is not possible to fully open the bottom valves (Fig 10.2f). In unlined tunnels pulsating flow can cause tunnel failure. Such phenomena occurred in the Gökcekaya Dam diversion tunnel and the construction of the dam was delayed one year.

Example 10.01

It is required to determine the hydraulic efficiency of a culvert, the characteristics of which are given (Fig. 10.4).

Data

1. $S_b = 0.020$

2. Cross section: Rectangular $B = 1.0$ m, $D = 3.00$ m in the tunnel and in the downstream open channel.

3. Downstream flow depth $= 3.10$ m

4. $L = 50.00$ m

5. $q = 6.25$ m³/sec.m

6. $n = 0.014$

7. $H_1 = 3.05$ m

Solution

1. Draw $q = f(h)$ diagram at section 1-1

$$H_1 = h_1 + \frac{U_1^2}{2g}$$

$$U_1 = \frac{Q}{A_1} = \frac{q}{h_1} \qquad\qquad H_1 = h_1 + \frac{q^2}{h_1^2\, 2g} = 3.05 \text{ m}$$

$$h_1 = 2.80 \text{ m}$$

$$q = [2g\,(3.05 - h_1)\, h_1^2]^{1/2}$$

2. Compute q_c as

$$h_c = \left(\frac{q_c^2}{g}\right)^{1/3} \qquad\qquad q_c = (gh_c^3)^{1/2}$$

$$h_c = \frac{2}{3} \cdot 3.05 = 2.03 \text{ m} \qquad\qquad q_c = (9.81 \cdot 2.03^3)^{1/2} = 9.06 \text{ m}^3/\text{sec.m}$$

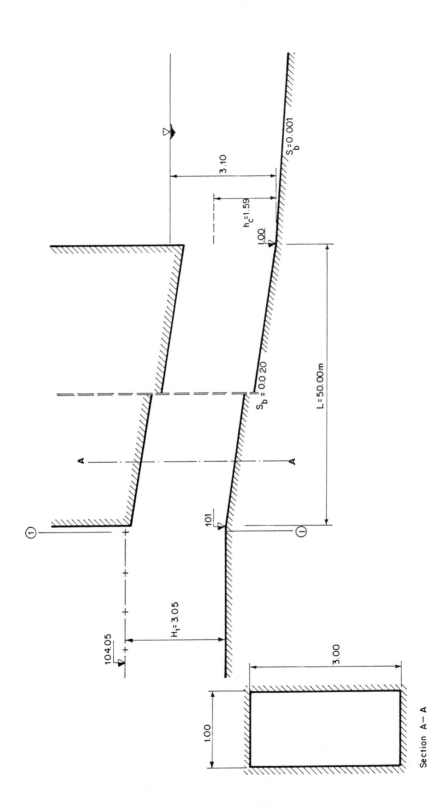

Figure 10.4. The culvert in Example 10.01.

3. The entrance loss is:

$$\Delta H_1 = 0.2 \frac{U^2}{2g} \qquad\qquad U = \frac{q}{h} = \frac{6.25}{2.80} = 2.23 \text{ m/sec}$$

$$\Delta H_1 = 0.2 \frac{2.23^2}{19.62} = 0.05 \text{ m}$$

4. The hydraulic head at the section immediately upstream of the tunnel and in the tunnel is

$$H_2 = H_1 - 0.05 = 3.05 - 0.05 = 3.00 \text{ m}$$

Then the q = f(h) curve can be drawn for this section

$$q = [2g(3.00 - h_2) \, h_2^2 \,]^{1/2}$$

5. The culvert is too short for uniform flow to take place, so gradually varied flow will prevail in the tunnel.

6. Downstream condition,

   Computation of the downstream channel critical depth:

$$h_c = (\frac{q^2}{g})^{1/3} = (\frac{6.25^2}{9.81})^{1/3} = 1.59 \text{ m}$$

7. Gradually varied flow in the tunnel

   Assume h_2 the depth of flow at the exit section, applying the energy equation for gradually varied flow yields

$$S_b L + h_1 + \frac{U_1^2}{2g} = h_2 + \frac{U_2^2}{2g} + S_E L$$

where

$S_b = 0.02$ \qquad $L = 50.00$ m \qquad $h_1 = 0.981$ (flow depth of supercritical flow)

$$U_1 = \frac{6.25}{0.981} = 6.37 \text{ m/sec} \qquad n = 0.014$$

and

$$U_2 = \frac{6.25}{h_2} \qquad S_E = \frac{U^2 n^2}{R^{4/3}} \qquad U = \frac{U_1 + U_2}{2}$$

$$R = \frac{R_1 + R_2}{2} \qquad R_1 = \frac{h_1}{1 + 2h_1} = \frac{0.981}{1 + 2 \cdot 0.981} = 0.33$$

$$R_2 = \frac{h_2}{1 + h_2} \qquad U = \frac{6.37 + \dfrac{6.25}{h_2}}{2} = \frac{6.37h_2 + 6.25}{2h_2}$$

$$R = \frac{0.33\,(1 + 2h_2) + h_2}{2\,(1 + 2h_2)} = \frac{0.33 + 1.66h_2}{2\,(1 + 2h_2)}$$

Substituting these values in the above equation yields

$$0.02L + 0.981 + \frac{6.37^2}{19.62} = h_2 + \frac{6.25^2}{2gh_2^2} + 0.014^2 \frac{(6.37h_2 + 6.25)^2}{4h_2^2} \frac{L}{\left[\dfrac{0.33 + 1.66h_2}{2\,(1 + 2h_2)}\right]^{4/3}}$$

$$h_2 = 1.167 \cong 1.20 \text{ m}$$

With critical depth equal to 2.00 m the flow is supercritical in the channel

$$h < h_c \qquad \text{and} \qquad \frac{H}{D} = \frac{3.00}{3.00} = 1.00 < 1.20$$

and the water surface line takes the form shown in Figs. 10.2 and 10.5. The flow follows an S2 curve upstream of the tunnel, the control is at the entrance and the outlet is submerged but a jump can take place.

10.5.2 Hydraulics of Tunnels of Circular Cross Section

Tunnels used for diversion purposes differ from culverts by,

- Their length which is greater than the length of culvert, and

- Many restrictive features, such as stop logs slots, bends, etc., which causes extra head losses.

The flow in a diversion tunnel can be free surface flow and after priming it can be a pressure flow. Each case may be studied separately.

For relatively small discharges, a free flow takes place in the tunnel. When the discharge increases, the flow depth in the tunnel increases also. There is critical value of the flow depth above which priming occurs, and for greater discharge values pressure flow is expected in the tunnel.

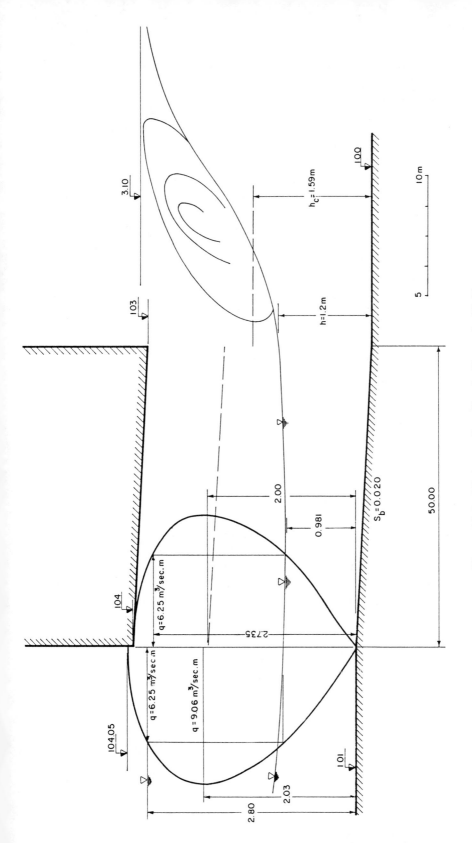

Figure 10.5. Water surface line in Example 10.01.

The determination of the discharge or the hydraulic head at which a diversion tunnel is primed can be made if all factors affecting the flow can be determined. A rough approximation can be made, but it must be checked by model studies because during priming, with head and discharge stationary pulsating flow taking place, it can be of some duration.

Priming will occur in diversion tunnels at a discharge value lower than the one obtained by the theoretical approach. It can be expected that the flow depth will be higher than indicated by computations because the flowing water evacuates the air trapped near the crown of the tunnel and causes a subpressure on the water surface. The tunnel passes its maximum discharge when the flow depth equals 0.975D. Discharge and depth increase together up to this point. Above this depth, there is a slight decrease in the discharge, the pipe fills, the system is abruptly primed and the tunnel flows under pressure.

Geometric factors have direct bearing on the priming of the tunnels, because they create local losses that are characteristic of the respective geometry.

The most important effect of priming is the development of subpressures in the tunnel so that the true discharge over a small range of heads cannot be readily calculated.

The priming and the depriming of the tunnel is a continuous source of change of the pressure from negative to positive values. If the tunnel is not properly lined, damage can occur.

Priming of tunnels and partially full tunnels is studied in more detail in the following paragraphs.

A. Pressure Flows Throughout Tunnel

Assuming H/D = 1.2 ~ 1.5 (see previous section), it is possible to say that pressure flow takes place in the circular tunnel. H is the hydraulic head at the entrance and D is the tunnel diameter.

It is also necessary to determine the downstream hydraulic condition in order to express the flow equation. For free flow at the exit the piezometric head can be taken equal to

$$\frac{2}{4}D \qquad (10.6)$$

and for a supported nappe

$$\frac{3}{4}D \qquad (10.7)$$

These two values were suggested by USBR (1967). A more detailed approach is given in Fig. 10.6, where the piezometric head at the exit is given as a function of the Froude number. Equations 10.6 and 10.7 were tested in DSI Hydraulic Laboratories, Turkey and found to be satisfactory. Under these conditions it becomes possible to write the pressure flow equation for the tunnel (Fig. 10.7).

The flow depth in the approaching channel is shown by h and the velocity head by $(U_o^2/2g)$. In general the velocity head is small and can be neglected.

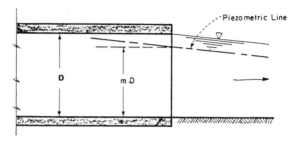

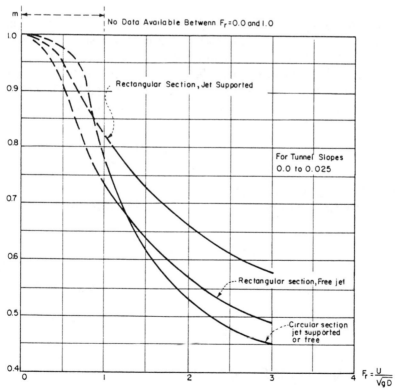

Figure 10.6. Effective piezometric head at pressure outlets (Roberts, E., 1960, Hydraulic Design of Outlet Works, DSI).

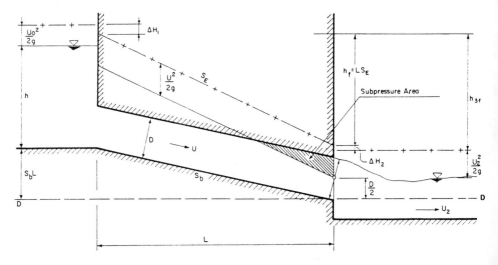

Figure 10.7. Pressure flow taking place in a tunnel.

Assume (ΔH_1) is the head loss at the inlet section, and (ΔH_2), the head loss at the exit. The friction head loss in the tunnel is shown by (h_f); it is assumed that H/D > 1.2 and $S_b < S_c$, then the flow is subcritical. An abrupt drop is added to the tunnel exit; then the jet is not supported. Equation 10.6 prevails and the piezometric line intersects the exit section at (D/2). Assume ($U_0^2/2g$) is small, then the energy equation can be written as

$$S_bL + h = \frac{D}{2} + \frac{U^2}{2g} + h_f + \Delta H_1 \qquad (10.8)$$

where L is the length of the tunnel.

Manning's equation is usually employed for computing h_f

$$h_f = S_EL = \frac{U^2 n^2}{R^{4/3}} L \qquad (10.9)$$

ΔH_1 can be written as

$$\Delta H_1 = k \frac{U^2}{2g}$$

Substituting these values in Eq. 10.8 yields

$$LS_b + h = \frac{D}{2} + \frac{U^2}{2g} + \frac{U^2}{2g} \frac{n^2}{R^{4/3}} L 2g + k \frac{U^2}{2g} \qquad (10.10)$$

and

$$LS_b + h = \frac{D}{2} + \frac{U^2}{2g} (1 + \frac{n^2 L}{R^{4/3}} \frac{2g}{} + k) \tag{10.11}$$

$$\frac{L}{D} S_b + \frac{h}{D} = \frac{1}{2} + (\frac{Q}{D^{5/2}})^2 (1 + 2g \frac{n^2 L}{R^{4/3}} + k) \frac{1}{2g} \frac{1}{(\frac{\pi}{4})^2} \tag{10.12}$$

$$\frac{h}{D} = \frac{1}{2} - \frac{L}{D} S_b + \frac{1}{2g (\frac{\pi}{4})^2} (1 + 2g \frac{n^2}{R^{4/3}} L + k) (\frac{Q}{D^{5/2}})^2 \tag{10.13}$$

The solution of this equation is given in Fig. 10.8, and the parameters n and k are listed in Table 10.3 and 10.4. Figure 10.7 shows how to determine the subpressure area in a tunnel.

B. Tunnel Flowing Partially Full

If the flow depth in the tunnel is high enough, the air confined between the free water surface and the tunnel will be evacuated as explained previously and priming will occur. If the discharge is insufficient, priming occurs intermittently, i.e. free air enters and is then evacuated repeatedly, then a pulsating flow occurs.

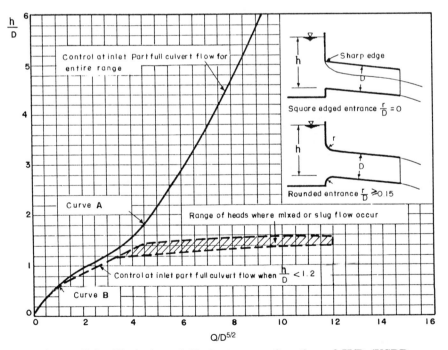

Figure 10.8. Variation of discharge as a function of H/D (USBR, Design of Small Dams).

525

TABLE 10.3

Values of Manning's (n) in a Lined Tunnel

	Nature of Lining	n
1.	Concrete lining, good quality, very smooth	0.013
2.	Concrete lining, prefabricated and transported from a central concrete plant (steel forms)	0.014
3.	Concrete lining, normal concrete prepared without any special care (rough forms)	0.017
4.	Low quality concrete lining, with rough joints	0.020

TABLE 10.4

Values of the Coefficient k

	Characteristics of the entrance section	k	k mean value
1.	Vertical, rectangular section	0.43 - 0.70	0.50
2.	Vertical (r/D≥ 0.15), circular section	0.08 - 0.27	0.10
3.	Concrete pipe, rounded wall, vertical	-	0.15
4.	Concrete pipe, rounded wall, upstream extended	0.10 - 0.33	0.20
5.	Metal pipe, corrugated, upstream extended	0.50 - 0.90	0.85

If partially full flow occurs in a tunnel, the flow in the free surface section is gradually varied. The partially full flow occurs in the tunnel when the entrance is submerged but the slope is steep enough so that a free surfaced flow can take place downstream. Assume that a free surface uniform flow has taken place in a tunnel as shown in Fig. 10.9.

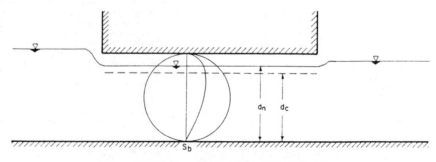

Figure 10.9. Free surface flow in a tunnel.

526

Assume that the discharge of the tunnel is Q_n and that the depth of the uniform flow which takes place in the tunnel is d_n. It is also assumed that some irregularities such as bend, plugs, etc. create local losses in the tunnel. The cross section of the tunnel is circular. Figures 10.10 and 10.11 define its characteristics. Figure 10.12 shows the form of the flow lines with similar cross sections.

Assuming that the flow is subcritical, the critical depth, d_c, corresponding to the bottom slope, S_b, is lower than the normal depth, d_n, of the flow. An examination of Fig. 10.11 reveals that Q_{max} does not correspond to a section flowing full. This is understandable due to the fact that the R value increases suddenly and more rapidly than the wetted area of a section flowing full.

The critical depth in closed conduits flowing partially full can be expressed as follows

$$\frac{U^2}{gD} = 1 \qquad (10.14)$$

where D is the hydraulic diameter of the tunnel

$$D = \frac{A}{T} \qquad (10.15)$$

Considering Eq. 10.14 it becomes possible to write

$$\frac{Q^2}{g} = \frac{A^3}{T} \qquad (10.16)$$

$$A = \frac{1}{8}(\theta - \sin\theta)D \qquad (10.17)$$

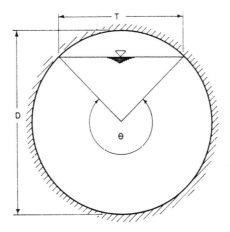

Fig 10.10. Geometric characteristics of a circular cross section.

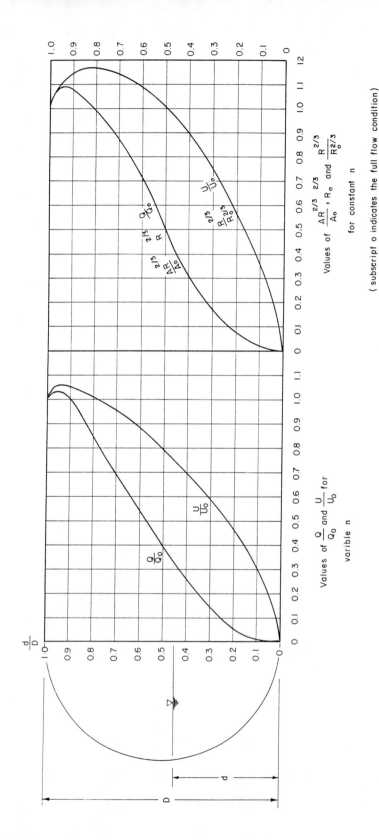

Figure 10.11. Discharge diagram in a closed conduit.

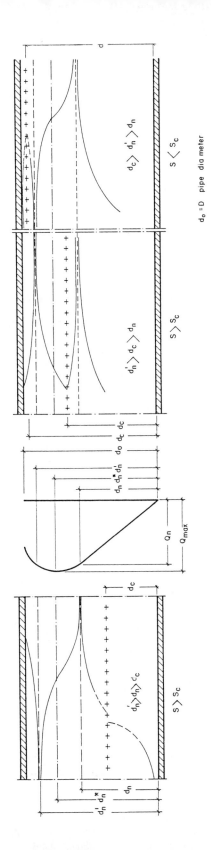

Figure 10.12. Flow lines in closed conduits.

$$T = (\sin \frac{\theta}{2})\, D \tag{10.18}$$

$$Q = \frac{(\theta - \sin\theta)^{\,3/2}}{8^{3/2}\,(\sin\frac{\theta}{2})^{1/2}}\, D^{5/2} \tag{10.19}$$

A function of Q is the angle θ. Then Q_{max} can be obtained by forming $dQ/d\theta = 0$. This operation yields

$$d' = 0.938D \neq d_c \tag{10.20}$$

when

for $S_b = S_w$ (uniform flow) and $d = d'$, $Q = Q_{max}$.

Where $Q_n > Q_{max}$, H increases, then the water surface level also increases upstream of the entrance section of the tunnel.

Assume now that a partially flowing discharge occurs in a tunnel as shown in Fig. 10.13. The mathematical definition of such a flow is

$$H_1 = S_b L + H = S_{1.2} L_{1.2} + S_{2.3} L_{2.3} + \frac{Q^2}{2gA_*^2} + d_* \tag{10.21a}$$

$$H_2 = S_b L_{2.3} + D + \frac{Q^2}{2gA_D^2} = S_{2.3} L_{2.3} + \frac{Q^2}{2gA_*^2} + d_* \tag{10.21b}$$

$$L = L_{1.2} + L_{2.3} \tag{10.22}$$

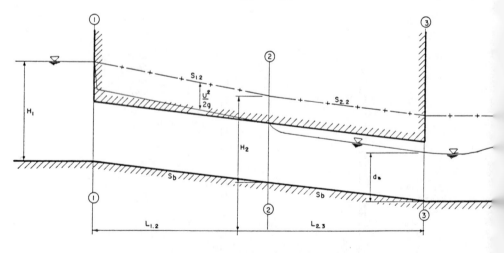

Figure 10.13. Partially full flow in closed conduits.

where   L   is the total length of the tunnel

S_b is the bottom slope of the tunnel

$S_{1.2}$ is the energy line slope for the flow flowing full

$S_{2.3}$ is the energy line slope for the free surface flow

d_* is the flow depth at the exit section, and

A_* is the wetted area at the exit section

In these three equations the unknowns are Q, $L_{1.2}$ and $L_{2.3}$, then the system can be solved for these three variables. The known values are:

- S_b the bottom slope of the tunnel, as chosen by the designer

- L   the total length of the tunnel, as chosen by the designer

- H_1 the total hydraulic head, as chosen by the designer

If H_1 is known, Q is computed as a function of H_1, and vice versa, if Q is given, H_1 can be obtained as its function

$$S_{1.2} = \frac{Q^2 n^2}{(\frac{\pi D^2}{4})^2 (\frac{D}{4})^{4/3}}$$

where

n is the Manning coefficient of resistance, and

D is the diameter of the tunnel flowing full

$$S_{2.3} = \frac{Q^2 n^2}{\left[\frac{A_* + \frac{\pi d^2}{4}}{2}\right]^2 \left[\frac{\frac{D}{4} + R_*}{2}\right]^{4/3}}$$

where A_* is the wetted area at the exit section and R_* is the hydraulic radius corresponding to d_*.

A better approach can be obtained by dividing the free flow reach between sections 2-2 and 3-3 into small reaches. It is assumed here that the variation of A_D is small so that Manning's relationship can be applied directly between sections 2-2 and the exit section. The value of A_D is

$$A_D = \frac{\pi [D(d)]^2}{4}$$

A_D, can be obtained from the table given in the Appendix. Increasing values of $S_{1.2}$ and $S_{2.3}$ correspond to increasing values of Q. Then d_x varies as a function of Q as indicated in Fig. 10.11. Also, increasing values of Q correspond to decreasing values of $L_{2.3}$; there is a critical value of Q above which the tunnel is completely filled. This is called priming of the tTunnel. It happens suddenly and H_1 increases accordingly.

Local head losses are not shown in Eqs. 10.21a, 10.21b and 10.22. The relationship expressing these losses are given in Chapter 11.

C. Subpressure in Tunnel

In cases previously outlined, the subpressure occurring in tunnels is capable of destroying the tunnel. Many examples throughout the world can be cited. The subpressure may be so high that it can nearly reach vacuum. Then the designer is obliged to reconsider the solution to be adopted. If the danger of subpressure exists, the design should be revised. Figure 10.14 shows a case in which $S_E < S_b$.

The pressure line being parallel to the energy line, the closed conduit shown in this figure is totally subject to subpressure. If (p_a/γ) is the atmospheric pressure

$$\frac{p}{\gamma} < \frac{p_a}{\gamma}$$

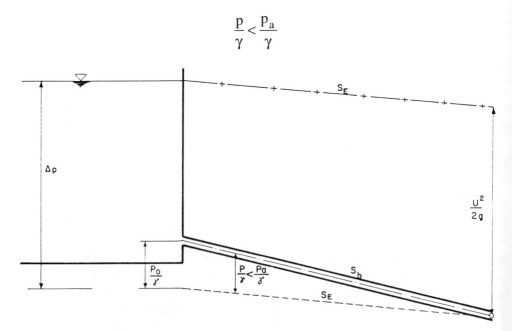

Figure 10.14. Subpressure caused by pressure flow in closed conduits.

The maximum value of the subpressure is (p_0/γ), which can in some instance reach vacuum (Problem 10.06). Figure 10.2e shows another example where sub-pressure occurs. In this particular case the sub-pressure is due to the convexity of the jet. In general, it is difficult to compute this subpressure. USBR suggests the use of the curves given in Fig 10.6. A better approach may be obtained by applying Eq. 3.26, for values of p/γ ($p/\gamma \neq p_a/\gamma$) smaller than a certain value, the water evaporates. Table 10.5 shows this limiting values of p/γ.

TABLE 10.5

Allowable Subatmospheric Pressures for Conduits Flowing Full

Elevation above sea level		Allowable subatmospheric pressures (p/γ)	
m	ft	m	ft
0	0	6.7	22
600	1970	6.1	20
1200	3940	5.5	18
1800	5900	4.9	16
2400	7880	4.3	14

10.6   DESIGN OF DIVERSION STRUCTURES

Following steps must be taken in designing a diversion system.

10.6.1 Layout

A diversion system consists of two cofferdams and one or more diversion tunnels. The problem to be solved includes location of the two cofferdams, determination of their height and the layout and number of the diversion tunnels. The location of the diversion tunnels and cofferdams is related to the characteristics of the dam and the topography of the site. Also, the number of tunnels, their diameter and the height of the upstream cofferdam are related to the hydraulic characteristics of the diversion. The hydraulic engineer, dam designer and geologist must all work together to develop a safe economical design. The layout of the tunnels necessitates a detailed geological investigation. Simple rules may be followed in preparing the preliminary design.

1. The tunnel must be in geologically sound material and must clear the nearest foundation excavation by a distance of at least

$$(4 \cong 5) \, D$$

2. The entrance to the tunnel must be located in a sound rock and tunnel construction must be more economical than the cut-and-cover.

3. When there is more than one tunnel, the distance separating the tunnels must be greater than

$$(4 \cong 6) \, D$$

10.6.2 Number of Tunnels

See section 10.4 for optimization analysis and section 10.2 for the determination of the number of tunnels.

10.6.3 Determination of the Crest Level of the Cofferdams and the Diameters of the Diversion Tunnels

See Section 10.4 for optimization analysis.

10.6.4 Detailed Hydraulic Computation of Diversion System

See section 10.5 and Example 10.02 for the detailed hydraulic computation of diversion systems.

10.6.5 Design of Diversion Systems

The hydraulic computation of the diversion system determines the dimensions of the hydraulic structures forming the system. The following steps may be taken for design purposes.

1. The tunnel layout is finalized by considering the length of the tunnels and the geology of the site. A decrease in the tunnel length is desirable because the cost of the tunnel excavation and hydraulic losses are decreased.

2. For preventing the bed load to enter the tunnel, the tunnel entrance may be located at the convex side of the river bend. The angle between the tunnel axis and the axis of the river at the entrance may be about 45°; at the exit, this angle may be about 30°.

3. The radius of the bends in plan may be equal to or greater than 100 m, to avoid helicoidal flow in the tunnel. At smaller radii the

helicoidal flow can be so strong that it can dama
downstream of the tunnel. In the Seyhan Dam Pow
butterfly valves were totally distorted and damaged b
flows; with the preventive measures taken for the pa..
damage to these valves were observed.

4. A horseshoe circular section is generally used for diversion tunnels. For relatively weak foundation material circular section is preferred to the horseshoe. For $D \geq 9.00$ m the horseshoe is preferable; in lined tunnels Manning's resistance coefficient, n, may be chosen to be 0.014 to 0.016, and the mean velocity

$$U \leq 12.00 \text{ m/sec}$$

The value of the resistance coefficient for new lining can be around 0.013 but with usage, solid material deposition on the concrete lining increases this value up to 0.016, or more.

5. If external hydrostatic pressure exists, the circular cross section is preferable to a horseshoe section.

6. If the foundation material is weak, contact injections are recommended. The characteristics of the grout required may be determined in-situ by experimental grouting.

7. Where the tunnel intersects the grout curtain of the dam, heavy high pressure grouting is necessary.

8. If a single tunnel is adopted, the tunnel can also be used later as a water intake tunnel. Where two tunnels are used, one can be blocked off and the other used as an intake tunnel. It is also possible to utilize one of the tunnels as a pressure tunnel for power generation. If so, an intake well is constructed at the entrance. Details of such a structure are given in Chapter 11.

9. The exit section of the diversion tunnel is generally equipped with stop log grooves so that stop logs can be placed to prevent the water from backing into the tunnel in case of an accident. In such a case the tunnel is blocked at both upstream and downstream ends and the water in it evacuated.

To prevent erosion downstream of the tunnel, a riprap blanket is generally installed.

A. General Principles for Design of Cofferdams

Cofferdams are temporary dams smaller than the main dam. Their purpose is to keep the water out of the structure foundation during struction. The upstream cofferdams are always much higher than the downstream. The upstream cofferdam for the Atatürk Dam in Turkey, for example, is a 42 m high structure. On large embankment dams, the upstream cofferdam is commonly incorporated into the main dam. When it is planned to incorporate the upstream cofferdam into the permanent embankment dam, it is built according to the main dam specifications. More details on cofferdams can be found in Dam Construction Treatise (USBR, 1957) (Figure 10.15).

10.6.6 Hydraulic Computation of Diversion Systems with Two Diversion Tunnels

A. Data Concerning Hydraulics of Diversion Systems

Diversion requires a complex system composed of two cofferdams and one or more diversion tunnels. The hydraulic definition of such a system requires the determination of the crest level of the upstream and downstream cofferdams and of the function

$$Q = f(d)$$

together with the geometric characteristics of the tunnel or tunnels. The function

$$d = f(D)$$

where

    d is the flow depth, and

    D is the tunnel diameter

is obtained by means of adequate flood routing.

There may be one or two diversion tunnels and occasionally three. Figure 10.16 shows a diversion system with two tunnels. Complete definition of the system is only possible when the following characteristics are known:

1. The length L_1 and L_2 of the tunnels, T_1 and T_2

The tunnel lengths are obtained by scaled dimensions from topographic maps for preliminary design. These are verified by field measurements.

2. $B_{1.1}$; $B_{1.2}$; $B_{2.1}$; $B_{2.2}$, are the bends in plan

3. $G_{1.1}$; $G_{1.2}$; $G_{2.1}$; $G_{2.2}$, are the characteristics of the entrance and the exit tunnels.

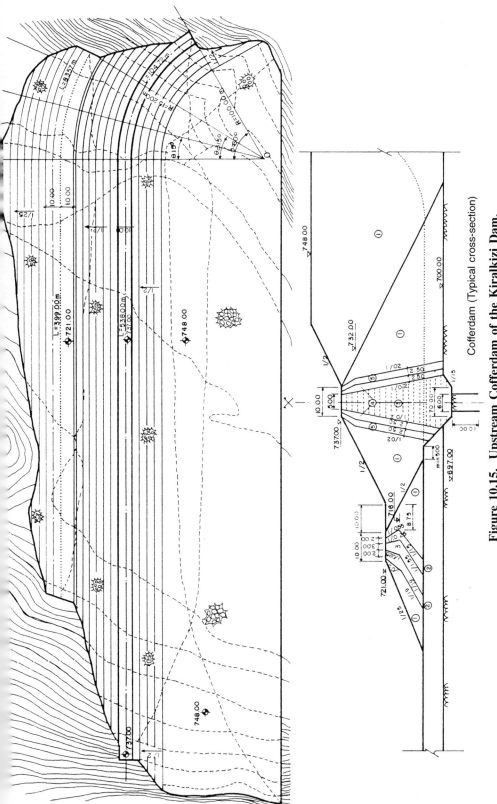

Figure 10.15. Upstream Cofferdam of the Kiralkizi Dam.

Cofferdam (Typical cross-section)

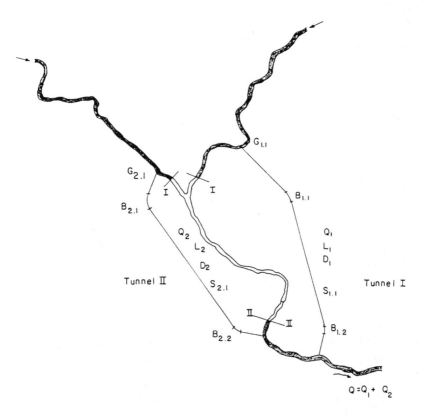

Figure 10.16. Diversion System with Two Diversion Tunnels.

4. The flood hydrograph.

5. $S_{1.b}$; $S_{2.b}$, are the bottom slope of the diversion tunnels and n_1; n_2 are the Manning coefficients of resistance relative to the tunnels 1 and 2, chosen by designer. If the bottom slopes change within the tunnels, $S_{1.b}$ and $S_{2.b}$, they then become $S_{1.1b}$; $S_{1.2b}$; $S_{2.1b}$ and $S_{2.2b}$.

6. The diameters, D_{1i} and D_{2i}, of the tunnel are computed by the engineer. The value of D_i generally constant within the tunnel, but occasionally changes for special conditions. If $i = 3$, then the diameter of the tunnel changes three times such as $D_{1.1}$; $D_{1.2}$; $D_{1.3}$ or $D_{2.1}$; $D_{2.2}$; and $D_{2.3}$.

7. Q_1 and Q_2 must be computed. The total discharge Q is $Q = Q_1 + Q_2$.

8. H is the hydraulic head of the diversion tunnel and is determined by the designer.

B. Hydraulic Computation of Diversion Systems

Assuming supported jet at the exit and the same invert elevation for the tunnels, the piezometric line crosses the exit section at (3/4 D). Applying the energy equation to the system yields

$$H = \frac{3}{4} D_i + \frac{Q_i^2}{2gA_i^2}(1 + \alpha_i) + L_i \frac{n_i^2 Q_i^2}{R_i^{4/3} A_i^2} - S_{ib} L_i \qquad (10.23)$$

where,

i   is the number of tunnels. For a single tunnel, $i = 1$; for two tunnels, $i = 1, 2$; and for three tunnels $i = 1, 2, 3$.

α   is a coefficient showing the local head losses. For example, in tunnel (i) there are the local head losses at the entrance $G_{i,1}$ and other losses at $G_{i,2}$, the bend losses at the bends $B_{i,1}$; $B_{i,i}$, etc. The entrance head loss is $0.1 U^2/2g$ and the local losses at bends $0.1\ U^2/2g$ (as an example). If the number of bends is 2 then

$$\alpha = 0.1 + 2 \cdot 0.1 = 0.3$$

A_i is the wetted area of the tunnel

Q_i is the total discharge diverted by the tunnel

R_i is the hydraulic radius of the tunnel, and

H   is the hydraulic head measured at the upstream entrance of the tunnel.

Writing Eq. 10.23 separately for tunnels T_1 and T_2 and equating then yields

$$\frac{3}{4}D_1 + \frac{Q_1^2}{A_1^2}\left(\frac{1+\alpha_1}{2g} + \frac{L_1 n_1^2}{R_1^{4/3}}\right) - S_{1b}L_1 = \frac{Q_2^2}{A_2^2}\left(\frac{1+\alpha_2}{2g} + \frac{L_2 n_2^2}{R_2^{4/3}}\right) - S_{2b}L_2 + \frac{3}{4}D_2$$

or

$$\frac{3}{4}(D_1 - D_2) + \left(\frac{Q_1}{A_1}\right)^2 \left(\frac{1+\alpha_1}{2g} + \frac{L_1 n_1^2}{R_1^{4/3}}\right) - S_{1b}L_1 - \frac{Q_2^2}{A_2^2}\left(\frac{1+\alpha_2}{2g} + \frac{L_2 n_2^2}{R_2^{4/3}}\right) + S_{2b}L_2 = 0$$

and

$$\frac{3}{4}(D_1 - D_2) + Q_1^2 \frac{\left(\frac{1+\alpha_1}{2g} + \frac{L_1 n_1^2}{R_1^{4/3}}\right)}{A_1^2} - Q_2^2 \frac{\left(\frac{1+\alpha_2}{2g} + \frac{L_2 n_2^2}{R_2^{4/3}}\right)}{A_2^2} + S_{2b}L_2 - S_{1b}L_1 = 0$$

Assume

$$C = \frac{3}{4}(D_1 - D_2)$$

$$E = \frac{\dfrac{1 + \alpha_1}{2g} + \dfrac{L_1 n_1^2}{R_1^{4/3}}}{A_1^2}$$

$$F = \frac{\dfrac{1 + \alpha_2}{2g} + \dfrac{L_2 n_2^2}{R_2^{4/3}}}{A_2^2}$$

and

$$G = S_{2b}L_2 - S_{1.b}L_1$$

Substituting these values in the above equation yields

$$EQ_1^2 - F Q_2^2 + G + C = 0$$

Then

$$E(Q - Q_2)^2 - F Q_2^2 + G + C = 0$$

where

$$Q = Q_1 + Q_2$$

and

$$(E - F) Q_2^2 - 2EQQ_2 + H = 0 \qquad (10.24)$$

where

$$H = G + C + EQ^2$$

The roots of this second degree equation are known

$$Q_2 = \frac{EQ \pm [E^2Q^2 - H(E - F)]^{1/2}}{E - F} \qquad (10.25)$$

Example 10.02

1. Compute the discharge Q_1 and Q_2 of the diversion tunnels (T_1, T_2) the characteristics of which are given:

    $$Q = 1200 \text{ m}^3/\text{sec}$$

$$D_1 = 8.00 \text{ m} \qquad\qquad D_2 = 8.00 \text{ m}$$

$$L_1 = 444.00 \text{ m} \qquad\qquad L_2 = 332.00 \text{ m}$$

$$n_1 = 0.015 \qquad\qquad n_2 = 0.015$$

$$R_1 = \frac{8}{4} = 2.00 \text{ m} \qquad\qquad R_2 = \frac{8}{4} = 2.00 \text{ m}$$

$$\alpha_1 = 0.4 \qquad\qquad \alpha_2 = 0.4$$

$$S_{1.b} = 0.007261 \qquad\qquad S_{2.b} = 0.008421$$

2. Compute the optimum solution assuming that Q, $S_{1.b}$, $S_{2.b}$, L_1 and L_2 are constant. Assume, furthermore, that (α) is constant and D_i and n_i variable.

Solution

1. The discharge Q_1 and Q_2 can be obtained directly by applying Eq. 10.25

$$Q_2 = \frac{EQ \pm [E^2Q^2 - H(E - F)]^{1/2}}{E - F} \qquad\qquad 10.25$$

and

$$Q_1 = Q - Q_2$$

$$E = \frac{(\frac{1 + \alpha_1}{2g} + \frac{L_1 n_1^2}{R_1^{4/3}})}{A_1^2}$$

$$A_1^2 = (\frac{\pi D_1^2}{4})^2 = \frac{\pi^2 D_1^4}{16}$$

$$E = \frac{\frac{16}{\pi^2}(\frac{1 + \alpha_1}{2g} + \frac{L_1 n_1^2}{R_1^{4/3}})}{D_1^4} = \frac{0.082267 (1 + \alpha_1) + 10.29359 \, n_2^2 \frac{L_1}{D_1^{4/3}}}{D_1^4}$$

$$F = \frac{0.082267\,(1+\alpha_2) + \frac{16}{\pi^2}\,n^2\,\frac{L_2}{R_2^{4/3}}}{D_2^4} = \frac{0.08267\,(1+\alpha_2) + 10.29359\,n_2^2\,\frac{L_2}{D_2^{4/3}}}{D_2^4}$$

$$H = G + C + EQ^2$$

$$C = \frac{3}{4}\,(D_1 - D_2)$$

$$G = S_{2b}L_2 - L_{1b}L_1$$

Substituting these values in Eq. 10.25 yields the solution of the problem which is shown in Table 10.6 and in Fig. 10.17.

A close review of Table 10.6 and Fig. 10.17 shows that,

1. The influence of the rugosity is relatively minor in affecting the velocity. However in order to obtain n = 0.013, it is necessary to use more costly framework. It may be advisable to use normal construction practice with forms that leave a surface for which the coefficient will be around 0.015.

TABLE 10.6

Search for the Optimal Diameter

D_1	D_2	n	E	F	C	Q_2	Q_1	U_2	U_1
m	m		10^5	10^5	m	m³/sec	m³/sec	m/sec	m/sec
8.00	8.00	0.017	4.84	4.330	0	612.75	587.25	12.19	11.68
8.00	8.00	0.015	4.39	3.997	0	609.90	590.10	12.13	11.74
8.00	8.00	0.013	4.00	3.710	0	606.94	593.06	12.07	11.80
8.00	8.00	0.014	4.19	3.850	0	608.82	591.58	12.10	11.77
8.00	7.75	0.015	4.39	4.596	0.18	590.99	609.01	12.53	11.76
8.00	7.50	0.015	4.39	5.311	0.38	571.10	628.90	12.98	12.51
8.00	7.75	0.014	4.19	4.420	0.19	589.21	610.21	12.50	12.14
8.00	7.50	0.014	4.19	5.100	0.38	570.16	629.84	12.91	12.53
7.50	7.40	0.014	5.58	5.410	0.08	601.88	598.12	13.62	13.90
7.40	7.40	0.014	5.93	5.410	0	610.69	589.31	14.20	13.70
7.40	7.40	0.015	6.24	5.640	0	612.16	587.84	14.23	13.67
7.00	7.00	0.015	8.02	7.210	0	613.56	586.44	15.94	15.24

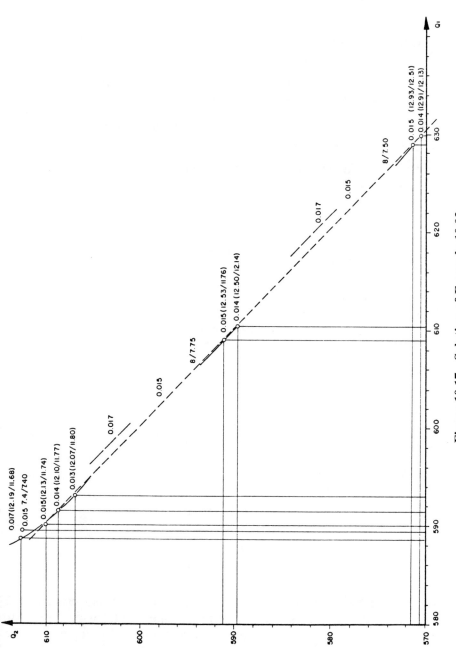

Figure 10.17. Solution of Example 10.02.

2. It might seem that $D_1 = D_2 = 7.40$ m is the cheaper solution. However, small diameters produce higher velocities ($U_{max} = 14.23$ m/sec), which may require special treatment to protect against erosion.

   If high velocities do occur for a short period (not more than 6 hours), then smaller diameters may warrant serious consideration.

3. For $D_1 = D_2 = 7.00$ m the maximum velocity attains 15.94 m/sec. Such a high velocity is generally avoided.

4. In preparing final design the case corresponding to $\alpha_1 \neq \alpha_2$ and $n_1 \neq n_2$ may also be checked.

5. The final decision concerning the choice of the diameter, D_i, is a function of the optimization analysis (section 10.4).

For $D_1 = D_2 = 8.00$ m the total head loss is

$$\Delta h = 1.4 \frac{U_2^2}{2g} + 332 \frac{0.015^2}{\left(\frac{8}{4}\right)^{4/3}} U_2^2 = 14.86 \text{ m}$$

where

$$U_2 = 12.13 \text{ m/sec}$$

and for

$$D_1 = D_2 = 7.40 \text{ m}$$

$$\Delta h = 1.4 \frac{U_2^2}{2g} + 332 \frac{0.015^2}{\left(\frac{7.4}{4}\right)^{4/3}} U_2^2 = 21.11 \text{ m}$$

where

$$U_2 = 14.23 \text{ m/sec}$$

Then

$$\Delta h' = 21.11 - 14.86 = 6.25 \text{ m}$$

In this case the upstream cofferdam crest may be further elevated for 6.25 m. If extra cost necessary for increasing the height of the upstream cofferdam by 6.25 m is less than the savings obtained by decreasing the diameter of the tunnels from 8.00 m to 7.40 m, this solution may be adopted. If not, a revision is necessary.

Example 10.03

Design the diversion system for the Dicle Dam, Turkey.

Data

The physical characteristics of the diversion system are listed in Table 10.7.

TABLE 10.7

Characteristics of the Diversion System of Dicle Dam, Turkey
$\alpha = 0.4$

Tunnels	Length of tunnels m	n	Apron Elevation		S	Design $Q_{i.25}$ m³/sec	Flood $Q_{i.10}$ m³/sec
			Entrance m	Exit m			
1	444	0.015	646.65	643.43	0.007261	2014	1601
2	332	0.015	646.76	643.96	0.008421	2014	1601

Solution

Preliminary design procedure

The diversion system is sized for $Q_{i.25}$ (flood discharge at the tunnel(s) entrance condition above the upstream cofferdam).

The dimension of the diversion tunnel or tunnels and the height of the cofferdam may be chosen in such a way that the flood is evacuated without any overtopping and with maximum velocities in the tunnel or tunnels smaller than the limiting value chosen by the designer according to local conditions. The following steps may be taken for solving the problem:

See Figs. 10.18 and 10.19 for design flood hydrographs of $Q_{i.25}$ and $Q_{i.10}$. Water surface elevation in the Dicle River is given by: h = $0.07379Q^{0.5805}$.

Step 1: Choose two diversion tunnels for the following reasons:

- Q is large enough to warrant two tunnels to carry $Q_{i.25}$

- The geology of the site is not suitable for the construction of one tunnel of large diameter.

- The topography allows for easy positioning of two tunnels with diameters of about 7.00 m.

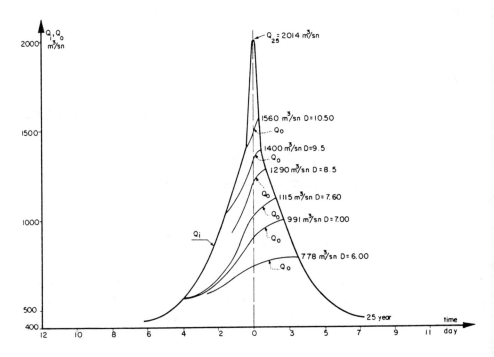

Figure 10.18. Design flood hydrograph $Q_{i.25}$ of the Dicle Dam Site.

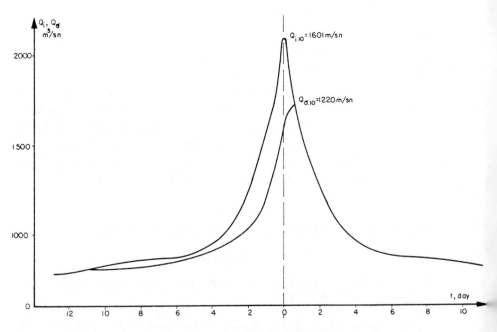

Figure 10.19. Design Flood Hydrograph $Q_{i.10}$ for the Dicle Dam Site.

Step 2: Compute Q = f(h) curve corresponding to the tunnels.

Step 3: Route the flood for D = 6.00 m, D = 7.00 m, D = 7.60 m, D = 8.50 m, D = 9.50 m and D = 10.50 m, and determine the hydraulic head at the entrance of the tunnels.

Step 4: Add 2.00 m (freeboard) to set the crest level of the upstream cofferdam.

Step 5: Compute the total cost of the cofferdam and the tunnels.

Step 6: Choose a solution corresponding to the minimum construction cost.

Computation of Q = f(h) Curves

For small discharges, free flow discharge takes place in the funnel. For Q greater than a critical value the tunnel primes suddenly. A discontinuity exists between the Q = f(h) curves for pressure flow and for free surface flow.

Pressure Flow

Figure 10.20 shows Q = f(h) curves for free surface and pressure flow. The application of Eq. 10.25 is sufficient for obtaining these curves. Assume $Q_o = 1200$ m³/sec and $D_1 = D_2 = 7.60$ m. Then

$$E = \frac{\frac{1.4}{19.62} + \frac{444 \bullet 0.015^2}{1.94^{4/3}}}{45.36^2} = 5.5312621 \bullet 10^{-5}$$

$$F = \frac{\frac{1.4}{19.62} + \frac{332 \bullet 0.015^2}{1.94^{4/3}}}{45.36^2} = 5.0108073 \bullet 10^{-5}$$

$$G = 0.008421 \bullet 332 - 0.007261 \bullet 444 = -0.4281$$

$$C = 0$$

$$H = G + C + EQ^2 = -0.4281 + 5.5312621 \bullet 10^{-5}Q^2$$

$$Q_2 = \frac{EQ \pm (E^2Q^2 - H(E - F))^{1/2}}{E - F}$$

$$H = \frac{3}{4}7.60 + \frac{1.4Q^2}{19.62 \bullet 45.36^2} - 332\frac{0.015^2\,Q^2}{1.94^{4/3} \bullet 45.36^2} - 332 \bullet 0.008421 \qquad 10.23$$

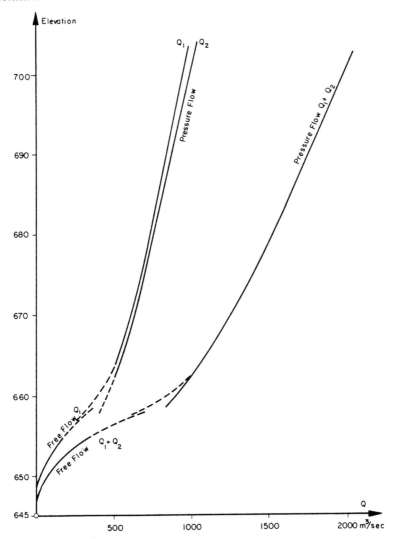

**Figure 10.20. Curves Q = f(h) for the Diversion System of the Dicle
Dam, Turkey.**

Then one point of Q = f(h) curve is obtained for $D_1 = D_2 = 7.60$ m and
$Q = 1200$ m³/sec. A different pair of values such as $D_1 = D_2 = 7.60$ m and Q
$= Q_o$ yields another point of the curve. Repeating the computation for $D_1 =$
$D_2 = D_o$ and $Q = Q_o$ yields a different Q = f(h) curve. Figure 10.20 is
obtained for $D_1 = D_2 = 8.00$ m. Regression technique gives the analytical
expression for the curve

$$h = 0.00017 \ Q^{1.66208}$$

(Table 10.10 for pressure flow computation)

Free Surface Flow

Step l: The regime of flow taking place in the tunnel:

$$S = \frac{Q^2 n^2}{A^2 R^{4/3}}$$

Using the table given in the Appendix, the critical depth can be determined as

$$\frac{d}{D} = 0.99$$

Then

$$Q_c = 8.8263 \ D^{5/2} \ = 1405.43 \ m^3/sec$$

$$A_c = 0.7841 \ D^2 \ \ = 45.29 \ m^2$$

$$R_c = 0.2666 \ D \ \ \ \ = 2.03 \ m$$

and

$$S_c = \frac{1405.43^2 \cdot 0.015^2}{2.03^{4/3} \cdot 45.29^2} = 0.08429 > 0.008421 > 0.007261$$

The regime of flow taking place in the tunnels is subcritical when $d/D = 0.99$. For smaller depth the regime of flow must be checked.

Step 2: Free surface flow in the tunnels

Assumptions:

- The approach velocity in the reservoir is assumed negligible, then

$$(U^2/2g = 0)$$

- The flow in the tunnels is uniform

- Local head losses:

$$\text{entrance loss} = 0.1 \ \frac{U^2}{2g}$$

$$\text{bend loss due to two bends} = 2 \bullet 0.1 \frac{U^2}{2g}$$

Friction loss

$$Q = \frac{A}{n} R^{2/3} S^{1/2}$$

The cross section of the tunnels being circular, the table given in the Appendix will be used for determining the variables in the Manning equation

$$\frac{A}{D^2} \cong 0.97801 \left(\frac{d}{D}\right)^{1.3721}$$

$$\frac{R}{D} \cong 0.4068 \left(\frac{d}{D}\right)^{0.83535}$$

These two equations are only approximately valid for d/D smaller than 0.80. For greater values, new relations may be derived. For a better approximation the computations may be repeated for each individual value of d/D.

Substituting these values in Manning's equation yields

$$Q = \frac{0.97801 \left(\frac{d}{D}\right)^{1.3721}}{0.015} D^2 \left[0.4068 \left(\frac{d}{D}\right)^{0.83535} D\right]^{2/3} S_i^{1/2}$$

For Tunnel no. 1, $S_{1b} = 0.007261$, then

$$Q = 3.0502 \left(\frac{d}{D}\right)^{1.929} D^{2/3} D^2 = 3.05 \left(\frac{d}{D}\right)^{1.929} D^{2.667} \qquad (10.26)$$

The case of this equation is shown schematically in Fig. 10.21. The energy equation applied to this flow yields

$$H = S_b L + d_e + \frac{U_2^2}{2g} = d_o + \frac{U_2^2}{2g} + 0.1 \frac{U_i^2}{2g} + 0.1 \frac{U_{i+1}^2}{2g} + 0.1 \frac{U_{i+2}^2}{2g} - \left(\frac{U^2 n^2}{R^{4/3}}\right) L \qquad (10.27)$$

where,

    H   is the hydraulic head

    d_o is the depth of flow at the exit section

    U_o is the velocity of flow at the exit section

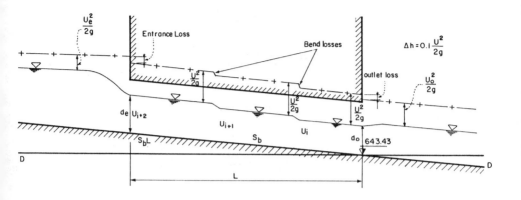

Figure 10.21. Free surface flow in the tunnel (schematic).

U_i is the velocity at the bend (1)

U_{i+1} is the velocity of flow at the bend (2)

U_{i+2} is the velocity of flow at the entrance section

U is the mean velocity

R is the mean hydraulic radius

d_e is the depth of flow at the entrance

U_e is the mean velocity of flow at the entrance section.

Each tunnel diameter corresponds to a particular energy equation of the form of Eq. 10.27. Assume that the tunnel diameter be chosen equal to $D = 7.6$ m

$$Q = 13.62 \, d^{1.929} \qquad (10.28)$$

Assume further on that the tunnel is long enough for the uniform flow to take place (this assumption must be checked) and $U_i \approx U_{i-1} \approx U_{i-2} \approx U$, then $S_E = S_b$, and

$$H = d_o + 1.3 \frac{U^2}{2g} + S_E L = S_b L + d_e + \frac{U_e^2}{2g}$$

Introducing the entrance apron elevation 646.65 m, this relationship takes the following form:

$$H = d_o + S_E L + 1.3 \frac{U^2}{2g} + 643.43 = S_b L + d_e + 643.43 + \frac{U_e^2}{2g} = 646.65 + d_e + \frac{U_e^2}{2g}$$

$$U = \frac{Q}{A}$$

$$A = 0.97801 \left(\frac{d_o}{D}\right)^{1.3721} D^2$$

$$U = \frac{Q}{0.97801 \, (d_o/D)^{1.3721} \, D^2}$$

$$\frac{U^2}{2g} = \frac{1}{2g} \left(\frac{Q^2}{0.97801^2 \left(\frac{d_o}{7.6}\right)^{2.7442} 7.6^4} \right)$$

$$H = d_o + \frac{1.3}{2g} \; \frac{Q^2}{0.9565\left(\frac{d_o}{7.6}\right)^{2.7442} 7.6^4} +$$

$$+ \, 43.43 + S_{EL} = d_o + 0.0055 \; \frac{Q^2}{d_o^{2.7442}} + 646.65 \qquad (10.29)$$

When d_o is known the elevation of the energy line at the entrance can be computed using Eq. 10.27, and the water depth at the entrance d_e can be obtained accordingly

$$H - 646.65 = d_e + \frac{U_e^2}{2g}$$

$$(H - 646.65) - d_e = \frac{Q^2}{2g} \; \frac{1}{0.9565 \left(\frac{d_e}{7.6}\right)^{2.7442}} \; \frac{1}{7.6^4}$$

$$H - 646.65 = d_e + 13.9234 \; \frac{Q^2}{d_e^{2.7442}} \; \frac{1}{7.6^4} \qquad (10.30a)$$

$$H - 646.65 = d_e + 0.0043 \; \frac{Q^2}{d_e^{2.7442}} \qquad (10.30b)$$

d_e, can be computed using this equation for the above given assumptions.
Substituting Q from Eq. 10.28 into Eq. 10.29 yields H, then the only unknown in Eq. 10. 30 is d_e

$$H = d_o + 0.0055 \; \frac{1}{d_o^{2.7442}} \, (13.62 \, d_o^{1.929})^2 + 646.65$$

$$H = d_o + 0.0055 \; \frac{185.50 \, d_o^{3.858}}{d_o^{2.7442}} + 646.65$$

$$H = d_o + 1.02 \ d_o^{1.1138} + 646.65 \qquad (10.31)$$

Equations 10.30 and 10.31 used together solve the problem (Table 10.8).

TABLE 10.8

Free Surface Flow taking place in Tunnel T_1,
$D = 7.60$ m

d_o	H	Q_1	d_e
m	m	m³/sec	m
1	2	3	4
0.50	647.62	3.58	0.90
1.00	648.67	13.62	1.88
1.50	649.75	29.78	2.88
2.00	650.86	51.86	3.91
2.50	651.98	79.76	4.96
3.00	653.12	113.38	6.08
3.50	654.27	152.65	7.17
3.75	654.84	174.38	7.71

Column 1:   Choose an arbitrary value for d_o

Column 2:   Use Eq. 10.31 to determine H

Column 3:   Use Eq. 10.28 to determine Q

Column 4:   Use Eq. 10.30b to determine d_e

For higher discharges, pressure flow takes place in the tunnel. Equation 10.28 is only approximate; for this reason and due to previous assumptions, higher approximations cannot be obtained from this kind of approach.

If the above mentioned approximations are not valid, then it is necessary to check for the existence of uniform flow corresponding to each discharge.

The flow taking place in tunnel No. 2 has the following characteristics:

1. The entrance apron elevation is 646.76 m.

2. The hydraulic head is the hydraulic head of tunnel No. 1.

The energy equation can be written as

$$H = d_o + 1.02d_o^{1.1138} + 646.76 \tag{10.32}$$

$$H = d_e + 0.0043 \frac{Q^2}{d_e^{2.7442}} + 646.76 \tag{10.33}$$

Simple transformation yields

$$H - 646.76 = d_o (1 + 1.02d_o^{1.1138}) \tag{10.34}$$

$$H - 646.76 = d_e + 0.0043 \frac{Q^2}{d_e^{2.7442}} \tag{10.35}$$

The solution of these equations are given in Table 10.9.

TABLE 10.9

Free Surface Flow Taking Place in Tunnel No 2

D = 7.60 m

H m	d_o m	Q_2 m³/sec	d_e m	Q_1 m³/sec	$Q=Q_1+Q_2$ m³/sec
1	2	3	4	5	6
647.62	0.45	2.92	0.80	3.58	6.50
648.67	0.95	12.34	1.78	13.62	25.96
649.75	1.45	27.89	2.78	29.78	57.67
650.86	1.95	49.39	3.80	51.86	101.25
651.98	2.45	76.72	4.85	79.76	156.48
653.12	2.95	109.77	5.91	113.38	223.15
654.27	3.45	148.47	6.98	162.65	311.12
654.84	3.70	169.92	7.49	174.38	344.30

Column 1:   From Column 2 of Table 10.8

Column 2:   From Eq. 10.34

Column 3:   From Eq. 10.28

Column 4:   From Eq. 10.35

Column 5:   From Column 3 of Table 10.8

Column 6:   Column 3 + Column 5 of Table 10.9.

In Fig. 10.20, Q = f(h) diagram is visualized, and the computation of pressure flow is shown in Table 10.10.

TABLE 10.10

Pressure Flow Taking Place in Tunnel No. 2
D = 7.60 m

Q m³/sec	Q_2	Q_1 m³/sec	H m	Elevation
1	2	3	4	5
600	300.62	299.38	7.43	654.19
650	326.76	323.24	8.25	655.01
700	352.82	347.18	9.14	655.90
800	404.78	395.22	11.11	657.87
900	456.60	443.40	13.35	660.11
1000	508.26	491.74	15.85	662.61
1100	559.86	540.14	18.61	665.37
1200	611.41	588.59	21.64	668.40
1300	662.93	637.07	24.93	671.69
1600	817.19	782.81	36.37	683.13
2000	1022.63	977.37	55.31	702.07

Column 1:   Discharge to be chosen by the designer

Column 2: Q_2 from Eq. 10.25

Column 3:   Column 1 - Column 2

Column 4:   H from Eq. 10.23; $\alpha_i = 0.4$; $L_i = 332.00$ m

$$R_i = \frac{\pi D^2}{4\pi D} = \frac{D}{4}$$

$$S_{ib} = 0.008421$$

Column 5:   Column 4 + 646.76

The pressure curve is also shown in Fig. 10.20.

A $h_p 11C$ type calculator is used for solving this particular problem.*

Tunnel diameter D and the upstream cofferdam crest elevation being computed as a function of D, a rough estimate of the diversion system can be done. The result is shown in Table 10.11. Table 10.11 is plotted in Fig. 10.22. The optimum tunnel diameter is found to be equal to 7.60. The dimensions of the diversion system are then

Tunnel diameter: $D_1 = D_2 = 7.60$ m

Flood Routing

The Bresse equation is used in flood routing computations

$$(Q_i - Q_o) \Delta t = \Delta V \qquad (10.36)$$

where

Q_i is the inflow discharge to the reservoir

Q_o is the outflow discharge from the reservoir

Δt is the time increment, and

ΔV is the incremental volume in the reservoir.

Equation 10.36 is actually a continuity equation. It states that the difference between the inflow and outflow for a given interval of time is

---

*   See Problem 10.07

equal to the variation of the volume in the reservoir during the same time interval

$$V = \frac{A_1 + A_2}{2} \Delta h \qquad (10.37)$$

A_1, and A_2 are the reservoir surface areas and Δh is the incremental depth. Introducing these values in Eq. 10.36 yields

$$Q_i \Delta t - Q_o \Delta t = \frac{A_1 + A_2}{2} \Delta h$$

where

$$Q_i = \frac{Q_{i.1} + Q_{i.2}}{2}$$

and

$$\frac{Q_{i.1} + Q_{i.2}}{2} \Delta t - Q_o \Delta t = \frac{A_1 + A_2}{2} \Delta h \qquad (10.38a)$$

where

$$\Delta h = h_1 - h_2$$

and $Q_{i.1}$ and $Q_{i.2}$ are two consecutive values of the inflow separated by a time interval equal to Δt.

If greater precision in the computation is required the time interval may be shortened. In the case of the Dicle Dam

$$A = 0.094 \, h^{1.8164}$$

introducing this value into Eq. 10.38a yields

$$(Q_{i.1} + Q_{i.2}) \Delta t - 2Q_o \Delta t = 0.094 \, (h_1^{1.8164} + h_2^{1.8164}) \, (h_2 - h_1) \quad (10.38b)$$

The solution of this equation is obtained by following the step-by-step method:

1. Choose a diameter, D, for the diversion tunnel

2. Choose a time interval, Δt

3. Determine $Q_{i.1}$ and $Q_{i.2}$ corresponding to the time interval, Δt, using the inflow hydrograph and compute

$$(Q_{i.1} + Q_{i.2}) \Delta t$$

TABLE 10.11

Optimization of Dimensions for the Diversion Structure

Tunnel Diameter	Height of the Upstream Cofferdam	Cost Tunnels	Estimates Cofferdam	Total
m	m*	10^6 unit**	10^6 unit**	10^6 unit**
6.00	26.28	302.1	432.1	734.8
7.00	22.40	361.4	366.2	727.6
7.60	18.50	397.3	316.8	714.1
8.50	18.44	453.1	287.5	740.6
9.50	16.40	517.4	246.5	763.9
10.50	15.18	585.6	221.6	807.2

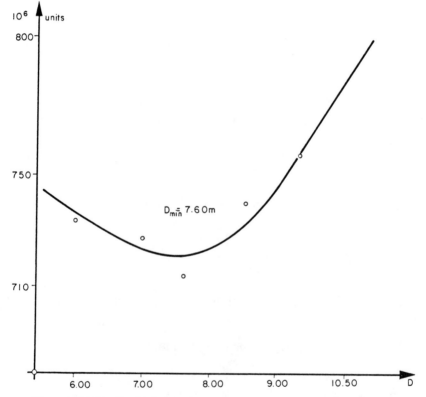

Figure 10.22. Optimization curve of the Dicle Dam Diversion System, Turkey.

---

* A 2.00 m freeboard is added.

** Monetary unit of the country where the optimization analysis is performed

4. Choose a value for Q_o; then the left side of the equation is determined

5. Q_o is the delivery of the diversion system. Equation 10.23 yields the value of h_2 as a function of Q_o; h_1 is the reservoir depth at the inflow start time. When h_1, and h_2 are known, the right side of the equation can be computed. If this value is different from the value of the left side of the equation a new value for Q_o is chosen and the computation repeated until the left side of the equation equals the right side.

A computer or a calculator can be used for performing this computation. The result is shown in Fig. 10.18 and in Table 10.10.*

Flood Routing for $Q_{i.25}$ Flood

The numerical application of the routing of Q_{i25} flood is given. The crest level of the upstream cofferdam is determined as a result of this computation. Q_o is chosen based upon the head difference between the centerline of the tunnel and the height of the cofferdam. The computation is performed using Eq. 10.38b

$$(Q_{i.1} + Q_{i.2}) \Delta t - 2Q_o \Delta t = 0.094 \, (h_1^{1.8164} + h_2^{1.8164}) \, (h_2 - h_1) \quad 10.38b$$

The results are listed in Table 10.12.

TABLE 10.12

Flood Routing for Q_{i25}

Q_{i1}	Q_{i2}	Δt	$(Q_{i1}+Q_{i2})\Delta t$	Q_o	h_1	h_2	Eq. 10.38b	
m³/sec	m³/sec	10^6sec	10^6 m³	m³/sec	m	m	right side	left side
1	2	3	4	5	6	7	8	9
550	900	0.1728	250.56	674.10	6.0975	8.551	17.53	17.59
900	1190	0.0864	180.58	813.10	8.5510	11.677	40.01	40.08
1190	2014	0.0864	276.48	991.90	11.6770	16.249	105.53	105.10
2014	1350	0.0432	145.15	1051.90	16.2490	17.915	54.37	54.27
1350	1190	0.0432	109.73	1068.70	17.915	18.393	17.39	17.39
1190	1050	0.0432	96.77	1072.53	18.392	18.503	4.10	4.10

* See Problem 10.09

Column 1:   From the hydrograph of Fig. 10.18

Column 2:   From the hydrograph of Fig. 10.18

Column 3:   The time increment Δt is chosen by the designer

Column 4: $(Q_{i1} + Q_{i2}) \Delta t$

Column 5: Q_o is chosen by the designer to begin the trial

Column 6:   Hydraulic head relative to $Q_{i1} = 550$ m^3/sec (for the first trial)

Column 7: $h_2 = 0.00017 \ Q_2^{1.66208}$ (This equation is assumed valid for $D = 7.60$ m also)

Column 8: $0.094 \ (h_1^{1.8164} + h_2^{1.8164}) \ (h_2 - h_1)$

Column 9: $(Q_{i1} + Q_{i2}) \Delta t - 2Q_o \Delta t$

If the value obtained in Column 8 is too different from the value in Column 9, change Q_o in Column 5 and repeat.

Conclusion:

The final dimensions of the diversion system are determined as:

1. $D_1 = D_2 = 7.60$ m

2. For Q_{i25} the $Q_o = 1072.53$ m^3/sec

3. Pressure flow takes place in the tunnels (Fig. 10.20) for

$Q_{o2} = 550$ m^3/sec

$Q_{o1} = 524$ m^3/sec

where

Q_{o2} is the outflow of tunnel 2

and

Q_{o1} is the outflow of tunnel 1

4. Reservoir water supply elevation upstream of the upstream cofferdam

$= 646.76 + 18.50 = 665.26$

5. Crest elevation of the upstream cofferdam

$= 646.76 + 18.50 + 2.00 = 667.26$

6. Freeboard $= 2.00$ m

7. Water surface elevation in the Dicle River for $Q_o = 1072.53$ m³/sec

$h = 0.07379Q^{0.5805} \cong 4.24$ m

Elevation $= 643.43 + 4.24 = 647.67$ m.

10.7   PROBLEMS

Problem 10.01

It is required to determine the hydraulic efficiency of a culvert the characteristics of which are given (Fig. 10.4).

Data

- Entrance invert elevation: 100.05

- Exit invert elevation: 100.00

- Cross section, $b = 1.50$ m, $D = 3.00$ m in the tunnel and in the downstream open channel

- $L = 50.00$ m

- $n = 0.014$

- $H_1 = 3.05$ m

- $H_2 = 3.00$ m

- $q = 6.25$ m³/sec m

- $S_b = 0.00886$

Problem 10.02

What happens to the flow described in Problem 10.01 when $q = 9.76$ m³/sec.m?

Data

$H_1 = 3.20$ m

Problem 10.03

What happens to the flow described in Problem 10.01 when $h = 2.00$ m; determine the length L of the culvert for this condition.

Problem 10.04

What happens to the flow described in Problem 10.01 when $q = 3.00$ m³/sec.m?

Data

$H_1 = 2.75$ m

Problem 10.05

What happens to the flow described in Problem 10.01 when $H_1 = 3.75$ m?

Data

The flow depth at the downstream end of the culvert is 1.75 m.

Problem 10.06

Determine the length of the closed conduit shown in Fig. 10.14 so that (p_o/γ) equals $p/\gamma = 5.45$ m. The conduit is built at elevation 1200 (m.s.d.).

Data

- Diameter of the closed conduit = 1.00 m

- $n = 0.013$

- $Q = 1.00$ m³/sec

- The slope of the closed conduit = 0.002

Problem 10.07

Compute the discharges, Q_1 and Q_2, of the diversion tunnels T_1 and T_2, the characteristics of which are given.

Data

- $Q = 1250$ m³/sec

- $L_1 = 500$ m

- $L_2 = 545$ m

- $\alpha_1 = 0.65$ (two bends only)

- $\alpha_2 = 0.67$ (two bends only)

- $S_{1b} = 0.0075$

- $S_{2b} = 0.0077$

D_1, D_2, n_1 and n_2 are variable. The invert elevation is the same for the tunnels. Determine the optimal solution under the following conditions

$$U_{imax} \leq 12.00 \text{ m/sec}$$

$$n_i \geq 0.014$$

The invert elevation is the same for the tunnels.

Problem 10.08

Draw the Q = f(H) curves for the tunnels described in Problem 10.07 and for the diameters chosen as result of the problem

Data

Entrance invert elevation = 0

Problem 10.09

Prepare a calculator or a computer program for computing Eq. 10.38a.

Chapter 11

HYDRAULICS OF OUTLET WORKS

11.1 INTRODUCTION

A dam is a hydraulic structure designed to retain water in the reservoir it creates. Flood water entering the reservoir is discharged over a spillway, through outlet valves or penstocks. Gradual discharges from the reservoir may be controlled by gates and uniform but constant small discharges by outlet works. The purpose of the outlet works is to direct the water to pipe lines, to irrigation areas or to the main river. The outlet works is also designed to empty the reservoir for emergency or maintenance reasons. Two types of intakes are used for outlet works. In order to supply water to pipe lines it is necessary to locate the intake at a relatively high elevation. For emptying a reservoir, the intake is placed at its lowest elevation.

The hydraulics of outlet works is basically pipe hydraulics in which local head losses play the basic part in the equilibrium relation. They are listed as follows:

a. Local head loss due to   the trashrack

b.   "   "   "   "   "   the entrance

c.   "   "   "   "   "   the bends

d.   "   "   "   "   "   the transitions

e.   "   "   "   "   "   the branching pipes

f.   "   "   "   "   "   the friction or flow resistance

g.   "   "   "   "   "   the gates and valves

The sum of head losses in each particular case, subtracted from the available hydraulic head, determines the net available head to be used for the effective discharge. The standards adopted for emptying the reservoir in case of an emergency varies from country to country.

Different problems concerned with outlet works are outlined in the following paragraphs and their solutions given. An example of the layout of the outlet work is shown in Fig. 11.1.

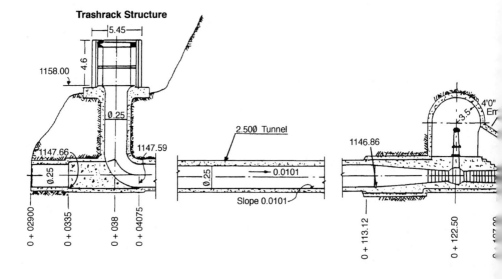

TUNNEL PROFILE

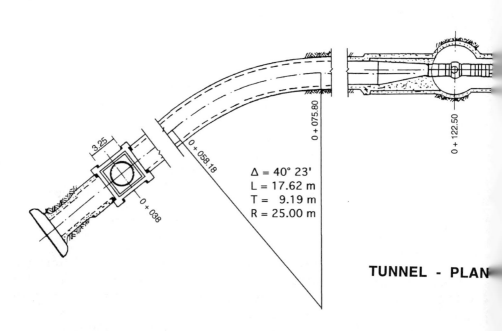

$$\Delta = 40°\ 23'$$
$$L = 17.62\ m$$
$$T = 9.19\ m$$
$$R = 25.00\ m$$

TUNNEL - PLAN

Figure 11.1. Outlet works: Layout.

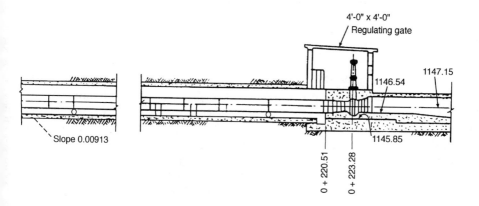

4'-0" x 4'-0"
Regulating gate

1147.15

1146.54

1145.85

Slope 0.00913

0 + 220.51

0 + 223.28

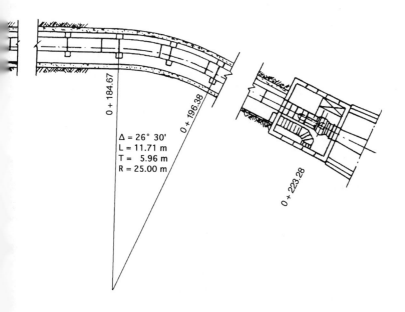

0 + 184.67

0 + 196.38

0 + 223.28

Δ = 26° 30'
L = 11.71 m
T = 5.96 m
R = 25.00 m

Figure 11.1. Continued.

11.2 HYDRAULICS OF OUTLET WORKS, LOCAL HEAD LOSSES

11.2.1 Local Head Loss due to Trashrack

Water intakes in an outlet system are equipped with trashracks to prevent floating debris such as branches, leaves and other debris from entering the system. In the case of power plants solid materials accompanied by solid particles in suspension can cause heavy damage to the mobile part of turbines. Trashracks can prevent, but only partially, the introduction of small floating solids to the system. Branches and leaves can obstruct the intake entrance at the head works; the trashrack is usually equipped with special devices for cleaning. The following steps can be taken in designing the trashrack and in estimating the total head loss through it.

The gross trashrack area, calculated from inlet sill to the reservoir design water surface elevation, should be such that the velocity through the trashrack for the design discharge does not exceed 0.75 m/sec. However if the trashrack is accessible for cleaning, a maximum velocity of 1.00 m/sec may be accepted. The normal type of trashrack structure used for diverting water for irrigation purposes is of the cage or box-type. The structural members are rectangular and sharp cornered, the volume inside the cage is small relative to the area of the trashracks and the trashbars are closely spaced. For this type of structure the trashrack head loss is

$$\Delta h_1 = 0.8 \, h_v \qquad (11.1a)$$

where h_v denotes the velocity head ($U^2/2g$) computed on the basis of the gross trashrack area.

When hydroelectric power is involved the trashrack or inlet structure will be streamlined. The volume inside the cage will be large relative to the trashrack area. In this case the trashrack head loss is

$$\Delta h_1 = 0.3 \, h_v \qquad (11.1b)$$

These two values represent the two extremes that are encountered in practice. The above coefficients are conservative and include an allowance for partial clogging.

The following relation suggested by USBR can also be used if the required information is available

$$\Delta h_1 = [1.45 - 0.45 \frac{A_N}{A_g} - (\frac{A_N}{A_g})^2] \, h_v \qquad (11.1c)$$

where

A_N is the net area through the trashbars, and

A_g is the gross area through the trashbars.

In carrying out the above calculations, allow for a 50% reduction in area due to clogging of the trashrack.

A second approach is suggested in Europe for computing trashrack loss (Şentürk, F., 1957)

$$\Delta h_1 = K_{tr} \, h_v$$

and

$$K_{tr} = B \left(\frac{S}{b}\right)^{4/3} \sin (d) \tag{11.2}$$

where

B   is a coefficient as defined in Fig. 11.2 and in Table 11.1

S   is the thickness of the trashbars

b   is the clearance between the trashbars, and

d   is the slope angle of the trashrack.

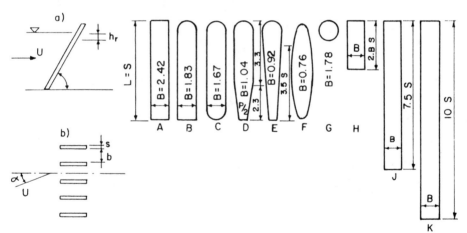

Figure 11.2. Definition of coefficient B.

TABLE 11.1

Corrections for S/b = 0.599 and θ = 90°

				K_{tr}					
α	A	B	C	D	E	F	H	J	K
30°	1.46	0.76	0.71	0.43	0.68	0.22	1.81	1.53	1.62
45°	2.05	1.29	1.29	0.94	1.29	0.67	2.72	2.32	2.12
60°	4.26	2.45	2.81	2.19	3.05	1.84	4.26	3.43	3.88

Figure 11.2 suggested by Krishmer is valid for a flow velocity vector perpendicular to the trashrack plan. If an angle, (α), exists between the two directions then S. Strangler suggests the corrections shown in Table 11.1.

11.2.2 Local Head Loss due to Entrances

Local head loss due to entrances is defined by USBR as follows

$$\Delta h_2 = K_e h_v \tag{11.3}$$

The values of K_e are listed in Table 11.2.

The velocity head, h_v, is computed on the basis of the circular section. If stop log or gate-guide grooves exist at the square section add 0.03 ~ 0.02 to the corresponding coefficient.

TABLE 11.2

Values of Coefficient K_e

Definition of the entrance	K_e		
	max	min	mean
1. Entrance equipped with a bulkhead gate, complete contraction	1.80	1.00	1.50
2. Entrance equipped with a bulkhead gate; contraction at the bottom and sides	1.20	0.50	1.20
3. Entrance equipped with a bulkhead gate; rounded walls	1.00	0.10	0.50
4. Square entrance	0.70	0.40	0.50
5. Slightly rounded corners entrance	0.60	0.18	0.23
6. (r/D) ≥ 0.15, completely rounded corners	0.27	0.08	0.10
7. Circular bell-mouth	0.10	0.04	0.05
8. Square bell-mouth	0.20	0.07	0.16
9. Tube bell-mouth	0.93	0.56	0.80

K_b is shown in Fig. 10.3.

11.2.3 Local Head Loss Due to Bends

Local head loss due to bends is given by

$$\Delta h_3 = K_b h_v \tag{11.4}$$

11.2.4 Local Head Loss Due to Transitions

Normally, the completed outlet works will incorporate gradual transitions throughout. However, where the maximum head is moderate, if hydraulic losses are relatively unimportant and the dimensions of control valves, gates and pipes are relatively small compared to tunnel dimensions, semi-abrupt, converging transitions are sometimes used. Abrupt, diverging transitions are rarely designed. If hydroelectric power is involved abrupt transitions should be avoided in order to reduce the local head losses and to prevent cavitation.

A. Abrupt Transitions

  a. Abrupt expansion

$$\Delta h_4 = K_{tra} (h_{v1} - h_{v2}) = K_{tra}\, \Delta h_v \qquad (11.5a)$$

For values of K_{tra} see Table 11.3.

  b. Abrupt contraction

$$\Delta h_4 = K_{tra} (h_{v2} - h_{v1}) = K_{tra}\, \Delta h_v \qquad (11.5b)$$

For values of K_{tra} see Table 11.3.

TABLE 11.3

Hydraulic Losses - Abrupt Expansion and Contraction

Larger area/ Smaller area	K_{tra}, expansion	K_{tra}, contraction
1	0.20	0.15
2	0.35	0.24
3	0.47	0.34
4	0.56	0.39
5	0.62	0.41
6	0.66	0.41
7	0.69	0.42
8	0.73	0.42
9	0.74	0.43
10	0.76	0.43
11	0.77	0.44
12	0.78	0.44
13	0.78	0.44
14	0.79	0.44
15	0.80	0.45

B. Gradual Transitions

a. Gradual expansion

$$\Delta h_4 = K_{tra} (h_{v1} - h_{v2}) = K_{tra} \Delta h_v \qquad (11.5c)$$

It is recommended that a minimum value of 0.15 and $\theta \geq 12°$ be used for design purposes.

b. Gradual contraction

$$\Delta h_4 = K_{tra} (h_{v2} - h_{v1}) = K_{tra} \Delta h_v \qquad (11.5d)$$

For a well designed transition of this type, $K_{tra} = 10$ and $\theta \geq 12°$ (Table 11.4 and 11.5 for K_{tra} values).

TABLE 11.4

K_{tra} Values for Gradual Contraction

Contraction angle		K_{tra}
Smaller than	12°	0.10
	90°	0.50

For angles between 12° and 90° the variation is assumed to be linear.

TABLE 11.5

K_{tra} Values for Gradual Expansion

Expansion Angle	K_{tra}	
	After King (1954)	After Rouse (1950)
2°	0.03	0.02
5°	0.04	0.12
10°	0.08	0.16
12°	0.10	---
15°	0.16	0.27
20°	0.31	0.40
25°	0.40	0.55
30°	0.49	0.66
40°	0.60	0.90
50°	0.67	1.00
60°	0.72	---

11.2.5 Head Loss Due to Friction or Flow Resistance

Manning's formula is generally used for computing head losses due to friction

$$\Delta h_5 = L(nU)^2/R^{4/3} \tag{11.6}$$

where L is the length of the conduit

 n   is the coefficient of resistance

 U   is the mean velocity, and

 R   is the hydraulic radius

A roughness coefficient of n = 0.014 is usually used for well constructed concrete lined tunnels. For welded steel pipe with good alignment a value of n = 0.012 is generally used (Simons, D.B. and Şentürk, F., 1992, for different values of the coefficient, n).

In many outlet work systems, rectangular sections of short length may be required at gate sections. The friction loss through such sections is of minor consequence and the elimination of S_E is justified. Where such sections are quite long, and the determination of resulting friction losses is therefore important, S_E should be computed by employing the basic formula.

11.2.6 Head Losses and Head Required to Produce Flow through Gates and Valves

The head losses resulting from the presence of an emergency gate or valve fully open in the pressure system is of interest. In case of gates, losses are caused by the presence of a gate slot or change of shape in the hydraulic section. Both of these two singularities can be present jointly in the system. In the case of a valve the loss is caused by the change in shape of the hydraulic section with the obstruction caused by the valve's closing member even in the fully open position; usually butterfly valves or wedge gate valves are the only types of valves used for emergency closure. Such losses may be expressed as

$$\Delta h_6 = K_g h_v \tag{11.7}$$

For most types of gates the coefficient generally selected is $K_g = 0.03 \sim 0.02$; for the USBR outlet gate, where relatively small gate slots are used, $K_g = 0.02$; for the butterfly valve, $K_g \cong 0.17$. Actually little test data are available on butterfly valve losses. For large wedge gates (above 30 cm), use $K_g \cong 0.07$. For ring follower and ring seal gates, $K_g = 0$.

For regulating gates and valves, it is recommended that the hydraulic losses within the gate or valve passages plus the velocity head of the jet below the gate or valve be considered. The sum of these two values is known as the head required to produce flow, $\Delta h_6 + h_{vj} = \Delta h_7$

$$\Delta h_7 = K_{rpf} h_v{}^* \tag{11.8}$$

The velocity head is based on the area at the upstream flange of the gate or valve and K_{rpf} may be derived from the coefficient of discharge, C, of the gate or valve as follows

$$Q = CA (2g\, h_{rpf})^{1/2}$$

$$\Delta h_7 = (\frac{Q^2}{A^2})(\frac{1}{C^2})(\frac{1}{2g}) = K_{rpf}\, h_v$$

Values of C and K_{rpf} for various gates and valves are given in Table 11.6.

TABLE 11.6

K_{tra} values for local head loss for gates and values

Gate or LCV Valve type	K_{rpf} (wide open)*	C
Hollow jet	2.04	0.70
Butterfly valve	~1.56	~0.80
Howell-Bunger valve	1.385	0.85
Bulkhead gate	~1.56	~0.80
Fixed wheel gate	"	"
High pressure slide gate:		
• with slot deflector	1.42	0.84
• without slot deflectors (USBR design)	1.11	0.95
Outlet gate (USBR design)	1.04	0.98

Figure 11.3 shows a complete outlet system and various local head losses taking place in it. These head losses are shown again in Table 11.7.

---

* rpf: required to produce flow

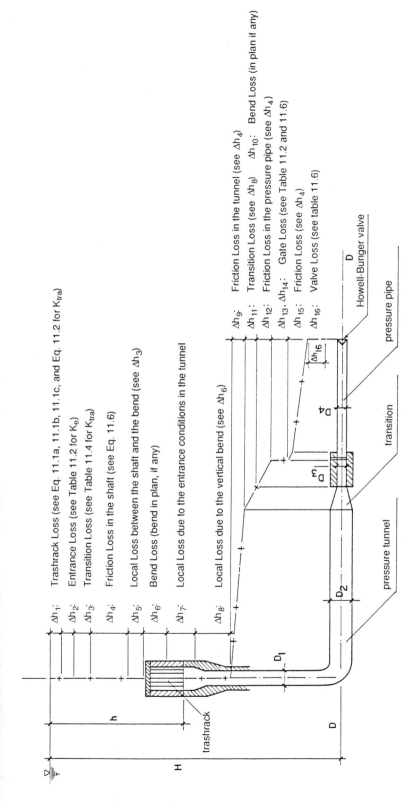

Figure 11.3. Outlet system head losses.

Δh_1:   Trashrack Loss (see Eq. 11.1a, 11.1b, 11.1c, and Eq. 11.2 for K_{tra})

Δh_2:   Entrance Loss (see Table 11.2 for K_e)

Δh_3:   Transition Loss (see Table 11.4 for K_{tra})

Δh_4:   Friction Loss in the shaft (see Eq. 11.6)

Δh_5:   Local Loss between the shaft and the bend (see Δh_3)

Δh_6:   Bend Loss (bend in plan, if any)

Δh_7:   Local Loss due to the entrance conditions in the tunnel

Δh_8:   Local Loss due to the vertical bend (see Δh_6)

Δh_9:   Friction Loss in the tunnel (see Δh_4)

Δh_{10}:   Bend Loss (in plan if any)

Δh_{11}:   Transition Loss (see Δh_8)

Δh_{12}:   Friction Loss in the pressure pipe (see Δh_4)

Δh_{13}, Δh_{14}:   Gate Loss (see Table 11.2 and 11.6)

Δh_{15}:   Friction Loss (see Δh_4)

Δh_{16}:   Valve Loss (see table 11.6)

TABLE 11.7

Head Losses Taking Place in an Outlet Work

Outlet Work elements	Local head losses
Δh_1, local loss due to trashrack	See Eq. 11.1a - 11.1b 11.1c - 11.2 for K_{tr}
Δh_2, local loss due to entrance	See Table 11.2 and Eq. 1.3 for K_e
Δh_3, local loss between the entrance and the shaft	See Tables 11.3, 11.4 and 11.5 for K_{tra}
Δh_4, friction loss in the shaft	See Eq. 11.6 for Δh_4
Δh_5, local loss between the shaft and the bend	See Δh_3
Δh_6, local loss due to bend	See Fig. 10.3
Δh_7, local loss due to the entrance conditions in the tunnel	In case that the entrance structure is coupled with the bend, $\Delta h_7 = K_{tra} (U^2/2g)$, U is the mean velocity in the bend
Δh_8, local loss due to the vertical bend	See Δh_6
Δh_9, friction loss in the tunnel	See Δh_4
Δh_{10}, local loss due to the bend in plan	See Δh_6
Δh_{11}, local loss due to the transition between the tunnel and the pressure pipe	See Δh_3
Δh_{12}, friction loss in the pressure pipe	See Δh_4
Δh_{13}, local head loss due to bulkhead gate	See Table 11.6
$\Delta h'_{13}$, local head loss for butterfly valve	See Table 11.6
Δh_{14}, local loss due to high pressure sliding gate	See Table 11.6
Δh_{15}, friction loss in the downstream pressure pipe	See Δh_4
Δh_{16}, local head loss due to Howell-Bunger valve	See Table 11.6

11.3   FLOW EQUATION AND BASIC COMPUTATIONAL APPROACHES

11.3.1   The Flow Equation for the Outlet Works

The flow equation for outlet works is given in Chapter 10, Eq. 10.23.

11.3.2   Determination of Sill Level

The sill level is the level of the dead storage in the reservoir.

The volume of dead storage corresponds to the active life of the reservoir. It is assumed to be around 100 years; and also that the dead storage volume will be completely filled with sediment during this period of time (Chapter 7, and Simons, D.B. and Şentürk, F., 1992).

Example 11.01

The allowable sediment storage in a reservoir is 500 million m^3. The known data are:

$$\gamma_S = 2.55 \text{ T/m}^3 \quad D_m = 0.3 \text{ mm} \quad S = 0.0009 \quad Q_{mean} = 750 \text{ m}^3/\text{sec}$$

and B = 90 m (the width of the river bed). (See Example 8.04).

Using Meyer-Peter and Müller formula, determine the necessary time required for the allowable storage to be completely filled with sediment.

Solution

The Meyer-Peter and Müller sediment transport formula is (Simons, D.B. and Şentürk, F., 1992)

$$\gamma dS = 0.047 \, \gamma'_s \, D_m + 0.25 \rho^{1/3} \, q_{bw}'^{2/3}$$

where
  $\gamma = 1000 \text{ kg/m}^3$ (density of water)

  $S = 0.0009$ (slope of the main river bed)

  $d = 2.50 \text{ m}$ (flow depth in the main river)

  $\gamma'_s = 1550 \text{ kg/m}^3$ (specific weight of the transported sediment measured under water), and

  $\rho = 102 \text{ kg} \cdot \text{sec.m}^4$ (specific mass of water).

Then

$$1000 \cdot 2.5 \cdot 0.0009 = 0.047 \cdot 1550 \cdot 0.0003 + 0.25 \cdot 102^{1/3} q'^{2/3}_{bw}$$

q'_{bw}, is the submerged weight of the transported sediment

$$q'_{bw} = 2.63 \text{ kg/sec.m}$$

The flow equation for the outlet works is given in Chapter 10 (Eq. 10.23)

$$Q'_{sw} = 2.63 \cdot 90 = 237 \text{ kg/sec.}$$

The dead storage will be filled in

$$\frac{500 \cdot 10^6}{\dfrac{237}{1550} \cdot 86400 \cdot 365} = 103.69 \text{ years}$$

11.3.3 Determination of the Driest and Wettest Years

In case of a failure of a fill type dam it is necessary to empty the reservoir in the shortest possible time period. The dimensions of the bottom gates or valves are a function of this emptying period. The dimensions of the outlet system may be chosen in such a way that the reservoir, assumed full, may be emptied in:

- 60 to 90 days in the wettest period
- 45 to 60 days in the driest period.

The wettest and driest periods and inflows during these periods are determined by the hydrologist.

Example 11.02

Design the outlet works for a storage reservoir with the following data:

- Diameter of the outlet tunnel   5.75 m
- Length of the tunnel   645.00 m
- Invert elevation at the entrance   709.45 m
- Bottom slope of the tunnel $4.65 \cdot 10^-$
- Basic discharge of the river   100.00 m³/sec

- Reservoir volume corresponding to elevation 715.20 of the reservoir, which is also the elevation of the highest point of the entrance section of the tunnel $0.41 \cdot 10^6 \text{ m}^3$

- Discharge for a two year return period flood Q_2 $450.00 \text{ m}^3/\text{sec}$

- Discharge for a five year return period flood Q_5 $734.00 \text{ m}^3/\text{sec}$

- Maximum dead storage elevation of the reservoir 740.00

- Volume of the dead storage $38.70 \cdot 10^6 \text{ m}^3$

- Volume of the reservoir: $V = 0.00049 \, (d+2)^{3.2379}$

- Minimum dead storage elevation of the reservoir, $V = 0, \quad 709.45$

- Flow equation for the main river, $d = 0.0647 \, Q^{0.4753}$

- Sill elevation of the energy intake 751.96

- Volume of the reservoir corresponding to elevation 751.96
   $92 \; 10^6 \text{ m}^3$

Characteristics of the intake (Fig. 11.3 and 11.4)

- Trashrack surface area $2.65 \cdot 5.75 = 37.49 \text{ m}^2$

- Distance between the entrance and the sloping tunnel   (Fig. 11.4) $= 7.96 \text{ m}$

- Dimension of the sloping shaft, cross sectional area, $3.58 \times 5.75 = 20.59 \text{ m}^2$

- Dimension of the vertical shaft, cross sectional area, $4.00 \times 4.00 = 16.00 \text{ m}^2$

- Length of the vertical shaft 8.42 m

- Angle between the vertical and sloping shafts $63°43'$

- Distance between the transition structure and the center line of the vertical shaft 2.00 m

- Length of the tunnel up to the concrete plug 310.86 m

- Length of the steel pipe in the concrete plug 39.60 m

- Diameter of the pipe in the plug 3.00 m

- Length of the transition structure relative to high pressure sliding gate 3.78 m

- Length of the Howell-Bunger valve block 16.62 m

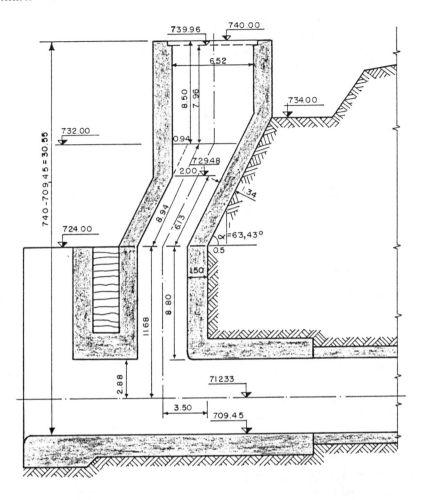

Figure 11.4. Characteristic dimensions of outlet work.

- Total length of the outlet structure (Fig. 11.6) $645 + 3.50 + 11.68 + 8.94 + 7.96 = 677.8$ m

- Wall thickness at the entrance 1.00 m
- Manning's coefficient (n) in the tunnel 0.016

  - "   "   " in the sloping shaft 0.015

  - "   "   " in the penstock 0.012

- Water surface elevation in the reservoir 815.75

- Crest elevation of the spillway 802.25

Solution

Assuming the water surface elevation in the reservoir, 815.75, the outlet system will work as follows:

1. Between elevations 815.75 and 802.25
   - the spillway
   - the energy system, and
   - the bottom valve

   discharge water.

2. Between the elevations 802.25 and 751.96
   - the energy system, and
   - the bottom valve

   are functioning.

3. Between the elevations 751.96 and 740.00
   - only the bottom outlet is functioning

Determination of the Driest and Wettest Years

For a first approximation, the method outlined below may be used.

- Consider the all existing discharge measurements at the nearest station
- A simple statistical analysis gives the results summarized in Table 11.8

Determination of the parameter (α) in flow equation

$$\alpha = \sum_{1}^{16} K_1$$

K_i values are computed according to Table 11.7

- Δh_1, local head loss due to trashrack

$$A_N = 0.60\ A_g$$

is the construction criteria for the trashrack

$$A_g = 6.52 \bullet 5.75 = 37.49\ m^2$$

$$A_N = 37.49 \bullet 0.60 = 22.494\ m^2$$

TABLE 11.8

Basic Flows to be Considered in the Design of the Outlet Works

1.	Fundamental discharge = 100 m^3/sec
2.	The driest year: 1973
3.	The wettest year: 1969

$$V(10^6 m^3)$$

Years	Month:											
	10	11	12	1	2	3	4	5	6	7	8	9
1973	6	15.9	8	11.3	66.0	53.0	64	29	9.8	4.0	2	2.1
1969	12.6	63.0	152	212	177	326	248	146	23	9	9	7.4

$$Q(m^3/sec)$$

Years	Month:											
	10	11	12	1	2	3	4	5	6	7	8	9
1973	2.24	6.11	2.99	4.22	26.9	19.8	27.4	10.9	3.8	1.4	0.8	0.8
1969	4.70	24.3	56.8	79.2	73.2	121.7	95.7	54.5	12.8	5.8	3.2	2.9

$$K_{tr} = 1.45 - 0.45 (A_N/A_g) - (A_N/A_g)^2 \qquad (11.1b)$$

$$K_{tr} = 1.45 - 0.45 \cdot 0.60 - 0.60^2 = 0.82$$

Assume that half of the trashrack can be blocked by flowing debris

$$U_N = \frac{Q}{A_N} = \frac{Q}{22.494} \text{ m/sec}$$

$$U = 2U_N = \frac{Q}{11.25} \text{ m/sec}$$

Then

$$\Delta h_1 = K_{tra} \frac{U_N^2}{2g} = 0.82 \frac{Q^2}{(22.494)^2} \frac{1}{19.62} = 8.26 \; 10^{-5} Q^2 \qquad 11.1b$$

or

$$\Delta h_1 = K_{tra} \frac{U^2}{2g} = 0.82 \frac{Q^2}{(22.494)^2} \frac{4}{19.62} = 3.30 \; 10^{-4} Q^2$$

- Δh_2, local head loss due to entrance

Assume $K_e = 0.0538$ (circular bell-mouth entrance, see Table 11.2)

$$\Delta h_2 = 0.0538 \frac{Q^2}{2gA^2} \qquad 11.3$$

$$A = 6.52 \bullet 5.75 = 37.49 \text{ m}^2$$

$$\Delta h_2 = 0.0538 \; \frac{Q^2}{19.62 \bullet 37.49^2} = 1.95 \; 10^{-6} \, Q^2$$

It is recommended that the entrance head loss be minimized. In this example a rather small coefficient is adopted. If adequate measures are not taken, K_e can increase rapidly up to $K_e = 1.00$ m.

- Δh_3, local head loss between the entrance and the shaft

$$U_1^2 = \frac{Q^2}{(37.49)^2} = 7.11 \; 10^{-4} \, Q^2$$

$$U_2^2 = \frac{Q^2}{(3.58 \bullet 5.75)^2} = 2.36 \; 10^{-3} \, Q^2$$

$$\Delta h_3 = 0.1 \, (2.36 \; 10^{-3} \, \frac{Q^2}{19.62} - 7.11 \; 10^{-4} \, \frac{Q^2}{19.62})^* \qquad \text{11.5c}$$

$$\Delta h_3 = 8.40 \bullet 10^{-6} \, Q^2$$

- Δh_4, friction loss in the shaft

$$K_f = 2g \, \frac{n^2}{R^{4/3}} \, L$$

K_f is the friction head loss coefficient

$$\Delta h_4 = SL = K_f \frac{U^2}{2g}$$

$$L = 8.80 + 8.94 + 7.96 = 25.70 \qquad\qquad R = \frac{A}{P}$$

$$A = \frac{4 \bullet 5.75 \bullet 8.80 + 3.58 \bullet 5.75 \bullet 8.94 + 6.52 \bullet 5.75 \bullet 7.96}{25.70} = 26.65 \text{ m}^2$$

$$R = \frac{\dfrac{20.59 \bullet 8.94}{2(3.58 + 5.75)} + \dfrac{23 \bullet 8.80}{2(4 + 5.75)} + \dfrac{37.49 \bullet 7.96}{2(6.52 + 5.75)}}{25.70} = 1.26 \text{ m}$$

$$K_f = 19.62 \, \frac{0.015^2}{(1.26)^{4/3}} \, 25.70 = 83.3657 \bullet 10^{-3}$$

$$\Delta h_4 = 83.3657 \bullet 10^{-3} \frac{U^2}{2g} = 83.3657 \bullet 10^{-3} \frac{Q^2}{A^2 \bullet 2g}$$

---

* See 11.2.4A for the value of K_{tra}

F. Sentürk

$$\Delta h_4 = 5.98 \cdot 10^{-6} Q^2$$

- Δh_5, local head loss between the shaft and the bend.

  This outlet system does not include such an element.

  Then

$$\Delta h_5 = 0$$

- Δh_6, local head loss due to the bend

  The bend angle is 63°43'

$$\Delta h_6 = 0.25 \frac{Q^2}{(37.5)^2} \frac{1}{2g}$$

$$\Delta h_6 = 9.06 \; 10^{-6} Q^2$$

- Δh_7, local head loss due to the entrance in the tunnel

$$\Delta h_7 = 0.42 \frac{Q^2}{[\frac{\pi 5.75^2}{4}]^2} \frac{1}{19.62} = 3.17 \; 10^{-5} Q^2$$

There is a 90° bend at the entrance, so $R_b/D = 0.50$;
Figure 10.3 yields $K_b = 0.42$.

- Δh_8, local head loss in the transition

$$A_1 = 5.75^2 = 33.06 \text{ m}^2$$

(the cross section of the shaft is rectangular)

$$A_2 = \frac{\pi 5.75^2}{4} = 25.97 \text{ m}^2$$

(the cross section of the tunnel is circular)

$$\Delta h_8 = 0.1 \frac{Q^2}{19.62} (\frac{1}{25.97} - \frac{1}{33.06})$$

$$\Delta h_8 = 4.21 \; 10^{-5} Q^2$$

- Δh_9, friction loss in the tunnel

$$\Delta h_9 = SL = \frac{n^2}{A^2} \frac{L}{R^{4/3}} Q^2$$

$$L = 645 + 3.50 - 60.00 = 588.50 \text{ m}$$

584

The reach, having a length of 3.50 m can be seen in Fig. 11.4. Obtain 60.00 m as follows:

- the length of the steel pipe in the blockage $=$ 39.00 m
- the length of the sliding gate structure $=$ 3.78 m
- the length of the Howell-Bunger valve structure $=$ 16.92 m

  Total $=$ 60.00 m

$$\Delta h_9 = \frac{0.015^2}{25.97^2} \frac{588.80}{(\frac{5.75}{4})^{4/3}} Q^2$$

$$\Delta h_9 = 1.21 \cdot 10^{-4} Q^2$$

- Δh_{10}, local head loss due to the bend in plan

  In this particular case there is no bend in plan

$$\Delta h_{10} = 0$$

- Δh_{11}, local head loss due to the transition between the tunnel and the pressure pipe

$$D_{pressure\ pipe} = 3.00\ m$$

$$U_1 = \frac{Q}{25.97}\ m/sec$$

$$U_2 = \frac{Q}{(\frac{\pi D^2}{4})} = \frac{Q}{7.07}$$

$$\Delta h_{11} = 0.15 \frac{Q^2}{2g} [\frac{1}{(7.07)^2} - \frac{1}{(25.97)^2}]^*$$

$$\Delta h_{11} = 1.41 \cdot 10^{-4} Q^2$$

- Δh_{12}, friction loss in the pressure pipe

$$\Delta h_{12} = SL = \frac{Q^2 n^2}{A^2} \frac{L}{R^{4/3}}$$

$$n = 0.012$$

---

* $K_{tr} = 0.15$ (See Table 11.4)

** See 11.2.6

$$L = 60.00 \text{ m}$$

$$A = \frac{\pi 3^2}{4} = 7.07 \text{ m}^2 \qquad R = \frac{3}{4} = 0.75 \text{ m}$$

$$\Delta h_{12} = 2.54 \cdot 10^{-4} Q^2$$

The actual length of the pressure pipe is 39.60 m but L = 60.00 m is chosen as a first approximation. The bend losses in different reaches can be computed separately for a more accurate computation.

- Δh_{13}, local head loss due to bulkhead gate[**]

  The dimension of the bulkhead gate are 2.60 • 2.25 m

  $$A = 2.60 \cdot 2.25 = 5.85 \text{ m}^2$$

  $$\Delta h_{13} = 0.19 \frac{Q^2}{19.65 \cdot 5.85^2} = 2.83 \cdot 10^{-4} Q^2$$

  $K_v = 0.19$ is assumed for this particular case

- Δh_{14}, local head loss at butterfly valve

  There is no butterfly valve in this system

  $$\Delta h_{14} = 0$$

- Δh_{15}, friction loss in the downstream pressure pipe

  Δh_{15} is included in Δh_{12}

- Δh_{16}, local head loss due to the Howell-Bunger valve

  The diameter of the valve is D = 3.00 m. Transition loss is a part of this loss and K_{HB} is chosen equal to 0.50

  $$K_{HB} = 0.50$$

  $$\Delta h_{16} = 0.50 \frac{Q^2}{(7.07)^2} \frac{1}{19.62} = 5.098 \cdot 10^{-4} Q^2$$

Then

$$\Delta h = \sum_1^{16} \Delta h_i = 1.50 \cdot 10^{-3} Q^2$$

$$\alpha = 1.5 \cdot 10^{-3} \cdot 19.62 \, A^2 \qquad \text{(Table 11.9)}$$

TABLE 11.9

Summary of Head Losses in the System

Definition of head losses	Head losses
• Trashrack	$\Delta h_1 = 8.26 \cdot 10^{-5}Q^2$
• Entrance	$\Delta h_2 = 1.95 \cdot 10^{-6}Q^2$
• Transition between the entrance and the shaft	$\Delta h_3 = 8.40 \cdot 10^{-6}Q^2$
• Friction in the shaft	$\Delta h_4 = 5.98 \cdot 10^{-6}Q^2$
• Transition from the shaft to the bend	$\Delta h_5 = 0$
• Bend in the shaft	$\Delta h_6 = 9.06 \cdot 10^{-6}Q^2$
• Tunnel entrance	$\Delta h_7 = 3.17 \cdot 10^{-5}Q^2$
• Transition to the tunnel	$\Delta h_8 = 4.21 \cdot 10^{-5}Q^2$
• Friction losses	$\Delta h_9 = 1.21 \cdot 10^{-4}Q^2$
• Bend in plan	$\Delta h_{10} = 0$
• Transition from tunnel to pressure pipe	$\Delta h_{11} = 1.41 \cdot 10^{-4}Q^2$
• Friction in the pressure pipe	$\Delta h_{12} = 2.54 \cdot 10^{-4}Q^2$
• Plate valve	$\Delta h_{13} = 2.83 \cdot 10^{-4}Q^2$
• Butterfly valve	$\Delta h_{14} = 0$
• Friction loss	Δh_{15} (included in h_{12})
• Howell-Bunger valve	$\Delta h_{16} = 5.098 \cdot 10^{-4}Q^2$
Total head losses	$\sum_i^{16} \Delta h_i = 1.5 \cdot 10^{-3}Q^2$
	$\alpha = 1.5 \cdot 10^{-3} \cdot 19.62\, A^2$

Computation of the emptying time of the reservoir

The emptying time of the reservoir can be obtained by flood routing.

Step 1: **Case 1**

Inflow =	100 m³/sec
Reservoir water surface elevation	750.00
Minimum reservoir water surface elevation	741.06

Elevation 741.06 corresponds to the hydraulic head of the tunnel for delivering Q = 100 m³/sec. For that case, the routing gives an emptying time equal to

$$8 + \frac{2}{24} \text{ days}$$

Case 2

Time necessary for the reservoir water level to drop from elevation 802.25 to elevation 741.06. The spillway crest is at elevation 802.25. The energy tunnel delivers water for reservoir elevations above elevation 750.00

$$Q_{energy} = 141.00 \text{ m}^3/\text{sec}$$

The energy tunnel is functioning and $Q = 121.7 \text{ m}^3/\text{sec}$, which is the inflow in March for the wettest year (the mean value for mid-April is 95.7 m³/sec).[*]

Result: Emptying time 54 days.

The detail of the last case is given as follows.

Data for flow routing:

1. Volume of the reservoir:

$$V = 0.00049 \ (d+2)^{3.2379}$$

The elevation corresponding to $d = 0$ is 709.45.

2. Flow equation for the main river

$$h = 0.0647 Q^{0.4753}$$

3. The flow equation

$$H = \frac{3}{4} D + \frac{Q^2}{2gA^2} (1 + \alpha) + L \frac{n^2}{R^{4/3}} \frac{Q^2}{A^2} + S_b L$$

is applied to the outlet works in the Example to obtain

$$S = 4.65 \bullet 10^{-3} \qquad\qquad L = 645.00 \text{ m}$$

$$S_b L = 3.00 \text{ m} \qquad\qquad D = 5.75 \text{ m}$$

$$H - 3.00 = \frac{3}{4} \bullet 5.75 + \frac{Q^2}{2gA^2} (1 + 1.5 \bullet 10^{-3} \ 2g \ A^2)$$

$(\frac{Ln^2}{R^{4/3}})$ is introduced in the parameter (α)

$$H = h + 740.00 - 709.45$$

then

---

[*]   If the emptying period is larger than 31 days $Q = 95.7 \text{ m}^3/\text{sec}$

$$\frac{3}{4} \cdot 5.75 + \frac{Q^2}{2gA^2}(1 + 1.5 \cdot 10^{-3}2gA^2) = 3.00 + h + (740.00 - 709.45)$$

$$A = \frac{\pi(5.75)^2}{4} = 25.97 \text{ m}^2$$

Substituting this value for A in the above equation yields

$$4.31 + 7.559 \cdot 10^{-5}Q^2 + 1.5 \cdot 10^{-3}Q^2 = 33.55 + h$$

$$h = 1.58 \cdot 10^{-3}Q^2 - 29.24$$

Using these relations the time computation relative to the emptying of the reservoir is summarized in Table 11.10.

Column 1: Time required for emptying the reservoir (chosen by the engineer)

Column 2: Time increment, Δt (10^6 sec)

Column 3: $V_i = Q_i \Delta t$ ($Q_i = 121.7$ m³/sec)

Column 4: Water surface elevation in the reservoir = 707.45 + depth (from Column 10).

The minimum depth storage elevation of the reservoir is 709.45. The lowest elevation in the reservoir from where the surface elevation is measured is 707.45.

Column 5: $V_{res} = 0.00049 \, d^{3.2379}$

Column 6: $V_{res} + \frac{V_i}{2}$

Reservoir volume increment $\Delta V = V_i/2$ is chosen to compensate the depth variation in the beginning and in the end of the month.

Column 7: $d = (\frac{V}{0.00049})^{0.3088}$ (V, is taken from Column 6)

Column 8: h = Column 7 + 707.45 - 740.00

740 is the entrance elevation of the intake

TABLE 11.10
Emptying Time of the Reservoir Using Outlet Facilities

t	Δt	Vi	Reserv. WSE	Vres	Vres + Vi/2	dres	h	Q	V	Vres	dr	Final WSE	Observations
day	10^6 sec	10^6 sec	m	10^6 m³	10^6 m³	m	m	10^6 m³	10^6 m³	10^6 m³	m		
1	2	3	4	5	6	7	8	9	10	11	12	13	14
1	0.0864	10.52	802.25	1232.8317	1238.0917	94.8658	62.3158	240.7212	34.8815	1203.2102	94.0323	801.4823	
2	0.0864	10.52	801.482	1200.7978	1206.0578	94.1009	61.5509	239.7136	34.7945	1171.2633	93.2541	800.7041	
3	0.0864	10.52	800.704	1168.9176	1174.1777	93.3257	60.7757	238.6880	34.7058	1139.4717	92.4650	799.9150	Original d_r = 94.80
4	0.0864	10.52	799.915	1137.1943	1142.4543	92.5397	59.9897	237.6436	34.6156	1107.8386	91.6646	799.1146	Q_{oe} = 163 m³/sec (energy
6	0.1728	21.03	799.115	1105.6274	1116.1424	91.8763	59.3263	236.7585	69.0783	1047.0641	90.0814	797.5314	flow discharge
8	0.1728	21.03	797.531	1044.9834	1055.4984	90.3049	57.7549	234.6488	68.7137	986.7841	88.4471	795.8971	Original water surface
10	0.1728	21.03	795.897	984.8317	995.3467	88.6833	56.1333	232.4516	68.3340	927.0127	86.7563	794.2068	elevation from where
12	0.1728	21.03	794.207	925.1860	935.7010	87.0071	54.4571	230.1583	67.9378	867.7633	85.0053	792.4533	the emptying begins=802.25
14	0.1728	21.03	792.455	866.0626	876.5776	85.2710	52.7210	227.7587	67.5231	809.0545	83.1861	790.6361	Minimum reservoir
16	0.1728	21.03	790.636	807.4838	817.9888	83.4687	50.9187	225.2407	67.0880	750.9008	81.2919	788.7419	elevation for energy
18	0.1728	21.03	788.742	749.4422	759.9572	81.5934	49.0434	222.5904	66.6300	693.3272	79.3139	786.7639	production=765.09
20	0.1728	21.03	786.764	691.9890	702.5040	79.6366	47.0866	219.7907	66.1462	636.3578	77.2245	784.6915	V_{res}, corresponding to
22	0.1728	21.03	784.692	635.1377	645.6527	77.5881	45.0381	216.8213	65.6331	580.0195	75.0517	782.5117	reservoir elevation
24	0.1728	21.03	782.812	573.9130	589.4280	75.4356	45.8856	213.6566	65.0863	524.3418	72.7586	780.2086	802.25=1232.83 10^6 m³
26	0.1728	21.03	780.209	523.3490	533.8640	73.1641	40.6141	210.6252	64.5002	469.3637	70.3120	777.7620	Q_i=121.7 m³/sec for the
28	0.1728	21.03	777.762	468.4818	478.9968	70.7545	38.2045	206.6069	63.8681	415.1288	67.6959	775.1460	wettest year (March)
30	0.1728	21.03	775.146	414.3461	424.8711.	68.1826	35.6326	202.6292	63.1807	361.6901	64.8757	772.3257	Q_i=95.7 m³/sec for the driest
32	0.1728	21.03	772.326	361.0240	369.2904	65.2938	32.7438	198.0663	62.3923	306.9018	61.6670	769.1170	year (April)
34	0.1728	21.03	769.117	306.3432	314.6132	62.1414	29.5914	192.9639	61.5106	253.1026	58.1039	765.5539	Limit WSE=740 + 5.7 • 0.45
					Energy Intake closed								=742.5975
36	0.1728	16.54	765.554	252.6483	260.9183	58.6522	26.1022	187.1542	32.3402	228.5781	56.3038	763.7538	WSE for the shaft
38	0.1728	16.54	763.754	228.1719	236.4419	56.8949	24.3449	184.1589	31.8227	204.6192	54.4111	761.8611	entrance <742.5875
40	0.1728	16.54	761.861	204.2581	212.5281	55.0521	22.5021	180.9645	31.2707	181.2574	52.4118	759.8618	C=3.5 Q=3.5 A $h^{1/2}$
42	0.1728	16.54	759.862	180.9406	189.2106	53.1115	20.5615	177.5385	30.6787	158.5319	50.2879	757.7379	A=25.97 m² (area of the shaft)
44	0.1728	16.54	757.738	158.2576	166.5276	51.0578	18.5078	173.8394	30.0395	136.4882	48.0159	755.4659	Q=95.7 m³/sec corresponds
46	0.1728	16.54	755.467	136.2548	144.5248	48.8717	16.3217	169.8132	29.3437	115.1811	45.5640	753.0140	to elevation 740.04
48	0.1728	16.54	753.014	114.9869	123.2569	46.5275	13.9775	165.3870	28.5789	94.6780	42.8877	750.3377	Reservoir water level
50	0.1728	16.54	750.338	94.5208	102.9708	43.9905	11.4405	160.4592	27.7274	75.0635	39.9209	747.3709	reaches this elevation
52	0.1728	16.54	747.371	74.9413	83.2113	41.2106	8.6616	154.8818	26.7636	56.4477	36.6574	744.0074	assymptotically
53	0.0864	8.27	747.007	56.3577	60.9428	37.3472	4.7972	146.7736	12.6812	47.8115	34.7302	742.1802	Emptying time (54) days
53	0.0432	4.13	742.180	47.7367	49.8017	35.1703	2.6203	142.0027	6.1345	43.6672	33.7713	741.2213	
					Control at the Shaft Entrance								
53	0.0216	2.07	741.221	43.5993	45.6643	34.2409	1.6908	118.1938	5.1060	40.5383	33.0098	740.4598	
54	0.0216	2.07	740.460	40.4957	42.5607	33.5047	0.9547	88.8108	3.8366	33.7241	32.5414	739.9900	
												740	

Column 9: $Q_o = (\dfrac{h + 29.24}{1.58 \ 10^{-3}})^{1/2}$ (See flow equation)

h, is taken from Column 8

Column 10: $V_o = Q_o\Delta t + 163 \ \Delta t = (Q_o + 163) \ \Delta t$

Column 11: Residual $V = V_{res} - V_o$ (Column 6 - Column 10)

Column 12: d_r = Relative depth = $(\dfrac{\text{Column 11}}{0.00049})^{0.3088}$

Column 13: Final water surface elevation = 707.45 + Column 12.

This outlet work is shown in Figs. 11.5, 11.6, 11.7 and 11.8. Figure 11.5 shows the curves of the emptying process.

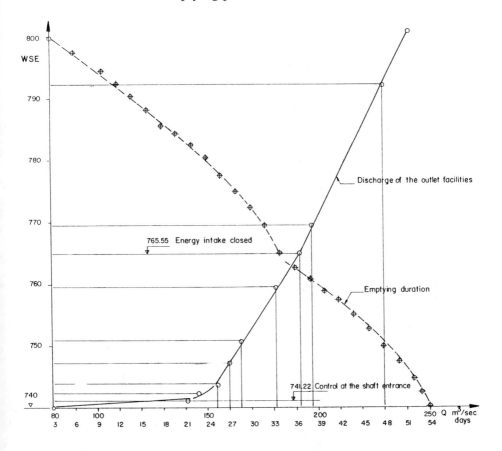

Figure 11.5. Discharge of the outlet system and emptying duration.

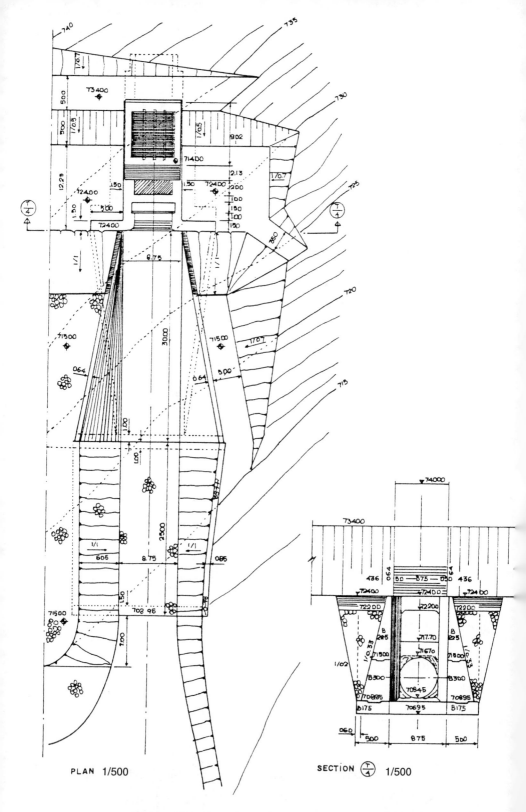

Figure 11.6. Outlet works: intake structure details.

PLAN 1/500

SECTION $\frac{T}{4}$ 1/500

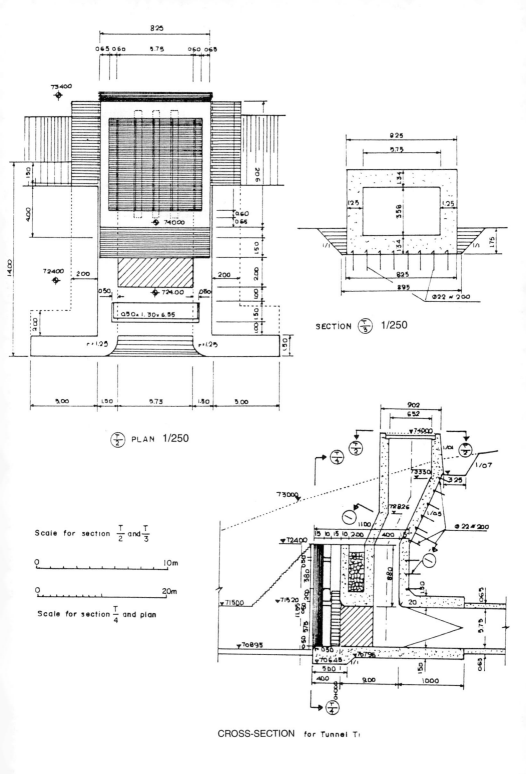

SECTION $\left(\frac{T}{3}\right)$ 1/250

$\left(\frac{T}{2}\right)$ PLAN 1/250

Scale for section $\frac{T}{2}$ and $\frac{T}{3}$

0 ————————— 10m

0 ————————— 20m

Scale for section $\frac{T}{4}$ and plan

CROSS-SECTION for Tunnel T₁

Figure 11.6. Continued.

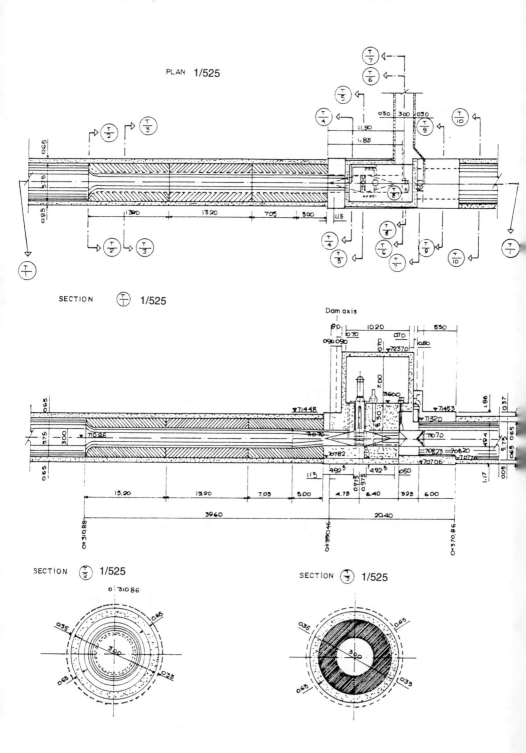

Figure 11.7. Longitudinal section and cross sections of tunnels T_1 and T_2.

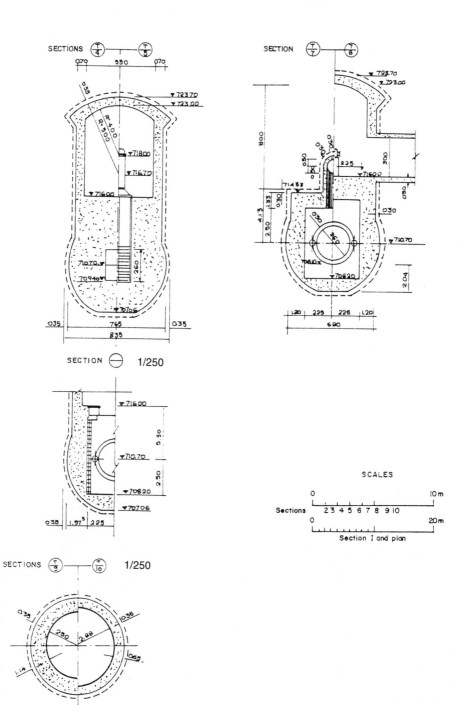

Figure 11.7. Continued.

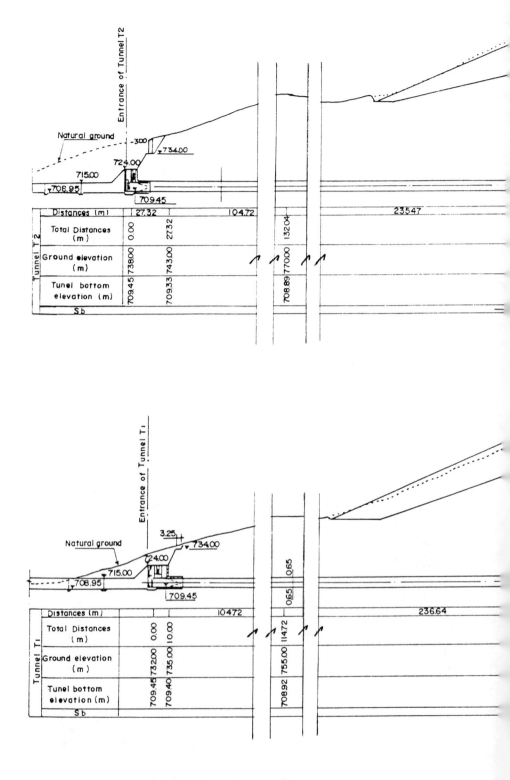

Figure 11.8. Outlet works: structure details.

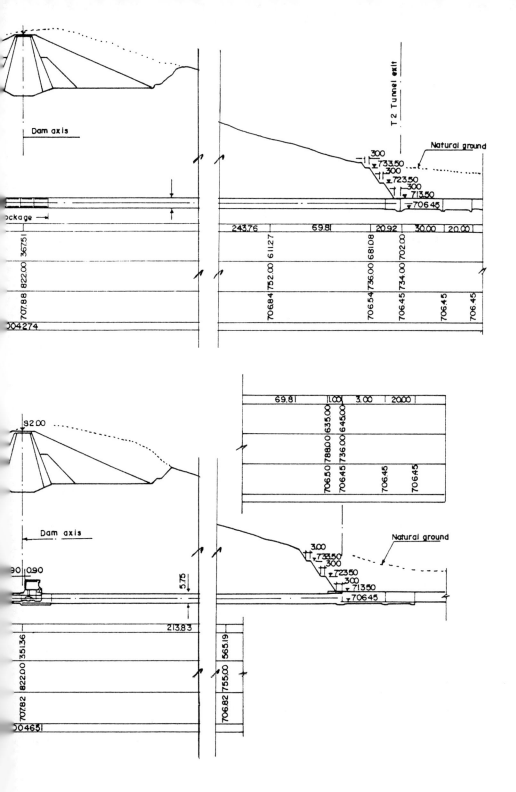

Figure 11.8. Continued.

11.4   PROBLEMS

Problem 11.01

Calculate the discharge curve for the pressure side of the outlet work shown in Fig. 11.1.

Data

• Maximum reservoir surface elevation	1182.50
• Minimum operating reservoir water elevation	1163.20
• The elevation of top of intake	1162.60
• Dead storage elevation	1158.00
• Invert elevation	1147.59
• n, concrete	0.014
• n, steel pipe	0.012
• Area of trashrack is on the basis of four side openings, and one top opening each 3.25 m square	52.80 m^2
• Diameter of the shaft	2.50 m
• Length of the shaft (1158.00 - 1147.66) (Fig. 11.1)	10.34 m
• Length of the tunnel (113.12 - 40.75)	72.37 m

(Fig. 11.1)

• Bend angle in plan	40°23'
• Bend radius in plan	25.00 m
• Bend length in plan (Fig. 11.1)	17.62 m
• Diameter of the pressure pipe	1.50 m
• Length of the pressure pipe	70.00 m

- Dimensions of the bulkhead gate: 1.65 • 1.85

- S_b 0.0101

Problem 11.02

Calculate the discharge for the lower section of the outlet work in Problem 11.01, assuming "weir control" at the sill of the inlet shaft (Fig 11.1).

Data

- Choose C

- Crest elevation of the entrance structure   1158.00

- Invert elevation   1147.59

- $D = 3.25$ m

- $P = 10.41$ m

- Diameter of the circular conduit   2.50 m

- Length of the shaft   10.34 m

- Diameter of the shaft   2.50 m

- n   0.014

- α 1.2

- Exit depth of flow $\dfrac{2.5}{2} = 1.25$ m

- Wetted area just downstream of the exit: $A_{ex} = (A_{con}/2)$ or $A_1 = \dfrac{A}{2}$

It is assumed that the circular conduit flows full but the flow is not supported as it is assumed in Problem 11.01, and the exit depth is equal to the half of the diameter.

F. Sentürk

Problem 11.03

Calculate the discharge curve for the lower section of the outlet work described in Problem 11.01, assuming orifice flow at the intake.

Problem 11.04

Show the complete variation of Q = f(H) for the intake described in Problem 11.01.

Chapter 12

DEGRADATION, AGGRADATION AND LOCAL SCOUR IN DOWNSTREAM CHANNELS AT DAMS *

12.1   INTRODUCTION

The construction of a dam on a river where a stable flow takes place for a long period of time causes a change in the regime of flow and a stability disturbance. Stable flow is a flow in which the total sediment load entering from an upstream section is discharged completely from a downstream section and in which the bottom elevations stay unchanged. The reservoir causes a reduction in the upstream transport capacity of the stream, and downstream of the dam the actual sediment load is reduced below the transport capacity of the stream due to the trap efficiency of the reservoir. In both regions the regime of sediment transport is disturbed and the stream reacts accordingly, so that aggradation takes place upstream of the reservoir and degradation downstream of it. Both conditions should be studied by the designer.

Problems arising as a consequence of aggradation upstream of the reservoir are described:

1. **Local effects**

    a. Aggradation of bed

    b. Loss of waterway capacity

    c. Change in river geometry

    d. Increased flood stage

2. **Upstream effects**

    a. Change in the base level for tributaries

    b. Deposition in tributaries near confluences

    c. Aggradation causing development of a perched river channel or a change in alignment of the main channel.

---

*   This Chapter is summarized from Simons, D.B. and Şentürk, F., 1992.

Problems arising as a consequence of degradation downstream of the dam:

1. **Local effects**

    a. Channel degradation

    b. Possible change in river form

    c. Local scour

    d. Possible bank instability

    e. Possible dam failure

2. **Downstream effect**

    a. Degradation of the reach up to a control section

    b. Reduced flood stage

    c. Reduced base level for tributaries, increased velocity and reduced channel stability causing increased sediment transport to main channel.

If water released from the reservoir is clear enough, it causes local erosion immediately downstream of the dam. Its sediment transport capacity is high enough to erode the downstream reach up to a control section where a new equilibrium condition is reached. Higher scour will then occur. This tends to deepen and widen the main river and at the same time, decrease the bottom slope. Fine material is removed from the bed and then coarse material builds up an armor coat, which tends to establish an equilibrium. This establishes a new regime in which scouring is slowed and the process of degradation is ended.

This new equilibrium condition must be predicted by the designer so that necessary measures can be taken in advance.

12.2   DEGRADATION PHENOMENA

As stated earlier, degradation occurs downstream of a dam immediately following the year of completion and the release of clear water from the reservoir. The mechanical action of flow causing degradation can be stated as follows:

a. Local scour caused by the turbulence of the flow around the hydraulic structure which crates vertical axis vortices.

Scour downstream of bridge piers and upstream of groins may be cited as an example of this kind of scour.

b. If the flow is confined, the velocity increases with constant discharge and the flow depth keeps increasing. This causes a continuous deepening of the downstream channel.

Scouring between levees is an example of this kind of scour.

c. Finally, bottom erosion along a long reach due to the change in the sediment transport regime must be considered.

12.2.1 Confined Flow

In confined flow with constant discharge, the velocity increases and the tractive force increases accordingly. As the bottom erosion takes place, the velocity decreases and the flow approaches its initial condition except when a wide cross section is replaced by a deep one. When the total load entering the confined reach leaves the exit section without any quantitative change, the erosion stops.

Considering a confined flow reach, assume that the depth of flow at the entrance section is d_1, the mean velocity U_1, and the linear sediment discharge, q_{B1}, Nordin, 1971). If the width of the cross section is B_1, then the total sediment discharge is

$$q_{B1} B_1$$

and the total liquid discharge

$$q_1 B_1$$

In the confined reach, the linear liquid discharge is shown by q_2, the mean velocity by U_2, the linear sediment discharge by q_{B2}, and the width of the cross section by B_2.

Then the total sediment discharge in the confined reach will be

$$q_{B2} B_2$$

which may be equal to

$$q_{B1} B_1$$

then

$$q_{B2} B_2 = q_{B1} B_1$$

$$q_{B2} = \frac{B_1}{B_2} q_{B1} \tag{12.1}$$

A similar computation yields

$$q_2 = \frac{B_1}{B_2} q_1 \tag{12.2}$$

Using one of the known sediment transport formulas, the variation of the sediment load can be obtained as a function of the representative particle size, the depth of flow and the mean velocity. The only unknown being the depth d, then the depth of erosion is

$$d_s = d_2 - d_1$$

Example 12.01

Determine the depth of erosion in a confined reach where

$$B_1 = 90 \text{ m} \qquad B_2 = 60 \text{ m} \qquad q'_{B1} = 1 \text{ kg/sec.m}$$

$$q_1 = 1.00 \text{ m}^3/\text{sec.m}$$

using the Meyer-Peter formula.

Data

$$\gamma'_s = 1700 \text{ kg/m}^3 \quad S = 0.001 \quad D_m = 0.0001 \text{ m} \quad \rho_s = 102 \text{ kg.sec.m}^4$$

Solution

The Meyer-Peter and Müller sediment transport formula can be written as

$$\frac{\gamma d S}{\gamma'_s D_m} = 0.047 + \frac{0.25 \ \rho^{1/3}}{\gamma'_s D_m} q'^{2/3}_{bw} \tag{12.3}$$

Substituting for the unknown by values given as data of the problem yields

$$\frac{1000 \cdot d_1 \cdot 0.001}{1700 \cdot 0.0001} = 0.047 + \frac{0.25 \cdot 102^{1/3}}{1700 \cdot 0.0001} 1^{2/3}$$

$$d_1 = \frac{0.047 \cdot 1700 \cdot 0.0001}{1000 \cdot 0.001} + 0.25 \cdot 102^{1/3} = 0.008 + 1.17 = 1.18 \text{ m}$$

and

$$q'_{B2} = \frac{90}{60} 1 = 1.50 \qquad\qquad 12.1$$

$$d_2 = \frac{0.047 \bullet 1700 \bullet 0.0001}{1000 \bullet 0.001} + 0.25 \bullet 102^{1/3} \bullet 1.5^{2/3}$$

$$= 0.008 + 1.531 = 1.54 \text{ m}$$

The deepening of the channel is

$$d_s = 1.54 - 1.18 = 0.36 \text{ m}$$

12.2.2 Overbank and Channel Flow

The study of overbank flow was performed by Laursen in 1960. If the flow can not be contained in the confining reach and overbank flow exists together with channel flow then Laursen's suggestion is

$$\frac{d_2}{d_1} = (\frac{Q_t}{Q_c})^{c_1} (\frac{B_1}{B_2})^{c_2} (\frac{n_2}{n_1})^{c_3} \qquad\qquad (12.4)$$

where

$$c_1 = \frac{6}{7} \qquad c_2 = \frac{6\,(2 - f)}{7\,(3 - f)} \qquad c_3 = \frac{6f}{7(3 - f)}$$

Q_c: the discharge upstream of the confined reach

Q_t: the discharge in the confined reach

$Q_t > Q_c$; the discharge of the overbank flow is introduced in Q_t, (n_1) and (n_2) are the Manning's roughness coefficient in and out of the channel and the values of (f) are listed below.

$\dfrac{U_*}{w}$	f
< 0.5	0.25
1.00	1.00
> 2.00	2.25

The relation is written in English units, U_* is the shear velocity and w is the fall velocity of the particle with D_{50} diameter.

Laursen (1963) defined the erosion around a bridge pier in an inundated area as follows

$$\frac{d_2}{d_1} = \left(\frac{B_1}{B_2}\right)^{6/7} \left(\frac{U_1^2}{120 d_1^{1/3} D_{50}^{2/3}}\right)^{3/7} - 1 \qquad (12.5)$$

This relation is obtained assuming a clear water flow. The variables are defined as follows:

d_2 is the limiting value of erosion in feet

d_1 is the depth of flow in feet

B_1 and B_2 are the width of the cross section in and out of the contracted and uncontracted sections respectively, in feet

D_{50} is the mean diameter of the sediment, in feet

U_1 is the velocity of the flow in the uncontracted section, in fps.

12.3   LOCAL SCOUR

Local scour is a consequence of an abrupt change in flow direction. The change of direction is accompanied by vortices causing the scour. Some scour examples encountered in engineering practice are:

• scour around bridge piers

• scour at a head structure equipped with gates

• scour at the nose of groins and embankments

• scour around guide walls

• scour downstream of stilling basins

• scour downstream of canal bends

Scour causes the lowering of a defined small portion of the channel bed below its normal level. The term "depth of scour" defines the depth of a scour hole upstream or downstream of an obstacle. Figure 12.1 shows such a scour. The depth of upstream scour can sometimes be greater than the depth of the downstream scour.

Scour may be localized at the nose of a pier or it may extend continuously around it. In the former, the scour hole is "horseshoe shaped."

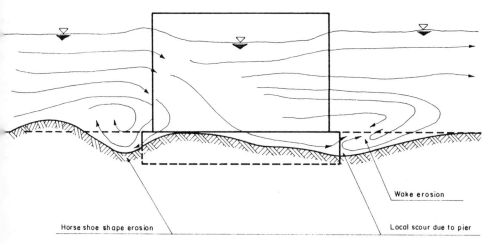

Figure 12.1. Local scour around a bridge pier.

Downstream scour at a pier is called "wake erosion." If the flow direction is not parallel to the pier axis, the resulting scour is called "skewed erosion." Figure 12.2 shows different types of erosion.

Local scour may occur in conjunction with, or in the absence of degradation, aggradation and scour due to contractions. For practical purposes three cases of scour are considered.

1. Stable scour: The sediment entering the scour hole is equal to the sediment expelled. In general, the local disturbances caused by a pier result in local scour around the pier.

2. Clear water scour: The flow of sediment into the scour hole is zero. The erosion is continuous and the depth of scour increases with time until a limiting value is reached.

3. Scour with varying sediment inflow: The flow of sediment from upstream may be smaller or greater than the rate of sediment discharge from the scour hole. If the amount of sediment entering the scour hole is greater than the amount of sediment discharged from the hole, the depth of the scour hole decreases with time; in the reverse situation the depth of the hole increases with time.

12.3.1 Mechanism of Local Scour

The problem has been investigated by many researchers such as White, (1940), Urbonas (1968), Simons and Stevens (1971). It is believed that the scour is a consequence of the combined influence of shear and lift forces.

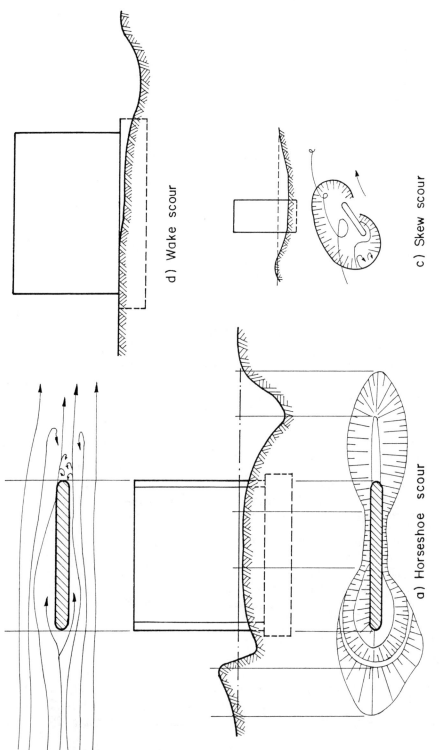

d) Wake scour

c) Skew scour

a) Horseshoe scour

Figure 12.2. Scour around bridge piers.

The influence of one or both of these two forces depends on local conditions and results in the development of the scour hole in a vertical or horizontal direction. The following paragraphs explain the procedure for performing computations.

12.3.2   Evaluation of Local Scour

The bed of a river is subject to a general degradation downstream of a newly constructed dam. In addition to the overall scouring effect additional scouring occurs downstream or at local obstacles such as spurs, noses, bends, and downstream of stilling basins. The magnitude of such local scour is a function of:

- the slope of the river bottom

- the geometric and hydraulic characteristics of the cross section

- the characteristics of bed material such as the granulometry, specific weight, etc.

- the characteristics of water turbidity and density - the characteristics of the flood hydrograph and history of floods

- the characteristics of floating debris such as logs, branches, ice, etc.

- the characteristics of man-made hydraulic structures, weirs, stilling basins, and similar engineering works (constrictions, bridge piers, spurs, lining, revetments, training works).

Many investigations of scour have been conducted in the field and in the laboratory; but the diversity of the variables involved have prevented the derivation of a world-wide general scour formula. Only relations using a limited number of parameters have been suggested.

The basic factors involving the scour phenomena are the velocity of flow and the shear imposed by the flow. These two parameters can be evaluated and measured. The effect of turbulent fluctuation cannot be measured, it can only be analyzed qualitatively. Then the one or the other of the measurable quantities or both may be introduced in scour formulas together with measurable characteristics of the bed material. In most cases attempts can be made for evaluating the scour by using physical models. The accuracy of model investigation is dubious (see Chapter 13) and subject to subjective interpretation.

The resistance of the bed material to scouring is a function of the grain size, size distribution, specific weight of the solid particles, shape factor of the grains and their packing arrangement. The influence of the cohesion on the scour has also been investigated but a comprehensive result has not been reached. For non-cohesive bed material, variables such as shape factor and packing factor are not taken into account due to the complexity of their composition. It is also true that in model analysis these factors are also excluded, preventing the hydraulic engineer from reaching a conclusive judgment on the scour phenomenon. Under these circumstances a complete theoretical approach of the scour problem seems illusionary. However, Nell (1973) suggested a trial and error procedure capable of guiding the designer to the first step in the evaluation of scour between bridge piers. Laursen and Flick (1983) more recently suggested an approximate formula for predicting scour depth. Carstens (1966) defined a sediment number N_s to define scour phenomena

$$N_s = \frac{u}{[(\frac{\gamma_s}{\gamma} - 1) gD_g]^{1/2}} \qquad (12.6)$$

where   u is the fluid velocity adjacent to the bed

γ_s is the specific weight of the sediment particles

and D_g is the typical grain diameter of the surface solid material.

LeFeuve (1965), seeing the difficulty of computing or measuring the fluid velocity adjacent to the bed suggested substituting the mean velocity (U) for (u) since Eq. 12.6 is rather approximate, this substitution facilitates its use greatly. LeFeuve conducted research showing that the results from Carstens's equation are not very much altered by his suggestion. Then, substituting U for u yields

$$N_s = \frac{U}{[(\frac{\gamma_s}{\gamma} - 1) gD_g]^{1/2}} \qquad (12.7)$$

where D_g is the geometric mean size of the sediment mixture. Note that D_g is defined as:

$$D_g = (D_{84.1} \bullet D_{15.9})^{1/2}$$

Then, from a well known S grainsize distribution curve, $D_g = D_{50}$; its influence on the variation of N_s is in the second order.

Altinbilek (1971) and Altinbilek and Okyay (1973), starting from the general form of the continuity equation

$$t = \int_o^s \frac{dV}{Q_s} \tag{12.8}$$

have suggested another approximate formula. In Eq. 12.8

V   is the scour hole volume

t   is the time

S   is the depth of scour, and

Q_s is the sediment discharge from the scour hole.

A.   Scour Around Controlled Waterway Openings

Şentürk (1972) stated that the first step in solving a scour problem involves the choice of the right location for the hydraulic structure and the dimensions of openings between the piers in the case of bridges and gated diversion dams and head structures. If this choice is erroneous, the river response is prompt; either it changes its course or it erodes its bed so greatly that the hydraulic structure is deeply buried in it. Two interesting examples can be cited from Turkey.

1.   The Gediz River in the Western region of Anatolia refused to flow between the piers of Emiralem Weir, left its original course and excavated a new bed 1500 m away from the hydraulic structure.

2.   In the same region, after a large flood, the drops on the Milas Tekfürambari drainage canal were so deeply buried in the sand that engineers had difficulty finding them.*

Today's knowledge of the hydraulics of erodible channels allows the designer to bring a better solution to similar problems.

---

*   These drops and the drainage channel were designed in 1955 by the author of this book. The errors involved in the design were the lack of
  • a full understanding of the flow characteristics
  • an adequate analysis of the channel bed material, and
  • the wish of the designer to inflict his desire on the river, to which the river's reply was terrible.

F. Sentürk

Many bridges all over the world have failed because of an improper location, an incorrect axis orientation, inadequately sized openings between the piers. As a result of such failures the old bed of river is dry and a new bridge has been constructed near the old one to serve traffic.

A good approach for solving such problems is to study the history of floods coupled with a physical model study of exceptional floods to determine the flow direction relative to the axis of the proposed structure for immediate and for high discharges (Article 13.2.5.E). The selection of the direction of the axis of the hydraulic structure will be discussed in more detail together with river characteristics.

Bridge piers are directly subject to scour. Similar scouring occurs downstream of gated weirs also, although these structures are often assisted by the addition of downstream stilling basins. Scour is a function of

- the opening between piers

- the shape of the piers

- the flow characteristics, and

- the bed material properties of the waterway.

The opening between bridge piers can be determined by using d'Aubuisson's formula (Chow, 1953) or by direct application of basic hydraulic principles (Şentürk, F., 1957).

The chart of Fig. 12.3, prepared by using Lacey's regime formula, can also be used for the determination of the pier opening

$$P = CQ^{1/2}$$

where

P is the wetted perimeter

Q is the water discharge, and

C $= \dfrac{8}{3}$ or   C = 1.8

Nell. (1973) suggested C = 8/3 for shifting channels in sandy materials and C=1.8 for channels formed in more stable material. When the value of P is selected according to Fig. 12.3, a physical model study is helpful to check the validity of the assumption adopted.

The location of the hydraulic structure in a selected reach of the river, if properly made, resolves many problems (also Chapter 13, subsection

13.2.5E). Local scours will occur but their magnitude will be smaller; these scours can be defined as follows

- Degradation of the river bed due to the retention of sediment by the dam upstream of a weir, and downstream of the weir, if it also traps sediment.

The choice of the location of the weir is a function of the overall degradation. But if the degradation begins downstream of the stilling basin, adequate measures may be taken to preserve the stability of the structure. Riprap lining has often been chosen for preserving the stability of the streambed. Weirs constructed on tributaries in Italy have been built to form a series of steps to restrain the progressive erosion of the bed. River degradation is studied in section 12.4.

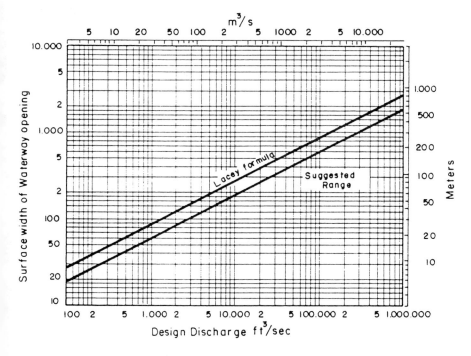

Figure 12.3. Chart for selecting a trial pier opening width.

- Scour due to constriction. This kind of scour can be prevented by riprap protection.

- Scour around bridge, abutments, approach embankments and guide walls. Often, engineering measures needed to prevent this kind of scour can be implemented only after some scour has taken place (subsection 13.2.5).

Nell's (1973) approach is summarized for the evaluation of scour due to constriction:

1. Determine the mean velocity U, between the bridge piers corresponding to Q_{max}. No scour is considered in computing U.

2. Determine the competent velocity.

   The competent velocity is, after Nell., a limiting velocity defined in Fig. 12.4 for a bottom covered with loose material. Table 12.1 can be used for determining the competent velocity for cohesive material.

3. Compare the mean velocity U, with the competent velocity U_{comp}. If the mean velocity is greater than the competent velocity, scour should be expected.

4. The only solution may be to decrease the mean velocity so that $U < U_{comp}$. Different ways of decreasing the mean velocity are:

   • Increasing the water depth between the piers by lowering the bottom elevation

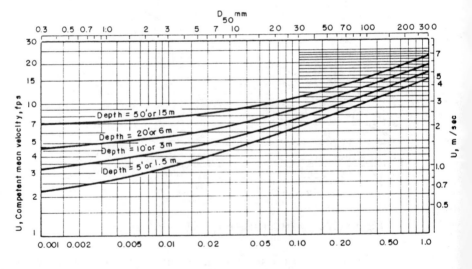

Figure 12.4. **Suggested competent mean velocity for significant bed movement of cohesionless material, in terms of grain size and depth of flow, after Nell (1973).**

TABLE 12.1

Estimate of U_{comp} for Cohesive Soil after Nell, (1973)

Depth of flow		Competent mean velocity					
		Easily erodible material		Average erodibility		Resistant material	
ft	m	fps	m/sec	fps	m/sec	fps	m/sec
5	1.5	1.9	0.60	3.4	1.0	5.9	1.8
10	3.0	2.1	0.65	3.9	1.2	6.6	2.0
20	6.0	2.3	0.7	4.3	1.3	7.4	2.3
50	15.0	2.7	0.8	5.0	1.5	8.6	2.6

- Increasing the clearance between piers if feasible.

B. Local Scour around Piers

Local scour around piers is a function of:

1. The geometric characteristics of the pier:

   - the length (ℓ) of the pier

   - the width (a) of the pier,

   - the angle of attack (α) in the vertical plane

   - the angle of attack (β) in the horizontal (Fig. 12.5)

2. The characteristics of the flow; the upstream depth d_1, the Froude number F_{r1}, upstream of the hydraulic structure.

3. The characteristics of the bed material

   - The standard deviation (σ) of the sediment mixture
   - the mean size D_{50} of the sediment

4. The characteristics of the scour hole

   - the depth S of the scour, and

615

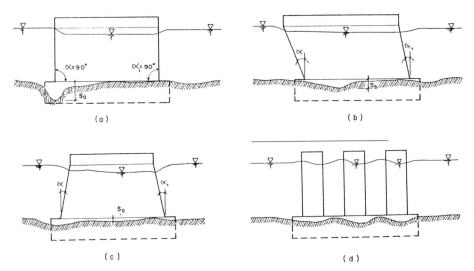

Figure 12.5. Characteristics of the piers.

- the terminal depth S_b of the scour hole.

Pier characteristics will be used as a parameter in deriving scour formulas.

a. Scour Formulas

Many authors have suggested different relations to approximate the depth of the scour hole; some will be summarized here.

The continuity equation stating that the quantity of sediment entering the hole may be equal to the quantity of sediment discharging from it, in order that the scour volume remains constant is given by Eq. 12.8. More explicitly this relation can be written as

$$\frac{df(S)}{dt} = Q_{si} - Q_{so}$$

where

$f(S)$ is the mathematical expression of the scour hole volume

t is the time

Q_{si} is the sediment discharge entering hole,

and Q_{so} is the sediment discharge leaving the hole.

This general form is difficult to derive due to the difficulties in expressing the boundaries of the scour hole mathematically. However it is possible to write

$$f(S) = (\frac{U}{\sqrt{\frac{\tau_c}{\rho}}}, \sigma, \frac{Ut}{a}, \frac{Ua}{v}, \frac{U^2}{ga}, \frac{b}{a}, Q_{so})$$

where

f(S) is a dimensionless function of the scour hole

U   is the mean velocity

a   is a characteristic length describing the size of the system

τ_c is the critical tractive force applied to the sediment subject to scouring

σ is the standard deviation of the particle size distribution

t   is the time involved in the scouring process

v   is the kinematic viscosity

g   is the gravitational acceleration

ρ is the density of water, and

b   is a typical length of the flow pattern or scour geometry

An explicit relation including the total of these variables is not available as of this date, but approximate formulas starting from this general form have been suggested.

Carstens starting with Eq. 12.8 proposed Eq. 12.9 for computing the scour around a vertical cylinder

$$4.1 \cdot 10^{-6} (N_s^2 - N_{sc}^2)^{5/2} (\frac{D_g}{D})^{5/2} (\frac{D_g}{D})(\frac{Ut}{D}) = \frac{(\frac{S}{D})^5}{\tan\varnothing} + \frac{(\frac{S}{D})^4}{16} -$$

$$\frac{\tan\varnothing(\frac{S}{D})^3}{24} + \frac{\tan\varnothing(\frac{S}{D})^2}{32} - \frac{(\tan\varnothing)^3(\frac{S}{D})}{32} + \frac{(\tan\varnothing)^4}{64} Ln(\frac{2(\frac{S}{D})}{\tan\varnothing} + 1) \quad (12.9)$$

where

N_s   is the sediment number

N_s is the sediment number

N_{sc} is the sediment discharge at zero transport

D   is the diameter of the cylinder

Ø   is the angle of repose of the sediment mixture (submerged)

t   is the time, and

S   is the depth of the scour hole.

Figure 12.6 illustrates the variation of (S/D) defined by Eq. 10.9. The terminal scour depth is defined as

$$\frac{S_t}{D} = 0.546 \left(\frac{N_s^2 - 1.64}{N_s^2 - 5.02} \right)^{5/6}$$ (12.10)

where S_t denotes the scour depth.

Maza Alvarez and Sanchez (1964) modified Jaroslovitsiev's (1960) equation and proposed

$$\frac{S}{a} = K_1 K_2 \frac{U^2}{ga} - \frac{30 D_g}{a}$$ (12.11)

where K_1 and K_2 are parameters defined in Table 12.2 and 12.3.

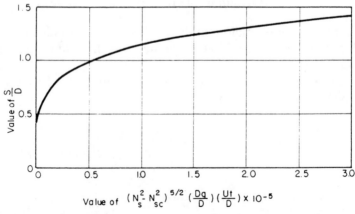

Figure 12.6. Scour depth versus time (vertical cylinder).

TABLE 12.2

Values of coefficient, K_1

Shape of pier		Values of K_1				
β =	$0°$	$10°$	$20°$	$30°$	$40°$
Circular	10.0					
Prismatic	8.5	8.7	9.0	10.3	11.3

β is the skew angle

TABLE 12.3

Values of coefficient, K_2*

d_1/a	$U^2/2g$ =	0.25	0.15	0.10	0.04
1.8		0.75	0.95	1.15	1.40
2.2		0.75	0.95	1.10	1.35
2.6		0.75	0.90	1.10	1.30
3.0		0.75	0.90	1.10	1.30

d_1 is the flow depth   a is the pier thickness

- Breusers's suggestion (1965)

$$S_b = 1.4 \, a \qquad (12.12)$$

where a is a variable geometric characteristic of the pier defined in Fig. 12.7.

- Larras's suggestion

$$S_b = 1.42 \, K \, a^{0.75} \qquad (12.13)$$

where

$$K = 1.0 \text{ for cylindrical pier}$$

---

* approximate

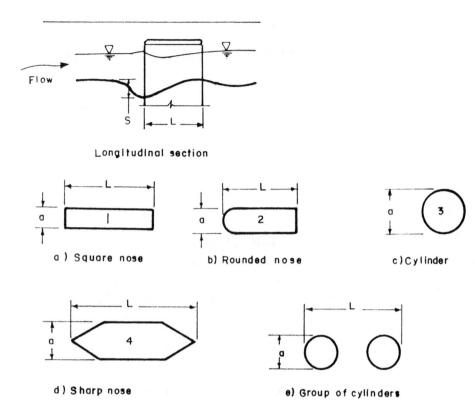

Figure 12.7. Common pier shapes.

K = 1.4 for rectangular pier and for $\beta = 0$

β is the skew angle

- Nell's suggestion (1973):

$$S = ka \qquad (12.14)$$

with k given in Table 12.4.

- Colemans' s suggestion (1971)

$$\frac{U}{(2gS)^{1/2}} = A \left(\frac{\rho Ua}{\mu}\right)^{9/10}$$

where

μ is the dynamic viscosity of water

TABLE 12.4

Local scour allowances for a pier aligned parallel to the flow.

Pier shape in plan	Suggested allowance for local scour
	(k)
Pier 1	2.0
Pier 2	1.5
Pier 3	1.5
Pier 4	1.2
Pier 5	1.2

ρ is the density of water

a is the pier thickness

$$A = 0.6 \left(\frac{\mu}{\rho g^{1/2} a^{3/2}}\right)^{9/10}$$

Substituting its value for A yields

$$S = 1.49 \, a^{9/10} \left(\frac{U^2}{2g}\right)^{1/10} \tag{12.15}$$

For $U^2/2g = 1$, and $a^{9/10} \cong a$, and if 1.4 is substituted for 1.49 for rectangular piers, S reduces to

$$S = 1.4 \, a \tag{12.12}$$

which was given by Breusers in 1965.

b.   **Influence of geometry of cross section of the pier on the scour hole**

Figure 12.7 summarizes the geometry of piers commonly used in practice. The depth of scour at the nose of the pier no. 1 is given by Jain and Fisher

$$\frac{S}{d_1} = 2.2 \left(\frac{a}{d_1}\right)^{0.65} F_{r1}^{0.43} \tag{12.16}$$

where d_1 is the depth of flow upstream of the pier, and

F_{r1} is the Froude number upstream of the pier.

For Pier no. 3.2, Eq. 12.16 takes the form:

$$\frac{S}{d_1} = 2.0 \, (\frac{a}{d_1})^{0.65} \, F_{rl}^{0.43} \tag{12.17}$$

Table 12.5 outlines the reduction in scour depths due to the geometry of the cross section.

TABLE 12.5

Reduction in scour depth.

Type of pier	S/S_*	
Square nose	1.00	
Cylinder	0.90	
Round nose	0.90	S_* is the ratio of the
Sharp nose	0.80	modified pier nose to the square nose.
Group of cylinders	0.90	

Nell suggests that if a pier of the shape of Fig. 12.5b is chosen, the depth of scour hole increases according to the ratio

$$\frac{S_b}{S_a} = \frac{2.0}{1.5} \tag{12.18}$$

and if a pier of the shape of Fig. 12.5c is chosen the depth of scour hole decreases according to

$$\frac{S_c}{S_a} = \frac{1.0}{1.5} \tag{12.19}$$

c. **Influence of the geometry of the longitudinal section of pier on the scour hole**

The characteristics of the longitudinal section of the pier are:

- the angle of attack (α), and
- its length.

The angle (α) reduces or increases the pressure exerted by the flow on the bed and the length of the pier affects the flow profile.

For streamlined piers, the scour depth decreases slightly.

d. Influence of the Skew of the Pier on the Scour Depth

The skew is defined by the horizontal angle of attack (β). It is desirable to have $\beta = 0$ in general but it is not always possible. The influence of the skew of piers on the scour hole is shown in Table 12.6.

TABLE 12.6

Multiplying factor for depth of scour S for the skew.

β	Length to width ratio of pier			
	4	8	12	16
0	1.0	1.0	1.0	1.6
15	1.5	2.0	2.5	3.0
30	2.0	2.5	3.5	4.5
45	2.5	3.5	4.5	5.0
60	2.5	3.5	4.5	6.0

C. Local Scour around Embankment

Often, bridges with piers and embankments are located immediately downstream of a dam or upstream of a reservoir. Figure 12.8 shows the scour at the embankment and the adjacent pier. This scour can be so large that the pier can collapse and the bridge with it. Figure 12.9 is a typical example of this case. The horseshoe shape around the pier can be seen together with the deep scour at the upstream front of the side pier. The importance of the bridge location can also be seen on this figure. The bridge is located immediately downstream of a natural bend of the river.

The main force of the current was directed against the side pier, scouring the subgrade and failure occurred. A comprehensive mathematical approach of the phenomenon is not available, only laboratory investigations and some in-situ observations are available.

Liu et al. (1961) suggested that local scour equilibrium can be determined using Eq. 12.20

$$\frac{S}{d_1} = 1.1 \ (\frac{a}{d_1})^{0.4} \ F_{r1}^{0.33} \tag{12.20}$$

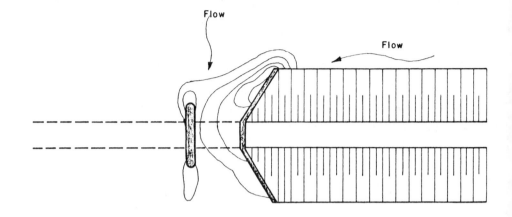

Figure 12.8. Scour at the embankment and the adjacent pier.

Figure 12.9. Typical bridge failure due to the scour at the embankment and the adjacent pier.

where

a   is the embankment length measured normal to the wall of the flume

d_1 is the upstream depth of flow, and

F_{r1} is the upstream Froude number

$$F_{r1} = \frac{U_1}{(gd_1)^{1/2}}$$

For a wing wall, Liu (1961) and Gill (1972)'s suggestion is

$$\frac{S}{d_1} = 2.15 \, (\frac{a}{d_1})^{0.4} \, F_{r1}^{0.33} \tag{12.21}$$

In-situ investigations on rock dikes on the Mississippi River indicate that the relationship

$$\frac{S}{d_1} = 4 \, F_{r1}^{0.3} \tag{12.22}$$

yields an acceptable estimate for scour depth.

Scour depth is difficult to evaluate and measure in nature due to the lack of a stable bottom surface. Large dunes are in motion on the bottom; they fill the scour hole momentarily then the scouring force of the flow reestablishes the scour hole. The researcher is unsure of the state of scour; he has to repeat measurements during the year for many years. As a first approximation it is recommended that Eq. 12.20 be used for $0 < a/d_1 < 25$ and Eq. 12.22 for $a/d_1 > 25$. Figure 12.10 illustrates this approximation.

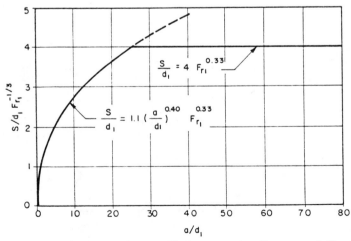

Figure 12.10. Recommended prediction equation for scour defined in Figs. 12.8 and 12.9.

F. Sentürk

Example 12.02*

Determine the clear distance between piers of a bridge that will allow a discharge of 500 m³/sec without any dangerous scour.

Compute the total scour around the piers and design the bridge.

Data

- Design discharge = 500 m^3/sec

- The stage-discharge curve is given in Fig. 12.11

- D_{50} = 1.00 mm

- γ_s = 2.7 T/m^3

- The bottom elevation of the riverbed between the piers = 0

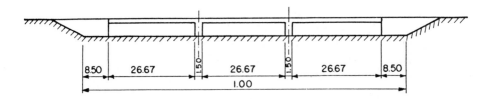

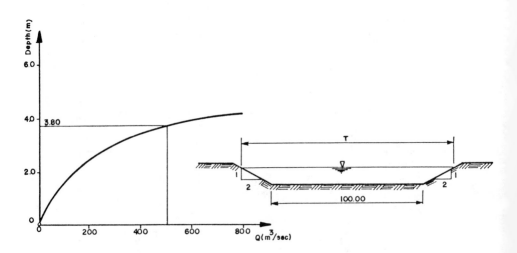

Figure 12.11. Data for Example Problem 12.02.

---

* This example is taken from Simons and Şentürk (1976) and reproduced here after revision.

- The cross section of the river where the bridge will be located is given in Fig. 12.11.

- $\alpha = 20°$.

Solution

Following steps are taken for solving the problem as suggested by Nell.

Step 1: Estimate the net surface width for a discharge of 500 m³/sec using Fig. 12.3. The obtained value is variable between 70 m and 100 m. T = 80.00 m is chosen for the first trial.

Step 2: Following Nell's fourth approach, estimate the general scour which can take place between the piers.

- The depth of flow corresponding to Q = 500 m3/sec is d = 3.80 m (Fig. 12.11).

- The wetted area is then

$$A = 3.80 \cdot 80.00 = 304 \ m^2$$

and

$$U = \frac{Q}{A} = \frac{500}{304} = 1.64 \ m/sec$$

- Using Fig. 12.4 determines the depth of flow with a 10% increase in the competent flow velocity U_{comp}= 1.20 m/sec,

$$1.1 \cdot 1.2 = 1.3 \ m/sec$$

The corresponding safe depth is 5.0 m, approximately.

For this condition the scour will be S = 5.00 - 3.80 = 1.20 m. The corresponding value of U is then

$$U = \frac{500}{5.00 \cdot 80} = 1.25 \ m/sec$$

Step 3: Determine the scour around the embankments using Fig. 12.10.

Assume an embankment length (a) equal to 8.50 m from both sides of the channel.

The computations in Step 3 is conducted using Eq. 12.21 and Fig. 12.10.

$$\frac{a}{d_1} = \frac{8.5}{3.8} = 2.24$$

and (Fig. 12.10)

$$\frac{S}{3.80} F_{r1}^{-1/3} = 1.5 \qquad S = \frac{1.5 \cdot 3.8}{F_{r1}^{-1/3}} = 1.5 \cdot 3.8 \, F_{r1}^{1/3}$$

The Froude number can be written as:

$$F_{r1} = \frac{U_1}{(gd_1)^{1/2}} = \frac{Q}{A_1 (gd_1)^{1/2}} = \frac{500}{A_1 (gd_1)^{1/2}}$$

where A_1 is

$$A_1 = (100 + 23.8) \, 3.8 \cong 409 \text{ m}^2$$

Substituting this value of A_1 in the Froude number relation yields

$$F_{r1} = \frac{500}{409 \, (3.8g)^{1/2}} = 0.20$$

and $\quad S = 1.5 \cdot 3.8 \cdot 0.20^{1/3} = 3.35 \text{ m}$

Step 4: Determine the scour depth at the pier assuming

1. no lateral flow

2. the shape of the pier in cross section is cylindrical with a circular nose, a = 1.50 m

3. the shape of the pier in the vertical plane is as given in Fig. 12.5b with $\alpha = 20°$.

The scour phenomena being very complicated, the mean value of the results of existing formulas are selected for the solution of Step 4.

1. Carstens' approach

$$N_s = \frac{U}{[(\frac{\gamma_s}{\gamma} - 1) gD_g]^{1/2}}$$
12.7

U is the mean velocity of the flow between the piers. It is not constant and varies with the scour. As a first step the mean velocity between the piers is assumed constant and this value is used in Eq. 12.7.

$$N_s = \frac{1.64}{(1.7 \cdot 9.81 \cdot 0.001)^{1/2}} \cong 12.699 \cong 13.00$$

$$\frac{S_t}{D} = 0.546 \left(\frac{N_s^2 - 1.64}{(N_s^2 - 5.02)}\right)^{5/6}$$
12.10

where D is the pier thickness, then

$$S_t = 1.5 \cdot 0.546 \left(\frac{13^2 - 1.64}{13^2 - 5.02}\right)^{5/6} = 0.83 \text{ m}$$

2. Maza Alvarez-Sanchez's approach

$$\frac{S}{a} = K_1 K_2 \frac{U^2}{ga} - \frac{30 D_g}{a}$$
12.11

Determine K_1 and K_2 from Table 12.2 and Table 12.3 respectively as

$$\frac{d_1}{a} = \frac{3.80}{1.50} = 2.54 \qquad \frac{U^2}{2g} = \frac{1.64^2}{2 \cdot 9.81} = 0.137$$

$$K_1 = 8.50 \qquad K_2 = 1.10$$

and

$$S = 1.5 \left(8.50 \cdot 1.10 \frac{1.64^2}{9.81 \cdot 1.5} - \frac{30 \cdot 0.001}{1.5}\right) = 2.53 \text{ m}$$

3. Breusers's approach

$$S_t = 1.4 \cdot 1.5 = 2.1 \text{ m}$$
12.12

4. Larras' approach

$$S_t = 1.42 \cdot 1 \cdot 1.5^{0.75} = 1.92 \text{ m}$$ 12.13

5. Nell's approach

Using Table 12.4

$$S_t = 1.5 \cdot 1.5 = 2.25 \text{ m}$$ 12.14

6. Using Eq. 12.16

$$S = 3.80 \, [2.0 \, (\frac{1.5}{3.8})^{0.65} \, (\frac{1.64}{(9.81 \cdot 3.8)^{1/2}})^{0.43}]$$

$$S = 2.36 \text{ m}$$

Eliminating the smallest value, the mean value of the scour hole depth becomes

$$S = \frac{2.53 + 2.10 + 1.92 + 2.25 + 2.36}{5} = 2.23 \text{ m}$$

Correcting for vertical skew yields

$$S_b = S \frac{2}{1.5} = 2.23 \cdot \frac{2}{1.5} = 2.98 \cong 3.00$$ 12.18

Then the scour due to the piers is estimated to be 3.00 m. This value must be added to the scour due to constriction (Fig. 12.12).

If the bridge is located near the sea, as high flows and low tides occur, an increase in average velocity at the bridge site may cause additional scour. The bridge may collapse if this condition persists. Many coastal bridges in Italy, France, Turkey, etc. have collapsed because of an inadequate depth of foundation resulting from inadequate scour analysis.

D. Local Scour Downstream of Hydraulic Structures

a. Flow conditions downstream of hydraulic structures

Scour downstream of hydraulic structures such as stilling basins, diversion works, etc., constitute an important field of investigation and

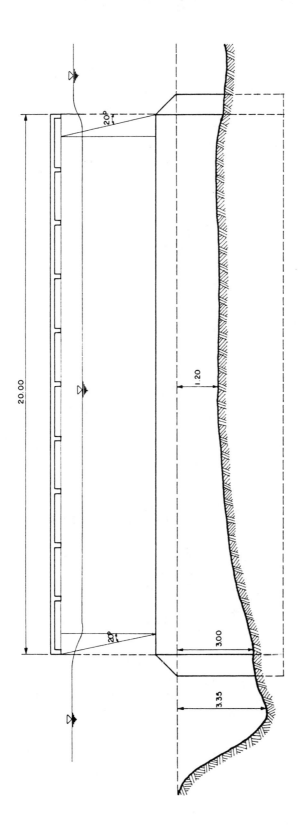

Figure 12.12. Solution of Example 12.02.

research because of its frequent occurrence in engineering applications. Several solutions to this problem were proposed in Europe by Schoklitsch (1932), Eggenberger (1943), Jäger (1939) and more recently by Altinbilek and Okyay (1973), etc., and in the USA by Simons and Stevens. (1971), Laursen (1980), etc.

Figure 12.13 summarizes the possible flow conditions downstream of hydraulic structures.

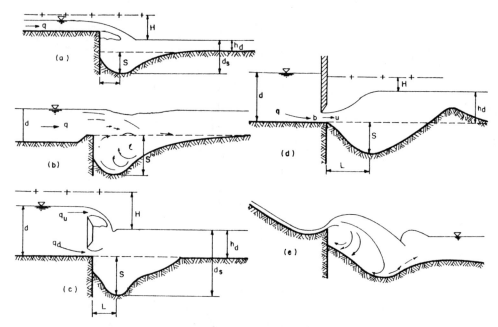

Figure 12.13. Different kind of erosion that can take place downstream of hydraulic structures and gates.

A universal flow formula does not exist that is capable of handling the different flows shown in Fig. 12.13. The existing approaches that can be used for estimating the local scour are summarized here.

b. Determination of depth of scour downstream of hydraulic structures

Early scour formulas proposed by Schoklitsch, Eggenberger, Jäger, Müller, etc. have only historical importance today.

Carstens (1966) using Laursen's (1952) approach defined the depth of scour, for the flow shown in Fig. 12.13d as

$$(\frac{S}{b})^6 = 2.85 \cdot 10^{-3} (N_s^2 - 4)^{5/2} \tan\varnothing (\frac{D_g}{b})(\frac{ut}{b}) \qquad (12.23)$$

where

S   is the depth of the scour hole

b   is the jet thickness in Laursen's experiments

N_s is Carstens' sediment number

Ø   is the angle of repose of the submerged bed material

t   is the time in seconds

D_g is the diameter of particle equal to the geometric mean of sediment mixture

and

u   is the reference velocity.

Valentine (1967) assumed that S is a function of the Froude number F_r and proposed for flows defined in Fig. 12.13d, the following relationship

$$\log \left(\frac{S}{b'}\right) = \frac{F_{rb} - 2.0}{4.7} + 0.55 \log \frac{b'}{D_{90}} \qquad (12.24)$$

where

F_{rb} is the Froude number computed at the vena contracta, and

b'   is the thickness of the vena contracta.

Altinbilek and Okyay (1973) stated that the terminal scour S_t, regenerated by a two-dimensional vertical jet, is independent of time. They proposed the following simple relation to compute it

$$\frac{S_t}{b} = \frac{30.5 \left(\frac{U}{w}\right)^2 \left(\frac{D_g}{b}\right)}{F_r \left(\frac{h}{b}\right)^{0.5}} \qquad (12.25)$$

where

S_t is the terminal depth of the scour hole

b   is the thickness of the jet

U   is the velocity of water issuing from the nozzle

w   is the fall velocity of the particle with the diameter D_g

D_g is the geometric mean diameter

F_r is the Froude number and is taken equal to $U/(gb)^{1/2}$, and

h is the depth of water.

A model test is necessary to determine the scour depth shown in Fig. 12.13e. Laboratory tests and experiments performed on prototypes have shown that in aerated flows the scour is smaller than in non-aerated flows. Dissipation of energy by means of flip buckets and deflectors is in part based on this assumption. The physical interpretation of the phenomena is simple. The specific weight of water in aerated flow is smaller than the specific weight of water in non-aerated flow. Air in the jet of water that issues from the end spill disperses the water particles and damps their energy. Figure 12.14 (G. Johnson, 1967) summarizes the results of tests conducted in laboratories. It is clear that the scour caused by a water-air mixture is less than the scour caused by non-aerated flow downstream of a free overflow. The influence of time is also indicated on the same figure. The depth of the scour hole as predicted by Franke and Dodiah's formula is shown together with the experimental results. To the best of our knowledge, no adequate mathematical approaches have been developed that will predict local scour caused by the aerated jet. To facilitate the comparison of scour formulas, they are listed in Table 12.7.

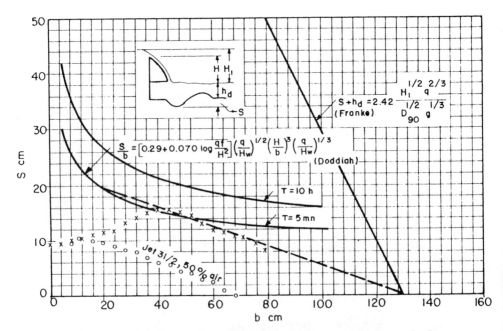

Figure 12.14. Comparison of formulas for scour at the base of a free overflow, with jet scour test results. (Simons, D.B. and Şentürk, F., 1992.)

TABLE 12.7

Formulas giving the depth of the local scour.

M: Metric system of units
E: English system of units
D: Dimensionless

Name of the author and Equations	Type of Formula (cf Fig. 12.13 for nomenclature) SCOUR DOWNSTREAM OF			Observations
	Vertical jet	Horisontal jet	End of Apron =	
1. Dodiah (1955) Eq. 12.26	$\dfrac{s}{b} = \left[0.29 + 0.07\log\dfrac{qt}{H^2}\right]\left(\dfrac{q}{Hw}\right)^{1/2}\left[\dfrac{H}{b}\right]^3\left(\dfrac{q}{Hw}\right)^{1/3}$			w: fall velocity D
2. Franke (1960) Eq. 12.27	$S = 2.42\dfrac{H^{1/2}q^{0.67}}{D_{90}^{1/2}g^{1/3}} - h_d$			M
3. Carstens (1966) Eq. 12.23		$\left(\dfrac{S}{b}\right)^6 = 2.85\,10^{-3}(N_s^2 - 4)^{5/2}\tan\emptyset\left(\dfrac{D_g}{b}\right)\left(\dfrac{U_t}{b}\right)$		cf Eq. 12.23 for nomenclature (D)
4. Valentine (1967) Eq. 12.24		$\log\dfrac{S}{b'} = \dfrac{F_{rb} - 2.0}{4.7} + 0.55\log\dfrac{b'}{D_{90}}$		cf Eq. 12.24 for nomenclature M
5. Zimmerman and Maniak (1973) Eq. 12.28			$S = \dfrac{2.89q^{0.82}}{D_{85}^{0.23}}\left(\dfrac{h_d}{q^{2/3}}\right) - h_d$	M
6. Altinbilek and Okyay (1973) Eq. 12.25	$\dfrac{S_T}{b} = \dfrac{30.5\left(\dfrac{U}{w}\right)^2\left(\dfrac{D_g}{b}\right)}{F_r\left(\dfrac{h}{b}\right)^{0.50}}$			cf Eq. 12.25 for nomenclature D

F. Sentürk

TABLE 12.7a

Old formulas giving the depth of local scour
Metric system

Name of the author and Equations	Type of Formula (cf Fig. 12.13 for nomenclature) SCOUR DOWNSTREAM OF	
	Sharp Crested Spillway	Observations
1. Schoklitsch (1932) Eq. 12.30	$S = 4.75 \dfrac{H^{0.2} q^{0.5}}{D_{90}^{0.32}} - h_d$	S: Depth of scour hole H: Vertical distance between the energy grade line and the downstream water surface
2. C. Jäger (1939) Eq. 12.31	$S = 6 H^{0.25} q^{0.5} \left(\dfrac{h_d}{D_{90}} \right)^{\frac{1}{3}} - h_d$	q: Linear water discharge h_d: Downstream water depth
3. Eggenberger (1943) Eq. 12.32	$S = 22.88 \dfrac{H^{0.5} q^{0.6}}{D_{90}^{0.4}} - h_d$	(*) Eq. 12.29
	$S = C \dfrac{H^{0.5} q^{0.6}}{D_{90}^{0.4}} - h_d$ (for sharp crested spillway and sluice gate)	
4. Müller (1944) Eq. 12.33	$S = \dfrac{C''}{g^{0.3}} \left(\dfrac{\gamma}{\gamma_s'} \right)^{4/9} \dfrac{H^{0.5} q^{0.6}}{D_{90}^{0.4}}$	$C' = C g^{0.3} \left(\dfrac{\gamma}{\gamma_s} \right)^{-4/9}$ Eq. 12.34

$$(*) \qquad C = 22.8 - \frac{1}{0.0049 \left(\dfrac{q_u}{q_d} \right)^3 - 0.0063 \left(\dfrac{q_u}{q_d} \right)^2 - 0.0029 \left(\dfrac{q_u}{q_d} \right) + 0.064} \qquad \text{Eq. 12.29}$$

Example 12.03

Determine the scour depth S which will take place downstream of the sluice gate shown in Fig. 12.15 for different gate conditions.

Data

- $D_{90} = 7$ mm $\qquad\qquad D_g = 5$ mm

- For $q_d = 2 m^3/sec.m$, the upstream water elevation is 10.0 m and the gate operates as a sluice gate, a = 0.33 (Fig. 12.15a). For $q_u = 4$ $m^3/sec.m$ the upstream water elevation is 12.00 m and with $q_d = 2 m^3/sec.m$; the gate operates as a sharp-crested spillway and as a sluice gate (Fig. 12.15b). For $q_u = 2 m^3/sec.m$, the upstream water elevation is 10.00 m, and the gate operates as a sharp-crested spillway (Fig. 12.15c)

636

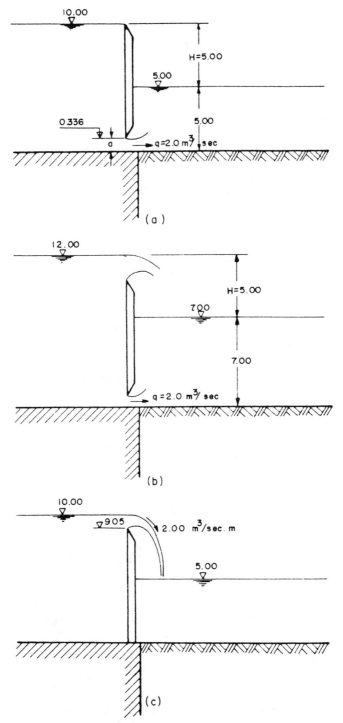

Figure 12.15. Flow conditions at the sluice gate for Example Problem 12.03.

- The tailrace characteristics are:

  For $q_d = 2$ m^3/sec.m, $d = 5.00$ m (sluice gate)

  $a = 0.336$ m

  For $q_u = 4$ m^3/sec.m and $q_d = 2$ m^3/sec.m, $d = 7.00$ m

- The flow is bidimensional

- The approach velocity is small so that the velocity head can be considered as zero; the opening of the sluice gate $a = 0.336$ m

- $\varnothing = 38°$, $\gamma_s = 2.65$ T/m^3 and $w_{50} = 52$ cm/sec*

Solution

Step 1: The gate is operating as a sluice gate Fig. (12.15a). Using Eggenberger's formula (Table 12.7a)

$$d_s = C \frac{(10-5)^{1/2} \, 2^{0.6}}{7^{0.4}}$$

The value of $C = 10.35$ as determined from Eq. 12.29, then

$$d_s = 10.35 \frac{5^{1/2} 2^{0.6}}{7^{0.4}} = 16.11 \text{ m}$$

Using Carstens' formula

$$\left(\frac{S}{b}\right)^6 = 2.85 \cdot 10^{-3} (N_s^2 - 4)^{2.5} \tan \varnothing \left(\frac{D_g}{b}\right) \left(\frac{u}{b}\right)t \qquad 12.23$$

with

$$b = 1.00 \text{ m}$$

$$N_s = \frac{U}{(\gamma_s' g D_g)^{1/2}} = \frac{2/0.336}{(1.65 \cdot 9.81 \cdot 0.005)^{1/2}} = 20.92$$

and

$$\left(\frac{S}{0.336}\right)^6 = 2.85 \cdot 10^{-3} (437.65 - 4.00)^{2.50} \left(\frac{0.005}{0.336}\right) \left(\frac{5.95}{0.336}\right) t \cdot 0.781$$

$$S = 1.22 \, t^{1/6}$$

---

* This example is taken from Simons and Şentürk (1976).

At the end of the first day

$$S = 8.11 \text{ m}$$

At the end of the tenth day

$$S = 11.91 \text{ m}$$

Using Valentine's formula

$$\log \frac{S}{b'} = \frac{F_{rb} - 2.00}{4.70} + 0.55 \log \frac{b'}{D_{90}} \qquad 12.24$$

with

$$F_{rb} = \frac{2/0.30}{(9.81 \bullet 0.30)^{1/2}} = 3.89$$

$$b' = 0.90 \quad a = 0.90 \bullet 0.336 = 0.30$$

$$\log S = \log 0.30 + \frac{3.89 - 2.00}{4.7} + 0.55 \log \frac{0.30}{0.007} = 0.78$$

$$S = 5.98 \text{ m}$$

Step 2:   The gate is operating as a sharp crested spillway, Fig. 12.15c.

$$S = 4.75 \frac{H^{0.2} q^{0.5}}{D_{90}^{0.32}} - h_d \qquad 12.30$$

where $H = 5.00 \text{ m}$

Then $S = 4.75 \dfrac{5^{0.2} 2^{0.5}}{7^{0.32}} - 5.00 = -0.03$

no scour is indicated. Using Jäger's formula

$$S = 6 H^{0.25} q^{0.5} \left(\frac{hd}{D_{90}}\right)^{1/3} - h_d \qquad 12.31$$

and

$$S = 6 \bullet 5^{0.25} 2^{0.50} \left(\frac{5}{7}\right)^{1/3} - 5.00 = 6.34 \text{ m}$$

F. Sentürk

Using Altinbilek - Okyay formula

$$\frac{S_t}{b} = \frac{30.5\left(\frac{U}{W}\right)^2\left(\frac{D_g}{b}\right)}{F_r\left(\frac{h}{b}\right)^{0.5}}$$

12.25

where b = 0.95 • 0.30 = 0.285 m

(It is assumed that the jet issued from the spillway crest loses 70% of its thickness 5.00 m below crest level)

$$U = \frac{2}{0.285} = 7.01 \text{ m/sec}$$

$$w = 0.32 \text{ m/sec}$$

$$D_g = 5 \text{ mm}$$

$$h = 5.00 \text{ m}$$

$$S_t = 0.285\left[\frac{30.50\left(\frac{7.01}{0.52}\right)^2\left(\frac{0.005}{0.285}\right)}{\frac{7.01}{(9.81 \cdot 0.285)^{1/2}}\left(\frac{5}{0.285}\right)^{1/2}}\right]$$

and

$$S_t = 1.58 \text{ m}$$

Step 3: The gate is operating as a spillway and as a sluice gate at the same time (Fig. 12.15b).

Using Eggenberger's formula

$$S = C \frac{(H)^{1/2} q^{0.6}}{D_{90}^{0.4}} - h_d$$

12.32

and with

$$\frac{q_u}{q_d} = \frac{4}{2} = 2, \text{ the coefficient C is obtained from Eq. 12.29. Then}$$

$$S = 15.55 \frac{5^{0.5}4^{0.6}}{7^{0.4}} - 7.00 = 29.68 \text{ m}$$

Step 4: The results are summarized in Table 12.8.

Discussion

Eggenberger's, Carstens', and Altinbilek-Okyay's formulas give very large values for the depth of scour S. These formulas may only be applied in laboratory conditions for small values of q and for coarse material. To

TABLE 12.8

Summary of results of Example 12.08

Formulas	Depth of scour Sluice gate	Spillway	Spillway and sluice gate
	m	m	m
Schoklitsch	-	no scour	-
Eggenberger	11.11	-	29.68
Jäger	-	6.34	-
Carstens	8.11	-	-
Valentine	5.98	-	-
Altinbilek-Okyay	-	1.58	-

some the extended Schoklitsch's and Jäger's formulas can be applied to engineering problems but, in general, empirical relations for determining the scour hole depth reveal inadequacies.

12.4   DEGRADATION OF STREAM BEDS

12.4.1 Mechanism of Degradation

The degradation process is a function of several variables such as:

1. The characteristics of flow released from the reservoir

2. The sediment concentration of the flow released from the reservoir

3. The properties of sediment forming the river bed

4. Natural and man-made irregularities and singularities in the river bed

5. The geometric and hydraulic characteristics of the river channel

6. The existence and location of controls in the downstream channel.

F. Sentürk

Common assumptions used in the analysis of degradation are:

- Water released from the reservoir is clear

- Irregularities in the river course are not important

- Characteristics of sediments in the degrading reach are constant

- Sediment is mainly transported as the bed load.

12.4.2 Evaluation of Degradation

General rules of hydraulics can be applied for evaluating bed degradation. The continuity equation

$$\frac{\partial y}{\partial x} + \frac{\partial Q}{\partial x} = 0 \qquad (12.35)$$

is applied for a first approximation in predicting the magnitude of degradation of the bed of a river subject to erosion. If (z) is the vertical distance between a point on the bed to a datum line, its variation in time will be (dz) and its rate of change with time will be (dz/dt). Assume the sediment transport, as bed load per unit width, is q_B. The suspended load in this study is eliminated which is a constraint on the theory as a whole. Assuming that the cross section of the river is rectangular and its bottom width equal to B, the total sediment load will be $q_B B$, and its variation along the distance (∂x) will be ($\partial(q_B B)/\partial x$). In a unit time period this quantity constitutes a volume and the vertical dimension of the volume will be

$$\frac{\dfrac{1}{B(1-\lambda)}\,\partial\,(q_B B)}{\partial x}$$

where (λ) represents the porosity of the bed material. Substituting the computed value of ($\partial y/\partial t$) as ($\partial z/\partial t$) and ($\partial Q/\partial x$) in Eq. 12.35 yields

$$\frac{\partial z}{\partial t} + \frac{1}{B(1-\lambda)}\frac{\partial(q_B B)}{\partial x} \qquad (12.36)$$

Practical uses of this general equation have been proposed by many researchers.

A. Computation of River Beds in Equilibrium-Gamal Mostafa Suggestion

Gamal Mostafa (1957) tried to introduce Einstein's relation into the equilibrium process

$$\frac{U}{U_*} = 5.75 \log \left(12.27 \frac{R}{k_s} x\right) \tag{12.37}$$

where

- U is the mean velocity

- $U_* = (gRS)^{1/2} = (\tau/\rho)^{1/2}$

- S is the slope of the energy line

- R is the hydraulic radius

- τ is the shear stress

- g is the gravitational acceleration

- ρ is the fluid density

- k_s is the characteristic height of the rugosity, and

- x is a correction factor defined by Einstein (Fig. 12.16)

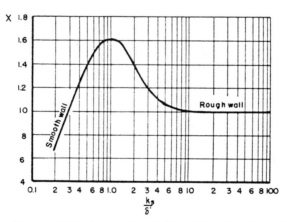

Figure 12.16. Correction factor defined by Einstein.

Shields relation can be expressed as

$$\frac{\tau_c}{\gamma_s' k_s} = 0.06 \tag{12.38}$$

where τ_c is the critical shear stress.

F. Sentürk

Gamal Mostafa combining Eq. 12.37 with Eq. 12.38 proposed

$$\frac{Q}{A} = 5.75 \, [\frac{0.06 \, \gamma'_s \, (k_s/y)}{\rho}]^{1/2} \, \log \, (12.27 \, \frac{R}{k_s} \, x) \qquad (12.39)$$

where

Q   is the discharge

A   is the wetted area

γ'_s is the specific weight of submerged solid particles

k_s is the characteristic sand rugosity, and

$$y = \frac{0.06}{K} \qquad\qquad K = \frac{\tau_c}{\gamma'_s k_s}$$

This formula leads to the slope relation

$$S = \frac{0.06 \, \gamma'_s \, (\frac{k_s}{y})}{\gamma \cdot R} \qquad (12.40)$$

defining the slope of a profile in equilibrium. Gamal Mostafa assumed

$$k_s = D_{98}$$

Equation 12.39 includes A and R as variables. It is possible to express A as a function of (R).* Introducing this value in Eq. 12.39 yields Eq. 12.40.

---

*   For a rectangular channel:

$$R = \frac{A}{B + 2(A/B)} \qquad \text{B: bottom width}$$

For a triangular channel:

$$R = \frac{1}{2} \left(\frac{\alpha}{1+\alpha^2} A \right)^{1/2}$$

where

$$\frac{\alpha}{1} = \frac{\text{horizontal}}{\text{vertical}} \text{ of side slopes}$$

For a trapezoidal channel:

$$R = \frac{A/B}{1 + \left(\frac{1+\alpha^2}{\alpha^2}\right)^{1/2} \left(\sqrt{1+\frac{4\alpha A}{B^2}} - 1\right)}$$

To solve the equations assume x and y equal to unity; then compute R for each value of Q and return to x and y using Fig. 12.17. Repeat the procedure until the calculated values of x and y equal the assumed values.

The choice of $k_s = D_{98}$ requires careful consideration because it has a direct influence on the potential armoring of the bed as degradation occurs.

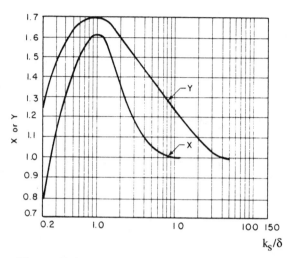

Figure 12.17 - Relationship between X or Y and k_s/δ.

Example 12.04

Determine the slope of a channel that will maintain an equilibrium flow of $Q = 15.00$ m³/sec (Fig. 12.18), with the following characteristics.

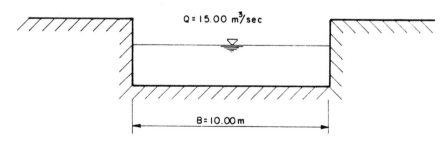

Figure 12.18

Data

- The cross section of the channel is rectangular
- The bottom width is equal to 10.00 m

F. Sentürk

- $\gamma'_s = 1.7 \text{ T/m}^3$
- $k_s = D_{98} = 0.04 \text{ m}$
- $\rho = 102 \text{ g.sec}^2/\text{m}^4$

and

$$n = 10^{-6} \text{ m}^2/\text{sec}$$

Solution

Gamal Mostafa's approach will be used for solving this problem.

- The hydraulic radius of a rectangular cross section is given as

$$R = \frac{A}{B + 2(A/B)} = \frac{10A}{100 + 2A}$$

- The wetted area of a cross section in which a flow in equilibrium takes place is, after Gamal Mostafa

$$\frac{Q}{A} = 5.75 \left(\frac{0.06 \, \gamma'_s \frac{k_s}{y}}{\rho}\right)^{1/2} \log\left(12.27 \frac{R}{k_s} x\right) \qquad 12.39$$

$$15 = A \, 1.15 \log 12.27 \frac{10A \, 0.04^{-1}}{100 + 2A}$$

This equation gives the value of A as

$$A = 5.90 \text{ m}^2$$

- The slope of a flow in equilibrium is

$$S = \frac{0.06 \cdot 1.7 \cdot 0.04}{\gamma R} \qquad y = 1 \qquad 12.40$$

then

$$S = \frac{0.06 \cdot 1.70 \cdot 0.04}{10A}(100+2A)$$

A, being known

$$S = 0.00773$$

- Checking for x and y gives

$$\delta = 11.6 \frac{\nu}{U_*} \qquad\qquad U_* = (gRS)^{1/2} = (9.81 \cdot 0.00773 \cdot 0.52)^{1/2}$$

$$= 0.000058 \qquad\qquad\qquad = 0.200$$

and

$$\frac{k_s}{\delta} = \frac{0.04}{0.000058} = 689.66$$

Using Fig. 12.17 x = y = 1, then for Q = 15.00 m^3/sec the slope of the energy line is computed, after G. Mostafa as:

$$S = 0.00773$$

The corresponding mean velocity is

$$U = 2.54 \text{ m/sec}$$

B. The Komura and Simons Approach for Computing Bed Degradation

Komura and Simons (1967), starting with Eq. 12.36, obtained a general expression for bed degradation (Fig. 12.19).

They used Kalinske and Brown's (1950) sediment transport formula

$$q_B = K \frac{S^{1.4} \, \gamma^2}{(1.486/n)^{1.2}} q^{3/5} (q^{3/5} - q_c^{3/5})$$

expressed as

$$q_B = \beta \, D_s^{(1-p)} U_* (U_*^2 - U_{*c}^2)^p \qquad\qquad (12.41)$$

where

$$(\beta) \text{ is } \frac{a_s}{[(\frac{\gamma_s}{\gamma} - 1)g] \, p}$$

a_s is a constant (equal to 10 in general)

p   is a dimensionless exponent (equal to 2 in general)

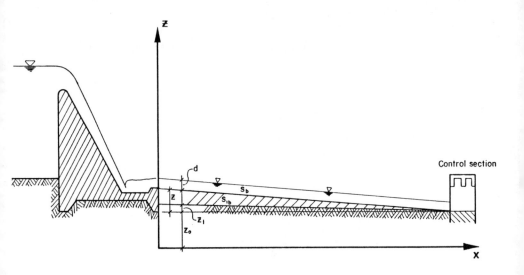

Figure 12.19. Definition sketch.

U_* is the friction velocity, and

U_{*c} is the critical friction velocity.

Differentiating Eq. 12.41 with respect to x, they obtained

$$\frac{\partial q_B}{\partial x} = \beta(1 - p) \; D_s^{-p} \; U_* \; (U_*^2 - U_{*c}^2)^p \; \frac{\partial D_s}{\partial x} + \beta \, D_s^{(1-p)} \, (U_*^2 - U_{*c}^2)^{\,(p-1)}$$

$$[(1 + 2p) \, U_*^2 - U_{*c}^2] \, \frac{\partial U_*}{\partial x} - 2\beta p \, D_s^{(1-p)} U_* U_{*c} (U_*^2 - U_{*c}^2)^{\,(p-1)} \frac{\partial U_{*c}}{\partial x} \qquad (12.42)$$

Then U_* can be derived from Manning's equation as

$$U_* = \frac{g^{1/2} Q n}{1.486 B y^{7/6}} \quad (*)$$

---

*   Manning's relation is

$$U = \frac{1}{n} R^{2/3} S^{1/2} \quad \text{or} \quad \frac{Q}{A} = \frac{1.486}{n} R^{2/3} \frac{R^{1/2} S^{1/2} g^{1/2}}{R^{1/2} g^{1/2}}$$

For very large cross sections R=y and A=By then

$$U_* = \frac{g^{1/2} Q n}{1.486 B y^{7/6}}$$

where

n   is the Manning's roughness coefficient and

y   is the depth of flow.

Equation 12.42 yields

$$\frac{\partial U_*}{\partial x} = \frac{U_*}{n}\left(\frac{\partial n}{\partial x} - \frac{7n}{6y}\frac{\partial y}{\partial x} - \frac{n}{B}\frac{\partial B}{\partial x}\right) \tag{12.43}$$

Komura (1963) expressed U_{*c} as

$$U_{*c} = (a_c(\frac{\gamma_s}{\gamma} - 1)gD_s)^{1/2} \tag{12.44}$$

where a_c is a function of the critical shear velocity Reynolds number D_s/ν. The term a_c can be obtained by measurement in nature. An example of the variation of this parameter is shown in Fig 12.20. Şentürk has conducted research in the field to determine a_c. He has seen that its variation is small; so the curve given in Figure 12.20 can be used as a first step. It is recommended for checking the validity of final computations.

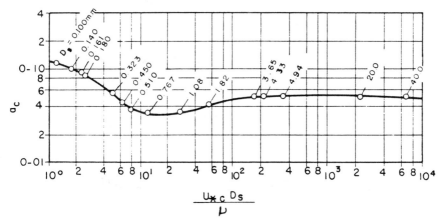

$$\frac{U_{*c} D_s}{\nu}$$

Figure 12.20. Variation of a_c with $U_{*c} D_s/\nu$.

Figure 12.20 is also used for computing the armoring effect. To do so, Komura and Simons defined a corrected value for a_c, $a'_c = C_a = a_c\varepsilon\sigma^r$, in which C_a, the armoring coefficient, is a constant, r is a dimensionless coefficient, e is a constant, and σ is the standard deviation of particle size distribution. Introducing these values in Eq. 12.44 yields

$$U_{*c} = (a'_c \gamma'_s gD_s)^{1/2} \tag{12.45}$$

Egiazaroff (1965) proposed the following explicit form for a_c'.

$$a_c' = \frac{0.1}{\log 19(D_{50}/D_m)}$$

Differentiating Eq. 12.44 leads to

$$\frac{\partial U_{*c}}{\partial x} = \frac{U_{*c}}{2D_s} \frac{\partial D_s}{\partial x} \qquad (12.46)$$

Manning's roughness coefficient can be written as

$$n = C_n \, D_s^{1/6}$$

then

$$\frac{\partial n}{\partial x} = \frac{C_n}{6} \, D_s - \frac{5}{6} \, \frac{\partial D_s}{\partial x} \qquad (12.47)$$

The equation of continuity (Eq. 12.36) can be transformed into

$$\frac{\partial z}{\partial t} = -\frac{1}{(1-\lambda)} \frac{\partial q_B}{\partial x} - \frac{q_B}{B(1-\lambda)} \frac{\partial B}{\partial x} \qquad (12.48)$$

and substituting Eqs. 12.41, 12.42, 12.43, 12.46, and 12.47 into Eq. 12.48 leads to:

$$\frac{\partial z}{\partial t} = \frac{\beta}{6(1-\lambda)y \, D_s^{(p-1)}} U_* (U_*^2 - U_{*c}^2)^{(p-1)} [4pU_*^2 - 7(U_*^2 - U_{*c}^2)] \frac{\partial D_s}{\partial x}$$

$$+ \frac{7\beta}{6(1-\lambda) \, y \, D_s^{(p-1)}} U_*^2 (U_*^2 - U_{*c}^2)^{(p-1)} [(1+2p) U_*^2 - U_{*c}^2)] \frac{\partial y}{\partial x}$$

$$+ \frac{2p\beta}{(1-\lambda) \, B D_s^{(p-1)}} U_*^2 (U_*^2 - U_{*c}^2)^{(p-1)} \frac{\partial B}{\partial x} \qquad (12.49)$$

For p = 2, $U_* (U_*^2 - U_{*c}^2) = 2U_*^3$ and $(5U_*^2 - U_*^2) = 4U_*^2$, Eq. 12.49 reduces to

$$\frac{\partial z}{\partial t} = \frac{2\beta}{(1-\lambda)D_s} U_*^3 (U_*^2 - U_{*c}^2) \left(\frac{1}{6D_s} \frac{\partial D_s}{\partial x} + \frac{7\partial y}{3y\partial x} + \frac{2\partial B}{B\partial x} \right) \qquad (12.50)$$

This relation denotes the bed degradation downstream of a reservoir which traps the sediment and releases clear water. At the final stage of degradation, $\partial z / \partial t = 0$, and

$$U_{*f}^2 = C_s g D_{sf} \tag{12.51}$$

Introducing these values in Eq. 12.50 yields the final equilibrium formula as proposed by Komura and Simons:

$$z_f = z_0 + \frac{14}{15}(\frac{C_s}{C})(\frac{D_{sfo}}{y_{fo}})(e^{(15C/14x')} - 1) + y_{fo}(1 - e^{-(C/14)x'}) -$$

$$\frac{y_{fo}}{2}(\frac{y_c}{y_{fo}})^3(e^{(c/7)x'} - 1) \tag{12.52}$$

where

z_f is the final profile of the river bed

z_0 is the bottom level at section 0 (Fig. 12.19)

C_s = a_c'

C =   a constant (Example 12.05)

x' is the distance along the x axis measured in the upstream direction along the river bed as shown on Fig. 12.19

D_{sfo} is the final representative particle size at section 0, and

y_{fo} is the ordinate of the depth at the origin of coordinates (section 0).

The final slope of the bed corresponding to z_f is proposed by Komura and Simons as

$$S_{fb} = C_s (\frac{D_{sf}}{y_f}) - \frac{1}{14}(1 - \frac{y_c^3}{y_f^3})(\frac{y_f}{D_{sf}})\frac{\partial D_{sf}}{\partial x} - \frac{1}{7}(\frac{y_f}{B})(6 + \frac{y_c^3}{y_f^3})\frac{\partial B}{\partial x} \tag{12.53}$$

Example 12.05 *

It is required to predict the degradation downstream of the Milburn Diversion Dam on the Middle Loup River in Nebraska. The initial conditions are:

• The water released from the dam is clear

---

† Komura and Simons (1967) presented a numerical example to illustrate their method. This example is reproduced here, after reconsideration.

F. Sentürk

- The river banks are not erodible

- Seasonal variation in discharge and temperature of water can be neglected

- Sediment injection by tributaries does not occur

- Meandering and the growth of vegetation does not occur.

Data for the problem:

- The variation of the bed material sizes downstream of the dam is given in Fig. 12.21.

- The bed material properties for the study reach are summarized in Table 12.9

- $\gamma_s = 2.65$ T/m³

- The width of river is assumed to vary as shown in Table 12.10

- The discharge of the river is 780 cfs.

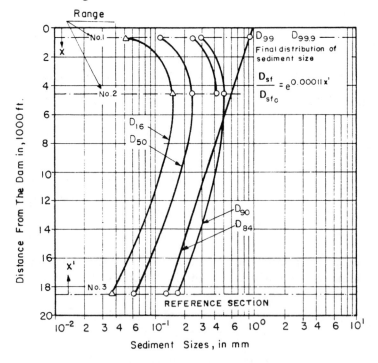

Figure 12.21. Sediment size distributions.

TABLE 12.9

SEDIMENT PROPERTIES OF RIVER BED

Range No. (1)	in feet (2)	D_{16} in mm (3)	D_{50} in mm (4)	D_{84} in mm (5)	D_{90} in mm (6)	σ (7)	a_c (8)
1	17,850	0.052	0.115	0.25	0.30	4.8	0.036
2	14,000	0.150	0.210	0.43	0.52	2.9	0.50
3	0	0.036	0.061	0.14	0.17	3.6	0.110
Mean			0.129	(0.000426 ft)		3.8	0.065

TABLE 12.10

n	x'	Δx'	B	ΔB$_n$	ΔD$_{sf}$ 10^3	ΔD$_{sfn}$ 10^3	y$_{fm}$	$y\left(\frac{3}{c}\right)_m$	$\left(\frac{y_c 3}{y_f}\right)$	$\left(\frac{y_f}{B}\right)_m$	$\left(\frac{y_f}{D_{sf}}\right)$	Δz$_{fn}$	z$_{fn}$
	feet	feet	feet	feet	feet	feet	feet	feet				feet	feet
1	2	3	4	5	6	7	8	9	10	11	12	13	14
7	18600	750	300	0	3.296	2.610	1.111	0.210	0.1533	0.0037	336.95	0.9246	10.75
6	17850	1750	300	-50	3.035	5.320	1.113	0.210	0.1523	0.0037	366.72	1.8201	9.83
5	16100	2100	350	-50	2.503	5.160	1.128	0.154	0.1073	0.0032	450.66	1.7997	8.01
4	14000	2900	400	-50	1.987	5.430	1.147	0.118	0.0782	0.0029	577.25	2.2274	6.21
3	11100	2500	350	+50	1.444	3.470	1.173	0.154	0.0954	0.0034	812.33	1.4373	3.99
2	8600	2500	300	0	1.097	2.640	1.196	0.210	0.1228	0.0040	1090.25	0.7901	2.55
1	6100	6100	350	0	0.833	4.060	1.220	0.154	0.0848	0.0035	1464.59	1.7590	1.76
0	0	0	350	0	0.427	0	1.280	0.154	0.0734	0.0037	3004.69	0	0

Solution

A step method is used for solving the problem.

Step 1: Compute C_s

$$a_c' = C_a = 3.80 \cdot 0.065 = 0.25$$

$$C_s = a_c' \gamma_c' = 0.25 \cdot 1.65 = 0.413 \ ^*$$

---

* ε is assumed equal to unity (see 12.42.B)

$a_c = 0.065 \qquad \sigma = 3.8$

Step 2: Assume $z_o = 0$ and compute the final equilibrium profile by applying

$$\Delta z_{fn} = \sum_{n=0}^{7} \{0.413 \, (\frac{D_{sf}}{y_f}) \, \Delta x_n' + \frac{1}{14} [1 - (\frac{y_c}{y_f})_m^3] \, (\frac{y_f}{D_{sf}}) \Delta D_{sfn} +$$

$$\frac{1}{7} \, (\frac{y_f}{B}) \, [6 + (y_c/y_f)_n^3] \Delta B_n \} \qquad (12.54)$$

derived from Eq. 12.52, and noting that

$$(y_c^3)_m = \frac{Q^2}{gB_m^2}$$

In this example, Eq. 12 52 will be used to develop the final equilibrium profile.

Step 3: Compute the sediment size distribution.

- Determine from Table 12.09, $D_{sfo} = 0.000426$ ft

- $D_{sf} = D_{sfo} e^{cx'}$ (Komura's assumption)

- $D_{sf} = 0.000426 \, e^{cx'} = 0.000426 \, e^{0.00011x'}$ (*)

- The final equilibrium profile is

- $z_f = 1.17(e^{0.000118x'} - 1) + 1.28(1 - e^{0.00000786x'}) - 0.0518(e^{0.0000157x'} - 1)$

Introducing these values in Table 12.10 yields the solution of the problem.

Column 1:   Divide the reach in sections

Column 2:   Distance to the beginning of each section from the origin of coordinates, x' (Fig. 12.21)

Column 3: $\Delta x_n' = x_n' - x_{(n-1)}'$

Column 4:   B, width of sections

Column 5: $\Delta B_n = B_n - B_{(n-1)}$

Column 6: $D_{sf} = D_{sfo} e^{0.00011x'}$, where $D_{sfo} = 0.000426$ ft $= 0.14$ mm (Table 12.9)

Column 7: $\Delta D_{sfn} = D_{sfn} - D_{sf(n-1)}$

---

*   See Fig. 12.21 for coefficient 0.00011

Column 8: $y_f = y_{fo}e^{-(c/14)x'}$ where y_{fo} is the depth of flow at section 0

Column 9: $(y_c^3)_m = \dfrac{\alpha Q^2}{gB^2}$, $\alpha = 1$, $B_m = B$

Column 10: $\left(\dfrac{y_c^3}{y_f^3}\right) m$

Column 11: $\left(\dfrac{y_f}{B}\right) m$

Column 12: $\left(\dfrac{y_f}{D_{sf}}\right) m$

Column 13: Δz_{fn} is obtained from Eq. 12.54

Column 14: $z_{fn} = z_{fn} + z_{f(n-1)}$

Table 12.10 summarizes the computation. The result is shown in Fig. 12.22.

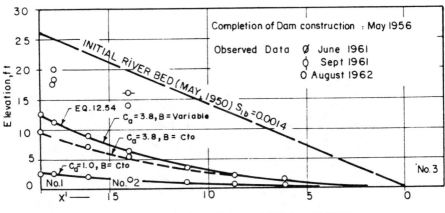

Figure 12.22. Final equilibrium profiles.

12.5 THE ARMORING OF RIVER BEDS

River beds are generally subject to erosion. When the bottom is covered by loose coarse material which covers small size particles varying in size from 0.06 mm to 5 cm erosion takes the form of severe scour. Fine sands are easily transported by the flow. If the tractive force of the flow is not large enough to move the coarse particles a kind of armoring takes place. The bed becomes stable and will not be eroded. As the tractive force becomes smaller the sediment will settle and stratification as shown in Figure 12.23 occurs. The armoring of this bed is shown in Fig. 12.24.

Figure 12.23. Stratification in bottom material in the Doğançay River, Turkey.

Figure 12.24. Armoring in the Doğançay River, Turkey.

12.5.1 The Armoring Process

Different methods have been suggested for determining the particle size of an armoring under given hydraulic conditions. The simplest of them all is the application of Shields method. Shields diagram is shown in Fig. 12.25.

For given flow characteristics it is possible to define the smallest diameter particle which cannot be transported. This diameter represents the size of the armor coat (Example 12.06). An exact mathematical approach is difficult because of the many parameters involved. As seen in Figure 12.24, the fine particles are intercalated into the larger ones to constitute the armor coat. The mathematical analysis of such a formation is very difficult. On the other hand, the initiation of motion for non uniform bed material is different from uniform bed material and therefore the application of Shields procedure may be erroneous. However Egiazaroff (1965) has taken the sheltering effect and using Shields's parameters has suggested the following expression

$$\tau_c = \frac{0.1}{(\log 19\frac{D}{D_m})^2} \tag{12.55}$$

where

τ_c is the initial shear value at which particle motion begins

D is the diameter of a particular grain, and

D_m is the mean diameter of the grain representing the sediment mixture

The definition of D_m is rather difficult. Egiazaroff suggests also including the grains in suspension when computing D_m.

Example 12.06

The mean diameter of the particle resting on the bottom of an infinitely large river is equal to $D_m = 10.00$ mm. The slope of the river in the reach is $S = 0.001$, $v = 10^{-6}$ m²/sec, $\gamma_s = 2.60$ T/m³. Determine the flow characteristics at which $D_m = 10.00$ mm constitutes the armor coat.

Solution

Use Shield's criterion for determining the start of motion (Fig. 12.25).

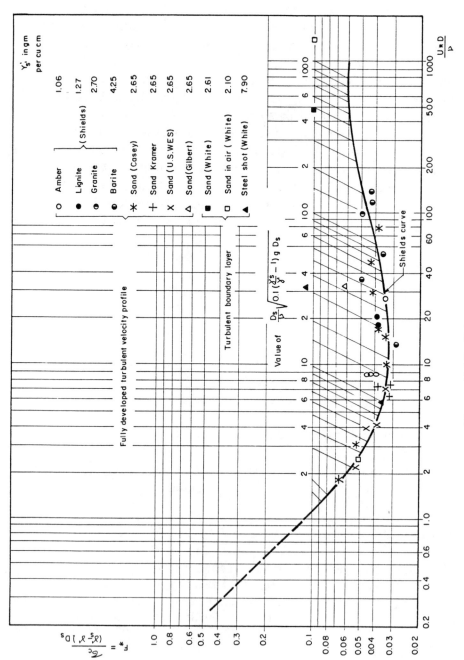

Figure 12.25. Shields diagram.

Step 1:   Determine

$$R_* = \frac{U_* D_m}{\nu}$$

$$U_* = (gRS)^{1/2}$$

and

$$R_* = \frac{(gRS)^{1/2} D_m}{\nu} = \frac{0.01 \cdot (9.81 \cdot 0.001)^{1/2} R^{1/2}}{10^{-6}} = 990.45 R^{1/2}$$

Step 2:   Determine the parameter

$$\frac{D_s}{\nu} (0.1 \, \gamma_s g D_s)^{1/2} = \frac{0.01}{10^{-6}} (0.1 \cdot 1.60 \cdot 9.81 \cdot 0.01)^{1/2} = 1253$$

This value of the parameter corresponds to:

$$R_* = 700 = 990.45 \, R^{1/2}$$

Assume R = 0.50 m $R \cong d$

Then for the depth of flow smaller than 0.50, D = 0.01 m constitutes an armor coat. Checking with Egiazaroff's relation yields

$$\tau_c = \gamma RS = \frac{0.1}{(\log 19 \frac{0.01}{0.01})^2} = 0.061 \qquad \qquad 12.55$$

$$R = \frac{0.061}{0.001 \cdot 1000} = 0.061 \text{ m}$$

which is different from 0.50 m. Note that in Shields criterion the sheltering effect is not taken into consideration.

Gessler (1970-1971) proposed the relation given by Eq. 12.56 for determining the size distribution of the armor coat

$$p_a = \frac{\int_{D_{min}}^{D} q p_o \, dD}{\int_{D_{min}}^{D_{max}} q p_o \, dD} \qquad \qquad (12.56)$$

F. Sentürk

where

   p_0 is the density function of the grain size distribution of mixture, and

   q is the probability that a grain size D will not be moved.

Then

$$q(D) = \frac{1}{\sigma\sqrt{2\pi}} \int_{\infty}^{\frac{\tau_{c(D)}}{\tau-1}} Exp\left(-\frac{x^2}{2\sigma^2}\right) dx$$

where

   σ is the standard deviation

   $\tau_c(D)$ is the critical shear stress for grain size D, and

   τ_c is the mean bed shear stress.

   Gessler reached q(D) = 0.50 for σ = 57 and Komura using Gessler's equation proposed

$$p_a(D) = \frac{\sum_{D_{min}}^{D} q(D)\,\Delta p_0(D)}{\sum_{D_{min}}^{D_{max}} q(D)\,\Delta p_0(D)} \tag{12.57}$$

Also Komura suggested a computational procedure using Eq. 12.57. This step method is summarized as follows:

Step 1: Compute the hydraulic radius R_b corresponding to the bed using the initial value of the flow

Step 2: Compute

$$U_* = (gR_bS)^{1/2}$$

Step 3: Considering the grain size distribution curve determine

$$U_{*b} = (a_c \gamma_s' \, gD_s)^{1/2} \tag{12.58}$$

corresponding to Δp_0 (D) taken from the curve shown in Fig. 12.26.

660

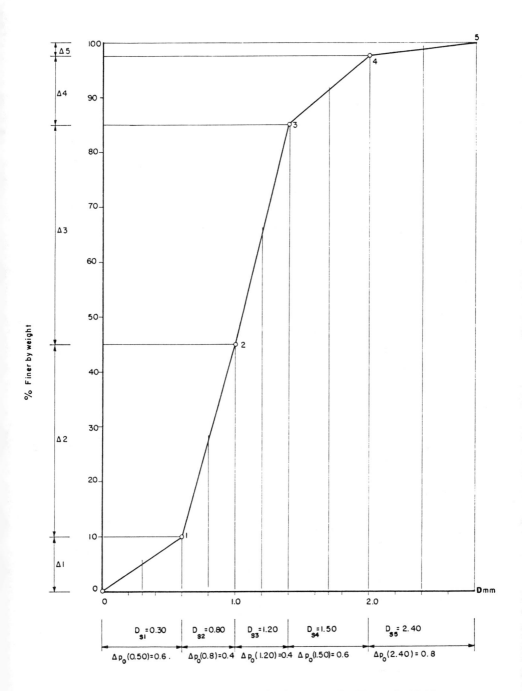

Figure 12.26. Size-frequency distribution curve for Example 12.07.

Step 4: Determine

$$\frac{\tau_c}{\tau_b} = \frac{U_{*c}^2}{U_{*b}^2}$$

Step 5: Determine

$$\xi = 1.754 \, (\frac{\tau_c}{\tau_b} - 1)$$

corresponding to each grain size.

Step 6: Determine q(D) using Table 12.11, for each geometric mean size.

Step 7: Using Eq. 12.57 compute $p_a(D)$ values.

Step 8: Plot $p_a(D)$ versus D (Fig. 12.27)

Step 9: Determine D_a, the mean diameter of the armor coat as indicated in Fig. 12.27.

Example 12.07

Determine the mean diameter, D_a, of the armor coat for a river, the characteristics of which are given:

- Use Fig. 12.26 to obtain the grain size distribution

- Use Fig. 12.28 to obtain $a_c = f(D_s, U_{*c}D_s/\nu)$

- The energy slope of the river is 0.0001

- $R_b = 0.35$ m and g = 9.81 m/sec^2

- $\gamma_s' = 1.7$ T/m^3

- The channel is wide so that d = R

Use the method suggested by Komura and Simons for solving the problem.

TABLE 12.11

Table of q(ξ)

For $\xi < 0$, q(-ξ) = 1 - q(ξ)

x	0.00	0.01	0.02	0.03	0.04	0.05	0.06	0.07	0.08	0.09
0.0	.5000	.5040	.5080	.5120	.5160	.5199	.5239	.5279	.5319	.5359
0.1	.5398	.5438	.5478	.5517	.5557	.5596	.5636	.5675	.5714	.5753
0.2	.5793	.5832	.5871	.5910	.5948	.5987	.6026	.6064	.6103	.6141
0.3	.6179	.6217	.6255	.6293	.6331	.6368	.6406	.6443	.6480	.6517
0.4	.6554	.6591	.6628	.6664	.6700	.6736	.6772	.6808	.6844	.6879
0.5	.6915	.6950	.6985	.7019	.7054	.7088	.7123	.7157	.7190	.7224
0.6	.7257	.7291	.7324	.7357	.7389	.7422	.7454	.7486	.7517	.7549
0.7	.7580	.7611	.7642	.7673	.7704	.7734	.7764	.7794	.7823	.7852
0.8	.7881	.7910	.7939	.7967	.7995	.8023	.8051	.8078	.8106	.8133
0.9	.8159	.8186	.8212	.8238	.8264	.8289	.8315	.8340	.8365	.8389
1.0	.8413	.8438	.8461	.8485	.8508	.8531	.8554	.8577	.8599	.8621
1.1	.8643	.8665	.8686	.8708	.8729	.8749	.8770	.8790	.8810	.8830
1.2	.8849	.8869	.8888	.8907	.8925	.8944	.8962	.8980	.8997	.9015
1.3	.9032	.9049	.9066	.9082	.9099	.9115	.9131	.9147	.9162	.9177
1.4	.9192	.9207	.9222	.9236	.9251	.9265	.9279	.9292	.9306	.9319
1.5	.9332	.9345	.9357	.9370	.9382	.9394	.9406	.9418	.9429	.9441
1.6	.9452	.9463	.9474	.9484	.9495	.9505	.9515	.9525	.9535	.9545
1.7	.9554	.9564	.9573	.9582	.9591	.9599	.9608	.9616	.9625	.9633
1.8	.9641	.9649	.9656	.9664	.9671	.9678	.9686	.9693	.9699	.9706
1.9	.9713	.9719	.9726	.9732	.9738	.9744	.9750	.9756	.9761	.9767
2.0	.9772	.9778	.9783	.9788	.9793	.9798	.9803	.9808	.9812	.9817
2.1	.9821	.9826	.9830	.9834	.9838	.9842	.9846	.9850	.9854	.9857
2.2	.9861	.9864	.9868	.9871	.9875	.9878	.9881	.9884	.9887	.9890
2.3	.9893	.9896	.9898	.9901	.9904	.9906	.9909	.9911	.9913	.9916
2.4	.9918	.9920	.9922	.9925	.9927	.9929	.9931	.9932	.9934	.9936
2.5	.9938	.9940	.9941	.9943	.9945	.9946	.9948	.9949	.9951	.9952
2.6	.9953	.9955	.9956	.9957	.9959	.9960	.9961	.9962	.9963	.9964
2.7	.9965	.9966	.9967	.9968	.9969	.9970	.9971	.9972	.9973	.9974
2.8	.9974	.9975	.9976	.9977	.9977	.9978	.9979	.9979	.9980	.9981
2.9	.9981	.9982	.9982	.9983	.9984	.9984	.9985	.9985	.9986	.9986
3.0	.9987	.9987	.9987	.9988	.9988	.9989	.9989	.9989	.9990	.9990
3.1	.9990	.9991	.9991	.9991	.9992	.9992	.9992	.9992	.9993	.9993
3.2	.9993	.9993	.9994	.9994	.9994	.9994	.9995	.9995	.9995	.9995
3.3	.9995	.9995	.9995	.9996	.9996	.9996	.9996	.9996	.9996	.9997
3.4	.9997	.9997	.9997	.9997	.9997	.9997	.9997	.9997	.9997	.9998
3.6	.9998	.9998	.9999	.9999	.9999	.9999	.9999	.9999	.9999	.9999

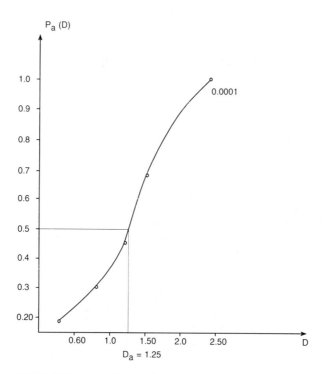

Figure 12.27. Determination of the mean diameter of the armor coat.

Solution

As suggested in the data, the step method presented by Komura and Simons is used for determining D_a.

Step 1: R_b is given as data of the problem. If R_b is not given, it must be determined using either Einstein's or Şentürk's approach (Simons, D. B., Şentürk, F., 1992). In this particular example $R_b = 0.35$.

Step 2: Determine U_{*b}

$$U_{*b} = (gR_bS)^{1/2} = (9.81 \bullet 0.35 \bullet 0.0001)^{1/2} = 0.01853 \text{ m/sec}$$

$$U_{*b} = 0.01853 \text{ m/sec}$$

Step 3: Compute U_{*c}

$$U_{*c,} = [a_c \left(\frac{\gamma_s}{\gamma} - 1\right) gD_s]^{1/2} \qquad\qquad 12.45$$

a_c is defined by Fig. 12.28.

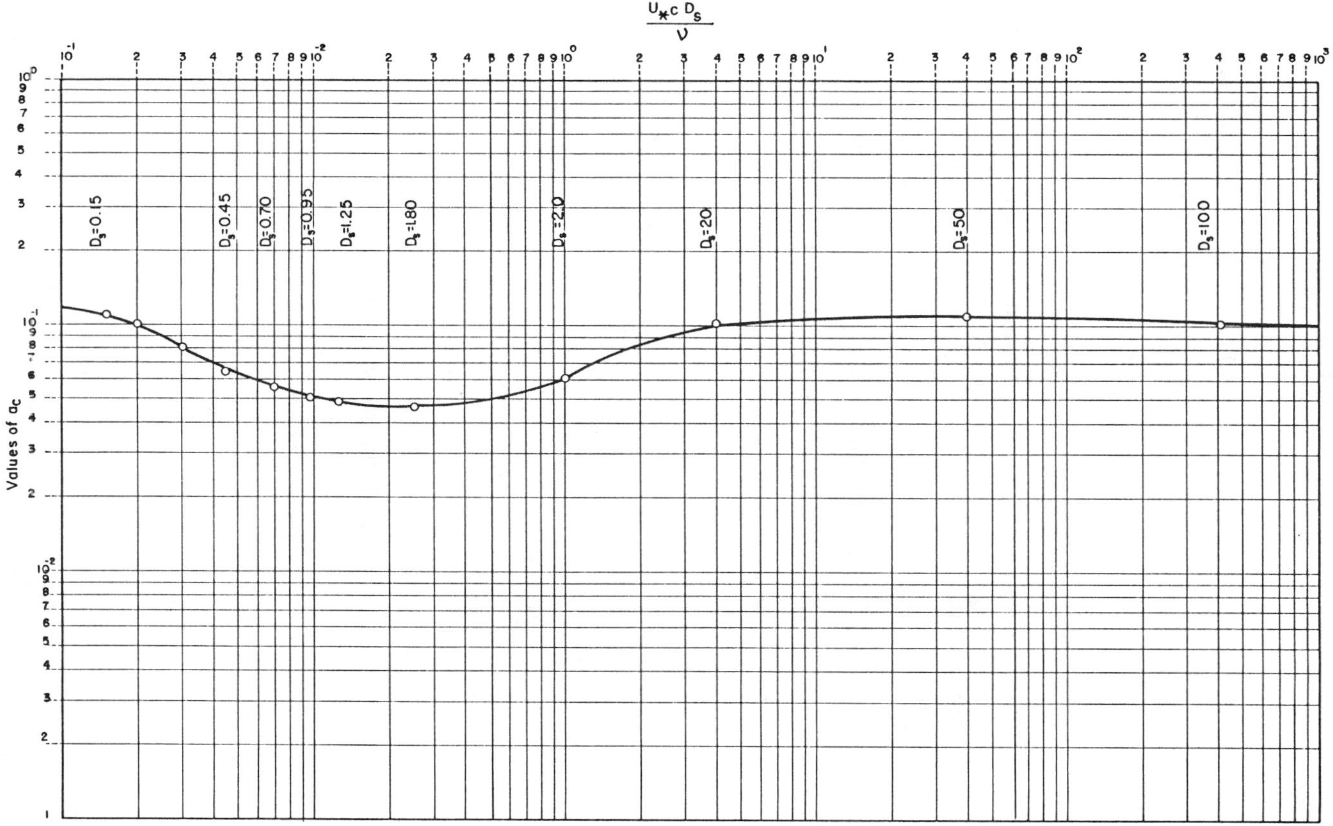

Figure 12.28. Curve $a_c = F(D_s, U_{*c} D/\upsilon)$.

Step 4: Prepare the grain size analysis as shown in Fig. 12.26:

- Determine the grain size intervals

  $\Delta 1$, from 0 mm to 0.6 mm

  $\Delta 2$, from 0.6 mm to 1.0 mm

  $\Delta 3$, from 1.0 mm to 1.2 mm

  $\Delta 4$, from 1.2 mm to 2.0 mm

  $\Delta 5$, from 2.0 mm to 2.8 mm

- Determine the representative diameter for each interval

  The representative diameter of interval $\Delta 1$, is $\quad D_{s1} = 0.30$ mm

  The representative diameter of interval $\Delta 2$, is $\quad D_{s2} = 0.80$ mm

  The representative diameter of interval $\Delta 3$, is $\quad D_{s3} = 1.20$ mm

  The representative diameter of interval $\Delta 4$, is $\quad D_{s4} = 1.50$ mm

  The representative diameter of interval $\Delta 5$, is $\quad D_{s5} = 2.40$ mm

- Determine the magnitude of chosen intervals

  $\Delta p_{o1} (0.3) = 0.6$

  $\Delta p_{o2} (0.8) = 0.4$

  $P_{o3} (1.20) = 0.4$

  $P_{o4} (1.50) = 0.6$

  $P_{o5} (2.40) = 0.8$

Table 12.12 shows different values of the variables as function of the chosen intervals.

TABLE 12.12

Intervals		D_s	$\Delta P_{oi} (D)$	a_c	U_{*c}	$\dfrac{\tau_c}{\tau}$	ξ	$q(D)$	$P_a(D)$
		mm	mm		m/sec				
1		2	3	4	5	6	7	8	9
0	- 0.6	0.3	0.6	0.080	0.020	1.16	0.28	0.61	0.14
0.6	- 1.0	0.8	0.4	0.055	0.027	2.12	1.96	0.97	0.30
1.0	- 1.4	1.2	0.4	0.050	0.029	2.45	2.54	0.99	0.45
1.4	- 2.0	1.5	0.6	0.048	0.034	3.37	4.16	1.00	0.68
2.0	- 2.8	2.4	0.8	0.062	0.049	6.99	10.51	1.00	1.00

Column 1:   See step 4

Column 2:   See step 4

Column 3:   See step 4

Column 4:   See step 3

Column 5: $U_{*c} = (a_c \gamma'_s g D_s)^{1/2}$

Column 6: $\dfrac{\tau_c}{\tau} = (\dfrac{U_{*c}}{U_{*b}})^2$

Column 7: $\xi = 1.754 \, (\dfrac{\tau_c}{\tau} - 1)$

Column 8:   q(D) values are given in Table 12.11

Column 9:   The values of $P_a(D)$ are obtained from

$$p_a(D) = \frac{\sum\limits_{D_{min}}^{D} q(D)\,\Delta p_o(D)}{\sum\limits_{D_{min}}^{D_{max}} q(D)\,\Delta p_o(D)}$$

Step 5:   Plot $P_a(D)$ versus D.

Step 6:   Determine D_a, the mean diameter of the armor coat from this curve (Fig. 12.27)

$$D_a = 1.25 \text{ mm}$$

D_a, corresponds to p_a (D) = 0.5, after Komura. Then for a given bed material the energy slope corresponding to the armor coat is $\xi = 0 = 1.754$ $(\tau_c/\tau) - 1)$

$$\frac{\tau_c}{\tau} = 1 \quad \tau_c = \tau$$

The greater value of τ_c being $\tau_c = \rho \, U_*^2$

$$\tau_c = 0.102 \cdot 0.049^2 = 0.00024 = \gamma dS$$

$$S = \frac{0.00024}{0.35 \cdot 1.00} = 0.0007$$

Then, $\xi = 0$ and $q(D) = 0.50$; the corresponding value of D_a is 2.32 mm.[*]

12.6   PROBLEMS

Problem 12.01

Determine the depth of erosion in a confined reach where

$$B_1 = 120 \text{ m} \quad B_2 = 75 \text{ m} \quad q'_{b1} = 2 \text{ kg/sec.m} \quad q_1 = 15.00 \text{ m}^3/\text{sec m}$$

using the Meyer-Peter and Müller formula.

Data

$$\gamma'_s = 1700 \text{ kg/m}^3 \quad S = 0.002 \quad D_m = 0.00015 \text{ m} \quad \rho = 102 \text{ kg.sec.m}^4$$

B_1 is the width of the cross section

B_2 is the width of the confined cross section

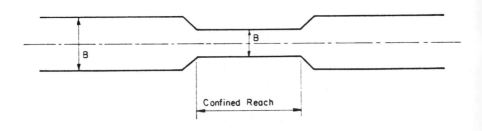

Fig. 1. Problem 12.01.

Problem 12.02

Determine the clear distance necessary to carry a discharge of 800 m³/sec through a specific cross section of a bridge crossing. Also determine the depth of possible scour, locate piers in the channel cross section and compute the magnitude of the local scour around piers and embankments.

---

[*] $U_{*c} = (a_c \gamma'_s g D_5)^{1/2}$ 12.45

$(0.049)^2 = 0.062 \times 1.7 \times 9.81 \text{ D}_s$

$D_s = 0.00232 \text{ m} = 2.32 \text{ mm}$

Data

- Stage-discharge curve at the selected section, see Fig. 1, Problem 12.02.

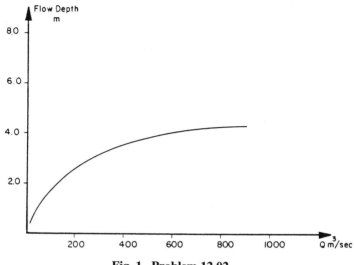

Fig. 1. Problem 12.02

- The representative size of the bed particles, $D_{50} = 3$ mm

- Design discharge = 800 m^3/sec

- Specific weight of the bed particles, $\gamma_s = 2.7$ T/m^3

Problem 12.03

Determine the scour depth S, which will take place downstream of the sluice gates shown in Fig. 12.15, for different gate conditions (Fig. 1, Problem 12.03).

Data

- $D_{90} = 9.00$ mm

- $D_g = 5.00$ mm

- For $q_d = 3$ m^3/sec.m, the upstream water elevation is 12.00 m and the gate operates as a sluice gate, a = 0.4 m (Fig. 1a, Problem 12.03).

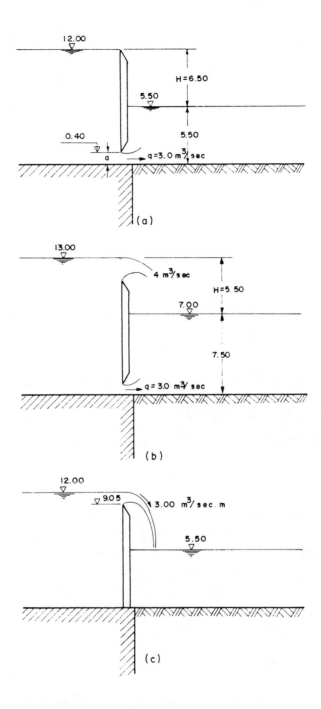

Fig. 1. Problem 12.03.

- For $q_u = 4.00$ m³/sec.m the upstream water elevation is 13.00 m; and with $q_d = 3.00$ m³/sec.m, the gate operates as a sharp crested spillway and sluice gate (Fig. 1b, Problem 12.03).

- For $q_u = 3.00$ m³/sec.m, the upstream water elevation is 12.00 m and the gate operates as a sharp crested spillway (Fig. 1c, Problem 12.03).

- The tailrace characteristics:

    For $q_d = 3.00$ m³/sec.m,   d = 5.5 m (sluice gate) a = 0.4 m.

    For $q_u = 4.00$ m³/sec.m   and $q_d = 3.00$ m³/sec.m,

    d = 7.50 m.

- The flow is bidimensional

- The approach velocity is small so that the velocity head can be taken equal to zero.

- The opening of the sluice gate is 0.4 m.

- $\varnothing = 38°$ $\gamma_s = 2.70$ T/m³,   and $w_{50} = 0.58$ m/sec

Problem 12.04

It is required to determine the slope of the channel, the characteristics of which are given for Q = 35.00 m³/sec, so that the flow is in equilibrium. Determine the value of S so that X = Y = 1.00 at the limit.

Data

The cross section of the channel is shown in Fig. 1, Problem 12.04.

$\gamma_s = 1.7$ T/m³

$k_s = D_{98} = 0.02$ m

$\rho = 102$ kg.sec²/m⁴, and

$\nu = 10^{-6}$ m²/sec.

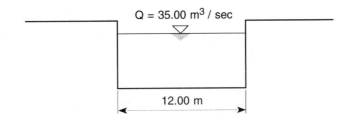

Fig. 1. Problem 12.04.

Problem 12.05

Check the results obtained in the solution of Problem 12.04 using the Shields criterion.

Problem 12.06

Study the variation of the flow characteristics as a function of D_{98} when the flow is in equilibrium, for the previous problem.

Problem 12.07

Compute the degradation of the bed of a river downstream of a storage dam assuming clear water is released from the reservoir, the river banks are not erodible, seasonal variation in discharge and temperature of water is negligible, there is no sediment injection by tributaries, meandering does not occur, and vegetation effects are negligible. Additional information:

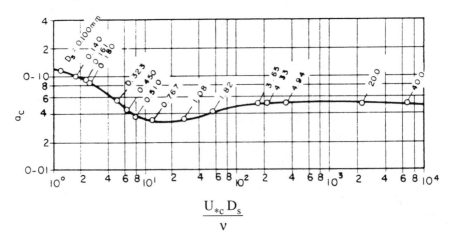

Fig. 1. Problem 12.07.

- The sediment properties of the reach are summarized in Table 1, Problem 12.07

TABLE 1, PROBLEM 12.07

D_{16}	D_{50}	D_{84}	D_{90}	σ	a_c	γ_s	c	D_{sfo}
mm	mm	mm	mm			T/m^3		m
0.5	0.7	1.2	1.3	3	0.035	1.65	0.00010	0.0005

- The discharge of the river is assumed equal to 2500 cfs,

- The variation of the width of the river is assumed as shown in Table 2, Problem 12.07.

TABLE 2, PROBLEM 12.07

Stations	x'	B	Y_{fo}	α
	ft	ft	ft	ft
5	35000	500	-	1
4	30000	400	-	
3	20000	450	-	
2	10000	380	-	
1	5000	400	-	
0	0	400	3.00	

Problem 12.08

Determine the mean diameter of the armor coat of a river, using Komura and Simons' approach.

Data

- The grain size distribution is given in Fig. 1, Problem 12.08

- $a_c = f(D_s, U_{*c}D_s/\nu)$ is given in Fig. 12.28

- The channel is wide so that d = R

- $R_b = 0.25$, g = 9.81 m/sec², and $\gamma'_s = 1.7$ T/m³

- S = 0.0002

Discuss the influence of the channel slope on the armor coat.

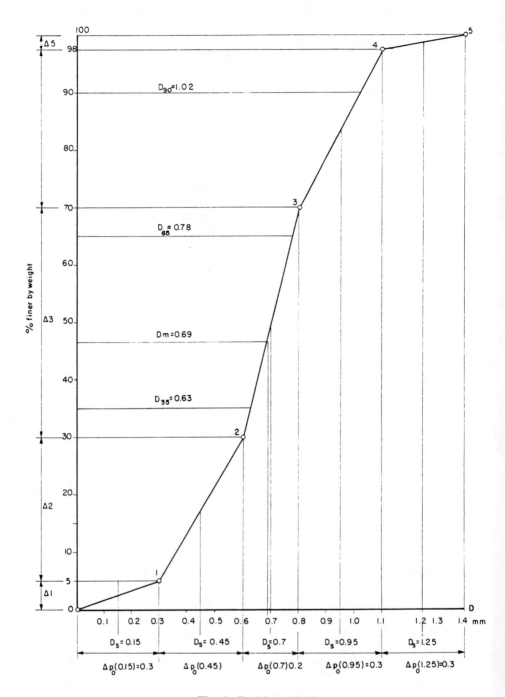

Fig. 1. Problem 12.08.

Chapter 13

MODEL TECHNIQUES USED FOR SOLVING SPECIAL FLOW PROBLEMS

13.1 INTRODUCTION

The hydraulic design of dams was discussed in previous chapters. It was seen that many hydraulic problems could not be solved satisfactorily by the use of the computational approach. The use of a scale model is a necessary helpful tool in solving them.

Scale models have been especially used since the first half of 20th century. Many hydraulic laboratories have been built throughout the world. A variety of hydraulic problems have been solved by these institutions. The golden era for the use of scale models lasted until the end of the first half of the 20th century. Since then, there has been a tendency to use computers more and laboratories less, especially since physical model studies are more time consuming and costly. It became apparent that similar accuracy was possible in a shorter time and at less cost by using computational methods.

Many hydraulics laboratories around the world have shifted their field of interest toward thermodynamics today. But even under these circumstances there are problems which can only be solved by means of hydraulic model studies. These problems are defined and enumerated in this chapter enabling the designer to decide which procedure is best for the resolution of the specific hydraulic problem to be solved.

There are many questions that cannot be answered by model studies conducted on industrial models; special models must be constructed for solving these kinds of problems. Some of these are cited below:

- Problems concerning boundary layer and rugosity

- Problems concerning atmospheric pressure which is a constant for both the model and the prototype;

- Problems concerning friction of water and air in the jet flow; and

- Generation of vortices.

These different cases are studied and the accuracy of obtained results are discussed in the following sections.

675

13.2   MODEL STUDIES OF SPILLWAYS

Hydraulic relations are derived, in general, by simplifying the natural phenomenon. The formulas obtained are used if they have the overall applicability and the results are accurate enough for practical application. The discharge formula pertaining to the spillway is a good example. This relation is obtained assuming that,

- the stream lines are orthogonal to the spillway axis;

- the approach velocity is uniform in cross section

- the approach velocity is uniform along the longitudinal section; and

- the helicoidal current does not exist.

In nature these conditions are in general not too clear cut. The stream lines are generally oblique and a certain angle exists between their direction and the spillway axis. The flow is not uniform across the cross section and helicoidal currents can generally be encountered. How can the discharge be calculated under these conditions? It is customary to say that the choice of the discharge coefficient can resolve the difference between the empirical and the actual. Here, the experience of the hydraulic engineer is assumed to be adequate to make the proper choice of the coefficient. Naturally such an approach cannot be accepted scientifically.

Scale models can solve these problems and since the early part of the 20th century engineers have tried to improve the model technique for achieving a better result (Şentürk, F., 1947). Their ceaseless efforts have shown that some hydraulic phenomena can indeed be approximated by this technique; others cannot.

Hydraulic problems concerning flow over spillways can be studied in four groups as shown below:

1   Flow approaching the hydraulic structure;

2.   Flow taking place over a hydraulic structure;

3   Flow in the downstream channel and in the energy dissipators; and

4.   Flow in the natural water course downstream of the dam.

The hydraulic problems concerning these different phases of the flow will be considered separately and the possibility to be solved by computational approaches or by scale model techniques are discussed in the following subsections.

13.2.1 Problems Concerning Flow Approaching the Spillway

Approaching flow is defined as the flow in the reservoir or in the approach channel, moving towards the hydraulic structure, which is in this case the spillway body. If the spillway is located on the dam body as in the example of Bangala Dam Spillway (Fig. 2.5), the spillway is in direct contact with the reservoir. Sometimes flood water is conveyed to the spillway by means of a channel called the *approach channel*. If the spillway blocks the approach channel (Fig. 2.1a), it is then situated at the far end of it.

Each case has its own special problem which may be solved by its own specific analysis.

A. Spillway is on the Dam Body

The reservoir is assumed large enough so that there is no particular current taking place in it. The water moves as a large block in approaching the spillway so that the stream lines are perpendicular to the spillway axis. If there is a deviation from the 90° approach, its effect is assumed to be negligible.

Many field studies have been made of flows at existing spillways. It has been remarked that although some of these assumptions are valid, many are not. Some of these observations can be described as follows:

• Stream lines follow a special path as a function of the geometry of the upstream valley; the water in the remaining area stays stagnant, without any motion or turns around as large vortices of small velocities. This special path has a direction which is not necessarily orthogonal to the spillway axis.

• The path followed by the approaching flow is not stationary, it changes its position as a function of the water level in the reservoir.

• In case of a V-type valley the mean velocity of the deep approach channel can be so large compared to the surface velocity that the flow hits the dam wall at the bottom of the valley, is projected up towards the water surface and then reflected back downstream by the spillway. In this case the flow in the spillway bays is erratic. If the reason for its variation is not properly detected, the designer will be obliged to use very costly solutions, as in the case of the Keban Dam Spillway (Fig. 2.20).

The following criteria may be taken into consideration for solving different aspects of flow which can be encountered in the approach phase.

a. The effect of skew of the approach flow

This problem can be solved only by the use of a physical model. Computational models are not efficient.

The cause of the skew is determined first. Then the model must be designed to reproduce the conditions causing the skew as accurately as possible. The results obtained from the model must then be scaled up to actual conditions in the field.

Skew can be corrected by the use of special shapes for the side piers and abatements adopted to each particular case. The effect of the skew is such that,

- the discharges, in the side bays particularly, are very different from each other,
- shock waves are generated in side and middle bays,
- shock waves are generated in the downstream discharge channel, and
- energy dissipators cannot function efficiently.

Figure 13.1 shows flow taking place in the side bay of a spillway due to skew. It is clear that the stream lines contraction in the side bay is quite different from that in the neighboring bays.

Measurements on the model and in nature show that the discharge coefficient can be decreased as much as 60%. In the case of weirs it has been observed in some cases that the side bay discharge has been negative, which means that the flow had been reversed and an inverse flow had taken place in the side bay.

Figure 13.1. Skewed flow at the side abatements of a spillway.

Different measures have been taken to correct the flow shown in Fig. 13.1. Figure 2.20 illustrates one solution adopted. This spillway has been operating for a decade and it seems that the solution adopted is adequate. But as can be observed in Fig. 2.20, the adopted solution required the construction of unusual forms that were costly. The shape of piers shown in Fig. 2.20 was developed from physical model studies without determining the actual cause of the skewed flow. Additional studies have been conducted to determine the cause. Figure 13.2 shows the cross section of the spillway.

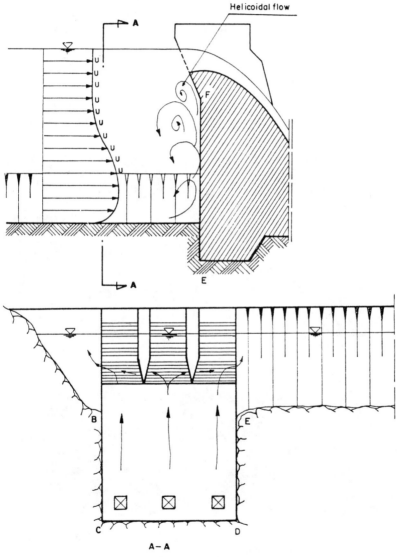

Figure 13.2. Generation of skewed flow due to the existence of the submerged approach channel.

The velocity diagram is indicated on the same figure. The power intake was placed in the submerged approach channel BCDE. The approach velocity was higher in the channel. The current, striking the upstream face (EF) of the spillway was projected with great velocity towards the reservoir surface. A helicoidal flow was generated beneath the upstream extension of the spillway and directed towards both sides. Flow was then reflected back into the approach channel, generating further disturbance.

A possible solution to the problem might be to modify the shape of the approach channel so that helicoidal flow would not develop at the entrance. It is possible that such construction would have been more economical than that used. However it is clear that a computational model could not have solved this problem either.

Prevention of shock waves and cross waves necessitates the construction of special scale models. Shock waves can be observed in Fig. 13.1 and cross waves in Fig. 13.3.

Figure 13.3.   Cross waves on the model of the Buldan Dam discharge canal, Turkey.

The characteristics of cross waves are enumerated below:

1. Cross waves are not steady; they travel in both cross sectional and longitudinal directions;

2. The height of the cross waves is also variable.

If these two characteristics of cross waves are not studied seriously enough, overtopping of the side walls of the downstream channel can occur causing heavy erosion on the outside of the channel walls.

The existence of cross waves is also a hazard for energy dissipation, devices. For instance, if a ski jump is used for energy dissipation the cross waves will so disturb the flow that only a portion will be projected in the air and the remaining portion will fall directly on the flip bucket foundation causing heavy erosion.

Figure 13.4 shows skew on the ski jump of Seyhan Dam.

Figure 13.4. Effect of skew on ski jump.

B. Model Study of Skew and Related Problems

It is evident that computational models cannot reproduce adequately the skew problem. The study of skew on a scale model is also very difficult. Field observation by the hydraulic engineer may detect the reasons for the development of this phenomenon. It takes time to collect sufficient field data and then to build the scale model. Designer and model engineer must work together to schedule the completion of the model tests and the design so that the construction schedule is not delayed.

Possible solutions include:

1. The correction of the direction of the spillway axis relative to the flow direction. If geologic and design conditions do not permit such a correction, this solution cannot be applied. For example, such a solution was adopted for Aslantaş Dam, Turkey.

2. The geometry of side piers and abatements can be studied on scale models for correcting the influence of skew. This method was used for the Keban Dam Spillway, Turkey and Ben-Metir Dam Spillway (North Africa).

This solution is not perfect, but is helpful (Figs. 13.5 and 13.6). A direct consequence of skew is the generation of cross waves.

Some elementary computational approaches exist for studying cross waves (Chow, V.T., 1959). The work of Ippen and Dawson (1951) in this field is also remarkable. They studied the cross waves generated within the parallel wall contractions which are similar to those generated in downstream discharge channels. Cross waves observed in circular channels

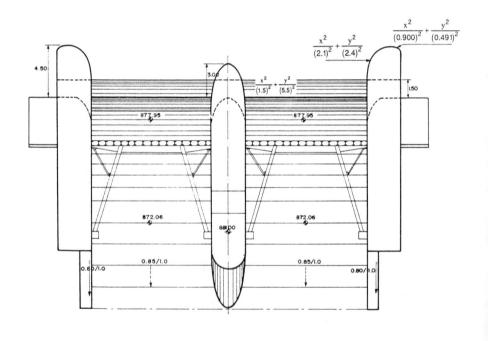

SECTION A-A

Figure 13.5.   Plan of a spillway designed for correction of skew (see abatements).

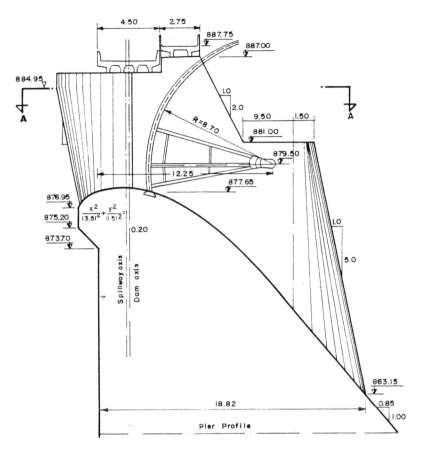

Figure 13.6. Longitudinal section of a pier designed for correcting the skew.

were studied by Ippen (1951) who has defined the superelevation caused by cross waves as follows:

$$y = \frac{U^2}{\gamma} \sin^2 (\beta - \frac{\theta}{2}) \qquad (13.1)$$

where y is the height of the cross wave

U is the mean velocity

β is $\sin\sqrt{\frac{gy}{U}}$, and

θ is the central angle of a pair of cross waves (Fig. 13.7)

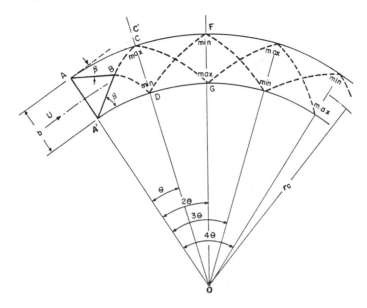

Figure 13.7. Cross waves pattern for supercritical flow in a curved canal.

This equation is valid under the following conditions,

- the flow is bidimensional
- U is considered as constant
- $S = 0$
- the flow is frictionless
- the cross section is rectangular

Ski jumps are also influenced by cross waves which are a result of skew. Correction of skew will also lead to perfect operation of the ski jump. The construction of a model for studying the cross waves in a steep channel, say the downstream channel of a spillway, necessitates particular study of the model scale.

13.2.2 Problems Concerning Flow Taking Place in the Approach Channel

Flow in the approach channel causes a variety of problems which should be solved before adopting a final solution (Fig. 13.8). Open channel problems can be solved by direct application of uniform flow formulas if uniform flow conditions prevail. The first thing to do is to check the state of

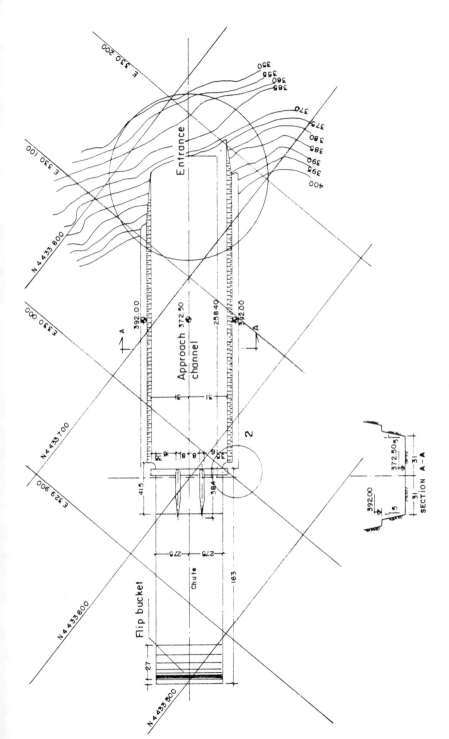

Figure 13.8. The approach channel of Gökçekaya Dam Spillway before correction.

flow. In general, approach channels are very short and uniform flow cannot be established because of this property. If the designer is sure that uniform flow takes place in the upstream approach channel, computational models can solve the problem easily (Example 3.01). Generally the perturbation caused by the entrance condition destroys the uniformity of the flow and if a curve exists in plan, the water surface becomes inclined in cross section so that the flow depth at the inner side of the curve becomes deeper than the flow depth at the outer side of it. Such perturbation can also be studied computationally if it is not accompanied by extra perturbations. The error that may be reached by the use of a computational model should be compared with the error that may be caused by a physical model. Based upon this type of evaluation, the computational model may be preferred to the physical model.

The following sources of perturbations are to be considered:

1. Evaluation of head losses at the entrance of the approach channel

2. Determination of the separation zones in the channel

3. Problem caused by the flow entering the approach channel from the reservoir by overtopping the guide walls

4. Problem related to the transition from the channel to the spillway.

In summary, if these perturbations and the channel boundaries can be expressed mathematically, open channel flow problems can be solved by computational models more quickly and economically than by scale models.

Problems that involve changes in channel depth to facilitate uniform flow or the determination of the type of channel lining to be used can also be solved by scale models.

These problems outline the diversity of flow types which can be encountered in the approach channel. They cannot be studied on a single scale model because their characteristics are different and the available measuring devices are not sensitive enough to measure very small perturbations. Models with a larger scale may be used to amplify the perturbations which may, in turn influence the flow in the approach channel. If an intermediate scale is used, it may be a source of error. The possible error must be evaluated as a function of the magnitude of the approach flow. A 10% error in a flood discharge of 17,000 m³/sec for example (Keban Dam Spillway) corresponds to 1,700 m³/sec which is significant.

A. Evaluation of Head Losses at the Entrance of the Approach Channel

The head losses at the entrance of the approach channel can be evaluated using head loss formulas given in previous chapters (see Chapters 3, 4, 5, 6).

These relationships have been derived as a function of the mean velocity U. But U cannot always be obtained at the entrance of the channel (Fig. 13.9). In some instances the layout of the entrance is so inadequate that the velocity at the inner side of the channel entrance is negative. This means that the flow enters along one side of the entrance channel, but a portion is deflected and reverses its direction, traveling backward along the opposite side of entrance channel. In such a case the head loss cannot be evaluated

Figure 13.9.   Flow at the entrance of the Gökçekaya Dam Spillway's approach channel (before correction).

by a formula of the form $k(U^2/2g)$. The mean velocity can be established in the channel far downstream of the entrance, if the channel is long enough. If not, a mean velocity, U, obtained by using the continuity equation cannot be used. If the design is based on such assumptions the risks of overtopping will be great. A scale model is necessary to solve this type of problem. The entrance should be corrected in such a way that the entrance flow is smooth as shown in Fig. 13.10. Figure 13.10 illustrates the results after the correction adopted in this particular case. The gain in available head was 0.55 m, which is considerable. This figure shows the solution adopted on the scale model of the channel entrance to the Gökçekaya Dam Spillway. Dye was used in the model to highlight the homogeneous flow that resulted from the corrected entrance.

    If skew exists in the approach flow, the solution used at Gökçekaya can be dangerous. The approach flow conditions were the subject of serious discussions between the designer and the model engineer before adopting

**Figure 13.10. Flow at the entrance of the Gökçekaya Dam
Spillway's approach channel (after correction).**

the configuration used. This flow condition was checked in the field during
the first spilling and it was seen that the similitude was perfect (Model Study
Report of Gökçekaya Dam Spillway DSI-HML Report).

In many cases the results obtained from a physical model can be checked
by a computational model. If there is an important discrepancy, it is
recommended that the similitude conditions of the physical model be
reconsidered. The accuracy in a scale model is a function of the sensitivity
of the measurements and the measuring devices used. A simple example is
cited here to illustrate this statement. In a model constructed at a scale of
1/100, 1.00 m is equivalent to 10.00 mm. Thus a reduction of 0.55 m in the
head loss obtained after the correction of the entrance in Gökçekaya Dam
Spillway approach channel is equal to 5.5 mm on the model. This length
can be measured accurately on the model, since the sensitivity of the
measuring device is high enough to satisfy such a measurement. Assume
now that the quantity in question is equal to 5.5 cm instead of 55 cm. In this
case the corresponding length on the model is:

$$\frac{55}{100} = 0.55 \text{ mm}$$

A point gauge is generally used for measuring water levels on models, and
the sensitivity of such an instrument is ± 1.0 mm. In this case 0.55 mm
cannot be measured accurately. The sensitivity of a vibrating gauge is

0.5 mm. This kind of measuring devices may be used on the model or the scale of the model must be corrected. A smaller scale corresponds to a higher cost of construction of the model. This cost may be so great that a computational model is substituted for a physical model. Error analysis can be used to govern the choice of model scale and the type of model.

B. Study of Separation Zones in the Approach Channel

Separation of flow in the approach channel is caused by design deficiency of the entrance. Separation decreases the effective flow area. The vortices caused by the helicoidal flow in the separation zone can cause erosion of the channel bottom. Fig. 13.9 shows such a separation zone. Correction in the design of the entrance structure can eliminate the separation of the flow from the boundary (Fig. 13.10). If there is a need to study the vortices resulting from skew, a distorted model can be built. Even with the distorted model, perfect similitude of the helicoidal flow is impossible.

This problem can be studied on a scale model only, because if skew exists the flow in the approach channel will be three-dimensional. Separation zones cannot be studied adequately by a computational model due to the complexity of the boundary conditions.

C. Study of Uniform Flow in the Approach Channel

Even if the above mentioned measures are taken at the entrance, uniform flow cannot be established in the approach channel due to:

- the perturbations at the entrance

- the influence of the bottom slope

- the influence of the cross section of the channel, and

- the insufficient length of the channel.

If the direction of the channel axis is appreciably different from the direction of the approaching flow the skew of the flow cannot be corrected. In general the flow direction is not constant in the reservoir. A small dissymmetry will always exist in the channel due to the entrance conditions. If the skew due to this dissymmetry is important enough, the alignment of the canal in plan may be corrected. This correction can only be obtained by careful studies; computational models are of no use in similar cases.

To obtain a uniform flow in the approach channel the following steps may be taken by the designer and the hydraulic engineer before constructing the physical model.

a. Select the bottom slope of the approach channel

The bottom slope can be chosen in such a way that uniform flow is established in the shortest channel length. If space is not available for obtaining such a length, then the bottom slope can be selected equal to zero. In many cases the bottom slope is negative in order that stagnant water does not remain in the channel during drawdown. Stagnant water causes pollution.

If the bottom slope is equal to zero or negative, then gradually varied flow takes place in the channel.

b. Consider the influence of bends in plan

In many cases bends can not be avoided in plan. The diameter and the central angle of the bend can be chosen in such a way that perturbances can be reduced to a minimum.

c. Select the cross section

The cross section adopted in approach channels is generally rectangular or trapezoidal. The bottom angle in a trapezoidal section may be chosen so that the flow taking place in it is stabilized within the shortest distance after it enters in the channel. This should be carefully checked by the designer and the hydraulic engineer and the ratio h/B be chosen accordingly.

If such precautions are taken and the establishment of uniform flow in the approach channel is assured, a computational model can solve the problem. Then a physical model in similar cases will not be necessary (see Chapter 3).

D. Problems Concerning the Lining of the Bottom and Side Slopes

Since the bottom and side slopes of the approach channel are subject to erosion by the flow, a protective lining is often required. If the approach channel is long, it is possible that some portions are less subject to wear than others. A scale model study will show whether portions may be left unlined to reduce the cost of construction.

A scale model can show the distribution of the shear stress on the bottom and side slopes of the channel. The reach where the shear is high will be lined. The direct measurement of shear stress is difficult and necessitates special measuring devices. However, there are two simple ways to calculate the shear,

1. The energy line slope is measured at different cross sections and at different longitudinal sections along the channel. The shear stress is computed as

$$\tau = \gamma RS \qquad (13.2)$$

and the mean value determined.

2. The local velocities are measured at different cross sections and at different longitudinal sections along the channel

$$\frac{u}{U_*} = 5.75 \log (30.2 \frac{y}{k_s} x) \qquad (13.3)$$

where u   is the measured local velocity (m/sec)

U_* is the shear velocity (m/sec)

y   is the depth of flow where u is measured (m)

k_s is the rugosity of the bottom or the side slope and

x   is a correction factor due to the viscosity defined by Einstein (Fig. 13.11).

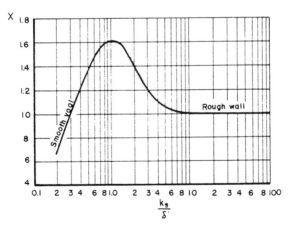

Figure 13.11. Values of x.

x is obtained from Fig. 13.11, k_s is measured in situ or previously measured and published values can be used. In an unlined channel, $k_s = D_{65}$ after

691

Einstein (Simons, D. B., Şentürk, F., 1992), y and u are measured in situ then U_* is obtained as

$$U_* = \frac{u}{5.75 \log (30.2 \frac{y_x}{k_s})}$$

and (τ) is then

$$\tau = \rho U_*^2 \qquad (13.4)$$

Another simple approach is to measure U, in situ, and adopt, according to this value of U, an adequate lining. It is sufficient to measure the local velocity at (1/6) of the depth for obtaining U. This particular problem cannot be solved by using a computational model. In situ measurements are necessary for determining the reach subject to heavy erosion.

E. Problems Caused by the Flow Overtopping the Guide Walls

Figure 13.12 shows the upstream guide wall of Hasan Uğurlu Dam Spillway's entrance channel and the velocities of the approaching flow

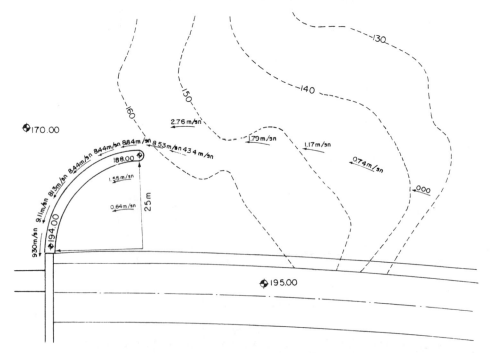

Figure 13.12. Measured velocities in the reservoir of the Hasan Uğurlu Dam Spillway near the upstream guide wall.

measured on the model. The top level of the guide wall is variable and increases from elevation 188 up to elevation 194. During a flood of $Q_{max} = 11,000$ m³/sec the maximum water level in the reservoir has been measured to be elevation 191.65. Figure 13.13 shows the model functioning for this flood discharge.

It is clear that the reservoir level being higher than the top of the guide wall, water will enter the approach channel by overtopping. Figure 13.13 shows the perturbations occurring in the side bay due to this phenomenon. The discharge capacity of the side gate decreases considerably. Taking into account that velocities observed in the approach channel increased from 4.34

Figure 13.13. Spillway of the Hasan Uğurlu Dam discharging Q=11,000 m³/sec on the model.

m/sec up to 9.30 m/sec, the difference between the flow level in channel and reservoir increases gradually from the upstream end of the guide wall to the spillway gate, and it reached

$$\frac{9.30^2}{19.62} = 4.41 \text{ m}$$

Adding the entrance losses to this value, it becomes easy to see that water drops into the approach channel from a height of around 4.50 m. If necessary measures are not taken in the approach channel, heavy and dangerous erosion can take place. In many cases the guide wall is endangered after such a flow. Furthermore, since the side bay discharge is diminished, the reservoir level rises and the freeboard decreases and the dam itself is endangered.

The problem can only be solved on a scale model and computational models cannot be used for solving similar problems. The model engineer must establish the following criteria:

1. Determine the decrease of freeboard as a function of the flood discharged from the spillway.

2. Determine the percentage of this decrease as a function of the correction adopted.

3. Determine the erosive capacity of the flow entering the channel by overtopping.

4. Determine the geometry of the guide wall in such a way that steady flow can take place in the approach channel.

Model studies can give satisfactory answers to these questions.

13.2.3 Flow Problems at and around Spillways

Flood water carried to the spillway poses a different kind of problem, enumerated as follows:

1. Problems pertaining to free flow spillways, such as

   - Ogee type spillways
   - Broad-crested spillways
   - Shaft spillways
   - Side spillways
   - Siphon spillways
   - Lateral spillways

2. Problems pertaining to gated spillways such as:

   - Determination of the gate seat position relative to the dam axis, the gate radius and the angle, θ, in case for radial gates (Fig. 4.1)

- Determination of subpressures and the coefficient of discharge (m) as a function of gate opening

- Determination of the spillway profile as a function of the gate characteristics.

- Study of gate vibration

- Problems concerning gate operation.

These problems cover a large field of engineering. Some can be solved easily using computational models. Others cannot be solved by model techniques. The questions which can be asked of the model engineer by the designer will be enumerated below, and the kind of solution expected from physical and computational models are described.

A. Free Flow on Ungated Spillways

The ungated or free flow spillways over which flow takes place freely are the type of hydraulic structures that correspond closely to the theoretical assumptions. Many problems on this type of hydraulic structures may be worked out on the physical model so that an acceptable solution can be reached. These problems are explained in the following subsections.

a. Ogee-type spillways

The profile of a free flow spillway may be worked out in such a way that two basic conditions are satisfied:

1. The value of the discharge coefficient must be about 0.52 while discharging the maximum flood

$$m \leq 0.52$$

2. The value of the subpressure must be about -0.4kg/cm^2 and less when discharging the maximum flood.

If the first condition is not satisfied, the nappe is broken, air is introduced between the face of the hydraulic structure and the lower nappe of the jet and pulsating flow can take place (see Chapter 3, Fig. 13.14). This phenomenon cannot be studied on a scale model due to the fact that the atmospheric pressure is usually not simulated on an industrial model. Scientific research can be undertaken on a particular model with reduced pressure. In this case vacuum may be applied to the model which may be

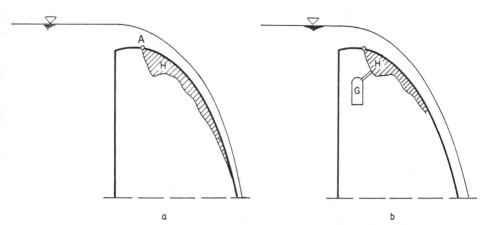

Figure 13.14.   **Vacuum area located between the hydraulic structure and the lower nappe.**

developed in a confined area. Then in the practice the accuracy of the result obtained from a scale model may be not higher then the accuracy which can be reached using a computational model.

As a general rule, it is possible to state that if the stream lines are perpendicular to the spillway axis the computational model may be preferred to the physical model. The answer to the problem is in the diagram

$$Q = f(h)$$

For a profile where the lower nappe is aerated, the problem presents some particularities. Figure 13.14 shows a free flow spillway and the flow taking place on it. The nappe plunges into the air downstream of point A. Air is confined between the nappe and the dam body and is entrained by the flow in such a way that vacuum is built up in the volume shown by H. It is costly to study a similar problem on a physical model as will be explained later. A solution for this condition may be the aeration of the volume by supplying air as shown in Fig. 13.14b. The study of air demand and the determination of dimensions of the gallery is also difficult on a scale model. Formulas derived from in situ measurement exist enabling the designer to reach a decision (subsection 4.3.2). These formulas should be used in the design. Then a mathematical analysis can solve the problem.

It is recommended that a computational model be used to obtain the answer to the problems enumerated below and encountered in free flow overfalls:

1. Aeration of the inner nappe of a flow taking place on a free flow spillway

2. Determination of the dimensions of the air gallery

3. Study of pulsating flow observed on free flow spillways

4. Air entrainment problems in the discharge channel downstream of free flow spillways and the variation of the flow depth due to the entrained air.

5. Study of cross waves.

In general all the problems connected with atmospheric pressure may be studied using computational models recognizing the mathematical relations obtained from in situ measurement. Also the subpressure buildup on the spillway profile are studied and its variation measured (see Chapter 3). The spillway profile may be shaped according to measurements suggested in Chapter 3. The profile is chosen in such a way that the subpressure is smaller than the critical value while discharging Q_{max}. In general physical models are used for solving this problem. Variation of pressure is obtained by using a measuring system as shown in Fig. 13.15.

Figure 13.15. Measurement of pressure on model using pressure tape.

The pressure buildup on the profile is measured by pressure tapes and a special measuring device. The connection between the tapes and the measuring device is provided by tubes. Piezometer tubes are generally used in such a system. The tapes are formed by small holes, 5 mm in diameter.

A plastic cylinder enters this hole. One extremity of the cylinder is open, the other extremity is closed by a porous membrane. The diameters of the pores are about 0.5 mm. This system is located on the spillway profile in such a way that its outer profile coincides perfectly with the spillway's model profile (Fig. 13.16).

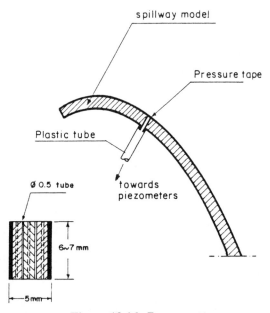

Figure 13.16. Pressure tape.

If the outer profile of the tape does not coincide exactly with the spillway profile, vacuum builds up immediately and this local artificial pressure does not represent the pressure which exists on the spillway profile (Fig. 13.17).

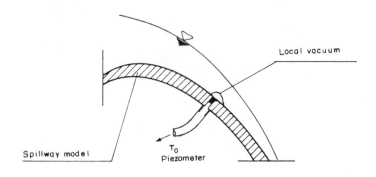

Figure 13.17. Local vacuum building up around the pressure tape.

This particularity may be checked carefully by the model engineer. Another item requiring great care is precision in reading the measurements of water levels in the piezometers. Because of the meniscus formation in the glass tube the accuracy may be within 1 mm. However, the water level in the piezometer is not constant, it vibrates and varies with time. The model engineer must determine the mean value of this variation and read the gauges at the proper moment. The involved error should not be higher than 5 mm, which corresponds to 0.50 mm in the 1/100 scale model.

The use of strain gauges is sometimes recommended for measurement of the pressure variation. Although the above mentioned source of error is eliminated with strain gauges, different other kinds of error are possible. The strain gauge works in a wet area. The accuracy of the measuring system should be checked before each new set of measurements are taken. This complicated procedure takes time and if not performed adequately, gives erratic results.

Measured pressures must be checked mathematically. Then a computational model must be used together with a physical model. In normal cases the designer may prefer the use of a computational model rather than a scale model. The scale model is recommended only for special profile such as a basket handle profile or the like and for skew problems.

b. Special Types Spillways

The special type spillways most often encountered in engineering practice are:

- broad-crested spillways

- shaft spillways

- side spillways

- lateral spillways, and

- siphon spillways.

The study of these hydraulic structures by using model engineering poses problems enumerated below.

- Broad-crested spillways

If the flow approaches the spillway orthogonally a computational model may be substituted for a scale model.

- Side spillways

The flow which takes place on a side spillway (Fig. 13.18) is not much different from the flow on an ogee spillway. The side spillway itself is also an ogee; but the flow taking place in the downstream channel poses problems which can be studied on a physical model. The flow on and around a side spillway is subject to Froude's law; it then can be studied on a scale model. In side spillway theory, it is assumed that the structure can be studied as a short structure (see Chapter 5). It is clear that all the head losses cannot be equal to zero. A scale model can include these losses, but it is difficult to introduce them in a computational model. It is suggested that side spillway problems be studied on physical models.

Figure 13.18. Side spillway in nature.

The problems which can be encountered on a side spillway are enumerated:

- Problems concerning approach flow conditions. These kinds of problems may be studied on physical models which can include skew problems.

- Problems concerning flow taking place on the spillway. The problems that may be studied on a physical model include:

- The determination of the flow profile in the attraction zone

- The variation of the discharge coefficients (m)

- The flow deviations and velocity profiles

- The evolution of subpressure, and

- Problems concerning flow in the downstream channel.

The last is a problem of a change in the direction of flow. It takes place as a function of Froude's law and is accompanied by high turbulence and head losses. The correction to be adopted is unique for each case and worked out individually for each case. The only technique to be used for reaching a satisfactory result is the technique of the scale model.

The scale model may be developed in such a way that it can resolve the following questions:

- determination of the mean water surface line in the downstream channel

- determination of the velocity profile at different cross sections of the downstream channel

- determination of the separation zones in the downstream channel.

The determination of these data is the basic factor to be used in the revisions to be adopted in the design for reaching a satisfactory flow.

The revision can be plotted on the preliminary drawings of the side spillway and channel. Designer and model engineers can then discuss the optimum shape and slope.

- Shaft Spillways

Problems similar to those enumerated above also exist with shaft spillways. The following additional problems may be investigated for an acceptable design (Fig. 13.19):

- Problems caused by the flow approaching the glory hole

Flow approaching the glory hole requires careful investigation. The topography may be such that the flow rotates around the glory hole. This phenomenon is rather rare, but if it happens, the only solution is to change the location of the shaft. The designer must first determine the characteristics of the approaching flow as a function of the spillway discharge, the reservoir level, and wind direction.

Figure 13.19. Flow taking place on a shaft spillway.

- Problems caused by the accumulation of water over the glory hole causing water level increase and pulsations

Flow approaching the glory hole from all sides is constricted causing a water level increase accompanied by pulsation. A pulsating flow is dangerous in all kind of hydraulic structures. Measures must be taken to eliminate it.

- Determination of the discharge capacity of the glory hole

When the problems caused by the approach flow are solved, the variation of the discharge as a function of the hydraulic head can be determined. A curve similar to the curve shown in Fig. 5.21 is the solution of the problem.

- Determination of the control section as a function of the hydraulic head

This problem can be solved on a scale model. The computational model can give an acceptable answer to this problem only if an odd section exists in the system.

- Study of the flow taking place in the shaft

This study includes the measurement of velocities, the measurement of subpressures, the determination of the factors causing vacuum, the determination of air supply required to avoid pulsations and the vacuum. Then the determination of the dimensions of air galleries, the measurement of head losses and the determination of the possibility of vortex formation is studied. Some of the hydraulic actions enumerated above can be studied on a physical model, some not.

- Velocity measurement

The velocity around the glory hole can be measured on a scale model using a micro propeller or a Pitot tube. But these devices cannot be introduced in the shaft. The only measurement technique available is laser beam gauges. In this case the model must be constructed of a transparent material. A second problem to be solved is the similitude of the rugosity. This problem is not particular to shaft spillways. In all cases, where rugosity acts as a determining factor, the reduction of the rugosity on the scale model causes unsolvable problems.

Assume that Manning's coefficient of roughness in a shaft is n = 0.015. The roughness coefficient of a model with the 1/100 scale may be

$$\frac{n_{prototype}}{n_{model}} = \frac{0.015}{n_{model}} = 100^{1/6}$$

$$n_{model} = 0.00698$$

Such a material does not exist in practice. This condition governs the choice of the scale and one may be obliged to choose a distorted model. If the flow taking place in the shaft is uniform a computational model can replace the physical model, but only within the limits of uniform flow.

- Measurement of subpressures

Either a physical or a computational model can be used for determining subpressures. The scale of the model is a function of the precision wanted in this kind of measurement.

- Measurement of vacuum

This can be done under the conditions outlined previously. But the designer may remember once more that the choice of scale has an important effect on these measurements. The model scale must be corrected according to the degree of accuracy desired.

- Measurement of the aeration zones

Real similitude does not exist. The model provides only qualitative data. Here a computational model can be substituted for a physical model. The existing formulas giving the dimensions of the air gallery are also approximate. But in this case the computational model is less costly.

- Measurement of head losses

The determination of head losses in shaft spillways using computational approaches is difficult due to the complexity of the flow. A scale model is unavoidable. The designer may require a step-by-step study of the head loss elements and specify the degree of accuracy he wants.

- Study of the downstream tunnel

A downstream tunnel can carry a free surface flow or a pressure flow. A scale model can provide the following information to the designer:

- $Q = f(h)$ diagram
- The existence of subpressure zones
- $U = f(h)$ diagram
- The pressure diagram

The subpressure may be investigated separately. For certain values of hydraulic heads, subpressure prevails along the top of the tunnel and pulsating flow can take place. These particularities can be determined computationally If the designer wants to see the evolution of subpressure on a scale model, he has to specify a precision analysis and a scale research from model engineer. The model can give the designer the highest subpressure value.

It was seen in previously described cases that scale analysis is very important in the model of a shaft spillway. It is also important to know that the scale must be adapted to each particular case. A universal model cannot provide all the data needed to the *precision desired*. The only adequate model for this purpose is the prototype.

Example 13.01

The flow taking place in the scale model of the downstream tunnel of a shaft spillway is to be studied. The chosen scale is 1/200. The designer wants to know, to an accuracy of ± 0.10 m, the variation of the hydraulic head at the tunnel exit. It is required to determine the accuracy needed on this model.

Data

- The data is shown in Fig. 13.20

- Point gauges and piezometric device are used for water level measurements.

- The accuracy obtained in the piezometer is 0.005 m due to the pulsation in the flow

- 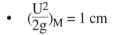 $(\frac{U^2}{2g})_M = 1$ cm

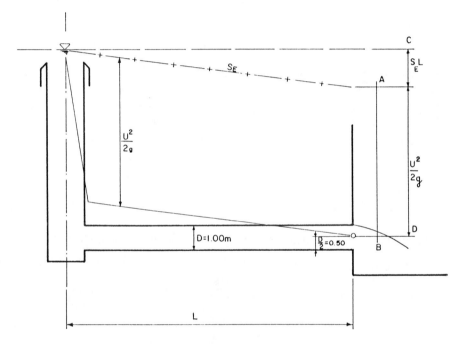

Figure 13.20. Flow in the exit tunnel of the downstream channel of the shaft spillway described in Example 13.01.

Solution

Assume that the datum line starts at point 0, which is the lowest point of the exit section.

- The total head is shown by CD

- The hydraulic head loss is shown by (CD-AB)

- CD is measured on the model with a precision of 0.001 m.

- AB is measured on the model with a precision of 0.005 m due to the pulsations in the piezometer.

Then the accuracy in the measurements of head losses is

$$0.001 + 0.005 = 0.006 \text{ m}$$

corresponding to:

$$0.006 \cdot 200 = 1.2 \text{ m} > 0.10 \text{ m}$$

An accuracy of 0.10 m can be obtained on a physical model with scale

$$\text{scale} = \frac{0.006}{0.10} = \frac{1}{16.7} \cong \frac{1}{15}$$

Note: It is assumed that point gauges are used for the measurement of water levels. If a vibrating point gauge is substituted for this device, it is possible to reach a higher degree of accuracy; then the accuracy of the pulsating level can be as low as 0.001 m and the total error will be 0.002 corresponding to 0.40 m on the prototype. Even here, the model is not accurate enough to measure the head loss with an accuracy of 0.10 m on the prototype. In this case a model with a scale of 1/50 is necessary.

- Siphon spillways

Hydraulic problems pertaining to siphon spillways are enumerated:

1. Problems concerned with self priming
2. Problems concerned with the coefficient of discharge and Q_{max}
3. Problems concerned with head losses
4. Problems concerned with pressures in the siphon
5. Problems concerned with flow at the exit section and downstream of the siphon
6. Problems concerned with vibration of the hydraulic structure during its operation.

- Problems pertaining to self priming

These problems can be solved on a scale model except for problems pertaining to the vibration, which are a function of the construction material used.

The most important hydraulic phenomenon observed on the siphon is self priming. If the priming does not occur, the siphon is transformed to a normal spillway; its discharge coefficient is reduced and the danger of overtopping is imminent. Self priming may occur at a given water elevation. The precision attained on the model is important due to the fact that the freeboard allowed is limited.

The study of self priming can be performed on a scale model with a scale that allows the measurement of subpressures with acceptable accuracy. When the water elevation in the reservoir reaches a given limit, water is discharged from the siphon throat, entraining the confined air between the siphon wall and the lower nappe. The local atmospheric pressure is then reduced and subpressure develops. When this subpressure attains a limiting value, the siphon is primed. The problem on the model is the accuracy in this similitude. The similitude of air entrainment is not perfect, the quantity of entrained air in the prototype cannot be similar to the air entrained in the model under the influence of a reduced flow velocity. The only transferable similitude, if the priming occurs at a given water elevation on the model, is that it will occur at a lower elevation in the field. This supplies a margin of safety. This margin is a function of the chosen scale. It is suggested that the designer and the model engineer work together on the problem.

- Problems pertaining to discharge

Since flow in the siphon is governed by Froude's law similitude becomes possible. If the hydraulic head losses can be computed with sufficient accuracy, a computational model can solve the problem (Example 5.06).

- Problems pertaining to head losses

The local head losses in a siphon are given by Eqs. 5.47, 5.48, 5.49, 5.50, and 5.51. The coefficients used in these relations are rather arbitrary, but they have been determined as a result of in situ measurements. If their precision satisfies the designer, a computational approach can solve the problem; if not, a scale model can be used. In this case the accuracy attained by the model must be carefully investigated. Entrance loss is especially difficult to simulate because of the geometric form of the entrance (Problem 13.03).

Example 13.02

It is required to determine the geometric characteristics of the siphon for $L_2 = 2.50$ m so that the maximum subpressure be equal to approximately 5.10 m. Compute the discharge of this siphon.

- Depth of water upstream of the siphon is d = 3.00 m

- L_2, L_3, L_4 are shown in Fig. 5.36; L_2 and L_3 are variable; L_4 is assumed constant and equal to 2.00 m.

- n = 0.014

- The cross section of the siphon is square-shaped with side length equal to 1.00 m

- Local head losses will be computed as follows:

Head loss at the entrance	$0.06\ (U^2/2g)$
Bend loss at the throat	$0.15\ (U^2/2g)$
Bend loss at the exit	$0.10\ (U^2/2g)$
Exit loss	$1.00\ (U^2/2g)$

Transition losses are assumed negligible.

- U, is the mean velocity in the siphon

- It is also assumed that the head loss due to bend at the throat occurs at the bend exit.

Solution

1. Compute the total local head loss

$$\Sigma \Delta h = (0.06 + 0.15 + 0.10 + 1.00)\ \frac{U^2}{2g} = 0.0667\ U^2$$

2. Total projection of the centerline on the datum line

$$AB = \frac{2\pi}{4}\ (L_2 + L_4) + L_3 = 1.5708\ (L_2 + 2.00) + L_3$$

3. Total head

$$H = 3 + L_3 + 2 - \Sigma \Delta h = 5 + L_3 - 0.067\ U^2$$

Energy line slope

$$S = \frac{5 + L_3 - 0.067U^2}{1.5708\ (L_2 + 2.00) + L_3}$$

4. Mean velocity

$$U^2 = \frac{1}{n^2} R^{4/3} \; S = \frac{1}{0.0142} \, 0.25^{4/3} \, [\, \frac{5 + L_3 - 0.067U^2}{1.5708 \, (L_2 + 2) + L_3} \,]$$

$$U^2 = 803.52108 \, [\, \frac{5 + L_3 - 0.067U^2}{1.5708 \, (L_2 + 2.00) + L_3} \,]$$

5. Discharge

$$Q = (1 \bullet 1) \, U$$

6. Subatmospheric pressure

The subatmospheric pressure head P is the distance between the piezometric line and the point of tangency on the CDF representative line (Fig. 13.21). The point of tangency is determined by drawing the tangent parallel to the piezometric line. The CD portion of the representative line is a cycloid with equation

$$x = L_2 \, (\theta - \sin\theta)$$

$$y = L_2 \, (1 - \cos\theta)$$

The slope of the tangent to this cycloid corresponding to angle θ is

$$S = \frac{\sin\theta}{1 - \cos\theta}$$

The coordinates of the point of tangency corresponding to θ is given by the equation of the cycloid. A graphical solution for obtaining the cycloid is shown in Fig. 13.21.

For given values of x, the corresponding values of θ are

$$\theta = \frac{x}{L_2} \frac{180}{\pi}$$

Then P can be computed as

$$P - \frac{U^2}{2g} - L_2 \, (1 - \cos0) + xS$$

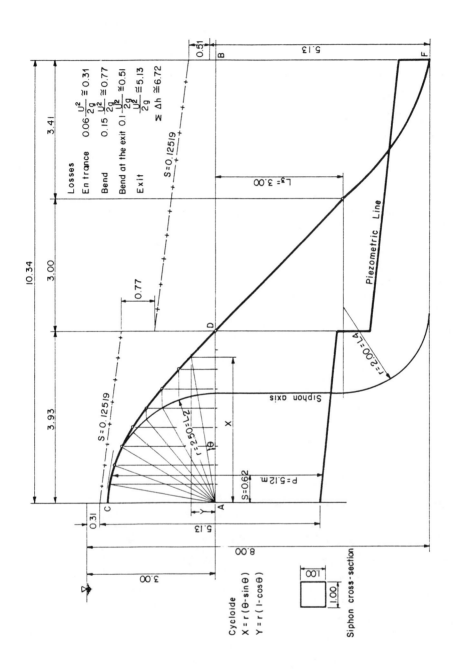

Figure 13.21. The siphon in Example 13.02.

The values of P corresponding to various values of L_2 and L_3 are listed in Table 13.1 and shown in Fig. 13.22.

To P = 5.10 m corresponds

| L_2 = 2.50 m | L_3 = 3.00 m |
| L_2 = 3.50 m | L_3 = 3.10 m |

The siphon, the characteristics of which are

L_2 = 2.50 m
L_3 = 3.00 m

is chosen due to its lesser length.
The discharge of this siphon is

$$Q \cong 10 \ m^3/sec$$

TABLE 13.1

Values of P corresponding to different values of L_2 and L_3

L_2 m	L_3 m	S 10^2	$U^2/2g$ m	x m	P m	U m/sec	Q m^3/sec	θ o
2.50	1.00	9.69	3.97	0.480	3.97	8.82	8.82	11.00
2.50	2.00	11.07	4.56	0.554	4.56	9.46	9.46	12.70
2.50	3.00	12.52	5.13	0.620	5.1	0.03	10.03	14.21
2.50	4.00	13.94	5.68	0.688	5.68	10.55	10.55	15.77
2.50	5.00	15.17	6.21	0.752	6.22	11.04	11.04	17.23
2.50	6.00	16.48	6.73	0.805	6.74	11.49	11.49	18.45
2.50	7.00	17.67	7.24	0.875	7.24	11.97	11.92	20.05
2.50	9.00	20.03	8.20	1.000	8.20	12.68	12.69	22.92
3.50	2.00	10.86	4.45	0.76	4.45	9.34	9.34	12.44
3.50	4.00	13.54	5.54	0.95	5.54	10.43	10.43	15.55
3.50	5.00	14.85	6.07	1.01	6.07	10.91	10.91	16.53
3.50	6.00	16.08	6.58	1.15	6.57	11.36	11.36	18.83
3.50	7.00	17.28	7.07	1.20	7.08	11.78	11.78	19.64
3.50	9.00	19.64	8.01	1.34	8.03	12.54	12.54	21.94

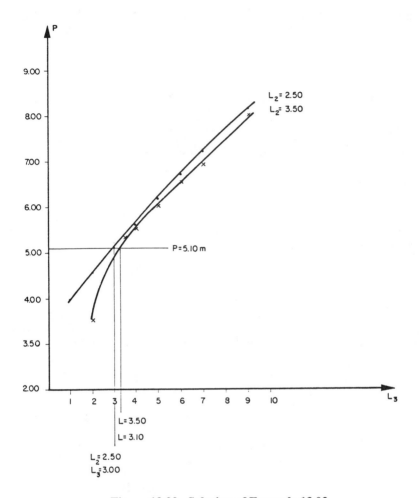

Figure 13.22. Solution of Example 13.02.

• Problems pertaining to the pressures in siphons

The pressure buildup in a siphon is a function of the hydraulic characteristics of the structure. In an S-shaped siphon with a circular cross section, the variation of pressure can be studied with acceptable accuracy using a computational model. If a siphon has a variable cross section, it is customary to work on a physical model. Since pressure variation causes vibration, a pressure analysis is necessary.

• Problems pertaining to the flow at the exit section and downstream of the siphon

This problem is similar to the flow conditions downstream of spillways (see subsection 13.2.4).

- Problems pertaining to the vibration of the siphon during its operation

The construction material used for a siphon is usually reinforced concrete or steel. Plastic material may be used for very small transportable siphons as used in irrigation. The material used for model construction is a kind of plastic called Plexiglas. It is difficult to simulate the vibration of reinforced concrete with Plexiglas. The model study of vibration, using a scale model is not satisfactory. Sometimes the maximum vibration in the model study is transposed to prototype without a careful review of the effects of scale. This casualty must be avoided because the relative magnitude of vibration in the model may be less than the actual magnitude in the field.

The most dangerous form of vibration is resonance which cannot be simulated using a physical model.

- Lateral spillways

These structures are used on levees and channels for water level regulation. The problems with these structures are:

- The determination of water line profile in the spillway

- The determination of the discharge of the structure, or if the discharge is known, the determination of the spillway length and the dimension, c (see Chapter 5)

Since the flow is governed by Froude's law, scale models can be used for studying the problems taking place on a lateral spillway. A computational model can also lead to acceptable solutions.

- Fuse Plugs

In a physical model investigation, the behavior of a fuse plug can lead to qualitative solution only. Models of fuse plugs have been constructed in DSI Laboratories in Ankara, Turkey. The similitude theory has been developed by Şentürk, F. (1965) and models constructed in the laboratory have been checked on prototypes. The parameters involved are various and numerous. The compaction of the earth body varies with time, a phenomenon which cannot be reproduced effectively on the model.

B. Problems Pertaining to Gated Spillways

Problems pertaining to gated spillways are numerous and complex; they are defined in the following paragraphs.

a. Determination of the gate seat position with respect to the dam axis, the gate radius, and the angle, θ, for radial gates.

When the type of gate is determined, the problem will be reduced to locating the gate on the spillway profile. The choice of gate type depends on:

• gate vibration and pulsating flow

• width of gate

• gate height

• hoisting procedure

• buildup of subpressures on the spillway profile due to gate characteristics, and

• buildup of the boundary layer around the piers.

The width of the gate versus its height is given in Table 4.1. The model engineer will study the gate for hydraulic efficiency, dimensions, and hoisting procedure. The type and location of the hoist, whether cable or chain will be used and the hoist mechanism will be examined. The scale of the model must be small.

If a radial gate is chosen, the following problems may be solved on a scale model:

• The influence of the location of the gate seat, gate radius and trunnion on the buildup of subpressure, the variation of the discharge and the vibration probability for partial gate openings.

• The diagram giving the discharge for partly open gates, and then the variation of the discharge coefficient, m.

• The upper and lower boundaries of the pressures, their location and the buildup of subpressure at the intersection of the piers with the spillway.

• The probability of pulsating flow.

A physical model can only give qualitative information on gate vibration and the occurrence of pulsating flow. The model is constructed of a transparent plastic material so that the model engineer can see the variation of the hydraulic phenomena such as those illustrated below (Fig. 13.23):

1. The tendency of vertical axis vortices to form at the gate groove

2. The tendency of separation zones to form downstream of gate

714

Figure 13.23. Vortex formation upstream of a gate.

3. The efficiency of the gate during overtopping

4. The tendency towards gate torsion while hoisting

b. Determination of the subpressures taking place on the spillway profile

The measurement of subpressure is done in the same way as described for ungated spillways (see subsection 13.2.3).

c. Gate vibration, pulsating flow;

As stated previously the similitude of gate vibration is difficult to obtain. The model can be costly. An industrial model can only give the guidelines to be followed to prevent vibration and the results obtained are only relatively accurate. The limits of the vibration range are different on the model and in nature. The helicoidal flow causing vibration can be studied on the model but even the similitude is minor in this situation.

The pulsating flow may be studied relative to gate vibration. It occurs only for certain openings of the gate. These openings can be observed and recorded from a physical model. This problem cannot be solved by a computational model.

d. Problems concerning gate operation

Problems concerning gate operation can be studied by computational models. Software is available for solving gate or reservoir operation problems. A physical model is costly and time consuming.

e. Flow around side piers (abutments)

Flow around piers and side piers (abatements) can be studied on a physical model. Results obtained from in situ measurements can also be transposed to prepare a computational model. In both cases the results must be discussed carefully. Often computational and physical models are used together for a comparison of obtained results.

Problems around piers and side piers (abutments) are enumerated.

1. Problems concerning the geometrical form of side piers.

It was previously shown that the skew of the approaching flow caused difficult problems (13.2.1.B). If macro measures cannot solve the problem perfectly, micro measures are taken by adjusting the geometric form of the side piers (Fig. 13.24). This operation can only be done on a physical model (Fig. 13.25). The influence of approaching flow and the resulting separation

Figure 13.24. Side piers of the Keban Dam Spillway for correcting the skew of the flow.

Figure 13.25. Flow around the side pier (abatement) of a spillway.

zones are shown in Fig. 13.25. This figure shows also the flow after the correction of the side pier shown in Fig. 13.24. The geometric form of the side pier after correction can be too difficult to execute in nature. For this reason the designer and the model engineer must work together.

Model engineer determines the flow conditions causing the skew and tries to reproduce them on the model. For example, if the skew is a result of a deep flow or a helicoidal flow and if surface flow only is reproduced on the model, the model does not reproduce nature accurately.

The following problems relative to side piers (abutments) can be investigated on scale models.

2. Vibration of the side pier due to the skew of the flow.

Only qualitative answers can be obtained from the model. Vibration characteristics cannot be obtained from the model.

Preventive measures to be taken cannot be fully investigated on the model. It is only possible to say that some corrections are better than none at all. The detection of vibration on a model necessitates complicated measuring devices and special materials for construction of the model.

3. Dynamic forces acting on the side pier.

The measurement of the dynamic forces caused by flowing water can be made by strain gauges. The scale of the model must be adequate to allow gauges to make precision measurements of the dynamic forces.

Physical model may be used for measuring dynamic forces.

717

4. Stream lines around the side pier.

The stream lines must be visualized in order to have an idea of the geometric form to be adopted for the side pier. A perfect solution will be the one where the stream lines coincide perfectly with the inner surface of the side pier.

The problem can only be studied on a physical model.

5. Determination of the increase in the discharge coefficient as a result of the adopted correction.

It is known that the discharge of a side bay is smaller than the discharge of a middle bay. This fact is shown by the decrease of the discharge coefficient in the side bay.

6. Measurement of the subpressure at the intersection of the side pier and the spillway.

The subpressure, if built up on the spillway profile, is greater at the intersection of the side pier or the pier with the spillway than at its center. Its determination is only possible on a model or in nature. The reinforcement of the concrete will be done according to these data.

The measurement of the subpressure at the intersection is not easy. The pressure tapes can be placed near the side pier or intermediate pier at a distance equal to the diameter of the tape. Then the measured subpressure will be smaller than the real subpressure.

The scale model is selected to allow for acceptable precision (Section 13.2.3).

7. Influence of the side pier on flow for partial gate openings; generation of vortices and pulsating flow, and accumulation of water around the side pier.

These hydraulic properties can be investigated only on physical models and the results are only qualitative. As an example, vortex formation occurs at a given water depth under atmospheric pressure. On a model and always under atmospheric pressure a vortex cannot be formed at a depth homologous to the depth in nature. A confined model must be used for assuring the similitude. Sometimes a distorted model is used for studying the vortex. Vortex problems have been studied in Chatou Laboratory, Paris, France and results have been published.

Similar considerations are true for pulsating flow. The accumulation of water immediately upstream of the gate can be studied on a physical model. The difficulty in this case is the separation of water accumulation from the superelevation of water level due to shock waves.

8. Perturbation of water due to stop log grooves in side piers.

This is a secondary problem which may be important. In some cases it can be investigated on a physical model.

9. Flow around piers.

The shape of the piers is worked out in such a way that the discharge coefficient can be increased without a sizable increase in subpressure.
Hydraulic problems to be solved on physical model are enumerated:

a. The upstream end of the pier is similar to a ship's prow (Fig. 13.26).

Here, the water surface line is shown in the figure by a dotted line. It is clear that the nappe at a pier with a ship's prow profile leads to smaller subpressure due to the flattening of the nappe. The gain in the subpressure can be measured on a scale model. This gain being rather small, the scale of the model must be chosen so that the measurement can be easily performed. It is possible that local defect in pressure tapes can obscure the gain, then the scale must be changed so that the gain in quantity of pressure is greater than the fluctuation in pressure due to the defects in the tapes.

The angle of attack, θ, must also be investigated. There is an optimum value of θ for which the flattening of the nappe attains its maximum value.

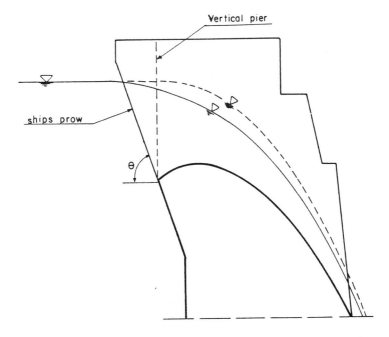

Figure 13.26. Longitudinal section of a pier in the form of a ship's prow.

b. The influence of the pier geometry, in plan, to the coefficient of discharge

This investigation has been made in laboratories previously and the results obtained have been covered in Chapter 4.

c. Study of flow around pier for partial gate opening (see Chapter 4)

d. Stream-lined downstream end of the pier

Various downstream end forms are used in practice (Fig. 13.27). Each of these piers presents a particularity which may be analyzed on the model. Special attention must be paid to elimination of the factors causing generation of cross waves.

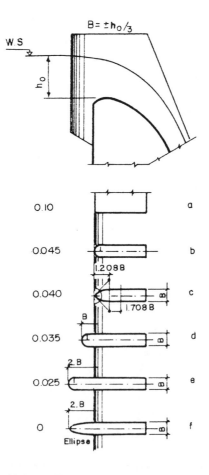

Figure 13.27. Different downstream end forms used in piers.

13.2.4 Model Studies of Restitution of Water to the Main Water Course

Problems involved with the restitution of flow to the main water course are enumerated:

- Problems involved with submergence

- Study of the water surface profile downstream of the control structure

- Study of the velocity distribution in the downstream discharge canal

- Study of the water pressure distribution in the downstream discharge canal

- Study of the cross waves in the downstream discharge canal

- Study of air entrainment in the downstream discharge canal

- Study of head losses in the downstream discharge canal

- Study of the energy dissipation with stilling basins

A. Problems Involved with Submergence

Submergence is encountered in low dams. The downstream water level may be so high that it influences the upstream water level. If the downstream water level can be determined, the submergence can be defined mathematically. For example, two head structures are located one below the other, on the same water course. If the upstream water elevation of the downstream structure is known, submergence for the upstream structure can be determined mathematically.

B. Study of Velocity Distribution in the Downstream Discharge Canal

The velocity distribution in the downstream discharge canal does not follow exactly the distribution of velocity on the model due to inability of the model to reproduce air entrainment. Therefore, a mathematical analysis is generally preferred to direct measurements on the model.

C. Study of Pressure Distribution in the Downstream Discharge Canal

It has been previously shown that the measurement of pressure and subpressure on the model were rather difficult. Therefore, computational models may be preferred to physical models for studying the variation of pressures in the downstream channel of spillways.

D. Study of Cross Waves in the Downstream Discharge Canal

Cross waves as stated earlier are not stationary. They migrate across the cross section of the downstream discharge canal. The migration of these waves cannot be studied mathematically. The simulation of this phenomenon is a function of the model scale. In general qualitative studies on models are found to be satisfactory.

The problem involved with cross waves is the overtopping likelihood of the flow taking place in the channel and the disturbances caused by the waves at the exit of ski jumps. This last possibility must be studied carefully, because it is the principal cause of the erosion taking place downstream of the ski jump. Figure 13.28 shows a flow at the ski jump of Seyhan Dam Spillway's energy dissipator.

Figure 13.28.   Influence of cross waves at the ski jump of an energy dissipator.

E. Study of Air Entrainment in the Downstream Discharge Canal

The study of air entrainment in the downstream discharge canal is difficult (Section 6.2.2). Physical models can solve the problem quantitatively. Many attempts have been made to measure air entrainment in situ. But these measurements were only approximate. The velocity of flow in the downstream channel is higher than 30 m/sec in many cases. The velocity head corresponding to it is approximately 16 m. The measurement of such a high head poses difficult problems. Liquid of higher density than the density of water has been tried. Strain gauges have also been used, and the results obtained for the velocity of the flow at the same location using the two methods separately have given differences higher than 100% for relatively shallow depths. A method of pure observation is also used for determining the water level in the canal. Strips of different colors are painted on the side walls of the discharge canal. Observation has shown that the water surface has not been stationary for a given discharge and that it varies unceasingly.

Thus, the evaluation of the air entrainment depends upon theoretical approaches. The results obtained on a physical model must be checked by a computational model. Hypercritical flow causes heavy erosion on the bottom of the spillway discharge canal (Fig. 13.29). In order to prevent this erosion a good approach is to artificially introduce air into the channel flow. This phenomenon can be studied on a physical model but only qualitatively. Some empirical formulas have been suggested to calculate the discharge of air necessary for preventing the erosion, but a theoretical basis does not exist for supporting these relations. They only can give the designer a general direction for solving the problem. Such a solution was used in the Karakaya Dam Spillway, Turkey. The result obtained is satisfactory.[*]

F. Study of Head Losses in the Downstream Discharge Canal

The available hydraulic head at the ski jump is the primary factor acting on the formation of the jet. The designer needs a minimum flow velocity at the exit from the ski jump to locate the impact of the jet at the location chosen. The exit velocity is a function of the hydraulic head which in turn is a function of the friction head loss and the local head losses. The friction head loss is important in hypercritical flow. The Manning coefficient of resistance which is about 0.014 for concrete is pushed up to 0.030 - 0.070 due to high wakes downstream of small roughness elements in the concrete. Therefore, extra roughness elements must be added to the physical model; but no one knows a quantitative approach for computing this extra roughness characteristics. Only rough measurements in situ exist for a qualitative approximation of the roughness of the downstream channel, which varies with the velocity.

Two examples are given below to illustrate the problem (Problem 13.06). Extra roughness has been added to the model of the Keban Dam Spillway downstream channel empirically. Observation in nature has shown that the model engineer has done a good approximation (Fig. 13.30).

The same method has not given a satisfactory result in the case of the Oymapinar Dam Spillway. The impact of the nappe striking the opposite slope of the valley destroyed it completely. The best method for solving the problem was to use both physical and computational models and compare the results obtained. It is again suggested that the model engineer and the designer cooperate closely in order to reach a good and acceptable solution.

---

[*] This solution is also applied in Turkey on the Keban and Aslantaş Dam Spillways.

Figure 13.29.   Erosion due to hypercritical flow on the bottom of the discharge canal of the Keban Dam.

Figure 13.30. Location of the impact of the Keban ski jump.

G. Study of Energy Dissipation

Energy dissipation at the exit of the downstream discharge canal and of the stilling basins is assured by means of ski jumps or by stilling basins. A sound criterion does not exist for comparing different solutions. Qualitative attempts have been made to obtain relatively satisfactory results on physical models. The problem cannot be solved using computational models. Problems involved with energy dissipators are enumerated as:

- Energy dissipators in form of ski jumps

725

F. Sentürk

The following geometric and hydraulic characteristics are investigated on physical models.

1. Study of the radius of curvature of the ski jump

2. Study of the exit level of the ski jump

3. Study of the geometric characteristics of the ski jump or the deflectors if they exist

4. Study of the water nappe issuing from the ski jump.

All the geometric characteristics of the ski jump must be studied in such a way that the energy dissipation is perfect. The criterion used for reaching the solution of the problem follows.

The downstream surface of the model is built up using homogeneous sand and the erosion caused by the impact of the jet is measured carefully (Fig. 13.31). The best solution is the one causing the least erosion on the downstream surface. This method has been criticized greatly during the past years, but a better and cheaper method does not yet exist.

- Energy dissipators in the form of stilling basins

Figure 13.31.   Illustration of erosion scour downstream of an energy dissipator.

The qualitative approach for comparing different solutions is used in this case also. The visualization of the turbulence can be observed using small floating objects and photographing their movement while flowing in the stilling basin or out of it. Figure 13.32 shows an example of the method used.

The measures to be taken for preventing heavy scour downstream of stilling basins can be summarized as follows.

- Each solution suggested by the model engineer must be discussed carefully with the designer. The most economical solution and resulting dimensions can be checked on the model. The construction feasibility of the geometric forms of different structural elements may

Figure 13.32.   Visualization of turbulence downstream of an energy dissipator.

be discussed with the designer, and if their realization are impractical, the solution may be eliminated. An interesting example is encountered in Turkey. The side walls of a stilling basin were worked out on the model and the required height was found to be 70 m. The solution was eliminated.

- Scouring downstream of stilling basins cannot be eliminated completely. A limiting scour depth must be suggested by the designer. The designer should know that the scouring phenomenon cannot be reproduced on the model and qualitative measurements are not possible, therefore it should not be expected that the hydraulic engineer be able to set the exact depth of the possible scour hole.

- Generally, riprap is used to reduce the scour depth. The dimensions of the riprap to be employed in this respect can be investigated on the model. The tetrahedron used in breakwaters for preventing erosion due to waves has been studied in laboratory and its geometric form has been improved as a result of model tests.

13.2.5 Model Studies of Flow in the Downstream Water Course

The investigation of the deformation of the natural bed of the water course downstream of the dam is a different field of model engineering. The problems can only be investigated on deformable bed models which takes time and effort. These problems are grouped as follows.

1. River bed deformation downstream of the dam

2. Investigation of the water surface profile in the river bed at the end of the scouring

3. Water surface profile resulting from the dam failure

A. Riverbed Deformation Downstream of the Dam

It is known that bed and banks degradations occur downstream of the newly constructed dams immediately after the closing of gates for storing water in the reservoir. The released water is free of sediment and is clear; its erosive capacity is at its highest. Consequently bed and banks degradation occur (see Chapter 12).

The study of bed degradation is possible on a scale model under certain conditions. The transposition of the results obtained from the model to the prototype demands a high degree of perfection in model engineering.

Data collection for determining the scale of the model takes time. Many years are necessary to obtain enough information on the characteristics of the river in its natural conditions. The model can give only qualitative answers to the scour problem. Figure 13.33 shows the delta formation downstream of the spillway at the Gökçekaya Dam. The problem to be investigated in this particular case was:

1. The spillway is located in a side valley. The scour and sediment transport from this valley by the flood discharged from the spillway may form an obstruction in the main river, causing an increase in the water level with a result in decrease of energy production, even a possibility of inundation of the power plant.

2. The duration of the flood hydrograph is 12 days

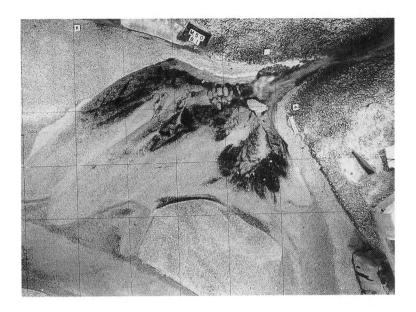

Figure 13.33. Delta formation by spilling of Q_{max} at the Gökçekaya Dam Spillway, Turkey.[*]

3. It is required to investigate the possibility of formation of a sediment bar in the main river and also the process of natural erosion of the bar by the flow.

The model had been constructed at the Hydraulic Research Center of General Directorate of the State Hydraulics Works at Ankara, Turkey (DSI). Pumice was used for simulating the transported sediment. A special distortion was applied for time simulation. The scale investigation, the construction of the model and its operation and calibration took more than two years. The results obtained were interesting but only qualitative. The dam had been constructed in 1973 and since then the gates of the spillway were never operated. A new large dam upstream of the Gökçekaya Dam will be constructed in the near future changing the existing conditions and all the money spent for the model were not necessary.

As a consequence it has been suggested that the bed deformation downstream of the spillway be studied using a computational model analysis if possible. A physical model is suggested only if urban areas are in the vicinity and hydraulic structures are necessary to prevent damages which can result from bed deformation. The model investigation takes time in this particular case and can be applied if it does not delay the dam construction.

---

[*] The black pumice with higher specific weight is used to represent large rock blocks.

B. Investigation of the Water Surface Profile in the River Bed at the End of Scouring

The problem can be reduced to the investigation of a man-made inundation downstream of the dam. The degradation, aggradation and deposition areas may be determined and the water surface profile be defined accordingly. If a physical model is constructed, the problem can be solved using this model. If the river bed will be lined a computational model can give the water surface profile accurately. The lining characteristics can be determined knowing,

- the shear stress buildup on the bottom and on the side slopes

- the shear stress around specific points such as bends, bridge piers, etc.

- the maximum water level around specific points.

Figure 13.34 shows the sediment deposition areas downstream of the Keban Dam resulting from the return current of a medium size flood.

Figure 13.34. Deposition areas downstream of the Keban Dam caused by the return current.

C. Investigation of Return Currents Caused by the Flood Discharge

One interesting and important problem is the investigation of return currents. The currents are generated by the nappe abruptly reaching the water surface downstream of the dam.

Water reflected by the immediate addition of a large quantity of extra water in an already existent flow begins to rotate about the impact point and its velocity can attain high values. In Keban Dam for example, the velocity measured in situ was as high as 10 m/sec. It has been observed that a good similitude existed between the model and the prototype (Fig. 13.34).

The return current is a local phenomenon. It can cause extra erosion, particularly on the side slope of the river bed. The material transported in the upstream direction can be deposited in front of the power plant if located in the area, causing a loss of energy and even an excess velocity of the turbine rotors because of pulsating flow. Such a problem can be examined only on a scale model.

D. Investigation of Water Surface Profile Resulting from Dam Failure

The problems of waves resulting from the catastrophic dam failures were actively investigated at the end of the Second World War. The accuracy obtained by a scale model is not greater than the accuracy attained by a computational model. It is, therefore, suggested that computational models be used to solve such problems.

E. Determination the Flow Direction after Construction

The location of the head structure in a selected river reach is an important choice. The silting of a water intake by a sediment laden stream is an immediate result of an erroneous choice of the intake location. The problem can be investigated only on a physical model. A computational model cannot be used for solving the problem because of difficulties in simulation of boundary conditions.

The construction of a diversion dam on a water course can change the direction of the current in the river bed. It has been observed that the sediment deposit upstream of the Berdan Dam was such that the current took a direction parallel to the axis of the structure and entered the river intake directly, transporting the sediment load into the irrigation canal.

To study this problem on the model it is necessary to reproduce on it the historical flood records. The model scale and characteristics should simulate the bed changes in such a way that the model reproduces the influence of time taking into consideration successive floodings. This is time consuming work but worthwhile, and if a good model is built, material transport, deposition and flashing problems can be solved on it.

An adequately calibrated model can establish the following:

- the location of the hydraulic structure in the river reach
- determination of the direction of the dam axis with respect to the main flow direction
- determination of the angle between the dam axis and the water intake axis
- the height of the entrance sill of the water intake
- the flushing technique to be adopted for a better evacuation of the deposited sediment.

13.3   MODEL STUDIES OF CLOSED CONDUITS

Many types of closed conduits exist in dam engineering. Closed conduits are used in diversion structures, in the downstream discharge facilities of spillways and in the downstream bottom valves for emptying the reservoir. The hydraulic problems involved can be divided into two categories:

1. Free flow in the closed conduit
2. Pressure flow in the closed conduit,

Physical and computational models may be used for solving such problems.

13.3.1   Free Flow in Closed Conduits

Problems involved with free flow in closed conduits are enumerated:

- uniform flow
- gradually varied flow
- rapidly varied flow
- aeration problem
- pulsating flow around bottom valves.

These problems are defined on the next page.

A. Uniform Flow in Closed Conduits

The closed conduit is, in general, a tunnel or a culvert. The hydraulic problems involved with a tunnel are cited:

- head loss problems: local head losses and head loss due to friction

- water surface superelevation around bends and the possibility of helicoidal flow generation

- investigation of downstream flow conditions

- air entrainment phenomenon

- hydraulic jump immediately downstream of the bottom valve and upstream of the uniform flow taking place in the closed conduit.

a. Head loss problems

These problems can be investigated on a scale model; the analogy is perfect. They can also be studied by computational models.

Because of the inadequacy in the similitude of friction, the friction head loss is rather difficult to reproduce on the model, but the local head losses can be simulated satisfactorily (Problem 13.01, 13.02, 13.03, and Subsection 13.2.3). If a physical model is chosen for investigation of head losses, the choice of the scale will be the main factor in obtaining the desired accuracy.

b. Water surface superelevation around bends and the possibility of helicoidal flow generation

When a bend is introduced in plan, different hydraulic phenomena are to be expected. The water surface, horizontal in the cross section shows a superelevation as shown in Fig. 13.35, the flow profile in the longitudinal section is also shown in the same figure. Each of these possibilities must be investigated by a physical or computational model. If the superelevation is greater than Δd critical, there is the danger in the flow filling the cross section temporarily. The upstream reach of the channel will carry a pressure flow, the discharge will decrease, the flow depth in the downstream reach will decrease and a drawdown curve will take place. The flow in the upstream reach will be a free surface flow once more. This is pulsating flow which must be eliminated.

This phenomenon can be studied on a physical model. A computational model is not suggested for studying such a case. If pressure flow subsides, a helicoidal flow can take place in the downstream reach.

For studying the influence of bends in a closed conduit the scale of the model must be chosen accordingly.

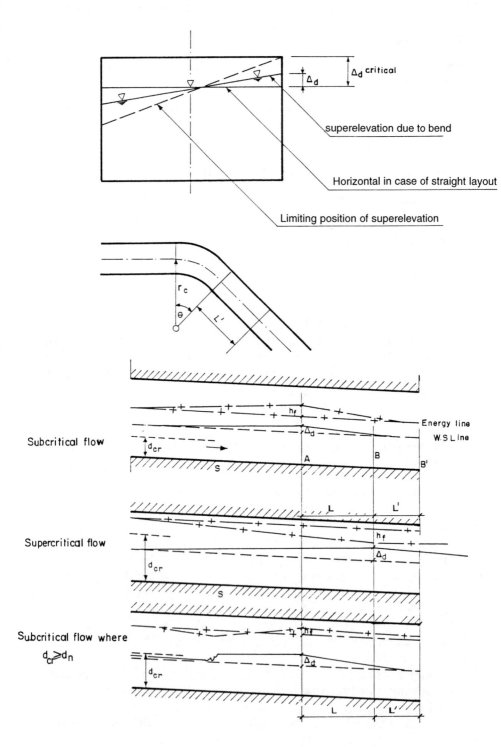

Figure 13.35. Flow around a curve.

Assume that the chosen scale is 1/40 and $\Delta d = 10$ cm in nature, corresponding to 2.5 mm on the model. If point gauges are used for the determination of superelevation, there is no possibility of detecting the phenomena accurately. If the length of the tunnel is 400 m, for example, the length of the model will be $10 + 5 = 15$ m (5.00 m is used for entrance and exit facilities on the model). This is cumbersome. Special techniques must be used to obtain a satisfactory model economically. An extensive scale study must be performed to solve the problem. If the designer can clearly specify the problem, the model engineer can usually find a satisfactory answer. If, for example, only the water surface superelevation is needed, and if secondary factors do not interfere with the flow, a partial model can be used or a computational model can be substituted for the physical model.

c. Investigation of downstream flow conditions

Downstream flow conditions play an important part in the performance of closed conduits.

Assuming a supercritical flow taking place in a tunnel, if the downstream conditions are such that a jump takes place downstream of the tunnel, this jump can be displaced upstream as a function of the downstream water level, and can enter the tunnel causing a pressure flow in it. This possibility can be investigated by a physical or by a computational model.

If the flow causes heavy erosion at the exit reach of the closed conduit, the supported nappe can be transformed to a free drop and the subcritical flow in the conduit can be transformed to a supercritical flow with high erosive capability. This problem can be investigated by a computational model, but if the scouring is to be introduced as part of the model investigation, a physical model is necessary.

Flow taking place in a closed conduit entrains the air, confined between the rigid boundaries and the moving water. If that air is not introduced from the upstream end, three phenomena can be observed:

d. Air entrainment phenomena

1. Water level is low enough (Fig. 13.36a). Air entering from the exit section is ejected from the same section.

2. Water level is higher than shown in Fig. 13.36a. Air is entrained by flowing water, but the air entrance from the exit section is not sufficient and cannot circulate through the whole length of the tunnel. Partial vacuum is built up, and beginning from the upstream sections, the water level is elevated above the normal water level by suction.

3. Water level is so high that the buildup of vacuum generates pressure flow which is not stable. In unlined tunnels the vacuum can cause slides and rockfalls blocking the tunnel section.

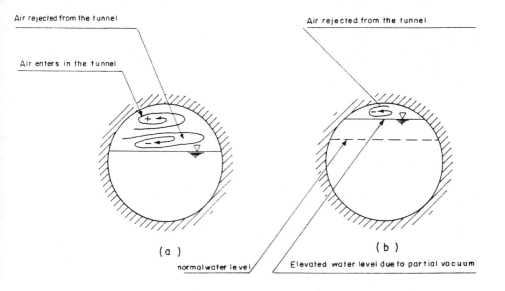

Figure 13.36. Air circulation in a closed conduit.

Pulsating flow causes imminent danger. This phenomenon cannot be reproduced entirely on a physical model due to the inadequacy of simulating vacuum formation. Qualitative simulation is only possible in particular cases. Computational models are not adequate for solving similar problems.

e. Problems pertaining to hydraulic jump

See Article 6.4.2.A and 13.3.2A

13.3.2 Gradually Varied Flow in Closed Conduits

Uniform flow can take place in a closed conduit if the conduit is long enough. In general, gradually varied flow is observed in relatively short tunnels. If the tunnel is straight and free of bends and obstacles, a computational model is usually suggested for solving gradually varied flow problems (see Chapter 10).

A. Rapidly Varied Flow in Closed Conduits, Aeration and Pulsating Flow Problems

Rapidly varied flow in closed conduits occurs under following conditions:

1. A pressure outlet valve is generally located at the upstream end of a tunnel. Its purpose is to deliver water for wild life, for irrigation, or for emptying the reservoir. This valve outlet can be a radial gate or a bulkhead gate. It is customary to locate a Howell-Bunger valve downstream of this valve for energy dissipation (see Chapter 11, Fig. 11.3). The hollow jet issuing from this valve strikes the tunnel wall with great energy (the velocity attaining 10-20 m/sec). Water moving rapidly causes a supercritical flow and a hydraulic jump occurs at some distance from the valve. The location and the dimensions of the jump being important in the tunnel design, the phenomenon is to be studied on a physical model.

The hydraulic jump, if not controlled, can reach the tunnel ceiling causing a pressure flow and then a pulsating flow. Vacuum builds up between the jump and the valve damaging the valve. To prevent this phenomenon air is introduced in the confined area by special air tubes. The dimensions of the air tubes are to be checked on the model. Also preventive measures may be taken to eliminate pulsating flow which causes high and dangerous vibrations in the valve.

As stated previously, the simulation of vacuum is not possible on industrial models. Special models may be prepared for solving the problem.

If the designer needs specific answers to the above three problems, three different models with adequate scales and different techniques must be constructed. Pressurized models are necessary to study the dimensions of air tubes and the buildup of the subpressure downstream of the bottom valve. Even the characteristics of the jump cannot be studied satisfactorily because of the vacuum upstream of it. In many instances the designer can ask the hydraulic engineer to study the above defined flow problems qualitatively. Models are prepared using transparent plastic material enabling the engineer to see the hydraulic phenomenon taking place. This type of physical model can also be prepared for evaluating various preventive measures.

2. If the designer wishes to eliminate supercritical flow in the tunnel, a controlled hydraulic jump can solve the problem. Different techniques can be used for generating and controlling the jump. The efficiency of these techniques can be studied in a scale model and/or by the computational model (Problem 13.08).

B. Pressure Flow in Closed Conduit

Pressure flow exists in the upstream reach of the bottom valve in the discharge tunnel. The pressurized energy tunnel is outside the scope of this book.

In case that a bend exists in plan in this reach, helicoidal flow can take place in the tunnel. A helicoidal flow is a flow which rotates around the tunnel axis up to the bottom gate. These currents may be so strong that they can damage the gate. This kind of current can be studied satisfactorily on a physical model and measurements can be performed if the model scale is chosen properly.

13.4   MODEL STUDY OF LEAKAGE

Leakage problems can be qualitatively studied on special models called Hele-Shaw models and membrane models. These models are constructed using two plastic plates parallel to each other in the first case. The space between them is filled with the material from which leakage may take place. A specially chosen oil circulates through the pores of this material. Perfect simulation of the foundation subgrade of the dam is not possible. This technique can be used for studying the circulation of water through the filter zone and the impervious core of an earth or rock fill dam.

13.5   MODEL STUDIES OF HYDRAULIC STRUCTURES

Hydraulic problems involved with dam construction are enumerated:

* diversion problems
* cofferdam failure problems
* water intakes problems
* water outlets problems.

These problems are defined in the following sections.

13.5.1 Diversion Problems

The diversion facilities in a dam construction are described in Chapter 10. The diversion system is composed of an upstream cofferdam, a downstream cofferdam and one or more tunnels. The upstream cofferdam plays the part of a small dam upstream of which water is stored temporarily. The diverted stream is conveyed by one or more conduits to a section downstream of the dam site. The downstream cofferdam is located upstream of the diversion conduit exit. Water between the two cofferdams is pumped out and the foundation of the dam is excavated in the dry.

Hydraulic problems involved with the diversion of water from the main water course and returned to the river bed downstream of the dam site are enumerated:

- location of the diversion conduit entrance
- investigation for the optimum dimensions of the diversion system
- operating procedure for the diversion system during high floods.

A. Location of the Diversion Conduit Entrance

The entrance to the diversion conduit which is generally a tunnel, may be located at an upstream section of the main river so that the sediment carried by the flow is deposited upstream and water, as free of sediment as possible, is delivered in the river downstream of the downstream cofferdam. This procedure requires first the choice of the adequate location and second, a sufficient storage volume to deposit the sediment. A physical model with a movable bed must be prepared to solve the problem. The solution may be satisfactory.

B. Optimum Dimensions of the Diversion System

The dimension system may divert a flood with a 25-year return period for economical dimension. The design is built on the model and the discharge of the system checked under different conditions. In large dams the upstream cofferdam may be a large dam impounding a reservoir holding millions of cubic meters. Flood routing will be necessary to determine the outlet discharge. It is not economical to simulate the entire upstream reservoir. Therefore, the outlet discharge is determined mathematically and the model tested accordingly. The $Q = f(d)$ curve is then determined experimentally. The model also provides information of upstream zones and of scouring zones at the exit section.

C. Operation of the Diversion System during Maximum Flood

The diversion system is designed generally for 25-year flood and checked for 50-year flood. The system should carry this discharge without overtopping. If a higher discharge occurs, the measures to be taken must be checked on the model.

A physical model can be used for solving this problem. A computational model can also be employed and is generally preferred to a physical model.

13.5.2 Cofferdam Failure Problems

If the flood discharge is greater than Q_{50}, overtopping occurs and flood water will be spilled over, with the cofferdam acting as a spillway. Since the cofferdam is an earthfill or rockfill structure, there is a probability of erosion, and failure is expected. The problem can be studied on a physical model and necessary preventive measures taken. A perfect similitude does not exist between the prototype and the model; only the time required for failure to take place and the preferred failure region can be determined qualitatively. Preventive measures to be taken at these locations can also be studied and evaluated.

13.5.3 Water Intake Problems

Special problems pertaining to water intakes are enumerated:

- study of head losses and precautions to be taken for minimizing them
- flow through partially open gates
- subpressure buildup at the entrance
- aeration immediately downstream of the gates.

These problems can be investigated on physical models; computational models are not to be used for solving such problems. The precision in measuring techniques is similar to that described previously.

13.5.4 Water Outlet Problems

These problems involve the study of erosion and scour taking place downstream of exit section. A movable bed model is necessary for solving them. In many cases failure of the outlet due to scouring has endangered the dam itself (Tarbela Dam, Pakistan).

Particular aspects of the problems to be investigated follow:

- study of the angle between the river axis and the outlet axis
- study of the erosion which can take place at the opposite side slope
- study of the local scouring immediately downstream of the outlet
- study of the possible change of the flow regime due to local scour.

These problems can be studied only on a physical model and only qualitative solutions may be expected. The model can serve only as a base for comparing and evaluating alternatives for preventive measures.

13.5.5 Head Structure Problems

In an irrigation system, for example, the head structure is a weir and turnout combined. The weir is either a gated or a free flow spillway; the turnout is a water intake.

A. Problems Pertaining to Weirs in a Head Structure

These problems are enumerated:

- variation of the discharge with time

- efficiency of the stilling basin downstream of the weir

- scouring downstream of the stilling basin.

If a heavy bed load exists in the river, a small reservoir is rapidly filled with sediment and vegetation begins to grow on the surface. This changes the original assumptions used for the design of the structure (Fig. 13.37).

In many cases the upstream reservoir has been completely filled with sediment and water has been seen flowing parallel to the axis of the weir.

Figure 13.37. Aquatic plants growing upstream of a diversion structure after partial deposition of sediments.

The reproduction of this phenomenon on a model is difficult but not impossible; only preventive measures are investigated on an industrial model.

B. Problems Concerned with Turnout

These problems are enumerated:

- design of the guide wall to direct the flow adequately (Fig. 13.38)
- Operation of the gates for flushing the sediment deposited in front of the turnout
- study of the water intake entrance
- study of the current entering the turnout
- study of the sediment entering the turnout and study of the preventive measures to be taken for solving this problem.

Figure 13.38. Guide walls upstream of a turnout.

13.6 PROBLEMS

Problem 13.01

It is required to measure on a physical model the flow velocity with an accuracy corresponding to 0.05 m/sec on the prototype. The velocity on the prototype is about 5.00 m/sec and the flow on the prototype and on the model takes place following the Froude's law.

Determine the precision of the micro flow meter to be used for the measurements of velocity on the model as a function of the model scale.

Problem 13.02

It is required to measure on the model the velocity at the exit section of the tunnel shown in Fig. 13.20. Determine the accuracy attained on the prototype as a function of the model scale.

Data

• A point gauge is used and $U^2/2g$ is measured on the model with a precision of 0.002 m because of the pulsating flow.

Problem 13.03

Assuming that a 1/100 scale physical model is used for simulating the flow in the siphon described in Example 5.06, determine the accuracy of the result obtained.

Data

• $U_M^2/2g$ is measured equal to 2.50 cm

• Point gauges are used for water level measurements (precision 0.001 m)

• Pressure gauges are used for pressure measurements (precision 0.0025 m)

• The subpressure is measured equal to 5.50 m.

Problem 13.04

It is required to determine the geometric characteristics of the siphon for $L_2 \leq 2.50$ m, so that the maximum subpressure is equal to approximately 6.00 m. Compute the discharge of the siphon.

Data

• Depth of water upstream of the siphon, $d = 3.00$ m

• L_2, L_3 and L_4 are shown in Fig. 1, Problem 13.04. L_2 and L_3 can be varied, L_4 is assumed constant and equal to 2.00 m.

F. Sentürk

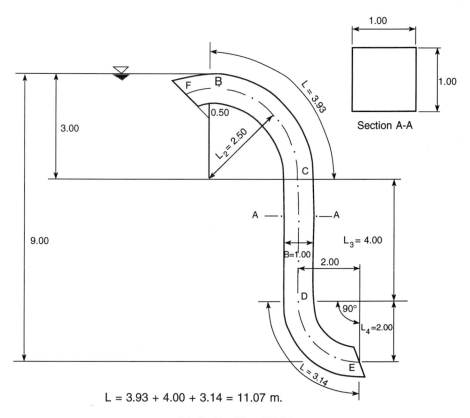

Fig.1. Problem 13.04.

- n = 0.014

- The cross section of the siphon is square shaped with side length equal to 100 m.

- Local head losses will be computed as follows:

Head loss at the entrance	$0.06 \, (U^2/2g)$
Bend loss at the throat	$0.15 \, (U^2/2g)$
Bend loss at the exit	$0.10 \, (U^2/2g)$
Exit loss	$1.00 \, (U^2/2g)$
Transition losses	negligible

- U is the mean velocity in the siphon

- It is also assumed that the head loss due to bend at throat occurs at the bend exit.

744

Problem 13.05

Prepare a computational model for the flow in Example 13.01 (Fig. 1, Problem 13 .05).

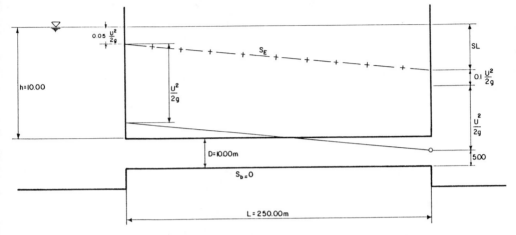

Fig. 1. Problem 13.05.

Data

- Local head losses: Entrance loss = 0.05 $U^2/2g$
  precautions are taken to minimize the loss

  Exit loss = 0.1 $(U^2/2g)$

- The nappe at the tunnel exit is not supported, then the pressure line passes through point 0 situated at the center of the cross section.

- $n = 0.014$

- The cross section is circular with diameter equal to 10.00 m.

- The bottom slope of the tunnel is equal to zero.

Problem 13.06

Determine the distance to the impact point of the jet issuing from the ski jump of the Keban Dam Spillway for the following values of the flood discharge taking place from one bay only (Fig. 1, Problem 13.06): $Q = 1000$ m³/sec and $Q = 6000$ m³/sec.

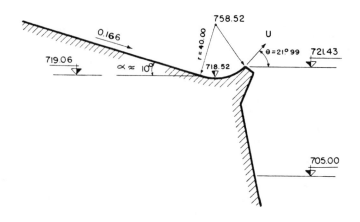

Fig. 1. Problem 13.06.

Data

- The mean velocity corresponding to Q = 1000 m³/sec and Q = 6000 m³/sec is assumed equal to the limiting value of 32 m/sec.

Problem 13.07

Prepare a computational model for the tunnel shown in Fig. 1, Problem 13.07 and investigate the flow taking place in it.

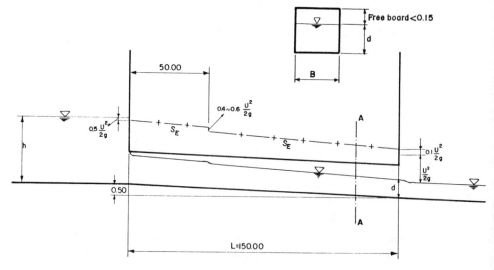

Fig. 1. Problem 13.07.

Data

- Cross section rectangular, with bottom width, B

- $Q = 100$ m³/sec

- Entrance loss $= 0.5\ U^2/2g$

- Bend loss $= (0.4 \cong 0.6)\ U^2/2g$

- Exit loss $= 0.1\ U^2/2g$

- A 22.92° bend, the entrance of which is located 40.00 m from the tunnel entrance, exists in plan $r_c = 50.00$ m (Fig. 2, Problem 13.07)

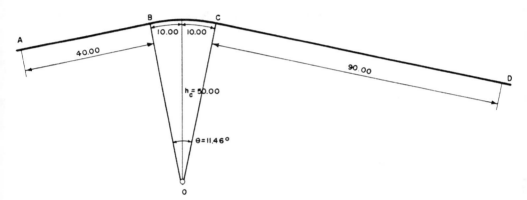

Fig. 2. Problem 13.07.

- (15) cm of freeboard is necessary to avoid pressure flow

- Any transition in plan or longitudinal section disturbs the flow

- n = 0.015; this value can be corrected to reach the most economical solution

- The dimensions of the downstream channel are similar to the dimension of the tunnel.

It is required to study,

- The variation of d, as a function of n (roughness coefficient) and B

- The determination of the most economical section

- The variation of the water surface superelevation in the most economical cross section at the bend (assume the water accumulation at the curve entrance to be negligible).

- Water surface profile in the longitudinal section for the most economical section

- The limiting value of d so that the pressure flow is excluded from all cross sections in case of the most economical solution.

Problem 13.08

A supercritical flow takes place in a tunnel. It is required to eliminate it by a hydraulic jump. Determine the characteristics of the tunnel and location of a downstream barrier so that the jump is located at least 10 m below a control section. Prepare a general mathematical solution of the problem.

Data

- $Q_1 = Q_{max} = 17.00$ m³/sec

- $Q_2 = Q_{min} = 12.00$ m³/sec

- Tunnel section: rectangular

- B: width of the tunnel is to be determined

- $n = 0.014$

- $S_b = 0$

- $d_2 = 0.75$ m corresponding to $Q_2 = 12.00$ m³/sec

- The distance between section 1-1 and the barrier is 125.00 m.

Appendix A

GEOMETRIC ELEMENTS FOR CIRCULAR CHANNEL SECTIONS

A: Area of flow
D: Diameter
d: Depth of flow
P: Wetted perimeter
R: Hydraulic radius
T: Top width

$\dfrac{d}{D}$	$\dfrac{A}{D^2}$	$\dfrac{P}{D}$	$\dfrac{R}{D}$	$\dfrac{T}{D}$
0.01	0.0013	0.2003	0.0066	0.1990
0.02	0.0037	0.2838	0.0132	0.2800
0.03	0.0069	0.3482	0.0197	0.3412
0.04	0.0105	0.4027	0.0262	0.3919
0.05	0.0147	0.4510	0.0326	0.4359
0.06	0.0192	0.4949	0.0389	0.4750
0.07	0.0242	0.5355	0.0451	0.5103
0.08	0.0294	0.5735	0.0513	0.5426
0.09	0.0350	0.6094	0.0574	0.5724
0.10	0.0409	0.6435	0.0635	0.6000
0.11	0.0470	0.6761	0.0695	0.6258
0.12	0.0534	0.7075	0.0754	0.6499
0.13	0.0600	0.7377	0.0813	0.6726
0.14	0.0668	0.7670	0.0871	0.6940
0.15	0.0739	0.7954	0.0929	0.7141
0.16	0.0811	0.8230	0.0986	0.7332
0.17	0.0885	0.8500	0.1042	0.7513
0.18	0.0961	0.8763	0.1097	0.7684
0.19	0.1039	0.9020	0.1152	0.7846
0.20	0.1118	0.9273	0.1206	0.8000
0.21	0.1199	0.9521	0.1259	0.8146
0.22	0.1281	0.9764	0.1312	0.8285
0.23	0.1365	1.0003	0.1364	0.8417
0.24	0.1449	1.0239	0.1416	0.8542
0.25	0.1535	1.0472	0.1466	0.8660
0.26	0.1623	1.0701	0.1516	0.8773
0.27	0.1711	1.0928	0.1566	0.8879
0.28	0.1800	1.1152	0.1614	0.8980
0.29	0.1890	1.1373	0.1662	0.9075
0.30	0.1982	1.1593	0.1709	0.9165

$\dfrac{d}{D}$	$\dfrac{A}{D^2}$	$\dfrac{P}{D}$	$\dfrac{R}{D}$	$\dfrac{T}{D}$
0.31	0.2074	1.1810	0.1755	0.9250
0.32	0.2167	1.2025	0.1801	0.9330
0.33	0.2260	1.2239	0.1848	0.9404
0.34	0.2355	1.2451	0.1891	0.9474
0.35	0.2450	1.2661	0.1935	0.9539
0.36	0.2546	1.2870	0.1978	0.9606
0.37	0.2642	1.3078	0.2020	0.9656
0.38	0.2739	1.3284	0.2061	0.9708
0.39	0.2836	1.3490	0.2102	0.9755
0.40	0.2934	1.3694	0.2142	0.9798
0.41	0.3032	1.3898	0.2181	0.9837
0.42	0.3132	1.4101	0.2220	0.9871
0.43	0.3229	1.4303	0.2257	0.9909
0.44	0.3328	1.4505	0.2294	0.9928
0.45	0.3428	1.4706	0.2331	0.9950
0.46	0.3527	1.4907	0.2366	0.9968
0.47	0.3627	1.5108	0.2400	0.9982
0.48	0.3727	1.5308	0.2434	0.9992
0.49	0.3827	1.5508	0.2467	0.9998
0.50	0.3927	1.5708	0.2500	1.0000
0.51	0.4027	1.5908	0.2531	0.9998
0.52	0.4127	1.6108	0.2561	0.9992
0.53	0.4227	1.6308	0.2591	0.9982
0.54	0.4327	1.6509	0.2620	0.9968
0.55	0.4426	1.6710	0.2649	0.9950
0.56	0.4526	1.6911	0.2676	0.9928
0.57	0.4625	1.7113	0.2703	0.9902
0.58	0.4723	1.7315	0.2728	0.9871
0.59	0.4822	1.7518	0.2753	0.9837
0.60	0.4920	1.7722	0.2776	0.9798
0.61	0.5018	1.7926	0.2797	0.9755
0.62	0.5115	1.8132	0.2818	0.9708
0.63	0.5212	1.8338	0.2839	0.9656
0.64	0.5308	1.8546	0.2860	0.9600
0.65	0.5404	1.8755	0.2881	0.9539
0.66	0.5499	1.8965	0.2899	0.9474
0.67	0.5594	1.9177	0.2917	0.9404
0.68	0.5687	1.9391	0.2935	0.9330
0.69	0.5780	1.9606	0.2950	0.9250
0.70	0.5872	1.9823	0.2962	0.9165
0.71	0.5964	2.0042	0.2973	0.9075
0.72	0.6054	2.0264	0.2984	0.8980
0.73	0.6143	2.0488	0.2995	0.8879
0.74	0.6231	2.0714	0.3006	0.8773
0.75	0.6318	2.0944	0.3017	0.8660

$\dfrac{d}{D}$	$\dfrac{A}{D^2}$	$\dfrac{P}{D}$	$\dfrac{R}{D}$	$\dfrac{T}{D}$
0.76	0.6404	2.1176	0.3025	0.8542
0.77	0.6489	2.1412	0.3032	0.8417
0.78	0.6573	2.1652	0.3037	0.8285
0.79	0.6655	2.1895	0.3040	0.8146
0.80	0.6736	2.2143	0.3042	0.8000
0.81	0.6815	2.2395	0.3044	0.7846
0.82	0.6893	2.2653	0.3043	0.7684
0.83	0.6969	2.2916	0.3041	0.7513
0.84	0.7043	2.3186	0.3038	0.7332
0.85	0.7115	2.3462	0.3033	0.7141
0.86	0.7186	2.3746	0.3026	0.6940
0.87	0.7254	2.4038	0.3017	0.6726
0.88	0.7320	2.4341	0.3008	0.6499
0.89	0.7380	2.4655	0.2996	0.6258
0.90	0.7445	2.4981	0.2980	0.6000
0.91	0.7504	2.5322	0.2963	0.5724
0.92	0.7560	2.5681	0.2944	0.5426
0.93	0.7612	2.6061	0.2922	0.5103
0.94	0.7662	2.6467	0.2896	0.4750
0.95	0.7707	2.6906	0.2864	0.4359
0.96	0.7749	2.7389	0.2830	0.3919
0.97	0.7785	2.7934	0.2787	0.3412
0.98	0.7816	2.8578	0.2735	0.2800
0.99	0.7841	2.9412	0.2665	0.1990
1.00	0.7854	3.1416	0.2500	0.0000

Appendix B

GEOMETRIC ELEMENTS FOR HORSESHOE CHANNEL SECTIONS

A: Area of flow
D: Diameter
d: Depth of flow
P: Wetted perimeter
R: Hydraulic radius
T: Top width

$\dfrac{d}{D}$	$\dfrac{A}{D^2}$	$\dfrac{R}{D}$	$\dfrac{d}{D}$	$\dfrac{A}{D^2}$	$\dfrac{R}{D}$
0.01	0.0019	0.0066	0.26	0.2013	0.1662
0.02	0.0053	0.0132	0.27	0.2107	0.1710
0.03	0.0097	0.0198	0.28	0.2202	0.1758
0.04	0.0150	0.0264	0.29	0.2297	0.1804
0.05	0.0209	0.0329	0.30	0.2393	0.1850
0.06	0.0275	0.0394	0.31	0.2489	0.1895
0.07	0.0346	0.0459	0.32	0.2586	0.1938
0.08	0.0421	0.0524	0.33	0.2683	0.1981
0.09	0.0502	0.0590	0.34	0.2780	0.2023
0.10	0.0585	0.0670	0.35	0.2878	0.2063
0.11	0.0670	0.0748	0.36	0.2975	0.2103
0.12	0.0753	0.0823	0.37	0.3074	0.2142
0.13	0.0839	0.0895	0.38	0.3172	0.2181
0.14	0.0925	0.0964	0.39	0.3271	0.2217
0.15	0.1012	0.1031	0.40	0.3370	0.2252
0.16	0.1100	0.1097	0.41	0.3469	0.2287
0.17	0.1188	0.1161	0.42	0.3568	0.2322
0.18	0.1277	0.1222	0.43	0.3667	0.2356
0.19	0.1367	0.1282	0.44	0.3767	0.2390
0.20	0.1457	0.1341	0.45	0.3867	0.2422
0.21	0.1549	0.1398	0.46	0.3966	0.2454
0.22	0.1640	0.1454	0.47	0.4066	0.2484
0.23	0.1733	0.1588	0.48	0.4166	0.2514
0.24	0.1825	0.1560	0.49	0.4266	0.2545
0.25	0.1919	0.1611	0.50	0.4366	0.2574

$\dfrac{d}{D}$	$\dfrac{A}{D^2}$	$\dfrac{R}{D}$	$\dfrac{d}{D}$	$\dfrac{A}{D^2}$	$\dfrac{R}{D}$
0.51	0.4466	0.2602	0.91	0.7943	0.2988
0.52	0.4566	0.2630	0.92	0.7999	0.2969
0.53	0.4666	0.2657	0.93	0.8052	0.2947
0.54	0.4766	0.2683	0.94	0.8101	0.2922
0.55	0.4866	0.2707	0.95	0.8146	0.2893
0.56	0.4965	0.2733	0.96	0.8188	0.2858
0.57	0.5064	0.2757	0.97	0.8225	0.2816
0.58	0.5163	0.2781	0.98	0.8256	0.2766
0.59	0.5261	0.2804	0.99	0.8280	0.2696
0.60	0.5359	0.2824	1.00	0.8293	0.2538
0.61	0.5457	0.2844			
0.62	0.5555	0.2864			
0.63	0.5651	0.2884			
0.64	0.5848	0.2902			
0.65	0.5843	0.2920			
0.66	0.5938	0.2937			
0.67	0.6033	0.2953			
0.68	0.6126	0.2967			
0.69	0.6219	0.2981			
0.70	0.6312	0.2994			
0.71	0.6403	0.3006			
0.72	0.6493	0.3018			
0.73	0.6582	0.3028			
0.74	0.6671	0.3036			
0.75	0.6758	0.3044			
0.76	0.6844	0.3050			
0.77	0.6929	0.3055			
0.78	0.7012	0.3060			
0.79	0.7094	0.3064			
0.80	0.7175	0.3067			
0.81	0.7254	0.3067			
0.82	0.7332	0.3066			
0.83	0.7408	0.3064			
0.84	0.7482	0.3061			
0.85	0.7554	0.3056			
0.86	0.7625	0.3050			
0.87	0.7693	0.3042			
0.88	0.7759	0.3032			
0.89	0.7823	0.3020			
0.90	0.7884	0.3005			

Appendix C

TABLE OF CONVERSION FACTORS

To convert from	To	Multiply by
Inches	feet	0.083333
"	angstrom	$2.54 \ 10^8$
"	microns	25400
"	millimeters	25.4
"	centimeters	2.54
Feet	inches	12.0
"	angstrom	$3.048 \ 10^9$
"	microns	304800
"	millimeters	304.80
"	centimeters	30.48
Microns	inches	$3.9370079 \ 10^{-5}$
"	feet	$3.2808399 \ 10^{-6}$
"	angstrom	$1 \ 10^4$
"	millimeters	$1 \ 10^{-3}$
"	centimeters	$1 \ 10^{-4}$
Millimeters	inches	$3.9370079 \ 10^{-2}$
"	feet	$3.2808399 \ 10^{-3}$
"	angstrom	$1 \ 10^7$
"	microns	$1 \ 10^3$
"	centimeters	$1 \ 10^{-1}$
Centimeters	inches	0.39370079
"	feet	0.032808399
"	angstrom	$1 \ 10^8$
"	microns	$1 \ 10^4$
"	millimeters	10

To convert from	To	Multiply by
Meters	inches	39.370079
"	angstrom	$1 \ 10^{10}$
"	micron	$1 \ 10^{6}$
"	millimeters	$1 \ 10^{3}$
Square meters	sf	10.763910
" "	sc	10000.0
" "	s.inches	1550.0031
Square feet (sf)	s. meters	$9.290304 \ 10^{-2}$
" "	sc	929.0304
" "	S inches	144
Square centimeters(sc)	s.meters	$1 \ 10^{-4}$
" "	sf	$1.0763910 \ 10^{-3}$
" "	s. inches	0.15500031
Square inches (s. inches)	s. meters	$6.4516 \ 10^{-4}$
" "	sf	$6.9444 \ 10^{-3}$
" "	sc	6.4516
Cubic centimeters (cc)	cm	$1 \ 10^{-6}$
" "	cf	$3.5314667 \ 10^{-5}$
" "	c. inches	0.061023744
Cubic meters (cm)	cf	35.314667
" "	cc	$1 \ 10^{6}$
" "	c. inches	61023.74
Cubic inches (c.inches)	cm	$1.6387064 \ 10^{-5}$
" "	cf	$5.7870370 \ 10^{-4}$
" "	cc	16.387064
Cubic feet (cf)	cm	0.028316847
" "	cc	28316.847
" "	c. inches	1728

Appendix C

To convert from	To	Multiply by
Pounds, avdp (lb)	dynes	$4.44822 \ 10^5$
"	gr	453.59237
"	kg	0.45359237
"	T (long)	$4.464286 \ 10^{-4}$
"	T (short)	$5 \ 10^{-4}$
"	kips	$1 \ 10^{-3}$
"	T (metric)	$4.5359237 \ 10^{-4}$
Kips	lb	1000
"	T (short)	0.500
"	kg	453.59237
"	T (metric)	0.45359237
Tons, short (T. short)	kg	907.18474
"	lb	2000
"	kips	2
"	T (metric)	0.907185
Kilograms (kg)	dynes	980665
"	gr	1000
"	lb	2.2046226
"	T (long)	$9.8420653 \ 10^{-4}$
"	T (short)	$11.023113 \ 10^{-4}$
"	kips	$2.2046226 \ 10^{-3}$
Tons, metric (T)	gr	$1 \ 10^6$
"	kg	1000
"	lb	2204.6226
"	kips	2.2046226
"	T (short)	1.1023113
Milliseconds	sec	10^{-3}
"	min	$1.6666 \ 10^{-5}$
"	hr	$2.7777 \ 10^{-7}$
"	days	$1.1574074 \ 10^{-8}$

To convert from	To	Multiply by
Seconds (sec)	milliseconds	1000
"	min	$1.6666 \ 10^{-2}$
"	hr	$2.7777 \ 10^{-4}$
"	days	$1.1574074 \ 10^{-5}$
Minutes (min)	millisecond	60000
"	sec	60
"	hr	0.0166666
"	days	$6.9444444 \ 10^{-4}$
Hours (hr)	milliseconds	3600000
"	sec	3600
"	min	60
"	days	0.0416666
Days	millisecond	86400000
"	sec	86400
"	min	1440
"	hr	24
Pound/square foot (psf)	psi	0.0069445
" " "	feet of water	0.016018
" " "	kips/sf	0.003
" " "	kg/scm	0.000488243
" " "	T/sm	0.004882
" " "	at	$4.72541 \ 10^{-4}$
Tons (short)/square foot		
T (short)/sf	kg/sm	9764.86
" "	T/sm	9.76487
" "	lb/s.inch	13.8888
" "	lb/sf	2000
" "	kips/sf	2.0
Feet of water at 39.2°F	lb/s.inch	0.43352
" "	lb/sf	62.427
" "	kg/s.cm	0.0304791
" "	T/sm	0.304791
" "	at	0.029499
" "	inches of Hg	0.88265

Appendix C

To convert from	To	Multiply by
Kips/sf	lb/s.inch	6.94445
"	lb/sf	1000
"	T (short(/sf	0.50000
"	kg/sc	0.488244
"	T/sm	4.88244
Kilograms/square centimeter	lb/s.inch	14.223
" "	lb/sf	2048.1614
" "	ft. of water (32°F)	32.8093
" "	kips/sf	2.0481614
" "	T/sm	10
" "	at	0.96784
Tons/square meter	kg/sc	0.10
" "	lb/sf	204.81614
" "	kips/sf	0.20481614
" "	T (short)/sf	0.102408
Grams/cubic centimeter (gr/cc)	T/cm	1.00
" "	kg/cm	1000.00
" "	lb/c.inch	0.036127292
" "	lb/cf	62.427961
Tons (metric)/cubic meter (T/cm)	gr/cc	1.00
" "	kg/cm	1000.00
" "	lb/c. inch	0.036127292
" "	lb/cf	62.427961
Kilograms/cubic meter (kg/cm)	gr/cc	0.001
" "	T/cm	0.001
" "	lb/c. inch	$3.6127292 \ 10^{-5}$
" "	lb/cf	0.062427961
Pounds/cubic inch (lb/c. inch)	gr/cc	27.679905
" "	T/cm	27.679905
" "	kg/cm	27679.905
" "	lb/cf	1728
Pounds/cubic foot (bl/cf)	gr/cc	0.016018463
" "	T/cm	0.016018463
" "	kg/cm	16.018463
" "	lb/c. inch	$5.78703704 \ 10^{-4}$

REFERENCES

Ahmed, M., 1953, "Experiments and Design and Behavior of Spur Dikes," Proc. IAHR, ASCE Joint Meeting, Univ. of Minnesota August.

Acatay, T., l967, "Karakteristikler Metodu," DSI, Teknik Bülten, no:1.

Aksoy, S., 1966, "Büyük Barajlarin Mansap Tarafindaki Nehir Tabani Alçalmalari," DSI Proceedings, no. 350.

Aksoy, S., 1971, "River Bed Degradation Downstream of Dams," IAHR, XIV Congress.

Aksoy, S., 1973, "The Influence of the Relative Depth on Threshhold of Grain Motion," Proc. Inter. Symp.on River Mechanics, Bangkok.

Altinbilek, H.D., 1971, "Similarity Laws for Local Scour with Special Emphasis on Vertical Circular Pile in Oscillatory Flow," XIV Congress of IAHR, Paris.

Altinbilek, H.D., and Okyay, S., 1973, "Localized Scour in a Horizontal Sand Bed Under Vertical Jets," XV Congress of IAHR, Istanbul.

Allen, J.R.L., 1969, "Physical Processes of Sedimentation," American Elsevir, New York.

Anderson, A.G., et al., 1970, "Tentative Design Procedure of Riprap Lined Channels," Project report no. 6, St.Antony Falls Hyd. Lab, Minneapolis, Minnesota NCHRP, Rep. 108

Aristowski-Beyer, 1959, "Entwurfsgrundlagen Zum Wehrbau," VEB Verlag Technik, Berlin.

Ashida, K., and Michiue, M., 1971, "An Investigation for River Bed Degradation Downstream of a Dam," Trans. XIV Congress Paris.

Army, Waterways Experiment Station, Vicksburg Mississipi, February, 1966.

ASCE Task Committee on Preparation of Sediment Manual, 1972, "Sediment Control Methods: B. Stream Channels," Journal of Hydraulics Division, ASCE, Vol. 98, no. HY7.

AWWA, 1964, "Spillway Design Practice," M.13 Spillway Design Manual.

Bata, G., Knezevich, B., 1953, "Some Observations on Density Currents in the Laboratory and in the field," IAHR, Minnesota Convention.

References

Bata, G., 1960, "Scour Around Bridge Piers," in Serbish, Institut za Vodoprivreder, Yugoslavia.

Bauer, W.J., 1954, "Turbulent Boundary Layer on Steep Slopes," TASCE V.119 pp. 1212-1234.

Beichly, G.S., 1963, "Stilling Basin Chute Block Pressures Basin II," Progress Report no. VI HYD-514.

Beichly, G.L., 1969, "Stilling Basin for Pipe or Open Channel Outlets - Basin IV," Report No. HYD 572, USBR.

Bell, Hugh Stefens, 1942, "Stratified Flow in Reservoirs and its use in the Prevention of Silting," U.S. Dept. of Agriculture, Pub. no. 491.

Beken, Şakir, 1970, "Derivasyon, Dipsavak, Dolusavak," in Turkish DSI Publications.

Blaisdell, F.W. and Al., 1951, "Hydraulic Design of the Box Inlet Drop Spillway," U.S. Dept. of Agriculture.

Blaisdell, F.W., 1958, "Hydraulics of Closed Conduit Spillway," U.S. Dept. of Agriculture.

Blaisdell, F.W., 1959, "The SAF Stilling Basin," St. Antony Falls.

Bogardy, J.L., 1965, "European Concepts of Sediment Transportation," J.H.D., ASCE Vol. 91, no HY1.

Bonnefille, R. and Goddet, J., 1959, "Study of Density Currents in a Canal," AIRH Montreal, Vol. 2.

Borland, M., 1971, "River Mechanics," edited by H.W. Shen, Chapter 29, Colorado State University, Colorado.

Bos., M.G., 1976, "Discharge Measurement Structures," L.R.I., Netherland.

Bouvard, 1958, "Barrages Mobiles et Prises d'Eaux en Riviére," Eyrolles.

Bradley, J.N., 1952, "Discharge Coefficients for Irregular Overfall Spillways," Engineering Monographs no. 9, USBR, Denver.

Bretschneider, C.L., 1958, "Revision in Wave Forecasting: Deep and Shallow Water," Proc. of Sixth conference on Coastal Engineering, Council on Wave Research, University of California.

Breusers, H.N.C., 1965, "Scour Around Drilling Platforms," Bulletin, Hydraulic Research 1964 and 1965, IAHR, Vol. 19, p. 276.

Brooks, N.H., 1963, "Boundary Shear Stress in Curved Channels," a discussion, Proc. ASCE, Vol. 89, no. HY3.

Brown, C.B., 1943, "The Control of Reservoir Silting," U.S. Dept. of Agriculture, Pub. no. 521.

Brown, C.B., 1944, "Discussion of 'Sediment in Reservoir' by B.J. Witzig," Transaction ASCE, Vol. 109.

Brown, C.B., 1950, "Engineering Hydraulics," edited by Rouse, J. Wiley and Sons, Inc.

Brudenell, R.N., 1935, "Flow Over Rounded Crests," Engineering News Record, V.115, no. 3.

Bruk, S., and Milorodov, V., 1971, "Bed Deformation Due to Stilling of Nonuniform Sediments in Backwater Affected Rivers," Trans. IAHR XV Congress, Paris.

Brune, G.H., 1953, "Trap Efficiency of Reservoirs," Am. Geophysical Union Trans., Vol. 34, no. 3.

Bruschin, J. and Dysli, M., 1973, "Erosion des Rives due aux Oscillations du Plan d'eaux d'une Retenue-Le Rhône à L'aval de Genéve," EPFL, no. 27.

Calhoun, C.C., Jr., and Compton, J.R. and Strohm, W.E., Jr., 1971, "Performance of Plastic Filter Clothes as a Replacement for Granular Filter Material," Paper presented at the annual meeting of Highway research Board, January.

California Division of Highways, 1970, "Bank and Shore Protection in California Highway Practice," Dept. of Pub. Works.

Campbell, F.B., 1966, "Hydraulic Design of Rock Riprap," Misc. Paper no. 2-777, Office, Chief of Engineers, U.S. Army.

Carlson, E.J., and Sayre, W.W., 1961, "Progress Report: 1-canal bank erosion due to wind-generated water waves," USBR.

Carstens, M.R., 1966, "An Analytical and Experimental Study of Bed Ripples under Wake Waves," Quart. Reports 8 and 9, Georgia Inst. of Technology, School of Engineering, Atlanta.

Carstens, L.R., 1966, "Similarity Laws for Localized Scour Journal of the Hydraulics Division," ASCE, Vo. 92, no HY3, Proc. Paper 4818, May pp. 13-16.

Casagrande, A., 1937, "Seepage Through Dams," J. New Eng. Water Works Assoc., June.

Castillo, J. and Del Campo, F., 1970, "La Auscultacion de la Presa de Valdecanas," Revista de Obras Publicas - Spain.

Chabert, J., and Engeldinger, P., 1956, "Etudes des Afouillements Autour de Piles

References

des Ponts," Lab. National d'Hydraulic, 6 Quai Watier, Chatou - Paris.

Charlton, F.G., 1966, "A Study of a Self Priming, Self Regulating Saddle Siphon," Inst. Water Eng.

Cheng, D.H., 1969, "Incipient Motion of Large Roughness Elements in Turbulent Open Channel Flow," Dissertation, Utah State University.

Chien, N., 1954, "The Present Status of Research on Sediment Transport," Proc., ASCE, Vol. 80.

Chow, V.T., 1935, "Open Channel Hydraulics," Mc Graw-Hill Book Company.

Churchill, M.A., 1948, "Discussion of 'Analysis and Use of Reservoir Sedimentation Data'," by L.C. Gottschalk, Federal Interagency Sedimentation Conf., Denver, Colorado, 1947, Proc., pp. 139-140.

Citrini, D., 1938, "Recherche d'une Formule Simplifiée Pour le Dimensionnement des Déversoirs Latéraux," Energia Electrica.

Coleman, N.L., 1969, "A Theoretical and Experimental Study of Drag Lift Forces Acting on a Sphere Resting on a Hypothetical Stream Bed," IAHR, Fort Collins, Colorado.

Coleman, N.L., 1971, "Analysing Laboratory Measurements of Scour at Cylindrical Piers in Sand Beds," Proceedings IAHR XIV, Paris, Vol. III.

Coleman, N.L., 1972, "The Drag Coefficient of a Stationary Sphere on a Boundary of Similar Spheres," La Houille Blanche, no. 1

Comision Federal de Electricidad, 1976, "Projecto Hidroelectri co Chicoasen," Mexico.

Coyne et Bellier, 1962, "Les Grands Barrages," Paris.

Craya, A. and Jaeger, C., 1948, "Hauteur d'eaux à l'Extremité d'un Long Déversoir," La Houille Blanche, pp. 185-187.

Creager, W.P. -Justin, J.D.-Hinds, J., 1950, "Engineering for Dams," Wiley, USA.

Crump E.S., 1954, "A Wortex-siphon Spillway for Maintaining a Constant Water Level Upstream of a Structure," Ing. C.E. no. 6108.

Davis, C.V., 1952, "Handbook of Applied Hydraulics," Mc Graw-Hill.

Demiröz, E., 1985, "Dolusavaklarin Boşaltim Kanallarindaki Büyük Hizli Akimlar Için Kullanilan Havalandiricilarin Bazi Proje Kriterleri," DSI (in Turkish).

Bureau of Yards and Docks, 1961, "Design Manual, Soil Mechanics, Foundations and Earth Structures," Nordocks DM.7 Dept. of the Navy.

Doddiah, D., Albertson, M., and Thomas, R., 1953, "Scour From Jets," IAHR Proc., Minnesota International Hydraulic Convention, September.

Doeringfeld-Barker, C.S., 1941, "Pressure Momentum Theory Applied to the Broad-crested Weir," TASCE, Vol. 106 pp. 934 - 946.

Dressler, R.F. and Yevjevich, V., 1984, "Hydraulic Resistance Terms Modified for the Dressler Curved Flow Equations," Jour. of Hydraulic Research, Vol. 22.

DSI, 1965, "Gökçekaya Baraji Savak Tesislerine Ait Model Deneyleri," HI 348, Ankara.

DSI, 1967, "Model Studies on Keban Dam," HI-426, Ankara.

DSI, 1967, "Keban Baraji Dolusavaği Model Deneyleri," HI-426.

DSI, 1968, "Aslantaş Baraji Dolusavaği Model Deneyleri," HI-500 (in Turkish).

DSI, 1968, "Model Studies on Aslantaş Dam," HI-500, Ankara.

DSI, 1970, "Hydraulic Computation of Spillways, Medik Dam Spillway."

Dubs, 1947, "Angewandete Hydraulic," Rascher Verlag, Zürich.

EDF, 1951, "Hydraulique et Electricité Française."

FDF, 1953, "Bort."

Edwards, R.H., 1952, "Model Tests on the Bellmouth Spillway," Waterway Eng.

Eggenberger, W., 1943, "Die Kolkbildung beim einen Überströmen beider Kombination Überströmen-Unterströmen," Dissertation ETH Zürich.

Egiazarof, I.V., 1950, "Coefficient (f) de la Force d'Entrainement Critique des Matériaux par Charriage," Proc. of Ac. of Sc., Armenian S.S.R. Transl. EDF.

Einstein, H.A. and Barbarossa, N.L., 1952, "River Channel Roughness," Transaction, ASCE, Vol. 117, pp. 1121 - 1146.

Elevatorski, E.A., 1959, "Hydraulic Energy Dissipators," McGraw-Hill Book Co.

EPDC, 1972, "Hydraulic Design of Spillways."

Escande, L., "Recherche sur le Fonctionnement Simultané des Barrages Mobiles en Deversoirs et en Vannes de Fond."

Escande, L., 1952, "Les Barrages Déversoirs à Fente Aspiratrice," Génie Civil.

Fair, G.M. and Hatch, L.P., 1933, "Fundamentals Governing the Stream Line Flow of Water Trough Sands," J.A. Water Works Association.

References

Faure, J., and Pugnet, L., 1959, "Etude de l'Allimentation d'un Evacuateur en Puit," VI İtalien Hydraulic Congress.

Fietz, T.R. and Wood, I.R., 1967, "Three-Dimensional Density Current," PASCE Vol. 93 no. HY2.

Fortier, S. and Scoby, F.C., 1926, "Permissible Canal Velocities," Transactions, ASCE, Vol. 89, pp. 940 - 956.

Franke, P.G., 1960, "L'Affouillement: Mécanisme et Formes," Österreichische Wasserwirtschaft, January.

Gaillard, D.D., 1904, "Wave Action in Relation to Engineering Structures," Prof. Paper, no. 31 Corps of Engineers.

Garde, R.J., and Swamee, P.K., 1943, "Analysis of Aggradation Upstream of a Dam," Inter. Sympo. on River Mechanics Bangkok , Thailand.

Gessler, J.,1965, "The Beginning of Bedload Movement of Mixtures Investigated as Natural Armoring in Channels," W.M.K.M. Keck Lab. of Hydraulics and Water Resources, Calif. Ins. Techn., Pasadena.

Gessler, J., 1970, "Self-Stabilizing Tendencies of Alluvial Channels," Journ. of the Waterways, Harbors and Coastal Engineering Division, ASCE, Vol. 96, no WW2, May.

Gessler, J., 1971, "Critical Shear Stress for Sediment Mixtures," Trans. IAHR, XIV Congress, Paris, Vol. III.

Gessler, J., 1971, "Beginning and ceasing of sediment motion," River Mechanics, edited by H.W.Shen, Chapter 7, Fort Collins, Colorado, 22pp.

Gill, M.A., 1972, "Erosion of Sand Beds Around Spur Dikes," Jour. Hyd. Div. ASCE, Vol. 98, no. HY9, September, pp. 1587-1602.

Graf, W.H., 1971, "Hydraulics of Sediment Transport," McGraw Hill Book Company.

Graf, W.H., 1977, "Reservoir Sedimentation," EPF de Lausanne Pub. n35.

Gould, H.R., 1951, "Some Quantatative Aspects of Lake Mead Turbidity Current," Soc. of Economic Paleontologists and Mineralogists, Pub. no.2.

Harleman, R.F., 1954, "Effect at Baffle Piers on Stilling Basin Performance," Boston S.C.E.

Harrison, A.J.M., 1966, "Hydraulic Model Investigation of a Siphon Bellmouth," Institution Water Engineering.

Harvey, B., Vershuren, J.P., and Stolte, J.V., 1971, "Exhaussement du lit à l'amont d'un Reservoir," Trans. IAHR XIV Congress Paris.

Hickox, G.H., 1944, "Aeration of Spillways," TASCE Vol. 109 pp 537-556.

Hinds, J., 1936, "Side Channel Spillway: Hydraulic Theory, Economic Factors and Experimental Determination of Losses," TASCE, Vol. 89.

Hinze, J.D., 1960, "On the Hydrodynamics of Turbidity Currents," Geol. Mijnbous, 39th Ing.

Hough, B.K., 1957, "Basic Soil Engineering," Ronald Press Co.

Hudson, R.Y., 1961, "Laboratory Investigation of Bubble-Mound Breakwaters," TASCE Paper no 3213 Vol. 126.

ICOLD, 1960, 1964, 1967, 1973, 1976, "Dams in Japan," Tokyo.

Icold, 1967, "U.S. Dams," Denver.

Ippen, A.T., 1950, "Channel Transitions and Controls," Engineering Hydraulics, editor: H. Rouse.

Ippen, A.T., 1961, "Mechanics of Supercritical Flow 1st paper in High Velocity Flow in Open Channels," TASCE Vol. 116 pp. 268-295.

Ippen, A.T. and Dawson, J.H., 1951, "Design of Channel Contractions," 3d paper in High Velocity Flow in Open Channels' A symposium TASCE Vol. 116 pp. 326-346.

Ippen, A.T., Harleman, D.R.F., 1952, "Steady State Characteristics of Subsurface Flow," Circular no. 521, Gravity waves symposium, National Bureau of Standards, Washington D.C.

Ippen, A.T., and Verma, R.B., 1953, "The Motion of Discrete Particles Along the Bed of a Turbulent Stream," Proceedings 5th Congress of IAHR, Minneapolis.

Ippen, T., and Drinker, P.A., 1962, "Boundary Shear in Curved Trapezoidal Channels," PASCE no. 3273.

Jäger, C., 1939, "Über die Aenlichkeit bei Flussaulichen Modelversuchen," W.W 34, no. 23-24, 269.

Johnson, G., 1957, "The Effect of Entrained Air on the Scouring Capacity of Water Jets," IAHR, XII Congress, Fort Collins, Colorado.

Jopling, A.V., 1964, "Laboratory Study of Sorting Process Related to Flow Separation," Journal of Geophysical Research, 69: 3403-3418.

Jumikis, A.R., "Thermal Soil Mechanics."

Keulogan, G.H., 1949, "Interfacial Stability and Mixing in Stratified Flow," National Bureau of Standards, Vol. 32 no. 303, RP 1591.

References

Khosla, A.R., 1953, "Silting of Reservoirs," Government of India Press.

King, D.l., 1965, "Stilling Basins for High Head Outlet Works with Slide-gate Control," USBR HYD-544.

King, H.W. and Brater, E.F., 1953, "Handbook of Hydraulics," Wiley.

Klein, H.J., 1963, "Esher-Wyss Schützen," E. Wyss Mitteilungen no. 6.

Knoroz, V.S., 1971, "Natural Armoring and its Effect on Deformation of Channel Beds Formed by Materials Nonuniform in Size," IAHR, XIV Congress, Paris, Vol. III.

Komura, S., 1963, "Discussion: Erosion of Sediments," Task Committee on Preparation of Sediment Manual, Journal Hyd. Div., ASCE, vol. 89, no. HY1, Proc. Paper 3405.

Komura, S., and Simons, D.B., 1967, "River Bed Degradation Below Dams," Proceedings, ASCE, Proc. Paper 5335, HY4, July.

Komura, S., 1971, "Prediction of River Bed Degradation Below Dams," Transaction IAHR XIV Congress, Paris, Vol. III.

Korzhavin, K.N., 1972, "General Research Results Obtained in the URSS on Ice-Thermal Conditions in the Vicinity of Hydraulic Structures," IAHR symposium, Leningrad.

Krynine, D.P., and Judd, W.R., "Principles of Engineering Geology and Geotechnics."

Kuenen, P.H., 1952, "Estimated Size of the Grand Banks Turbidity Current," American Journ. Sc.

Läufer, 1936, "Ecoulement Dans les Canaux à Radier Incurvé," Wasserkraft und Wasserwirtschaft (French Translation).

Lane, E.W., and Koelzer, V.A., 1943, "Density of Sediment Deposited in Reservoirs," Report no. 9, published at St. Paul, U.S. Engineer District sub-office, Hydraulic Laboratory, University of Iowa, Iowa City, Iowa.

Lane, E.W., and Koelzer, V.A., 1953, "Density of Sediments Deposited in Reservoirs," Report no. 5 of a Study of Methods Used in Measurement and Analysis of Sediment Loads in Streams, St. Paul, U.S. Eng. Distr. sub-office Hydraulic Lab. University of Iowa, Iowa City, Iowa.

Larras, J., 1963, "Profondeurs Maximales d'Erosion des Fonds Mobiles Autour des Piles en Riviére," Annales des Ponts et Chaussées, Vol. 133, no. 4, pp. 411-424.

Laursen, E.M., 1952, "Observations on the Nature of Scour," Proceedings of the 5th Hydraulic Conf., Bulletin 34, University of Iowa Studies in Engineering, Iowa City, Iowa.

Laursen, E.M., 1960, "Scour of Bridge Crossings," Journal Hyd., ASCE, Vol. 89, no. HY2, February, pp. 1-54.

Laursen, E.M., 1962, "Scour at Bridge Crossings," TASCE, 127, Part I.

Laursen, E.M., 1963, "An Analysis of Relief Bridge Scour," Journal Hyd. Div., ASCE, Vol. 89, no. HY3, May, pp. 93-118.

Laursen, E.M., and Flick, M.W., 1983, "Final Report Predicting Scour at Bridges, Questions not Fully Answered-Scour at Sill Structures," Report ATTI-83-6, Arizona Dept. of Transportation.

LeFeuve, A.R., 1965, "Sediment Transport Functions with Spherical Emphasis on Localized Scour," Thesis presented to the Georgia Institute of Technology, Atlanta, Georgia.

Linslay, R.K., and Franzini, J.B., 1945, "Water Resources Engineering," McGraw-Hill Book Company.

Liu, H.K., et al., 1961, "Effect of Bridge Construction on Scour and Backwater," Dept of Civil Engineering, Colorado State University, Res. no CER60HKL22, February.

Maître, R., and Obolensky, S., 1958, "Etude de Quelques Characteristiques de l'Écoulement dans la Partie Avale des Évacuateurs de Surfaces," La Houille Blanche.

Maître, R., 1960, "Note sur les Dépressions aux Ouvertures Partielles des Vannes dans les Évacuateurs de Crue et Leur Repercussion sur le Choix du Profil Déversant et de la Position des Vannes," Lab. Nation. d'Hydr. de Chatoux.

Mallet, G., and Pacquant, J., 1951, "Les Barrages en Terre," Eyrolles, Paris.

M.A.N., 1955, "Schützen."

Maza Alvarez, J.A., and Sanchez, B.J.L., 1964, "Contricion de le Socabocion Local en Pilas de Puente," Univ. Federal do Reo Grande de Sul.

McHenri. J.R. and Dendy, F.E., 1964, "Measurement of Sediment Density by Attenation of Transmitted Gamma Rays," Proc. of Soil Science Soc. Vol. 25.

McHenri, J.R. et al., 1969, "Consolidation of Sediment in a Small Reservoir in North Mississippi Measured in situ with a Gamma Probe," Proc. of the Mississippi W.R. Conf.

McHenri, J.R., 1971, "Discussion of Sediment Measurement Techniques: Reservoir Deposits," PASCE HD Vol. 97 no. HY8-8267.

Ministerio de Obras Publicas, 1973, "Grandes Presas en Mexico."

References

Molitor, D.A., 1935, "Wave Pressures on Sea-walls and Breakwaters," Trans. ASCE, Vol. 100, p. 984.

Mostafa, G., 1957, "River Bed Degradation Below Large Capacity Reservoir," TASCE, vol. 122, pp. 688-695.

Müller, R., 1944, "Experimentelle und Theoretische Untersuchungen über das Kolkproblem," Mitt. Versuchanschtalt Wasserbau ETH, no. 5, Zürich.

Neil, C.R., 1973, "Guide to Bridge Hydraulics," Road and Transportation Association of Canada, University of Toronto Press.

Nordin, C.F., 1971, "Graphical Aids for Determining Scour Depth in Long Contraction," Unpublished note (obtained from private communication).

Ortloff, C.R., 1988, "Canal Builders of Pre-Inca Peru," Scientific American, December.

Oziş, Ü., 1987, "Ancient Water Works in Anatolia," London, Water Resources Development, V.3.

Oziş, Ü., 1987, "Ancient Water Works in Anatolia per Respect to Hydraulic Engineering," Dokuz Eylül University, Izmir (in Turkish).

Oziş, Ü. and Harmancioğlu,, N., 1980, "Some Ancient Water Works in Anatolia," International seminar on Karst Hydrology, Antalya 1979, Proceedings, Ankara DSI-UNDP.

Oziş, Ü.; Benzeden, E., 1977, "Historical Hydraulic Works in Anatolia," IAHR, Baden-Baden XVII Congress.

Partensky, H.W., 1964, "Forces Dynamiques Agissant sur les Portes des Ecluses lors dela Dérivation Partielle de Crues," PHD Thesis, ENSEEHT.

Partensky, H.W., 1965, "Vibrations des Vannes lors de l'Utilisation des Ecluses Comme Évacuateur de Crue," IAHR.

Peterka, A.J., 1963, "Hydraulic Design of Stilling Basins and Energy Dissipators," Eng.Mono. no. 25, USBR.

Pinto, Nelson L. de S. and al., 1981, "Prototype and Laboratory Experiments on Aeration at High Velocity Flow," Water Power and Dam Construction.

Pinto, Nelson L. de S., 1985, "Bulking, Effects and Entraining Mechanism in Artificially Aerated Spillway Flow," Anales de la Universidad de Chile no. 8.

Pinto, Nelson L. de S., 1986, "Basic Hydraulics of Shooting Flows Over Aerators," ASCE, Special Conference, Minneapolis, Minnesota.

Pinto, Nelson L. de S., 1989, "Designing Aerators for High Velocity Flow," Water Power and Dam Construction.

Poggi, B., 1956, "l'Évacuateur à Déversoir en Cascade," Energia Elettrica.

Post, G. and Londe, P., 1953, "Les Barrages en Terre Compactée," Gothier-Villars, Paris.

Peers, E.F., 1938, "Low Dams," National Resource Committee U.S., Government Printing Office.

Press, H., 1958, "Stauanlagen und Wasserkraftwerke, l. Teil: Talsperren," W. Ernst and Sohn, Berlin.

Rafay, T., 1964, "Analysis of Change in Size of Bed Material Along Alluvial Channels," M.S. Thesis, Colorado State University, Fort Collins, Colorado.

Rana, S.A., and Simons, D.B., and Mahmood, K., 1973, "Analysis of Sediment Sorting in Alluvial Channels," Published in the Journal of Hydraulic Division, Vol. 99, no. 11.

Randolph, R.R., 1937, "Hydraulic Tests on the Spillway of the Madden Dam," PASCE, Paper no. 2001.

Raynaud, J.P., 1951, "Study of Currents of Muddy Water Through Reservoirs," ICOLD, New Delhi, India.

Richardson, E.V., 1971, "Sediment Properties," River Mechanics Chapter 6, Edited by W. Shen, Fort Collins.

Richardson, E.V., and Simons, D.B., 1957, "Discussion of Liu, H.K., 1957 Paper," PASCEHD, September.

Rhone, T.G., 1959, "Hydraulic of Energy Dissipators," McGraw-Hill Book Company.

Rose, E., 1947, "Thrust Exerted by Expanding Ice Sheet," TASCE, Vol. 112.

Rouse, H., 1956, "Discharge Characteristics of the Free Overfall," Civil Engineering, Vol. 6 no. 7, pp. 257-260.

Rouse, H., 1963, "Handbook of Hydraulics," Wiley.

Roy, 1949, "Note sur la Profondeur d'Eau à l'Extrémité d'un Déversoir Horizontal à Chute Libre," La Houille Blanche.

Saville, T. Jr., 1954, "The Effect of Fetch Width on Wave Generation," U.S. Army Corps of Engineers, Beach Erosion Board T.M. no. 70.

Saville, T. Jr., and McClendon, E.W., and Cochran, A.L., 1962, "Freeboard Allowances for Waves in Inland Reservoirs," Journal of Waterways and Harbors Div. ASCE, pp. 93-124.

References

Schauman, H., 1950, "Calcul des Déversoirs Latéraux," Ing. für Wasserbau-Berlin.

Scheidegger, A., 1970, "Theoretical Geomorphology," Springer Berlin.

Schmidt, M., 1954, "Ecoulement Sur les Déversoirs Latéraux," T.U. Berlin (French Translation).

Schlichting, E., 1960, "Boundary Layer Theory," McGraw-Hill Book Company.

Schoklitsch, A., 1914, "Über Schleppkraft und Geschie Bewegung," Engelmann, Leipzig (in German).

Schoklitsch, A., 1932, "Kolkbildung Unter Überfallstrahlen," Wasserwirtschaft, p. 343.

Schoklitsch, A., 1950, "Handbuch Des Wasserbaues, I," Springer Verlag Wien.

Schults, S., and Hill, R.D., 1968, "Bed Load Formulas, Part A. - A Selection of bed Load Formulas," The Pennsylvania State University, College of Engineering, University Park, Pennsylvania.

Scimemi, E., 1937, "Il Profilo Delle Dighe Sifioronti," Energia Elettrica.

Scimemi, E., 1946, "Sur la Forme à Donner aux Digues Déversantes," Energia Elettrica (French translation).

Searcy, J.K., 1967, "Use of Riprap for Bank Protection," Hydraulic Engineering Circular no. 11, Hydraulics Branch, Bridge Division, Office of Engineering and Operations, Bureau of Public Roads, Washington, D.C., June.

Secretaria de Recursos Hidrolicos, 1976, "Grandes Presas de Mexico," Mexico.

Sverdrup, H.U., and Munk, W.H., 1947, "Wind, Sea and Swell: Theory of Relationships for Forecasting," Publication no. 601, Hydrographic Office, U.S. Navy Dept.

Shalash, S., 1974, "Facts about Degradation," Arab Republic of Egypt Ministry of Irrigation, Dept of Hydrology.

Shields, A., 1936, "Anwendund der Aehnlichkeitsmechanik und Turbulentz-forschung auf die Geschiebewegung," Mitteilung Preussischen Versuchanstalt Wasser, Erd, Schiffbau, Berlin no. 26 (in German).

Simons, D.B., 1957, "Theory and Design of Stable Channels in Alluvial Material," PhD. Dissertation, Dept of Civil Engineering Colorado State University.

Simons, D.B., and Richardson, E.V., 1966, "Resistance to Flow in Alluvial Channels," USGS Professional Paper 422-J.

Simons, D.B., and Stefens, M.A., 1971, "Scour Control in Rock Basins at Culvert Outlets," River Mechanics, Colorado State University, Fort Collins, Colorado.

Simons, D.B. and Sentürk, F., 1976 & 1992, "Sediment Transport Technology," WRP, Littleton, Co.

SRH, 1976, "Presas construidas en Mexico," Mexico.

State of Illinois, 1970, "Model Test Result of Circular, Square and Rectangular Forms of Drop Inlet Entrance to Closed Conduit Spillways."

Stevens, M.A., and Simons, D.B., 1971, "Stability Analysis for Coarse Granular Material on Slopes," Ch. A., River Hydraulics, Vol. 1 (edited by H.W. Shen), Fort Collins, Colorado.

Stefenson, T.,1874, "The Design and Construction of Harbors," A.C .Block, Edinbourg.

Straub, L.G., 1935, "Some Observations for Sorting of Rivers Sediments," American Geophysical Union, Trans., 16:463-467.

Straub, L.G., and Lamb, O.P., 1953, "Experimental Studies of Air Entrainment in Open Channels," IAHR, Minneapolis.

Shui-Bo, Yingying, S., Qisuir, S., Xing-Lin, D., 1980, "Self Aeration Capacity of a Water Jet Over an Aeration Ramp," Journal of Hydraulic Engineering, Beijing, China no. 5.

Sunborg, A., 1956, "The River Klaralven, A Study of Fluvial Processes," Bulletin no. 52, Ins. of Hydraulics, Roy. Inst. of Tech., Stockholm, Sweden.

Şentürk, F., 1947, "Hidromekanik Benzeşim ve bu Metod ile Bazi Akim Olaylarinin Incelenmesi," ITU Publication no. 350 (in Turkish).

Şentürk, F., 1957, "The New Elmali Dam," DSI Publication, (in Turkish).

Şentürk, F., 1957, "Bağlamalar," DSI Publication, (in Turkish).

Şentürk, F., 1959, "Hydraulics of Water Workes," Mühendislik Haberleri, Andara, (in Turkish).

Şentürk, F., 1964, "Nehir Hidroliğinde Benzeşim," DSI Publication, (in Turkish).

Şentürk, F., 1965, "Güven Savaklarinin Benzeşimi," Teknik Bülten, DSI Publication, (in Turkish) no. 2.

Şentürk, F., 1965, "Similitude of Fuse Plugs," Technical Bulletin no. 2, Ankara.

Şentürk, F., 1966, "Similitude of Fuse-Plugs," golden Jubilee hydraulic Lab, Poona, India.

References

Şentürk, F., 1966, "Gökçekaya Dolusavağinin Mansabinda Yeralmasi Mümkün Olan Oyulmalarin Model Üzerinde Incelenmesi," Teknik Bülten no. 7, DSI Publication, (in Turkish).

Şentürk, F., 1969, "Nehir Hidroliği," DSI Publication (in Turkish).

Şentürk, F., 1969, "Sualma Yapilarinin Hidrolik Çalişaişlarina Dair," T. Mühendislik Haberleri Degisi, (in Turkish).

Şentürk, F., 1972, "Contribution to the Hydraulic Computation of Spillways," RCD Symposium, Ankara.

Şentürk, F., 1972, "A Propos du Coefficient de Decharge des Vannes Secteurs," DSI Publication.

Şentürk, F., 1972, "Dolusavaklarin Hidroliği ve Projelendirilmesi," DSI Pub., (in Turkish).

Şentürk, F., 1979, "Karakuz Baraji Raporlari," DSI Pub.

Şentürk, F., 1986, "Physical and Mathematical Model in Solving Flow Problems," DSI Publication.

Talvey, H.T., 1974, "Prediction of Wind Wave Heights," Journal of the Water Ways, Harbors and Coastal Engineering Div., WW1 no. 10324.

Task Committee on Sedimentation Manual, 1966, "Sediment Transportation Mechanics: Initiation of Motion," Proc. ASCE, no. HY2, March, p. 4738.

Task Committee on Sedimentation Manual, 1972, "Sedimentation Engineering," Vito A.Vanoni, Edit. ASCE.

Tison, L.J., 1953, "Recherche Sur la Tension Limite d'Entrainment Constitutif du lit," IAHR, 5th Congress, Minneapolis.

Taylor, K.V., 1973, "Slope Protection on Earth and Rockfill Dam," ICOLD, Madrid.

Terzaghi, K., 1943, "Theoretical Soil Mechanics," J. Wiley.

Terzaghi, K., and Peck, R.B., 1948, "Soil Mechanics in Engineering Practice," J. Wiley.

Terzaghi, K., and Bertram, G.E., 1948, "An Experimental investigation of Protective Filters," Soil Mechanics, Serie no. 7, Graduate School of Engineering, Harvard Universtiy.

Urbonas, B.R., 1968, "Forces on a Bed Particle in a Damped Rock / M.S.Thesis, Colorado State University, Fort Collins, Colorado.

Urrutia, C.M., 1955, "The effect of radial gates upon pressure distribution on overflow spillways", Test performed in USBR labs.

U.S. Army Corps of Engineers, 1948, "Laboratory Investigation on Filtrs for Enid and Grenada Dams," USWES Tech. Memo. no. 3245.

U.S. Army Corps of Engineers, "Hydraulic Design Criteria-Spillway Energy Losses," Manual 111-18.

U.S. Corps of Engineers, 1949, "Slope Protection of Earth Dams," Preliminiary Reportt - Vicksburg.

U.S. Army Corps of Engineers, 1965, "Hydraulic Design of Spillways," Manual EM no. 1110-2-1603.

U.S. Army Corps of Engineers, 1970, "Stone Stability-Velocity vs. Stone Diameter," Sheet 712-1, Civil Works Investigation Hydraulic Design Criteria, revised, August 1970.

USBR, 1941-1947, "Lake Mead Density Current Investigations," Vol. 1 and 2, Vol. 3.

USBR, "A Treatise on Dams."

USBR, 1951, "Stable Channel Profiles," Lab. Report no. Hyd. 325, September 27.

USBR, 1957, "Design of Small Dams," Denver.

U.S. Bureau of Standards, 1938, "Report on Investigation of Density Currents," Requested by the Geological Survey on March 10, 1936.

USSR, 1936, "The Max Permissible Velocity in Open Channels," Gidrotekhniches koiestroitel'stvo, Moskow, no. 5.

Valentine, H.R., 1959, "Applied Hydrodynamics," Butteworts.

Valentine, F., 1967, "Considerations Concerning Scour due to Flow Under Gates," IAHR, Fort Collins, 12th Congress.

Von Thun, J.L., 1985, "Use of Risk-based Analysis in Making Decision on Dam Safcty," NATO Advanced Study Institute on Engineering, Reliability and Risk, Tucson, Arizona.

rences

Wagner, W.E., 1954, "Morning Glory Shaft Spillways," PASCE no. 432.

White, C.H., 1940, "The Equilibrium of Grains on the Bed of a Stream," Proc. Roy. Soc. London, Vol. 174A.

Wood, I.R., 1967, "Horizontal Two-Dimensional Density current," PASCE, Vol. 93, HY2.

Yücel, Ö., and Graf, W.A., 1973, "Bed Load Deposition in Reservoirs," İAHR, İstanbul Congress.

Woodburn, J.G., 1932, "Tests of Broad Crested Weirs," TASCE Vol. 96.

Zimmermann, F., and Maniak, U., 1967, "Scours Behind Stilling Basins with Endsills of Baffle Piers," IAHR, XII Congress Fort Collins, Colorado.

AUTHOR INDEX

Author Index

Pinto, N.L. de S. 311, 313, 314, 315
Post, G. 375
Preece, E.F. 458
Pugnet, L. 244, 249

R

Randolph, R.R. 93, 96
Reeves, A.B. 418
Reynaud, J.P. 446
Roberts, E. 523
Rohne, T.J. 133, 153, 156, 157, 158, 159
Rose 386
Rouse, H. 93, 201, 205, 572
Roy, S.K. 205

S

Safranetz 317
Sanchez 618, 629
Saville, T. 413, 420, 422
Scheidegger, A. 443
Schlichting, H. 304
Schmidt, H. 218
Schoklitsch, A. 435, 632, 636, 641
Scimemi, E. 79, 85, 93
Shui-bo 315
Simons, D.B. 57, 301, 305, 379, 446, 461, 573, 577, 601, 607, 626, 632, 634, 638, 647, 649, 651, 662, 664, 673, 692
Smetana 85, 317
Smith, N.A.F. 3
Stark, H. 12
Stefan 386
Stefenson, T. 417, 419, 424
Stevens, M.A. 607, 632
Strangler, S. 570
Straub, G.L. 341
Strickler 436

Sverdrup, H.V. 412, 416

Ş

Şentürk, F. 48, 57, 92, 267, 268, 301, 305, 340, 379, 446, 461, 569, 573, 577, 601, 611, 612, 626, 634, 638, 649, 664, 676, 692, 713

T

Terzaghi, K. 481, 483, 485
Thomas, C.W. 418
Tyler 482

U

Urbonas 607
Urrutia, C.M. 153

V

Valentine 472, 474, 633, 635, 641
Viparelli 226
Von Thun, J.L. 59, 60

W

Wagner, W.E. 245, 251, 252, 253, 258, 261
Weissbach, J. 445, 446
Westergaard 400
White, C.H. 245, 607
Wood, I.R. 442
Woodburn 203
Woycicky 317

Y

Yevjevich, V. 100
Yücel, Ö 435, 437

Z

Zimmerman 635

SUBJECT INDEX

Subject Index

ecologic 58
general 53
geologic 58
hydrologic 55
reservoir 59
topographic 54

Deflector 316, 350, 368

Degradation 15, 316, 340, 602, 641

Delta 438

Demirköprü Reservoir 439, 468, 493, 496

Demre 10

Density
current 411, 430, 435, 438, 441, 442, 443, 447
of air-water solution 168
of deposit 440
of water 101

Dentated sill 335, 344, 348

Depth
of flow 115, 204, 310, 627
of the cross section 115

Design
discharge 506
flood 39, 40, 48
flood hydrograph 47
head 80, 97

Detroid Dam 134

Dicle Dam 545, 546, 558

Discharge 205, 212, 218, 244, 261
canal 721
coefficient 69, 76, 78, 79, 80, 87, 92, 95, 97, 98, 100, 108, 121, 127, 128, 133, 153, 156, 157, 160, 161, 176, 179, 198, 200, 205, 212, 242, 244, 284, 313, 371, 676
Decreasing 230
gradually increasing 226
increasing 226
linear 204, 218
maximum 97, 99

Diversion
facilities 738
structure 533
tunnel 507, 508, 509, 533

Djizveh Dam 3

Dogancay reservoir 431

Douma formula 168

Driest year 578, 581, 582

Drowned jump 283

DSI Research Lab. 93

Dumanli spring 10

Dynamic equation 230

E

Echo sounding device 430

Economic conditions 49

Economic study 50

Eggenberger's formula 636, 638, 640, 641

Einstein's formula 436

Einstein's relation 643

Elche Dam 5

Elephant Butte Dam 427

Embankment 491
dam 455

Energy
coefficient 218
dissipation 315, 634, 681, 721, 726
dissipator 316, 676, 678, 725
equation 539
flux 304, 305
flux loss 305
line 71, 72
loss 327, 368

Equation of continuity 248

Equipotential lines 396, 474

Erosion 281, 316

Subject Index

Subject Index

Mean size 615

Mean velocity 68, 69, 161, 269, 271, 305

Measuring device 203, 490, 491, 492, 493, 500, 501

Mehmet Sumra Dam 44

Mesopotamia 1, 3

Meyer-Peter formula 436

Middle East 3

Mississippi River 625

Model

    computational 678, 686, 688, 692, 695, 702, 704, 715, 723, 731, 735, 739, 740

    Hele-Shaw 738

    physical 612, 688, 690, 695, 696, 703, 717, 718, 719, 723, 731

    scale 161, 288, 335, 337, 340, 690, 694, 702, 703, 719

Momentum 203, 204

    change of 204, 227

Monteynard Dam Spillway 353

N

Nappe 136, 171, 255, 351, 360

Natillas Dam 6

National Bureau of Standards 441

Naugarh Dam 427

Navajo Dam 459

Negative pressure 97, 107, 131, 134, 137

Netzahualcoyotl 6

New Dam 17

O

Ocean formula 419

Ogee spillway 87

Operating bridge 38

Optimization

    analysis 508

    study 558

Orifice 10, 11, 13, 14, 136, 242

Oroville Dam 459

Örükaya Dam 10, 11, 12

Osceola Dam 134

Osman II Dam 17

Ottoman

    Dam 16

    Turks 15

Ottoman Empire (Turks) 1, 15

Overtopping 265

Owyhee Dam 427

Oymapinar Dam 10, 427, 723

Oymapinar Dam Spillway 117, 723

P

Pecos River 434

Penstock 565

Persia 2

Phreatophyte 434

Pier 623, 720

    characteristics 615

    end 132

    geometry of 160

    intermediate 132

    side 37, 161, 717, 719

Piezometer 491, 497

Pine Flat Dam 134

Pivot point 438

Pollution 6

Pore water pressure 491

Pre-Inca 7

Precision 705

Pressure 100

Subject Index

Subject Index

set-up 411, 412, 413, 423, 425, 426

Shock wave 678, 680

significant wave height 413, 419, 420

velocity 421

Weight of water 101, 204, 227

Weir 201, 246, 316, 613, 678

side 7

Wetted area 228

Wettest year 578

Whitney Dam 134

Wind

duration 412

speed 412

velocity 412, 413, 417, 419, 421, 422, 423

Y

Yagisawa Dam Spillway 48

Yuriraga Dam 6

Z

Zuider-Zee formula 423